Applied Mineral Inventory Estimation

Applied Mineral Inventory Estimation presents a comprehensive applied approach to the estimation of mineral resources/reserves with particular emphasis on

- The geological basis of such estimations
- The need for and maintenance of a high-quality assay data base
- The practical use of a comprehensive exploratory data evaluation
- The importance of a comprehensive geostatistical approach to the estimation methodology.

The text emphasizes that geology and geostatistics are fundamental to the process of mineral inventory. Practical problems and real data are used throughout as illustrations: each chapter ends with a summary of practical concerns, a number of practical exercises and a short list of references for supplementary study. The topics emphasized include estimation concepts, integration of geology into the estimation procedure, monitoring and maintaining the high quality of the assay database, appreciation and application of basic statistical procedures to the estimation process, exploratory data analysis as a means of improving confidence in the estimation process and applied geostatistical estimation methods. In addition, individual chapters are devoted to other important topics including grade–tonnage curves, bulk density, simulation, dilution, resource/reserve classification, and metal accounting reality checks of estimation procedures.

This textbook is suitable for any university or mining school that offers senior undergraduate and graduate student courses on mineral resource/reserve estimation. It will also be valuable for professional mining engineers, geological engineers, and geologists working with mineral exploration and mining companies.

Alastair J. Sinclair obtained his B.A.Sc. and M.A.Sc. degrees in geological engineering at the University of Toronto and a Ph.D. in economic geology at the University of British Columbia. He taught in the Department of Geology, University of Washington from 1962 to 1964 and at the University of British Columbia from 1964 to 1999. He is presently Emeritus Professor at the University of British Columbia. His teaching and research activities have focused on mineral deposits, mineral exploration data analysis, and mineral inventory estimation. He continues to be active in applied research in these fields. Over the past three and one-half decades he has given a wide range of short courses to industry and has consulted internationally on matters relating to data quality and mineral inventory estimation.

Garston H. Blackwell was educated in Cornwall, United Kingdom and graduated from Camborne School of Mines in 1968. He worked as a mining engineer for Falconbridge Nickel Mines in Sudbury for three years and then attended Queen's University, graduating with a M.Sc. (Eng.) in Mining Engineering in 1973. He then joined Brenda Mines Ltd. in British Columbia as senior mining engineer, working throughout the Noranda organization, becoming Chief Mine Engineer at Brenda in 1978. While at Brenda Mines, he oversaw development of computer applications in mining for ore reserve estimation and grade control, long- and short-term mine planning, financial evaluation, slope stability, and information reporting and technology. He applied many of these developments within the Noranda group of companies. In 1986, he joined Queen's University as Associate Professor, where he teaches open-pit mining, reserve estimation and grade control, computer applications, open-pit rock mechanics and surveying. He has taken extended industrial leaves with Barrick in Nevada and IMDI in Chile.

Applied
Mineral Inventory
Estimation

ALASTAIR J. SINCLAIR
The University of British Columbia

GARSTON H. BLACKWELL
Queen's University

CAMBRIDGE
UNIVERSITY PRESS

CAMBRIDGE UNIVERSITY PRESS
Cambridge, New York, Melbourne, Madrid, Cape Town, Singapore, São Paulo

Cambridge University Press
The Edinburgh Building, Cambridge CB2 2RU, UK

Published in the United States of America by Cambridge University Press, New York

www.cambridge.org
Information on this title: www.cambridge.org/9780521791038

First published 2002
This digitally printed first paperback version 2006

A catalogue record for this publication is available from the British Library

Library of Congress Cataloguing in Publication data
Sinclair, Alastair J. (Alastair James), 1935–
   Applied mineral inventory estimation / Alastair J. Sinclair, Garston H. Blackwell.
      p.   cm.
   Includes bibliographical references and index.
   ISBN 0-521-79103-0
   1. Ores – Sampling and estimation.   I. Blackwell, Garston H., 1944–   II. Title.
   TN560 .S56   2001
   669′.92 – dc21   2001035710

ISBN-13 978-0-521-79103-8 hardback
ISBN-10 0-521-79103-0 hardback

ISBN-13 978-0-521-02182-1 paperback
ISBN-10 0-521-02182-0 paperback

Contents

Preface   *page* xiii
Acknowledgments   xvii

**1   MINERAL INVENTORY:
      AN OVERVIEW**   1
1.1   Introduction   1
1.2   Mineral Inventory Estimates   2
1.3   Some Essential Concepts in Mineral
      Inventory   4
      1.3.1   Ore   4
      1.3.2   Cutoff Grade   5
      1.3.3   Continuity   7
      1.3.4   Reserves and Resources   8
      1.3.5   Dilution   9
      1.3.6   Regionalized Variable   10
      1.3.7   Point and Block Estimates   11
      1.3.8   Selective Mining Unit   13
      1.3.9   Accuracy and Precision   14
1.4   A Systematic Approach to Mineral
      Inventory Estimation   15
1.5   Traditional Methods of Mineral
      Inventory Estimation   16
      1.5.1   Method of Sections   17
      1.5.2   Polygonal Methods   17
      1.5.3   Method of Triangles   19
      *1.5.4   Inverse Distance Weighting
               Methods*   19
      1.5.5   Contouring Methods   20
      1.5.6   Commentary   22
1.6   Mine Revenues   23
1.7   Mining Software – Applications   26
1.8   Practical Considerations   27

1.9   Selected Reading   28
1.10 Exercises   28

**2   GEOLOGIC CONTROL OF
      MINERAL INVENTORY
      ESTIMATION**   31
2.1   Introduction   31
2.2   Geologic Mapping   32
2.3   General Geology   36
2.4   General Geometry of a
      Mineralized/Ore Zone   37
2.5   Geometric Errors in Geologic
      Modeling   39
2.6   Ore Deposit Models   45
      2.6.1   General Concepts   45
      *2.6.2   Volcanogenic Massive
               Sulphide Deposits*   46
      2.6.3   Besshi-Type Cu–Zn Deposits   47
      *2.6.4   Porphyry-Type Deposits (see
               also Sinclair and Postolski,
               1999)*   49
      2.6.5   General Summary   50
2.7   Mineralogy   51
2.8   Geologic Domains   55
2.9   Practical Considerations   56
2.10 Selected Reading   58
2.11 Exercises   58

3   CONTINUITY   59
3.1   Introduction   59
3.2   Geologic Continuity   59
3.3   Value Continuity   63

3.4	Continuity Domains	65
3.5	Continuity in Mineral Inventory Case Histories	66
	3.5.1 Silver Queen Deposit	66
	3.5.2 JM Zone, Shasta Deposit	68
	3.5.3 South Pit, Nickel Plate Mine	69
	3.5.4 Discussion	71
3.6	Practical Considerations	72
3.7	Selected Reading	73
3.8	Exercises	73

4	**STATISTICAL CONCEPTS IN MINERAL INVENTORY ESTIMATION: AN OVERVIEW**	76
4.1	Introduction	76
4.2	Classic Statistical Parameters	77
	4.2.1 Central Tendency	77
	4.2.2 Dispersion	78
	4.2.3 Covariance	80
	4.2.4 Skewness and Kurtosis	80
4.3	Histograms	80
4.4	Continuous Distributions	83
	4.4.1 Normal Distribution	83
	4.4.2 Standard Normal Distribution	84
	4.4.3 Approximation Formula for the Normal Distribution	85
	4.4.4 Lognormal Distribution	86
	4.4.5 Binomial Distribution	88
	4.4.6 Poisson Distribution	88
4.5	Cumulative Distributions	90
	4.5.1 Probability Graphs	90
4.6	Simple Correlation	94
4.7	Autocorrelation	96
4.8	Simple Linear Regression	97
4.9	Reduced Major Axis Regression	98
4.10	Practical Considerations	100
4.11	Selected Reading	100
4.12	Exercises	100

5	**DATA AND DATA QUALITY**	104
5.1	Introduction	104
5.2	Numeric Data for Mineral Inventory Estimation	105
	5.2.1 Types of Samples	105
	5.2.2 Concerns Regarding Data Quality	107
	5.2.3 Location of Samples	108
5.3	Error Classification and Terminology	108
	5.3.1 Definitions	108
	5.3.2 Relation of Error to Concentration	110
	5.3.3 Bias Resulting from Truncated Distributions	112
5.4	Sampling Patterns	113
	5.4.1 Terminology and Concerns	113
	5.4.2 Sample Representativity	115
5.5	Sampling Experiments	116
	5.5.1 Introduction to the Concept	116
	5.5.2 Comparing Sampling Procedures at Equity Silver Mine	117
	5.5.3 Sampling Large Lots of Particulate Material	118
5.6	Improving Sample Reduction Procedures	120
	5.6.1 The Mineralogic Composition Factor (m)	123
	5.6.2 The Liberation Factor	123
	5.6.3 The Particle Shape Factor	123
	5.6.4 The Size Range Factor	123
	5.6.5 Applications of Gy's Equation	124
	5.6.6 Direct Solution of Gy's Equation (Simplified Form)	124
	5.6.7 User's Safety Line	124
5.7	Assay Quality Control Procedures	124
	5.7.1 Introduction	124
	5.7.2 Using the Correct Analyst and Analytical Methods	125
	5.7.3 Salting and Its Recognition	127
5.8	A Procedure for Evaluating Paired Quality Control Data	129
	5.8.1 Introduction	129
	5.8.2 Estimation of Global Bias in Duplicate Data	129
	5.8.3 Practical Procedure for Evaluating Global Bias	130

5.8.4 *Examples of the Use of Histograms and Related Statistics* 131

5.8.5 *A Conceptual Model for Description of Error in Paired Data* 132

5.8.6 *Quantitative Modeling of Error* 133

5.9 Improving the Understanding of Value Continuity 139

5.10 A Generalized Approach to Open-Pit-Mine Grade Control 140

5.10.1 *Initial Investigations* 140

5.10.2 *Development of a Sampling Program* 140

5.10.3 *Sampling Personnel and Sample Record* 141

5.10.4 *Implementation of Grade Control* 142

5.10.5 *Mineral Inventory: Mine–Mill Grade Comparisons* 142

5.11 Summary 143

5.12 Practical Considerations 143

5.13 Selected Reading 144

5.14 Exercises 144

6 **EXPLORATORY DATA EVALUATION** 146

6.1 Introduction 146

6.2 File Design and Data Input 148

6.3 Data Editing 149

6.3.1 *Composites* 149

6.4 Univariate Procedures for Data Evaluation 151

6.4.1 *Histograms* 152

6.4.2 *Raw (Naive) versus Unbiased Histograms* 152

6.4.3 *Continuous Distributions* 152

6.4.4 *Probability Graphs* 153

6.4.5 *Form of a Distribution* 154

6.4.6 *Multiple Populations* 154

6.5 Bivariate Procedures for Data Evaluation 155

6.5.1 *Correlation* 155

6.5.2 *Graphic Display of Correlation Coefficients* 158

6.5.3 *Scatter Diagrams and Regression Analysis* 159

6.6 Spatial Character of Data 160

6.6.1 *Introduction* 160

6.6.2 *Contoured Plans and Profiles* 160

6.7 Multivariate Data Analysis 162

6.7.1 *Triangular Diagrams* 163

6.7.2 *Multiple Regression* 164

6.8 Practical Considerations 165

6.9 Selected Reading 165

6.10 Exercises 166

7 **OUTLIERS** 167

7.1 Introduction 167

7.2 Cutting (Capping) Outlier Values 168

7.2.1 *The Ordinary Case* 168

7.2.2 *Outliers and Negative Weights* 169

7.3 A Conceptual Model for Outliers 170

7.4 Identification of Outliers 170

7.4.1 *Graphic Identification of Outliers* 170

7.4.2 *Automated Outlier Identification* 171

7.5 Multiple Geologic Populations 172

7.6 Probability Plots 172

7.6.1 *Partitioning Procedure* 173

7.7 Examples 176

7.8 Structured Approach to Multiple Populations 177

7.9 Incorporation of Outliers into Resource/Reserve Estimates 178

7.10 Practical Considerations 178

7.11 Selected Reading 179

7.12 Exercises 179

8 **AN INTRODUCTION TO GEOSTATISTICS** 181

8.1 Introduction 181

8.2 Some Benefits of a Geostatistical Approach to Mineral Inventory Estimation 183

8.3   Random Function   183
8.4   Stationarity   185
8.5   Geostatistical Concepts
      and Terminology   185
8.6   The Variogram/Semivariogram   186
8.7   Estimation Variance/Extension
      Variance   186
8.8   Auxiliary Functions   188
8.9   Dispersion Variance   189
8.10 A Structured Approach to
      Geostatistical Mineral
      Inventory Estimation   189
      8.10.1 Applications of Geostatistics
              in Mineral Inventory
              Estimation   190
      8.10.2 Why Geostatistics?   191
8.11 Selected Reading   191
8.12 Exercises   191

9   SPATIAL (STRUCTURAL)
      ANALYSIS: AN INTRODUCTION
      TO SEMIVARIOGRAMS   192
9.1   Introduction   192
9.2   Experimental Semivariograms   193
      9.2.1   Irregular Grid in One
              Dimension   195
      9.2.2   Semivariogram Models   196
9.3   Fitting Models to Experimental
      Semivariograms   198
9.4   Two-Dimensional Semivariogram
      Models   199
      9.4.1   Anisotropy   201
9.5   Proportional Effect and Relative
      Semivariograms   204
9.6   Nested Structures   205
9.7   Improving Confidence in the Model
      for Short Lags of a Two- or
      Three-Dimensional Semivariogram   207
9.8   Complexities in Semivariogram
      Modeling   208
      9.8.1   Effect of Clustered Samples   208
      9.8.2   Treatment of Outlier Values   208
      9.8.3   Robustness of the
              Semivariogram   209

      9.8.4   Semivariograms in Curved
              Coordinate Systems   210
      9.8.5   The "Hole Effect"   211
9.9   Other Autocorrelation
      Functions   212
9.10 Regularization   212
9.11 Practical Considerations   213
9.12 Selected Reading   214
9.13 Exercises   214

10   KRIGING   215
10.1 Introduction   215
10.2 Background   216
      10.2.1 Ordinary Kriging   216
      10.2.2 Simple Kriging   217
10.3 General Attributes
      of Kriging   218
10.4 A Practical Procedure
      for Kriging   218
10.5 An Example of Kriging   219
10.6 Solving Kriging Equations   220
10.7 Cross Validation   221
10.8 Negative Kriging Weights   224
      10.8.1 The Problem   224
      10.8.2 The Screen Effect   225
10.9 Dealing with Outliers   227
      10.9.1 Restricted Kriging   227
10.10 Lognormal Kriging   228
10.11 Indicator Kriging   229
      10.11.1 Kriging Indicator Values   230
      10.11.2 Multiple Indicator Kriging
              (MIK)   230
      10.11.3 Problems in Practical
              Applications of Indicator
              Kriging   232
10.12 Conditional Bias in Kriging   233
      10.12.1 Discussion   235
10.13 Kriging with Strings of Contiguous
      Samples   236
10.14 Optimizing Locations for Additional
      Data   237
10.15 Practical Considerations   239
10.16 Selected Reading   240
10.17 Exercises   241

**11 GLOBAL RESOURCE
      ESTIMATION**   242
11.1   Introduction   242
11.2   Estimation with Simple Data Arrays   243
       *11.2.1   Random and Stratified
                 Random Data Arrays*   243
       11.2.2   Regular Data Arrays   243
11.3   Composition of Terms   244
       11.3.1   An Example: Eagle Vein   244
11.4   Volume–Variance Relation   245
11.5   Global Estimation with Irregular
       Data Arrays   246
       *11.5.1   Estimation with Multiple
                 Domains*   247
11.6   Errors in Tonnage Estimation   248
       11.6.1   Introduction   248
       *11.6.2   Sources of Errors in Tonnage
                 Estimates*   248
       11.6.3   Errors in Bulk Density   248
       *11.6.4   Errors in Surface (Area)
                 Estimates*   249
       *11.6.5   Surface Error – A Practical
                 Example*   250
       11.6.6   Errors in Thickness   251
11.7   Estimation of Co-Products
       and By-Products   251
       *11.7.1   Linear Relations and
                 Constant Ratios*   251
       *11.7.2   A General Model for
                 Lognormally Distributed
                 Metals*   252
       11.7.3   Equivalent Grades   253
       11.7.4   Commentary   253
11.8   Practical Considerations   253
11.9   Selected Reading   254
11.10 Exercises   254

12 GRADE–TONNAGE CURVES   255
12.1   Introduction   255
12.2   Grade–Tonnage Curves Derived from
       a Histogram of Sample Grades   257
12.3   Grade–Tonnage Curves Derived from
       a Continuous Distribution
       Representing Sample Grades   258

12.4   Grade–Tonnage Curves Based
       on Dispersion of Estimated
       Block Grades   259
       12.4.1   Introduction   259
       *12.4.2   Grade–Tonnage Curves from
                 Local Block Estimates*   261
12.5   Grade–Tonnage Curves by Multiple
       Indicator Kriging   262
12.6   Example: Dago Deposit, Northern
       British Columbia   263
12.7   Reality versus Estimates   265
12.8   Practical Considerations   266
12.9   Selected Reading   266
12.10 Exercises   266

**13 LOCAL ESTIMATION OF
      RESOURCES/RESERVES**   268
13.1   Introduction   268
13.2   Sample Coordinates   268
13.3   Block Size for Local Estimation   269
13.4   Robustness of the Kriging Variance   271
13.5   Block Arrays and Ore/Waste
       Boundaries   272
13.6   Estimation at the Feasibility Stage   274
       13.6.1   Recoverable "Reserves"   274
       13.6.2   Volume–Variance Approach   275
       13.6.3   "Conditional Probability"   276
13.7   Local Estimation at the Production
       Stage   276
       *13.7.1   Effect of Incorrect
                 Semivariogram Models*   276
       *13.7.2   Spatial Location of
                 Two-Dimensional Estimates*   278
       13.7.3   Planning Stopes and Pillars   279
13.8   Possible Simplifications   280
       *13.8.1   Block Kriging with Bench
                 Composites*   280
       *13.8.2   Easy Kriging with Regular
                 Grids*   280
       *13.8.3   Traditional Methods
                 Equivalent to Kriging*   280
13.9   Treatment of Outliers in
       Resource/Reserve Estimation   281
13.10 Practical Considerations   282

13.11 Selected Reading   282
13.12 Exercises   282

**14   AN INTRODUCTION TO
       CONDITIONAL SIMULATION**   284
14.1   Introduction   284
14.2   Aims of Simulation   285
14.3   Conditional Simulation as an
       Estimation Procedure   286
14.4   A Geostatistical Perspective   286
14.5   Sequential Gaussian Simulation   286
14.6   Simulating Grade Continuity   287
14.7   Simulation to Test Various Estimation
       Methods   287
       14.7.1   Introduction   287
       14.7.2   Procedure   287
       *14.7.3   Verifying Results of the
                 Simulation Process*   288
       *14.7.4   Application of Simulated
                 Values*   289
       *14.7.5   Sequential Indicator
                 Simulation*   292
14.8   Practical Considerations   292
14.9   Selected Reading   292
14.10 Exercises   292

15   BULK DENSITY   294
15.1   Introduction   294
15.2   Impact of Mineralogy on Density   295
15.3   Impact of Porosity on Bulk Density   296
15.4   Impact of Errors in Bulk Density   296
15.5   Mathematical Models of Bulk
       Density   297
15.6   Practical Considerations   299
15.7   Selected Reading   299
15.8   Exercises   299

**16   TOWARD QUANTIFYING
       DILUTION**   301
16.1   Introduction   301
16.2   External Dilution   301
       *16.2.1   Vein Widths Partly Less Than
                 Minimum Mining Width*   302
       16.2.2   Silver Queen Example   303

       16.2.3   Dilution from Overbreaking   304
       16.2.4   Contact Dilution   304
16.3   Internal Dilution   306
       16.3.1   A Geostatistical Perspective   306
       *16.3.2   Effect of Block Estimation
                 Error on Tonnage and Grade
                 of Production*   307
16.4   Dilution from Barren Dykes   311
       16.4.1   Snip Mesothermal Deposit   311
       *16.4.2   Virginia Porphyry Cu-Au
                 Deposit*   313
       *16.4.3   Summary: Dilution by Barren
                 Dykes*   313
16.5   Practical Considerations   314
16.6   Selected Reading   314
16.7   Exercises   315

17   ESTIMATES AND REALITY   316
17.1   Introduction   316
17.2   Recent Failures in the Mining
       Industry   317
17.3   Resource/Reserve Estimation
       Procedures   318
17.4   Geostatistics and Its Critics   319
17.5   Why Is Metal Production Commonly
       Less Than the Estimate?   322
17.6   Practical Considerations   323
17.7   Selected Reading   323
17.8   Exercise   324

**18   RESOURCE/RESERVE
       CLASSIFICATION**   325
18.1   Introduction   325
18.2   A Geologic Basis for Classification
       of Mineral Inventory   327
18.3   Shortcomings to Existing
       Classification Systems   327
18.4   Factors Traditionally Considered
       in Classifying Resources/Reserves   328
18.5   Contributions to Classification from
       Geostatistics   330
18.6   Historical Classification Systems   333
18.7   The Need for Rigor
       and Documentation   334

18.8 Examples of Classification
Procedures 335
18.9 Practical Considerations 335
18.10 Suggested Reading 336
18.11 Exercises 336

**19 DECISIONS FROM
ALTERNATIVE SCENARIOS:
METAL ACCOUNTING** 337
19.1 Introduction 337
19.2 Definition 337
19.3 Metal Accounting: Alternative
Blasthole Sampling Methods 338
19.4 Metal Accounting: Effect
of Incorrect Semivariogram
Model on Block
Estimation 340

19.5 Metal Accounting: Effect of Block
Estimation Error on Ore Waste
Classification Errors (After Postolski
and Sinclair, 1998; Sinclair, 1995) 341
19.6 Summary Comments 342
19.7 Practical Considerations 344
19.8 Selected Reading 345
19.9 Exercises 346

APPENDICES 347
A.1 British and International Measurement
Systems: Conversion Factors 349
A.2 U.S. Standard Sieves 350
A.3 Drill Hole and Core Diameters 351

Bibliography 353
Index 377

Preface

"... geostatistics is of real potential if it is reconciled with the geology of the deposit" (King et al., 1982).

Resource/reserve estimation is too often viewed as a set of recipes when it is in reality an intellectual undertaking that varies from one case to another. Even the estimation of two different deposits of the same type can involve significantly different procedures. We undertake an estimation study with *limited information* that represents perhaps 1/100,000 of the deposit in question. The information base, however, is larger than we sometimes think, including not just general geology and assay data, but information from *other sources* such as sampling practice, applied mineralogical studies, geophysical and geochemical survey data, and a range of engineering information – all of which contribute to the development and implementation of an estimation procedure. The use of such varied data demands that the opinions of a *range of experts* be taken into account during estimation. Increasingly, it is becoming more difficult for a single person to conduct a comprehensive resource/reserve estimation of a mineral deposit. Clearly, *all* sources of information must be considered in reasonable fashion if a mineral inventory study is to be professionally acceptable; it is morally unacceptable to provide an estimate that is seriously in conflict with pertinent data. Of course, it must be recognized that even with the best of intentions, procedures and abilities, errors will be made. After all, 99,999/100,000 of the deposit must be interpreted from widely spaced control sites. Moreover, continuing property exploration and evaluation ensures that new information will be added periodically to the database; *an estimate, once completed, might already be somewhat out of date!* The estimator must maintain an open mind to new information. A sense of humor will contribute to the success of a long resource/reserve team study.

This text was designed with a conscious consideration of the needs of senior undergraduate students who lack the advantage of years of experience in mineral inventory estimation. Those who are active in the field will appreciate that a text cannot possibly cover all possible nuances of the topic. In the end, experience is the best teacher. The authors both feel strongly that the best way to learn mineral inventory estimation is to become immersed in a real study where the practical situation requires solutions to problems so that a product is obtained within a reasonable time frame. A real life situation such as this is difficult to achieve in a university environment, and we have not attempted to do so. Instead, we have organized the text to supplement such a comprehensive study and have paid special attention to several aspects of estimation that our experience indicates have been underappreciated in recent decades.

Geology is widely recognized as the underpinning of any mineral inventory estimation and an extensive literature exists on the topic; nevertheless, modern texts that emphasize the importance of geology are lacking. Our experience suggests that all too often either geology is minimized or excellent geological work is not integrated adequately into the estimation procedure. The excellent text by McKinstry (1948) remains as valid today as when it was published, but geological understanding has advanced immeasurably since that time, and the impact on estimation procedures of new geological concepts can be momentous. A new look at the geological basis of estimation is essential.

The understanding of sampling theory and control of assay quality also has advanced substantially in recent decades, leading to improved understanding of, and confidence in, the underlying database. Much of this theory and quality control information is scattered widely through the technical literature and needs to be gathered in a concise, practical way so students can be introduced to both the need for high-quality procedures in data gathering and efficient procedures for ensuring high-quality data. "Garbage in, garbage out" is a computer adage that applies equally well to resource/reserve estimation.

The advent of cheap, efficient, multielement analytical techinques (chemical and instrumental) has led to an abundance of data that could barely be imagined as recently as 30 years ago. The availability of such data demands thorough evaluation; exploratory data analysis techniques are easily adapted to this aim and invariably lead to useful information for improving the quality of estimates. Both the quantity and variety of available data require a computer base for mineral inventory studies. Many data analysis procedures are impossible or difficult without access to computing facilities; fortunately, the revolution in personal computers has provided a level of computer access unimaginable a few decades ago.

Estimation procedures, too, have evolved substantially in recent years, even to the point of revolution. The advent of the computer age has resulted in automation of many traditional estimation procedures and the development of new approaches that were impractical or impossible to implement manually. Perhaps most significant has been the development and widespread use of a great range of "geostatistical" estimation procedures that provide a theoretical base to what had previously been an empirical undertaking. Over the past 30 years or so, geostatistics has evolved and proved its worth many times over. Too

often, though, geostatistical applications have been carried out without sufficient attention to the geology of the deposit for which a resource/reserve estimate is being prepared. Integration of geology and geostatistics is essential.

We have tried to incorporate much of the foregoing philosophy into this text. To summarize, some of our specific aims are to

- Present a comprehensive understanding of the geological base to mineral inventory estimation (e.g., domains, boundaries, and continuity.)
- Emphasize exploratory data evaluation as an essential component of a mineral inventory estimation
- Provide a relevance to simple classical statistical methods in mineral inventory estimation
- Integrate geology and geostatistical methodologies into mineral inventory estimation procedures in a practical manner
- Use as examples real deposits for which a substantial amount of information is publicly available
- Emphasize the importance of high-quality data by demonstrating design methodology and elements of monitoring data quality
- Document the practical approach to mineral inventory estimation at the exploration phase, as well as during feasibility and/or production stages.

Throughout, our emphasis has been on applied aspects of the estimation procedure, limited, of course, by our personal experience. We consider this text to be supplemental to a range of excellent texts of related but different themes (e.g., Annels, 1991; David, 1977; Deutsch & Journel, 1992; Isaaks & Srivastava, 1989; Journel & Huijbregts, 1978; and Rendu, 1978). In particular, with availability of such thorough and timeless geostatistical texts by Journel & Huijbregts (ibid), Deutsch & Journel (ibid), and Isaaks & Srivastava (ibid) we have avoided geostatistical theory entirely and restricted ourselves to reporting significant geostatistical relations and their practical applications. Moreover, a variety of highly specialized, complicated geostatistical techniques that find minor use in practice are ignored, including such topics as

nonparametric geostatistics, universal kriging, probability kriging, and cokriging, among others.

From a student's perspective we have aimed for
several additional goals. First, we use the concept of
ideal models throughout the text, standards against
which reality can be compared or measured. In our
experience this has been a productive way to get ideas
across to the inexperienced. Secondly, at the end of
each chapter we provide a selection of exercises that
are meant to serve both as assignments and as a basis
for developing new questions/assignments related to
a particular data set that an instructor is using in a
formal course. Also, we provide a few selected references at the close of each chapter as an aid to both
instructor and student in researching some of the main
themes of the chapter. The text assumes that appropriate software is available to the student. However,
several programs and data sets that relate to the text
are available through the publisher's website.

The reader will quickly find that the use of units
is not standardized in this text. The real world is not
standardized! In the various examples that we quote,
we have used the units of the original authors. An
ability to move between British and metric systems
is essential in the global environment in which estimations are conducted. A table of conversion factors is
included in an appendix to aid students in their ability
to "converse" between systems of units.

One of the more serious misgivings we have about
much of the published literature on mineral inventory
estimation relates to the all-too-common practice of
using artificial data or camouflaging real data, either
by not naming the mineral deposit represented or by
purposely omitting scales on diagrams and by multi-
plying data by an unknown factor to hide true values,
among other actions. Such procedures cloud the validity or interpretation of what is being presented, if
for no other reason than it is impossible to appreciate the impact of geology on the interpretation of the
numeric data. We have avoided altering data in any
way in this text unless a specific transformation is an
inherent part of an estimation procedure.

Errata

p. 109, col. 2, line 11---in both places on this line, in the quantity 's/x' the x should have a bar over it (standard notation for average value).

P. 122, col. 2, 2nd para—Gy's equation should have '1/' added at the front to read as follows: $1/(1/M_s - 1/M_L) = Cd^x/s^2$

p. 186, col. 1, 2nd line following equation 8.6. The reference to 'equation 8.8' should be to 'equation 8.6'.

4th line following equation 8.6. The reference to 'equation 8.7' should be to 'equation 8.6'.

p. 267, col. 2, 3rd line from end. Omit the word 'with'

p. 298, col. 2, first line following equation 15.5. should read 'where bs are constants'

p. 332, column 1, line 14. The expression should read
'Measured ≤ 0.3 < Indicated ≤ 0.5 < Inferred
(Note: the symbol $\leq$ means 'less than or equal to')

p. 353, The reference to Agricola should read "Agricola, 1556, de re Metallica, see English translation, Hoover and Hoover, 1950"

Acknowledgments

A book of this nature has evolved over a long period with contributions from many sources. We are indebted particularly to the many short course and university students who, over the years, have been guinea pigs for various parts of the material in this text. Questions and comments from students have greatly improved our appreciation of the material needed in a text book and the manner by which it could be presented most effectively.

This book could not have been completed without the dedicated assistance of longtime friend and colleague Asger Bentzen, who single-handedly computerized all the diagrams in addition to providing extensive computational assistance. Those who have attempted to produce original drawings in the computer will understand the magnitude of his invaluable contributions. We are equally indebted to Dr. Gerald K. Czamanske, who took on the onerous task of introducing order and editorial consistency throughout the text. His monumental editorial efforts are deeply appreciated. Similarly, we very much value the considerable efforts of the publisher to improve the quality of this work.

Personal Acknowledgments – AJS

The enthusiasm and dedication of Dr. S. H. Ward started me on a "quantitative" route in the earth sciences as an undergraduate student. In the intervening years my involvement in the study of mineral deposits benefited from my association with many, but particularly W. H. Gross and C. I. Godwin. I have had the good fortune to work with some outstanding

contributors to the field of resource/reserve estimation, beginning in 1972 when I was exposed to the ideas of G. Matheron in Fontainebleau by A. G. Journel, Ch. Huijbregts, and J. Deraisme. That, for me, was the start of an exciting and invigorating career focusing on matters related to mineral inventory. Since then, I have been fortunate to have worked at various times with H. M. Parker, M. Vallée, D. Francois-Bongarcon, T. Barnes, R. M. Srivastava, G. Raymond, G. H. Blackwell, and C. R. Stanley – all of whom have contributed to my education. To be useful, ideas require implementation: I have appreciated technical support over the years from A. Bentzen, Z. Radlowski, M. Nowak, and T. A. Postolski; all have been contributing partners in various aspects of my work.

Without interested and enthusiastic students, 36 years in academia would have been far less rewarding: I thank, in particular, the following former students for their efforts and the many informative discussions we have had together: T. A. Postolski, C. R. Stanley, C. Craig, G. H. Giroux, M. S. Nowak, M. Poloquin, A. Shahkar, M. Tilkov, and L. Werner.

Throughout my professional career I have had the opportunity to interact with many associates in the mining industry. I particularly appreciate ongoing association over many years with W. Green, J. Brunette, E. Lonergan, D. A. Sketchley, and G. H. Giroux. The many deposit evaluations that I have worked on with these people and others in the industry represent a long, continuous learning experience that could not have come from books. Over the years I have benefited from the financial support of the Science

Council of British Columbia, the Natural Science and Engineering Research Council of Canada, and members of the Canadian Mining Industry.

Personal Acknowledgments – GHB

I take this opportunity to thank the management (J. A. Hall, G. R. Montgomery, J. A. Knapp, G. R. Harris, and J. A. Keyes) and the staff of Brenda Mines Ltd. for their encouragement and support over a long period; and A. G. Journel and A. J. Sinclair for their assistance in the practical implementation of geostatistics, which has become essential for meaningful computerised mine planning. All the operating efficiency of Brenda Mines, a most successful mine despite head grades as low as 0.1% Cu and 0.025% Mo, was of no consequence without grade control and the ability to predict grade and its location. Mining engineers must remain involved in the grade modeling and estimation process from feasibility through to ongoing production to ensure the continuing profitability of their mines.

My industrial associates at Brenda, including P. D. D. Chick and N. I. Norrish and later T. Johnstone and G. Claussen, were a great assistance in the lengthy process of developing and testing mine-site procedures for resource/reserve estimation and grade control. My former students A. Chang, A. Bonham-Carter, and J. M. Anderson have continued to develop practical methods and software to enhance mine profitability and productivity through the application of geostatistics.

Mineral Inventory: An Overview

The life of a mine does not start the day that production begins, but many years before, when the company sets out to explore for a mineral deposit. A good deal of time and money is spent simply looking for, locating and quantifying a promising mineral occurrence. Not many will be found and not many of the ones found will have the potential to become mines. It is not unusual to spend five to ten years searching for a mineable deposit.
(Anonymous, Groupe de Reflexion, cf. Champigny and Armstrong, 1994)

> Chapter 1 introduces the setting within which mineral inventories are estimated, explains concepts and terminology important to the general problem of resource/reserve estimation, and describes a range of empirical estimation methods that are widely used in industry. Important concepts include the systematic and sequential nature of gathering data, three-dimensional block models of a mineral deposit, errors of estimation, cutoff grade, geologic and value continuity, and the structured nature of a mineral inventory estimation. Computers are an essential tool for modern requirements of mineral inventory estimation.

1.1: INTRODUCTION

The investment necessary to start a mine is on the order of tens to hundreds of millions of dollars. In order for the investment to be profitable, the potential product in the ground must be present in adequate quantity and quality to justify a decision to invest. Mining and processing systems used to extract the products must then operate so as to produce revenue to offset the planned investment and provide an acceptable profit. Clearly, all technologic and financial decisions regarding planned production are built on an understanding of the mineral assets available. Thus, the estimation of grade and location of material in the ground (in situ resources) must be known with an acceptable degree of confidence. This is especially true of certain large, low-grade deposits for which grade is only slightly above minimum profitable levels and for some precious metal deposits where only a small percentage of mineralized ground can be mined at a profit. Mining profits are strongly leveraged to product price and realized grade of material mined. A small difference between planned (estimated) and realized production grade or a small change in metal price can have a large impact on mine profitability.

To remain competitive, mining companies must optimize productivity of each mining operation. There are several ways to accomplish this goal. Moving and processing more tons with the same or less equipment is a common first goal, followed by inventory/materials management and control, purchasing new and better equipment, and so on. Each of these courses of action has an associated cost and a potential return on investment. Another method of increasing productivity is to increase the product content in the material being mined and processed (i.e., improved grade control during mining). This can be accomplished by increasing the grade for the same tonnage, increasing the tonnage while maintaining the same average grade, or some combination that involves improved selection of ore versus waste. Improved grade control arguably offers the best return for the least investment because of the leverage that grade mined has on revenue.

The three undertakings – ore estimation, mine planning, and grade control – are complementary in an efficient mining operation and are natural progressions. The integration of these three endeavors is important because the grade control system must balance with the ore reserve as well as with the final products of the operating plant, and both estimation and grade control are influenced by planned operational procedures. If this balance is not achieved, the original investment may be in jeopardy. Reappraisals of mineral inventories may be necessary many times prior to and during the life of a mine.

1.2: MINERAL INVENTORY ESTIMATES

Mineral inventories are a formal quantification of naturally occurring materials, estimated by a variety of empirically or theoretically based procedures. Inventories that are based on an *economic feasibility study* are commonly classed as *reserves*; inventories that are less well established are considered *resources*. These resource/reserve estimates, commonly determined from a two- or three-dimensional array of assayed samples, are applied to mineralized rock volumes that total many orders of magnitude larger than the total sample volume (Fig. 1.1). Thus, errors of estimation can be viewed as errors of extension (i.e., errors made in extending the grades of samples to a much larger volume [tonnage] of rock). For purposes of establishing a mineral inventory, a mineral deposit generally is discretized into an array of blocks, and the average value of each block is estimated in some manner from the nearby data. Thus, a mineral inventory can be viewed as a detailed breakdown of blocks whose individual sizes, locations, and grades are well established.

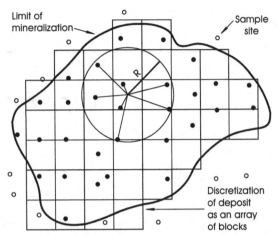

Figure 1.1: A two-dimensional representation of the general situation in mineral inventory estimation. A mineralized zone/deposit defined by geology is discretized by a number of blocks (commonly of uniform size, but not necessarily so). Each block is to be estimated using nearby data within a search area (volume), in this case defined by a circle centered on the block to be estimated. Small black dots are sample sites (for which there would be grade values) within the deposit; small open circles are samples outside the limits of the deposit.

Quantification of a resource/reserve is to a level of confidence (subjective or statistical) appropriate to the available data and the stated needs of the estimate. Volumes, tonnages, grades, and quantities of metals or minerals are the common attributes that are quantified. Their estimation must be optimal in the sense that they must be unbiased and the random error must not exceed an acceptable quality criterion. Mineral inventory estimates are used to determine economic viability that is relatively assured in the case of reserves. Volume (or tonnage) of ground classed as *resources* generally has not been evaluated rigorously for economic viability or has been found to lack immediate economic potential. Estimation procedures can differ substantially for deposits to be mined underground compared with deposits to be mined by surface pits. Similarly, methodology can vary depending on whether the mineral inventory in question is for short-term or long-term production planning.

Mineral inventories are determined at various times in the exploration, evaluation, and production of

Table 1.1 Staged sequence of data gathering in mineral exploration and evaluation

Phase of exploration	General description of work
Discovery	May result from a staged exploration program, prospecting wildcat invesigation, or by accident. This stage includes initial ground control by staking, options, etc.
Preliminary surface evaluation	Limited surface examination, including conceptual geologic appraisal, limited geochemical or geophysical responses are measured, sampling for assay and mineralogic studies, limited test pits, and stripping. This is the rapid property appraisal, or "scouting," stage of many major companies.
Detailed surface evaluation	Generally begins with the laying out of a regular grid on areas of interest to serve as a base for detailed geochemical and geophysical surveys and geologic mapping. Limited stripping, trenching, drilling, and systematic sampling are common at this stage as a guide to development of geologic concepts.
Subsurface evaluation	Involves various types of drilling, generally in a more or less systematic manner and initially with a relatively wide spacing of holes. Other methods, such as sinking exploratory shafts or declines and driving adits and other workings are useful for many deposit types.
Feasibility	Begins when a conscious decision is made to mount a detailed program to examine the possibility of economically viable production. It includes reserve estimates, mine planning, mill design, and costing of necessary infrastructure and environmental considerations, including mine reclamation. Several stages of prefeasibility work can be involved, i.e., studies that are close approaches to a true feasibility study, but with uncertainties that are unacceptable in a final feasibility study.
Development	Normally represents a halt in exploration efforts while deposit is prepared for production.
Production	An ongoing exploration program is common during the productive life of a mineral property. Both surface and underground techniques are used as needs arise. Work can be focused on extending the limits of known mineral zones or searching for new and discrete zones.
Reclamation	Exploration has normally been completed when reclamation begins.

Source: Modified from Champigny et al. (1980).

a mineral deposit (Table 1.1). At the exploration stage, a mineral inventory is useful in providing information concerning a target whose quantities are imprecisely known. The geologic setting of the mineralization may define features that provide limits to such targets, indicate possible directions of continuity, or help in constraining the extent of a target – that is, localizing the target for more detailed evaluation. Estimation errors, available quantitatively from some estimation methods, can be used to develop an appreciation of the effects of additional information (e.g., drill-hole density) on the quality of mineral inventory estimation.

Global estimates concerned with the average grade and tonnage of very large volumes of a deposit can be used to quantify a reserve or resource that will form the basis of continuing production. Thus, global resources/reserves represent a justification for long-term production planning. Global resources commonly are referred to as *in situ* or *geologic* because normally only very general economic factors have been taken into account. Increasingly, efforts are being made early in the exploration of a deposit to estimate the proportion of in situ resources/reserves that are recoverable under certain mining conditions.

Local estimation, both at the feasibility stage and in operating mines, commonly is used for short- and medium-range production planning. In particular, local block estimates generally serve as the basis for classifying blocks as ore or waste (see Fig. 1.1). Recoverable reserves are determined from a subset of local estimates (reserves actually recoverable by the planned mining procedures) and serve as a basis for financial planning.

In some cases, mineral inventories are approximated by numeric simulations. Conditional simulations (i.e., simulations that honor the available data) can be used to provide insight regarding grade continuity, mine planning, mill planning, and overall financial planning. Simulations are not used as widely as warranted.

Many governmental jurisdictions require that resource/reserve estimates be classified according to an accepted legal or professionally recognized system for purposes of formal publication, use by financial institutions, use in public fund-raising, and so on. In this respect, the resources/reserves represent an asset with an attendant level of risk related to the specifications (commonly poorly documented) for the corresponding classes of the classification system. In Canada, for example, reserves were classed historically in the following categories of decreasing reliability: proven, probable, or possible.

The process of mineral inventory estimation can be seen above to be an ongoing endeavor subject to continual refinement. King et al. (1982) note that "an Arizona porphyry copper mine achieved satisfactory ore reserve prediction only after 20 years of study and trial" (p. 18) and "it took a large Australian nickel mine six years to develop their computerized polygonal procedures to the point of yielding a 'planned mining reserve'" (p. 18). Optimal procedures for resource/reserve estimation are not cut and dried; rather, they contain an element of "art" based on experience that supplements technical routine and scientific theory.

In addition to preparing comprehensive mineral inventory estimations, geologic and mining professionals commonly are required to conduct a "reserve audit" or evaluation of a mineral inventory estimation done previously by others (cf. Parrish, 1990). A *reserve audit* is a comprehensive evaluation based on access to all geologic data, assays, and other pertinent information. An auditor does not repeat the entire study, but normally might take about one-tenth the time of the original study. The purpose is to provide confidence as to the quality of data and methodologies used and the general reliability of the reported estimates. An auditor's aim is to provide assurance that high professional standards have been maintained in decision making and that acceptable industrial practice has been followed in arriving at a resource/reserve estimate. Parrish (1990) provides a concise summary of the structure of a resource/reserve audit.

Exploration geologists are called on routinely to provide rough estimates of tonnages and grades of large volumes of rock based on a very limited amount of exploration data. Such "guesstimates" are not comparable to a mineral inventory study; rather, they are rough professional estimates as to the size of a likely target, based on limited geologic and assay information and/or a nonrigorous approach to estimation. These guesstimates should be viewed as attempts to define exploration targets that require verification. Such crude attempts should not be confused with more rigorous estimates of resources or reserves, unless the information is sufficient to satisfy the criteria of a formal classification (see Section 1.3.4).

1.3: SOME ESSENTIAL CONCEPTS IN MINERAL INVENTORY

As with so many professional undertakings, mineral inventory reports are filled with professional jargon and a range of usage not standardized everywhere. In some cases, the common use of terminology is lax relative to widely accepted technical definitions. In other cases, certain technical terms have specific meanings that are not widely appreciated either within or outside the industry because they have entered the mineral inventory literature from elsewhere; such is the case, for example, with a number of terms originating from the field of geostatistics. For the forgoing reasons, it is useful to define a number of terms and concepts that are now widely integrated into mineral inventory work.

1.3.1: Ore

The wide range of published definitions of the term *ore* has prompted Taylor (1986, p. 33) to propose the following definition: "the valuable solid mineral that is sought and later extracted from the workings of a mine; for the hoped or expected (though not always achieved) advantage of the mine operator or for the greater good of the community." The generality of this definition obscures the range of common usage.

The term *ore* is applied to mineralized rock in three senses: (1) a geologic and scientific sense; (2) quality control in ore reserves; and (3) for broken, mineralized ground in a mine, regardless of grade. In mineral inventory work, the second definition is of importance and implies the distinction of *ore* (mined at a profit) and *waste* (containing insufficient value to earn a profit). The recognition of such ore involves the consideration of three different categories of profit: (1) that relating to small increments of ore; (2) that referring to annual or other periodic outputs of ore; and (3) that expected from entire ore bodies. Note that for each of these types of profit, there is a different corresponding threshold (cutoff grade) that separates ore from waste. As indicated by Wober and Morgan (1993), the definition of the ore component of a particular mineral deposit is a function of time factors (metal price, technology, tax regime, etc.), place factors (relation to infrastructure), legal factors (safety, environmental, labor, etc.), profit, and discount rates.

In general, mines are put into production with the understanding that there will be an acceptable return on the necessary investment. Circumstances may dictate significant changes to the foregoing philosophy, or the concept of profit might change over time. A mine might operate at a loss for reasons of tax advantage, long-term planning, anticipation of short-term changes in metal prices or product sales, and so on. Moreover, government may impose regulations or incentives that affect operations normally expected to create losses. For example, historically, in the case of South African gold mines and the Alberta tar sands, some material was mined at a loss because it would otherwise not be recoverable and perhaps would be lost forever.

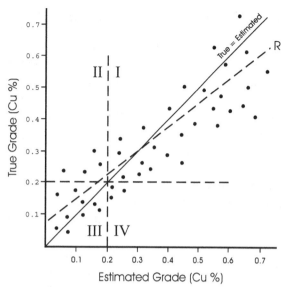

Figure 1.2: A plot of estimated grades versus true grades for many blocks (selective mining units) of mineralized ground (ore). A cutoff grade (x_c) applied to both axes divides the individual estimates into four quadrants that classify the estimates as follows: quadrant I = ore blocks correctly classed as ore; quadrant II = ore blocks incorrectly classed as waste; quadrant III = waste blocks correctly classed as waste; and quadrant IV = waste blocks incorrectly classed as ore. The fundamental concept inherent in this diagram is that random estimation errors necessarily require that an estimate can never be met by production (unless additional ore is encountered) because some ore is lost (i.e., incorrectly recognized as waste), and the remaining ore that is recognized is diluted by waste incorrectly classed as ore. A regression line (R) through the data indicates the common result for polygonal estimates (i.e., on average, estimates of high values overestimate the true grade, whereas estimates of low values underestimate the true grade). The alternate situation, in which high grades are underestimated and low grades overestimated, is common in situations in which groups of data are averaged to produce estimates.

1.3.2: Cutoff Grade

The concept of *cutoff grade and its practical applications* have invoked wide discussion in the technical literature (e.g., Lane, 1988; Taylor, 1972, 1985). For practical purposes, a *cutoff grade* is a grade below which the value of contained metal/mineral in a volume of rock does not meet certain specified economic requirements. The term has been qualified in many ways, particularly by accountants, and some ambiguity in its use has resulted (Pasieka and Sotirow, 1985). Cutoff grades are used to distinguish (select) blocks of ore from waste (Fig. 1.2) at various stages in the evolution of mineral inventory estimates for a deposit (i.e., during exploration, development, and production stages). Ore/waste selection is based on estimates (containing some error) rather than on true grades (which are unknown). Hence, in the case of block estimation it is evident that some ore blocks will be inadvertently classed as waste (quadrant II in Fig. 1.2) and that some waste blocks will be classed erroneously as ore (quadrant IV in Fig. 1.2). The volume of a predefined mining unit, on which mining selectivity is based and to which a cutoff grade is applied, can change during this evolution, as can the cutoff grade itself. As the cutoff grade increases, the tonnage of ore decreases and the average grade of that tonnage increases. As a rule, strip ratio (the units of waste that must be removed for each unit of ore) also increases with increasing cutoff grade. Generally, a fairly narrow range of cutoff grades must be considered in the process of optimizing the selection of a cutoff grade for a particular mining scenario.

The concept of cutoff grade is closely linked to the practical connectivity of blocks of ore at the production stage (Allard et al., 1993; Journel and Alabert, 1988). As the cutoff grade rises, the volume of ore decreases and becomes compartmentalized into increasing numbers of smaller, separated volumes (i.e., decreased "connectivity" with increasing cutoff grade) as illustrated in Fig. 1.3. Cutoff grades represent economic thresholds used to delineate zones of mineral/metal concentration for potential mining. This delimitation of ore/waste can be a *cutoff grade contour* or a series of straight line segments (steplike) separating blocks estimated to be above cutoff grade from those below cutoff grade. Consequently, the quality of block estimates to be contoured or otherwise grouped must be understood vis-á-vis grade continuity (Section 1.3.3) in order to appreciate the possible magnitude of errors in cutoff-grade contours or block estimates.

Estimation of cutoff grade, although a complex economic problem beyond the scope of this book, is tied to the concept of operating costs (per ton) and can be viewed simplistically as outlined by John (1985). Operating cost per ton milled, OC, is given by

$$OC = FC + (SR + 1) \times MC$$

where

FC = fixed costs/ton milled
SR = strip ratio
MC = mining costs/ton mined.

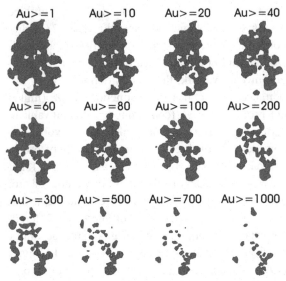

Figure 1.3: The concept of *connectivity* of ore as a function of cutoff grade. Data are 1,033 rock samples from the Mitchell–Sulphurets mineral district (Cheng, 1995), northern British Columbia, for which Au analyses (*g/mt*) have been contoured using different threshold values (cutoff grades). As the cutoff value increases, the high connectivity of Au deteriorates to increasing numbers of unconnected, isolated highs. Of course, where the cutoff value approaches the tail of the distribution, the number of high-grade patches decreases.

Cutoff grade, useful at the operational level in distinguishing ore from waste, is expressed in terms of metal grade; for a single metal, cutoff grade can be determined from operating cost as follows:

$$g_c = OC/p$$

where g_c is the operational cutoff grade (e.g., percent metal) and p is the realized metal price per unit of grade (e.g., the realized value from the smelter of 10 kg of metal in dollars, where metal grade is in percent). More exhaustive treatments of cutoff grade are provided by Taylor (1972, 1985) and Lane (1988).

Optimizing cutoff grade (selecting the cutoff grade that maximizes cash flow) is based on a confident mineral inventory estimate (e.g., as summarized in Table 1.2), where cash flow (*CF*) is given by

$$CF = \text{Revenue} - \text{Operating Costs}$$
$$= (g \times F \times P - OC)T$$

Table 1.2 Grade–tonnage relations that simulate a typical porphyry copper deposit

Cutoff grade	Tons of ore (millions)	Average grade ore	Strip ratio
0.18	50.0	0.370	1.00:1
0.20	47.4	0.381	1.11:1
0.22	44.6	0.391	1.24:1
0.24	41.8	0.403	1.39:1
0.26	38.9	0.414	1.57:1
0.28	35.9	0.427	1.78:1
0.30	33.0	0.439	2.03:1
0.32	30.0	0.453	2.33:1
0.34	27.2	0.466	2.68:1

Source: After John (1985).

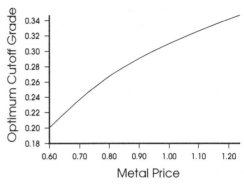

Figure 1.4: Optimum cutoff grade as a function of changing metal price for an information base used by John (1985) to illustrate the concept of cutoff grades. This diagram is based on the data presented in Tables 1.2 and 1.3.

where

g = average grade of ore mined

F = recovery proportion per ton milled

P = realizable value of metal per ton milled

T = tons milled.

The hypothetical mineral inventory data of Table 1.2 (John, 1985) simulate a porphyry copper deposit and are used to make the cash flow estimates shown in Table 1.3 for various potential cutoff grades. Clearly, for the situation assumed, cash flow is maximized for a cutoff grade of 0.28 percent metal. Changes in strip ratio, metal prices, percent recovery, and so on change the optimal cutoff grade. One useful concept emphasized by John (1985) is that the formulas presented here can be used to evaluate the effect that various alternatives (e.g., changing metal prices, different metal recovery) have on the selection of a cutoff grade; that is, a "sensitivity analysis" can be conducted for various parameters in which each parameter is varied independently in order to evaluate its impact on cutoff grade estimation. One such analysis (Fig. 1.4) shows variations in optimum cutoff grade with variation in metal price for the example used in Tables 1.2 and 1.3.

Parrish (1995) introduces the concept of *incremental ore* as

that material that in the course of mining "bona fide ore" must be drilled, blasted and moved but contains sufficient value to pay the incremental

costs of realizing that value and provide some profit as well. Incremental costs include the difference between delivery to the waste area and delivery to the feed bin, stockpile pad or crusher and the costs of crushing, processing, royalties, etc. (p. 986)

It is evident from this definition that mining costs are not included in incremental ore; hence, the cutoff grade used for its definition is less than the cutoff grade based on all costs and is consistent with the term *internal cutoff grade* (e.g., Marek and Welhener, 1985). The paradox of incremental ore is that in theory it cannot be classed as reserves (because all costs are not recovered), but in practice it makes sense to mine and process it. Owens and Armstrong (1994, p. 53) also recognize the paradox when they state, "The grade cut-off concept has a role for selection of stope or ore zone size units, but not for isolated blocks of low grade within ore zones intended for underground mining."

1.3.3: Continuity

Continuity is "the state of being connected" or "unbroken in space." (*Oxford English Dictionary*, 1985, p.186). In mineral deposit appraisals, this spatial definition commonly is used in an ambiguous way to describe both the physical occurrence of geologic features that control mineralization and grade values. Such dual use of the term *continuity* leads to

Table 1.3 Calculation of cash flow (dollars per ton milled) for example in Table 1.1[a]

Cutoff grade	Average ore grade	Strip ratio	Operating cost ($/t)	Total revenue	Operating cash flow
0.18	0.370	1.00:1	3.50	5.24	1.74
0.20	0.381	1.11:1	3.58	5.38	1.80
0.22	0.391	1.24:1	3.68	5.54	1.86
0.24	0.403	1.39:1	3.80	5.70	1.90
0.26	0.414	1.57:1	3.93	5.86	1.93
0.28	0.427	1.78:1	4.09	6.04	1.95
0.30	0.439	2.03:1	4.28	6.22	1.94
0.32	0.453	2.33:1	4.50	6.40	1.90
0.34	0.466	2.68:1	4.76	6.59	1.83

[a] Results in Table 1.3 can be obtained from information in Table 1.2 with $MC = 0.76$, $FC = 1.98$, recovery $= 0.83$, and a metal price of $0.85/lb in formulas 1.1 and 1.3.
Source: After John (1985).

ambiguity. To clarify this ambiguity, Sinclair and Vallée (1993) define two types of continuity that bear on the estimation of mineral inventories as defined in Table 1.4.

Distinction between the two types of continuity can be appreciated by the particularly simple example of a vein (a continuous geologic feature), only part of which (ore shoot) is mineralized with economically important minerals. Value continuity can be defined within the ore shoot. These two types of continuity are partly coincident in space, perhaps accounting for the ambiguity in past use of the unqualified term *continuity*. Understanding both types of continuity is essential in appreciating the implications of each to the estimation process. An example of the impact that an error in interpreting geologic continuity can have on mineral inventory estimation is shown in Fig. 1.5. Chapter 3 contains a detailed discussion of geologic continuity; Chapter 8 is concerned with a quantitative description of value continuity.

1.3.4: Reserves and Resources

Mineral inventory is commonly considered in terms of resources and reserves. Definitions currently vary from one jurisdiction to another, although there are increasing efforts being directed toward internationally acceptable definitions. In the absence of such international agreement, there is an increasing tendency

in both industry and technical literature for an ad hoc agreement centering on definitions incorporated in the "Australasian Code for Reporting of Identified Mineral Resources and Ore Reserves" (Anonymous, 1989, 1999). Thus, the Australasian terminology is summarized in Fig. 1.6.

A *resource* is an in situ (i.e., on surface or underground) mineral occurrence quantified on the basis of geologic data and a geologic cutoff grade only. The term *ore reserve* is used only if a study of technical

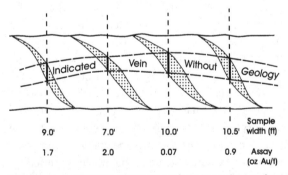

Figure 1.5: A simplistic illustration of the importance of geologic continuity (modified from Rostad, 1986). Interpretations concerning continuity clearly control the volume of ore (and therefore the tonnage) as well as the way in which sample grades will be extended. Detailed study of the geologic form and controls of mineralization constrain the geometric model of a deposit, which in turn has an impact on mine planning. In this case, vein intersections in a shear zone are shown misinterpreted as a simple vein rather than correctly as a series of sygmoidal veins.

Table 1.4 Two categories of continuity in mineral inventory estimation

Geologic continuity	Spatial form of a geometric (physical) feature such as a mineral deposit or mineral domain.
	    Primary: veins, mineralized shear, mineralized stratum
	    Secondary: postmineral faults, metamorphism, folding or shearing of deposits

Value continuity	Spatial distribution features of a quality measure such as grade or thickness within a zone of geologic continuity. Nugget effect and range of influence are quantified. Examine on-grade profiles (e.g., along drill holes) qualitatively in various directions. Quantify for each geologic domain using an autocorrelation function (e.g., semivariogram)

Source: After Sinclair and Vallée (1994).

and economic criteria and data relating to the resource has been carried out, and is stated in terms of mineable tons or volume and grade. The public release of information concerning mineral resources and ore reserves and related estimates must derive from reports prepared by appropriately qualified persons (i.e., a "competent person").

Prior to mineral inventory estimation, a variety of exploration information is available. As exploration continues, the information base increases and the level of detailed knowledge of a deposit improves. The estimation of reserves or resources depends on this constantly changing data and the continually improving geologic interpretations that derive from the data. Thus, the continuous progression of exploration information first permits the estimation of resources and, eventually, the estimation of reserves of different categories. Reserve estimation is thus seen as continually changing in response to a continually improving database. An indication of the wide range of data affecting mineral inventory estimation and classification is presented in Table 1.5.

On the international scene, it is becoming increasingly common to progress from resources to reserves by conducting a *feasibility study*. A feasibility study of a mineral deposit is "an evaluation to determine if the profitable mining of a mineral deposit is... plausible" (Kennedy and Wade, 1972, p. 70).

The term covers a broad range of project evaluation procedures yielding detailed insight into the geologic and quantitative database, resource/reserve estimation procedures, production planning, mining and milling technology, operations management, financing, and environmental and legal concerns. An exhaustive discussion of the classification of resources and reserves is provided in Chapter 18.

1.3.5: Dilution

Dilution is the result of mixing non-ore-grade material with ore-grade material during production, generally leading to an increase in tonnage and a decrease in mean grade relative to original expectations. A conceptual discussion of dilution during various mining operations is provided in Fig. 1.7 after Elbrond (1994). It is convenient to consider dilution in two categories: *internal* (low-grade material surrounded by high-grade material) and *external* (low-grade material marginal to high-grade material). Internal dilution can be subdivided into (1) sharply defined geometric bodies and (2) inherent dilution. *Geometric internal dilution* results from the presence of well-defined waste bodies within an ore zone (e.g., barren dykes cutting an ore zone, "horses"). *Inherent internal dilution* results from the decrease in selectivity that accompanies an increase in the block size (e.g., resulting from loss of equipment digging selectivity) used as the basis for discriminating ore from waste, even where no entirely barren material is present.

External dilution is the result of sloughing of walls, difficulty of sorting in open pits, or the inadvertent or purposeful mining of barren or low-grade material at the margin of an ore zone. Such dilution is generally significant in cases in which stope walls are physically difficult to maintain because of rock properties or where ore widths are less than the minimum mining width. External dilution can be of somewhat less significance in large deposits with gradational boundaries in comparison with small deposits because

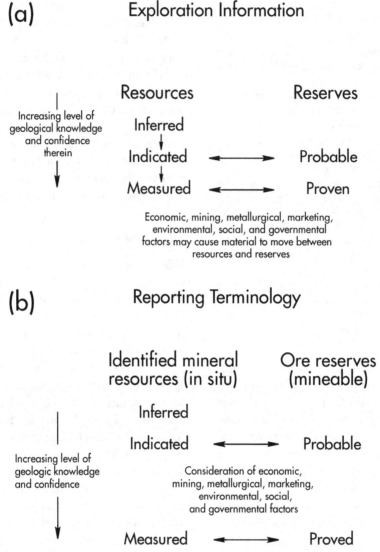

Figure 1.6: Examples of two published classification schemes for resources/reserves. (a) A proposed classification of the Society of Mining Engineers (U.S.). (b) Classification of the Australasian Institute of Mining and Metallurgy, in use in Australia since about 1980.

the diluting material (1) can be a small proportion of the mined tonnage and (2) contains some metal, possibly near the cutoff grade. In general, some uncertain proportion of waste must be taken along with ore during the mining operation. This form of dilution can be impossible to estimate with confidence in advance of mining; experience is probably the best judge. Accepted procedures for estimating dilution in underground mining operations are summarized by Pakalnis et al. (1995).

1.3.6: Regionalized Variable

A *regionalized variable* is a variable distributed in space in a partly structured manner such that some degree of spatial autocorrelation exists. The structured

Table 1.5 Examples of the information base required for a mineral inventory study

Location	Surveyed maps: cross sections showing locations of geologic control, mineral showings, sample locations, drill-hole locations/orientations, exploration pits, and underground workings; an indication of the quality of location data of various types.
Geologic	Detailed geologic maps and sections showing rock types, structural data, alteration, mineralization types, etc.; reliability of geologic information; description of supporting geochemical and geophysical survey data/intepretations; factual data distinguished from interpretation; documentation of drill-hole logging procedures (including scales and codes used); detailed drill logs; textural and mineralogic features of importance to mill design; spatial variability of geologic features that have an impact on mill and mine design, including effects on metal/mineral recovery and determination of cutoff grade; ore deposit model with supporting data; geologic influence on external and contact dilution; justification for separate domains to be estimated independently.
Sampling/assaying	Descriptions of all sampling methods, including quantification of sampling variability; sample descriptions; sample reduction procedures and their adequacy; bulk sampling design/results; duplicate sampling/assaying program.

character is called a *regionalization* and is characterized by the fact that nearby samples are, on average, more similar in value than are more widely spaced samples. A regionalization of grades is consistent with the occurrence of high-grade zones in a lower grade field, as is common in mineralized ground. In Fig. 1.8a, two samples with a spacing smaller than the dimensions of the "high-grade" mineralized zones will be more similar, on average, than two samples spaced more widely than the dimensions of the high-grade zones. Many mining-related variables are regionalized, including vein thickness, grades, fracture density, and metal accumulations (grade × length). In general, regionalized variables consist of at least two components: a random component and a structured component. The random component commonly masks the structured component to some extent (e.g., in contoured maps of the regionalized variable). Various autocorrelation (mathematical/statistical) functions can be used to characterize a regionalized variable and permit the autocorrelation to be incorporated into the mineral inventory estimation procedure. Statistics of independent (random) variables ignore the effects of spatial correlation, and thus might not take full advantage of the data in estimating average grades for parts or all of a mineral deposit.

Regionalized variables such as grades are defined on a particular support within a defined field or *domain* (e.g., Cu grades of 2-m drill-core lengths within a mineralized zone surrounded by barren rock). *Support* refers to the mass, shape, and orientation of the sample volume that is subsampled for assaying. One-m lengths of split core from vertical drill holes represent a regionalized variable with uniform support. If the length of split core in a sample is increased to, for example, 2 m, a new regionalized variable of different support is defined. A *smoothing* of values (decrease in variability) – that is, a *regularization* – accompanies the increase of support of a regionalized variable. The concept is illustrated numerically and graphically in Figs. 1.8b and 1.8c. It is evident from the foregoing discussion that the common procedure of constructing composite grades (uniform support) from grades of smaller supports leads to smoothing of the original data values.

1.3.7: Point and Block Estimates

It is essential to distinguish between the concept of estimating grade at a point (or a very small volume such as a sample) and estimating the average grade of a large volume such as a *production unit* or *block* (a volume that may be many orders of magnitude greater than the size of a sample). Point or punctual estimates are used in mining applications mainly as a means of validating or comparing estimation

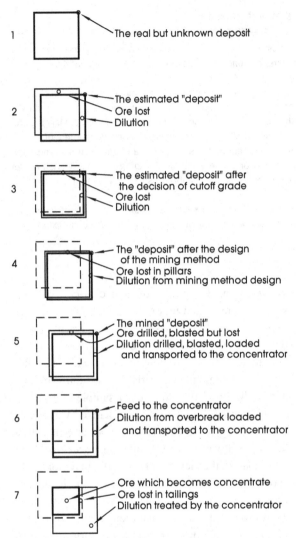

Figure 1.7: A conceptual representation of dilution of ore at various stages of mining and milling (after Elbrond, 1994). Note that at each of the stages represented, some ore is lost and some waste is included with ore.

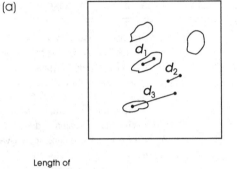

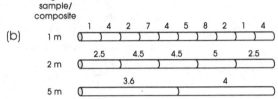

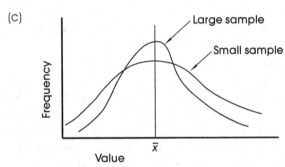

Figure 1.8: Regionalized variables and the concept of sample support. (a) The concept of a spatial dependency of a regionalized variable such as grade. Sample pairs separated by less than the dimensions of small high-grade (contoured) zones (d_1 and d_2) are more similar, on average, than are more widely spaced sample pairs (d_3). (b) Numeric demonstration of the smoothing effect of combining grades with small support into weighted grades of larger support. (c) General relation of grade dispersions illustrated by histograms of grades for relatively small volumes and relatively large volumes.

techniques and to estimate regular grid intersections as a basis for contouring (e.g., contouring the cutoff grade to define the ore/waste margin). The sample base (very small volumes) is the *point* base used to make point estimates. Block estimates are also made from the sample (point) database but are large volumes relative to sample volumes; generally, samples are measured in kilograms (a few hundreds of cubic centimeters), whereas blocks commonly are mea-

sured in hundreds or thousands of tons (hundreds to thousands of cubic meters).

Consider the example illustrated in Fig. 1.9 (data are listed in Table 1.6), in which the center of a large block is to be estimated from the available data shown. Estimation of the value at the block center is equivalent to determining the contour that passes through the center. Clearly, there are several ways to contour the

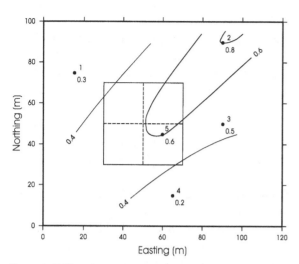

Figure 1.9: Five data points used to interpolate contours from which an estimated, average panel grade is obtained (see text). One possible estimate is the interpolated value at the panel center. A second estimate can be obtained by dividing the panel into four equal blocks by two dashed lines, interpolating the value at the center of each block, and averaging the four values. Data are summarized in Table 1.6.

Table 1.6 Coordinates and data values for estimation of the panel in Fig. 1.9

Sample no.	x Coordinates[a]	y Coordinates[a]	Grade (% Cu)
1	15	75	0.3
2	90	90	0.8
3	90	50	0.5
4	65	15	0.2
5	60	45	0.6

[a] Distances in meters.

data, each of which will provide a different estimate for the central point; for the contours shown, an interpolation provides a point estimate of about 0.59. The average grade of the block can be estimated from the available data; for example, a block grade can be estimated as the weighted average using areas of various contour intervals within the block as weights and the mean grades of the contour intervals selected. In the case presented, a weighted average of about 0.54 is obtained (compared with 0.59 for the block center). The two estimates differ by about 9 percent and, in general, there is no reason to expect a point estimate to be equivalent to a block estimate, even though the block contains the point in question.

1.3.8: Selective Mining Unit

A selective mining unit (SMU) is the smallest block on which selection as ore or waste is commonly made. The size of an SMU generally is determined from constraints associated with the mining method to be used and the scale of operations. For design purposes, a mineral deposit (and perhaps much of the surrounding waste) can be considered a three-dimensional array of SMUs (e.g., Fig. 1.10), each of which has assigned metal grades and other parameters. This three-dimensional array of SMUs is commonly the fundamental basis for feasibility studies; hence, SMUs are normally the minimum volume for which estimates are required. In a particular situation, an SMU could be the volume of broken material in a single truck that is directed to either the mill or the waste dump, depending on the assigned grade.

Selection of an SMU is also critical to mineral inventory estimation because it can be the basic unit classified, first as a resource (measured, indicated, or inferred) and second possibly as a reserve (proved or

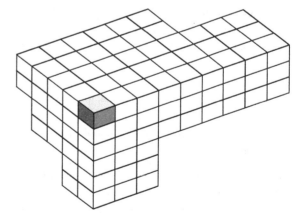

Figure 1.10: A three-dimensional array of blocks designed to approximate the geometry of an ore deposit. The block size is commonly taken as the selective mining unit (SMU), the smallest volume for which selection as ore or waste is possible and thus the smallest volume for which an average grade must be estimated.

probable). Consequently, the dimensions of an SMU must be determined carefully. Those dimensions depend on such interrelated factors as probable blasthole spacing during production, specifications of mining equipment, bench height, and blasting characteristics of the ground.

1.3.9: Accuracy and Precision

Accuracy is nearness to the truth, and significant inaccuracies can produce identifiable *bias* (departure from true value). Precision is a measure of reproducibility of a result by repeated attempts. It is possible to have good reproducibility and poor accuracy, so both must be considered in detail.

There are many potential sources of significant errors in the estimation of mineral inventory, including

1. Sampling error
2. Analytical error (including subsampling error)
3. Estimation (extension) error, that is, the error made in extending the grade of samples to a particular volume of mineralized rock
4. Bulk density (the all-too-common assumption that bulk density of mineralization remains constant throughout a deposit)
5. Geologic error, that is, errors in the assumptions of ore continuity and the geometry of a deposit
6. Mining method not adapted to the geometry of the deposit, that is, selectivity of ore and waste that is not optimal
7. Variable dilution from surrounding wall rock
8. Human error (misplotting of data, misplaced decimals, etc.)
9. Fraud (salting, substitution of samples, nonrepresentative data, etc.).

These factors can lead to both inaccurate estimates and imprecise estimates. Inaccurate results can be obtained very precisely, for example, if a bias (systematic error) is inherent in any one of the sampling procedures, the analytical method, or the data selection procedure. Regardless of how accurate assays or ore resource/reserve estimation procedures are, there is some level of random error in the data or estimates.

Concepts of accuracy and uncertainty are widely misused in the reporting of metal reserves and resources. King et al. (1982, p. 5) quote the following reported reserves for three mines of an unnamed mining company:

Mine A:	8,192,000 tons at 0.174 oz Au/t
Mine B:	27,251,000 tons at 1.87% Cu
Mine C:	152,533,400 tons at 7.11% Zn.

These examples are all given to more significant figures than are warranted and emphasize the importance of honoring the number of significant figures in presenting mineral inventory estimates. Mining companies traditionally report mineral inventory to more figures than are justified. A number with three significant figures (i.e., reproducible by test to better than the first two figures) indicates an uncertainty of between 1:100 and 1:1,000. Additional figures may be not only a waste of time but may be misleading, particularly to the uninitiated. Certainly, experience has shown that reserve estimation does not achieve an accuracy in excess of 1:1,000!

Errors in mineral inventory estimates are not entirely quantifiable. There are the obvious estimation errors that arise because a small sample is used to estimate a large volume. In addition, however, there is potential for very large error arising from the uncertainty of geologic interpretation of the geometry and internal continuity of ore. For a particular period of mining development in Australia, King et al. (1982, p. 3) state, "In Australia...some 50 new mining ventures...reached the production stage. Of these, fifteen were based on large, good grade deposits, relatively easily assessed. Of the remainder, ten suffered ore reserve disappointments more or less serious and some mortal." As examples of the level of error involved, they cite the following ratios of expectations (by estimation) to realizations (production) for a number of deposits:

100:75 major uranium mine (i.e., a 33% overestimate)

100:55 sizable copper mine (i.e., an 82% overestimate)

100:75 large gold mine (i.e., a 33% overestimate)
100:80 small nickel mine (i.e., a 25% overestimate).

These dramatic shortfalls of expectations "happened
to some of the most experienced companies in the
industry" (King et al., 1982; p. 3). Similar mishaps
in estimation have occurred internationally; for example, Knoll (1989) describes comparable cases of
initial estimates for some Canadian gold deposits not
being met as further evaluation or production ensued.
Champigny and Armstrong (1994, p. 23) summarize
several statistics relating to reserve estimation as follows: "In 1990 Graham Clow . . . presented an examination of 25 Canadian gold projects, at an advanced
stage . . . . Only three have lived up to expectations . . . .
He concluded the main reason for failure was
poor reserve estimation practice" "In 1991 . . .
Harquail . . . analyzed 39 recent North American gold
mining failures . . . . From the 39 failures, Harquail
attributed 20 failures to reserve issues: (1) basic
mistakes, (2) improper or insufficient sampling, and
(3) a lack of mining knowledge" "In early 1992 a
South African . . . provided a less than rosy picture
for small gold mines started in his country during the
1980s. For 13 mines which started out only 3 are still
working and all three have a marginal performance.
He correlated this lack of success to an overestimation
of reserves" It is one of the important aims of this text
to provide insight into how to minimize the effects
of the many sources of error that can lead to serious
mistakes in mineral inventory estimation.

1.4: A SYSTEMATIC APPROACH
TO MINERAL INVENTORY ESTIMATION

Mineral inventory estimation is a complex undertaking that requires a range of professional expertise. In
cases in which reserves (as opposed to resources) are
being estimated, a large team of professionals might
be required (cf. Champigny, 1989). Geologic, mining,
and milling engineers normally are involved, as well
as experts in mineral economics, financing, and any
number of specialized fields that might be pertinent
to a particular deposit or the local infrastructure re-

quired to mine a deposit. Fundamental to the mineral
inventory, however, is a knowledge of the spatial distribution of grade and the particular locations of volumes of mineralized rock that are above cutoff grade.
Whatever the purpose of a mineral inventory, the estimation should be done in an orderly or structured
manner. Weaknesses, especially important assumptions, should be flagged so that they can be improved
as new information permits. The eventual mineral inventory should not be seen as a fixture; rather, they
should be viewed as an evolving product, periodically
subject to improvement with the continuing flow of
new information. The more important topics relating
to mineral inventory estimation (see also Table 1.1),
roughly in their sequence of occurrence, include:

1. Geologic modeling
2. Continuity documentation, both geologic and
   value
3. Evaluation of quality of data and quality control
4. General data evaluation, i.e., implications of quantitative data (histograms, trends, correlations, etc.)
5. Global resources estimation
6. Local resources
7. Recoverable reserves
8. Simulation.

Of course, these undertakings are to some degree overlapping and should not be seen as mutually exclusive
or as all inclusive. Furthermore, although the order
listed is a relatively common order in practice, the specific aims of a mineral inventory study may necessitate
significant changes. Commonly, for example, geologic modeling, continuity, and data evaluation are
done partly contemporaneously. In some cases, simulation may be done early as an aid to understanding
continuity rather than later as a basis for optimizing
mining and/or milling procedures; only one of local
or global reserves might be required; and so on.

The early stage of geologic modeling should not
be confused with the more engineering-oriented terms
of deposit modeling, reserve modeling, and so on. Geologic modeling is concerned with the recognition,
spatial disposition, and interpretation of a wide range
of geologic features and how these are represented visually (perhaps simplistically). Terms such as *reserve*

modeling relate to how a deposit is compartmentalized geometrically for estimation purposes. Sides (1992b) summarizes a variety of geometric approaches to subdividing a mineral deposit for reserve estimation, methods that are not mutually exclusive, as follows:

1. Gridded seam – a "two-dimensional" or tabular body in which the planar surface is subdivided polygonally
2. Serial slices – commonly equispaced plans and/or sections showing polygonal outlines of "ore"
3. Block modeling – mineralized volume divided into a three-dimensional array of blocks that generally coincide in size with the SMU. This modeling procedure is widely used, particularly in computer-based approaches to reserve estimation.
4. Solid modeling – use of "geometric primitives defining volumes of common properties" adapted from CAD/CAM applications and being used increasingly
5. Boundary representations – modeling discontinuity surfaces within a large volume of ground. This procedure enables the distinction (and independent evaluation) of various ore types based on various criteria of a lithologic, structural, mineralogic, grade, or other nature.

1.5: TRADITIONAL METHODS OF MINERAL INVENTORY ESTIMATION

Several traditional and widely used mineral inventory estimation methods described, for example, by Patterson (1959), King et al. (1982), Annels (1991), and Stone and Dunn (1994), are illustrated in Fig. 1.11. These procedures include

1. Method of sections (plans); longitudinal, transverse
2. Polygonal methods
3. Triangular method
4. Regular grid, random stratified grid
5. Inverse distance weighting ($1/d$, $1/d^2$, $1/d^{2.7}$, $1/d^3$, etc.)
6. Contouring methods.

Methods 1 through 3 involve estimation of volumes; methods 4 through 6 concern point samples on which properties (e.g., grades) have been estimated and that are commonly used to estimate points on a regular grid. Each of these methods has found applications in industry for resource/reserve estimation; all are empirical and their use has depended largely on the experience of the user. When applied with close geologic control and adequate sample control, they

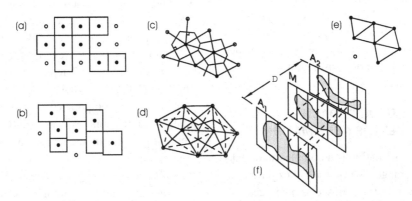

Figure 1.11: Examples of some common methods of grade estimation (after Patterson, 1959). (a) Polygonal: uniform rectangular blocks centered on uniformly spaced data. (b) Polygonal: nonuniform rectangular blocks centered on irregularly spaced data. (c) Polygonal: polygons defined by perpendiculars at midpoints between data points. (d) Polygonal: polygons about data points defined by bisectors of angles in a mesh of contiguous triangles. (e) Triangular: each polygon is assigned an average of grades at three vertices. (f) Method of sections: ore outlined on drill sections and weighted average grade determined on individual sections. Ore between sections is interpolated, commonly by a simple linear interpolation between neighboring sections.

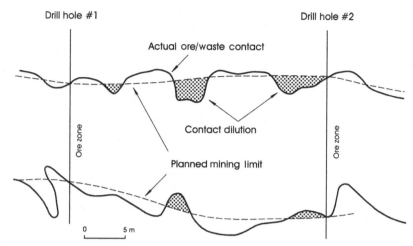

Figure 1.12: Depictions of an irregular ore–waste contact approximated by smooth dashed lines that are interpreted using information from drill holes (on sections), but that are interpolated between drill holes. With such irregular contacts, the method of sections must necessarily be incorrect (i.e., it can only coincidently result in the correct grade estimate for the interpolated zone because some waste is included within the interpreted zone of mineralization). Modified from Stone and Dunn (1994).

have commonly proved useful and reliable. However, there are characteristics inherent in their individual methodologies that can lead to drastic errors if the methods are applied inappropriately.

1.5.1: Method of Sections

The method of (cross) sections (cf. Fig. 1.11) is applied most successfully in the case of a deposit that has sharp, relatively smooth contacts, as with many tabular (vein and bedded) deposits. Assay information (e.g., from drill holes) commonly is concentrated along equispaced cross sections to produce a fairly systematic data array; in some underground situations, more irregular data arrays can result, for example, from fans of drill holes. A geologic or grade interpolation is made for each cross section (or plan), and the interpretation on each section is projected in smooth or steplike fashion to adjoining sections. Note that this two-stage interpolation (on and between sections) represents the principal underlying assumptions of the method of sections (or of plans) – that is, a smooth (or, alternatively, a one-step) interpolation between sections, as shown schematically in Fig. 1.11. A weighted-average grade on a section is normally

projected to a volume extending halfway to adjoining sections. Figure 1.12 clearly demonstrates how the irregularity of the ore–waste contact can influence mineral inventory; as a rule, grade is overestimated because an unknown quantity of waste is included in the interpreted ore. Moreover, an unknown amount of ore is lost to waste (i.e., some ore-grade material is outside the designed mining limits). The great strengths of the method of sections are the strong geologic control that can be imposed and the fact that it is a natural evolution from the plans and sections that are a standard industry approach to viewing and interpreting detailed mineral deposit data.

1.5.2: Polygonal Methods

Polygonal methods include a number of different approaches to the use of limited amounts of data to estimate individual polygonal volumes that have been defined geometrically in one of several ways. In the case of polygonal methods in which there is no smoothing of the raw data, a single sample grade is assigned as the mean grade of a large block (polygonal prism) of ore or waste (Figs. 1.11 and 1.13). Procedures are described by Annels (1991) and Stone and Dunn (1994),

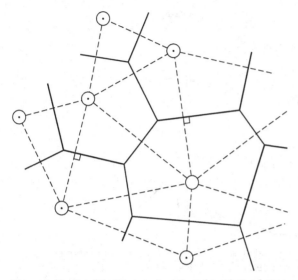

Figure 1.13: Details of the construction of a polygonal area to which a single contained grade is applied (i.e., the contained sample grade is extended to the polygon). Circles are data points; dashed lines join adjacent data points and form Delaunay triangles; thick lines defining a polygon (Voronoi tesselations) are perpendicular to the dashed lines and divide the dashed lines into two equal segments.

among others. On a two-dimensional projection of a deposit or a bench showing sample locations, polygons are defined in several ways, one of the more common being by a series of perpendicular bisectors of lines joining sample locations (see Fig. 1.13). This procedure is equivalent to a process known as *Voronoi tesselation*. Each polygon contains a single sample location and every other point in the polygon is nearer to the contained datum than to any external datum. There are arbitrary decisions that must be made as to how marginal prisms are bounded at their outer edge. The third dimension, the "height" of the polygonal prism, represents the thickness of the deposit or bench and is perpendicular to the projection plane. This process leads to a pattern of polygonal prisms that are assigned the grade of the contained datum. The method is simple, rapid, and declusters the data automatically. Other methods of determining individual polygonal prisms can result in data not being located optimally within prisms. For example, if a square grid is superimposed on a two-dimensional data field for one bench, each "square" might be estimated by the grade

of the nearest data value; this variation of the polygonal method is termed the *nearest neighbor estimation procedure.*

Previous reference has been made to the regularization of grades (decrease in variability) as the support (volume) increases (see Fig. 1.8). A corollary of this is that on average, the use of raw sample grades for mean grades of large volumes overestimates the grade of high-grade blocks (see Fig. 1.2) and, correspondingly, underestimates the grade of low-grade blocks (e.g., a conditional bias, in which the bias is dependent on the grade estimated). Royle (1979) discusses the factors that affect the magnitude of such bias in polygonal estimates, including variance of the data, the nugget effect, and the range of influence of samples. This bias is a particularly serious problem in cases in which cutoff grades are above the mean value of the grade distribution because strong biases toward overestimation of both "high"-grade tonnage and, particularly, average grade above cutoff, can arise, as illustrated in the grade-tonnage curve of Fig. 1.14. A further problem with polygonal methods is that spatial anisotropies in the grade distribution generally are not taken into account. In addition, margins of deposits

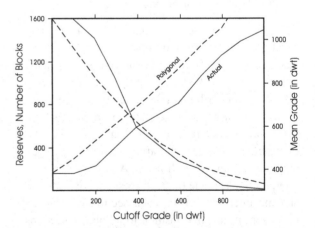

Figure 1.14: The magnitude of errors possible in applying a polygonal method to resource/reserve estimation (after Royle, 1979). Polygonal estimates (dashed lines) presented as "grade tonnage" (number of blocks replaces tonnage) for comparison with actual data (solid lines). Tonnage curves slope downward to the right; average grade curves slope downward to the left.

are generally known by relatively widely spaced assay information; hence, individual assays might be assigned to extremely large volumes of mineralized rock. Clearly, such a block array can be far from regular.

1.5.3: Method of Triangles

A triangular variation of the polygonal approach (Fig. 1.11) is more conservative than the assignment of single values to large blocks; triangular prisms are defined on a two-dimensional projection (e.g., bench plan) by joining three sample sites such that the resulting triangle contains no internal sample sites. The construction of the triangles may use "Delaunay trianglulation," the precursor to Voronoi tesselation (Section 1.2.11). The average of the three values at the apices of a triangle is assigned to the triangular prism (block). The principal advantage is that some smoothing is incorporated in the estimates of individual prisms; thus, estimation of the tail of the grade density distribution is more conservative than is the case with the traditional polygonal approach. Problems with the triangular method are (1) the smoothing is entirely empirical, (2) the weighting (equal weighting of three samples) is arbitrary and thus is not optimal other than coincidentally, (3) anisotropies are not considered, and (4) the units estimated do not form a regular block array.

1.5.4: Inverse Distance Weighting Methods

Inverse distance weighting methods generally are applied to a regular three-dimensional block array superimposed on a deposit; each block is estimated independently from a group of nearby data selected on the basis of distance criteria relative to the point or block being estimated and the amount of data desired for an individual estimate. Inverse distance methods must be done such that weights sum to one or the method is biased and therefore unacceptable. Thus, weights are defined as follows:

$$w_i = \left(1/d_i^x\right) \Big/ \left[\sum \left(1/d_i^x\right) \right] \qquad (1.4)$$

where x is an arbitrary power. In the coincidental sit-

uation that $d_i = 0$ the inverse weighting method theoretically (but not necessarily in practice) defaults to a polygonal estimate, i.e., $w_i = 1$ for $d_i = 0$. In fact, where data are roughly centrally located in individual blocks to be estimated, the inverse distance estimator more closely approaches a polygonal estimate as the power x (of $1/d^x$) increases. The choice of x is subjective and is commonly 1, 2, or 3 but can also be an intermediate value. In practice, anisotropies are not necessarily taken into account, although it is relatively easy to do so by coordinate transforms and greatly improves confidence in the quality of estimates. Such techniques as quadrant (or octant) search can also optimize the spatial distribution of data used to make a block or point estimate. Inverse distance weighting estimation procedures, although subjective, remain popular; they have been found in many cases to produce results that are relatively close to geostatistical estimates obtained by ordinary kriging. IDW methods tailored to deposits give estimates that compare well with production for very large blocks or volumes, a not surprising result where an abundance of evenly distributed data is available. An example of the inverse distance weighting procedure based on hypothetical data is given in Figure 1.15 and data are summarized in Table 1.7. In this example, a search radius is shown centered on a block B, to be estimated. Weights for the samples within the search radius for $1/d$, $1/d^2$, and $1/d^3$ are listed in Table 1.7 along with the corresponding estimates. Note that all the estimates assume isotropy (uniform weighting in all directions for a particular distance); this assumption is a potential weakness to the method as is the uncertainty as to which power of d to select. Note that the block estimate for a particular power of d will be the same regardless of the block size (e.g., B and B' in Figure 1.17). This is equivalent to the fact that exactly the same procedure is used in practice to make point estimates (e.g., the block centre) as block estimates. A nearest neighbour estimate (equivalent to a polygonal estimate) for the block gives 0.90 percent grade, about a 17 percent increase relative to the ISD estimate. As an indication of the arbitrariness of the procedure, one might well ask, "Which is correct?" or even, "Why not select $1/d$ or $1/d^3$ as the weighting procedure?"

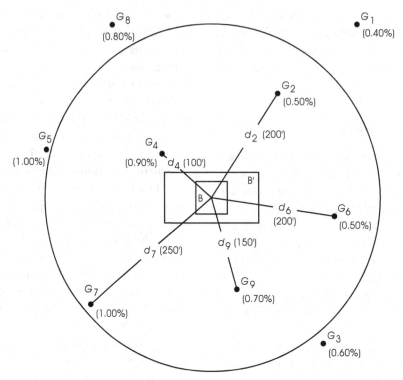

Figure 1.15: Illustration of block estimation using the method of inverse distance weighting (IDW) with hypothetical data (after O'Brian and Weiss, 1968). This is a common method of point interpolation (i.e., the central point of the blocks) but is used routinely for block estimation (e.g., B or B'). (Refer to Table 1.7.) Note that a given data array produces the same block estimate regardless of block shape and/or size.

In some definitions of inverse distance weighting (IDW) methods, in which a data value happens to coincide with the center of the block being estimated, that data value is assigned to the block (Fig. 1.15). This procedure avoids a zero divisor, but is a default to the polygonal method and does not take into account the presence of additional data or the magnitude of error inherent in a single value. Moreover, with highly regular grids of data, it is possible to contrive data geometries that give inconsistent results depending on how block estimation is conducted, as shown in Fig. 1.16. Estimation of the large square block by any isotropic distance weighting procedure gives equal weight to all samples and produces an estimated value of 63 percent. If the block is split into two rectangles along yy'; each rectangle is estimated at 60 percent by the value at the center, and the two blocks combine to provide a 60 percent estimate for the large block (clearly, this is wrong). Similarly, if the large block is divided into two rectangles along xx', each rectangle is estimated

at 66 percent, giving a large block estimate of 66 percent (also clearly wrong). This example demonstrates the inconsistency inherent in IDW methods. These errors emerge in cases in which a point or block center being estimated coincides with a data point, probably a fairly rare situation in which grids are not regular, but they can arise commonly in cases in which regular grids exist or data coordinates are rounded to very few significant figures. This problem can be avoided by assigning a nonzero, arbitrary distance to all samples that are very close (e.g., within one-quarter to one-half the block dimension) to the block center.

1.5.5: Contouring Methods

Contouring methods of reserve/resource estimation generally depend on estimation of a regular grid of points by some type of interpolation procedure (e.g., one of the traditional estimation methods described

Table 1.7 Grades, distance, and weights used for the inverse squared distance block estimate of Fig. 1.15[a]

Sample	Cooper (%)	d [b]	$1/d$	$1/d^2$	$1/d^3$
G1	0.4	360			
G2	0.5	200	0.005	0.000025	0.125 × 10-6
G3	0.6	290			
G4	0.9	100	0.01	0.0001	1.0 × 10-6
G5	1.0	275			
G6	0.5	200	0.005	0.000025	0.125 × 10-6
G7	1.0	250	0.004	0.000016	0.064 × 10-6
G8	0.8	320			
G9	0.7	150	0.0067	0.000044	0.195 × 10-6
Sum			**0.0307**	**0.00021**	**1.609 × 10-6**
[b]Estimated block grade			0.74	0.77	0.81

[a] Nearest-neighbor grade estimate = 0.9; local average (five nearest grades) = 0.72.
[b] Distance to block center.

previously) followed by contouring of the data. Contouring is normally invoked to avoid the jagged and commonly artificial ore/waste boundary that arises in estimating blocks. In addition, contouring is conceptually simple, at least superficially. In cases in which

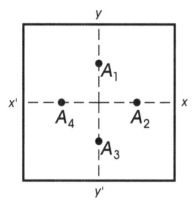

Figure 1.16: Example from David and Blais (1968) indicating inconsistent grade estimates of a large panel depending on the way in which an inverse distance weighting estimate is obtained. $A_1 = A_3 = 66\%$ Fe; $A_2 = A_4 = 60\%$ Fe. Values are equidistant from the panel center. If the large square panel is estimated directly, all values have equal weight because all data are equidistant from the block center (mean = 63% Fe). If the square block is divided horizontally into smaller rectangular blocks, each rectangular block is estimated at 66 percent Fe, as is the larger square panel. If the panel is divided vertically into rectangular blocks, each rectangular block is estimated at 60 percent Fe, as is the large square panel. Arbitrary decisions regarding block geometry control the panel estimate with no change in available data.

data are abundant (e.g., in the case of closely spaced blastholes in open-pit mining), they commonly are contoured directly without the intermediate stage of grid interpolation (e.g., manual contouring). In general, users should understand the rules that are the basis of an automatic contouring procedure to be used for estimation purposes. How are outliers treated? How are contours controlled at the margin of the field? What are the minimum and maximum data used for grid interpolation, and how are they selected? Automatic contouring procedures do not necessarily honor the data in detail because they commonly contain routines to produce "aesthetic" smoothed contours. An example of blasthole contouring is illustrated in Fig. 1.17. Grade and tonnage estimates by this method require that average grades (g_i) and volumes (v_i) between specified grade intervals (contours) be determined. Then, the average grade (g_p) of a large panel (panel volume $= \sum v_i$) can be estimated as follows:

$$g_p = \sum (g_i v_i) / \sum v_I.$$

As an estimation procedure, contouring of grades commonly is applied to control of grades in open pit mines (e.g., Fig. 1.17) where the controlling data are blasthole assays. Even in this situation involving a relatively high density of data, the method can lead to errors because the procedure corresponds to a variable, local smoothing of polygonal estimates.

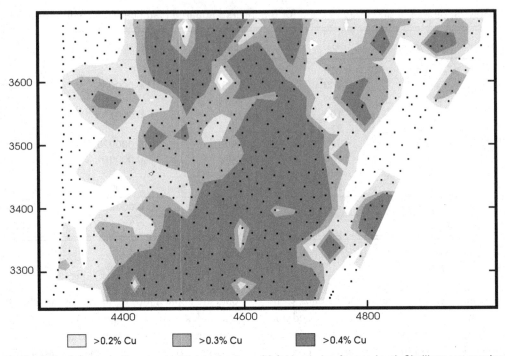

Figure 1.17: Contoured diagram of a subset of Cu grades from 30-ft blastholes for one level, Similkameen porphyry copper deposit, Princeton, B.C., illustrating the use of a contoured cutoff grade to define an ore/waste surface for mine planning purposes. Black dots are blasthole locations, mine coordinates are in feet (original mine units), contours are weight percent Cu. Cutoff grade (minimum contour) is 0.2 percent Cu; other contours are 0.3 percent Cu and 0.4 percent Cu. Contouring is based on a regular, interpolated grid with unit cell 29 × 29 ft^2; grid values were interpolated using $1/d^2$ and the nearest 16 data values.

Moreover, the method is commonly applied in two dimensions (i.e., using data from a single bench of an open pit operation), and thus, does not take advantage of data from the overlying bench. In addition, grid interpolation methods are commonly isotropic; hence, anisotropic (directional) geologic controls on mineralization might not be taken into account.

1.5.6: Commentary

Many variations on these traditional methods have been developed; in fact, most applications involve modifications to deal with peculiarities of a specific deposit, or that reflect an estimator's background. For example, Raymond (1979) found that an IDW of the form $1/(d^3 + k)$ and judicious choice of the constant k closely approximate kriged block estimates. The *ore zoning method* (Ranta et al., 1984) used by the Climax Molybdenum staff for many prophyry-type molybde-

num deposits is a particular variation of the method of sections (plans) adapted to large deposits with gradational margins. Various mineral inventory estimation procedures have been attempted but are not widely accepted, including a variety of multiple regression or trend analysis methods (cf. Agterberg, 1974; Fustos, 1982; Phillips, 1968); such methods have been ignored here.

The emphasis used has not been to describe in detail the detailed methodologies of the traditional approaches to mineral inventory estimation, but rather has been to document their general characteristics and some of their weaknesses. This should be construed as a demonstration of the need for a more theoretically based procedure for mineral inventory determination rather than an implication that the empirical methods are entirely unsatisfactory. Moreover, it has been the authors' experience that mineral inventory is done most usefully by more than a single method, and that a reconciliation of results by two or more methods is an important contributor to obtaining greater confidence in the results that are adopted. Detailed discussions of established traditional approaches to mineral inventory estimation are provided by Annels (1991) and Stone and Dunn (1994), among others.

Careful application of geostatistical procedures can minimize many or all of the problems inherent in subjective approaches to mineral inventory estimation because geostatistics has a rational theoretical basis for mine valuation. The recognition that data variations are due partly to a structured component (autocorrelation) and partly to a random component (nugget effect) is fundamental to the theory and implicit in the methodologies of geostatistics. The autocorrelation characteristics for various value measures (regionalized variables) can be determined specifically for a particular deposit or part of a deposit rather than depending on qualitative and subjective guesses. This autocorrelation can be expressed by an average mathematical function (model) that quantifies value continuity and the "average structure" of a deposit. This quantification of value continuity is the basis of geostatistical methods discussed in subsequent chapters.

Geostatistical applications in the English-speaking world are now well entrenched but were slow to develop through a general avoidance because of mathematical complexity, a limited literature in English, and the fact that geostatistics was viewed as a research tool rather than an applied tool. Now, however, the theory is well disseminated, a variety of texts are readily available, and many mining practitioners have been educated in both specific practical implementations and the broad potential applications of geostatistical methodologies; in particular, there are increasing numbers of case histories being reported in the scientific and technical literature. Specific features inherent in geostatistical methods are quantification of the area of influence of samples, consideration of the clustered nature of data used in an estimation, and variable support of samples relative to the size of blocks to be estimated (cf. Matheron, 1967).

1.6: MINE REVENUES

Superficially, mine revenues might not be viewed as directly influencing mineral inventory estimation. However, the selling arrangements for metal concentrates to smelters include important specifications that can determine what is or is not ore. The presence of deleterious minerals, for example, might override a grade that otherwise would be classed as ore, and thus has an impact on grade, tonnage, and various aspects of mine planning (cf. Goldie and Tredger, 1991).

Ore is defined in terms of expected profit based on mine revenues less expected costs (see Section 1.3.2). Operating costs are subject to a wide range of technical controls that are not necessarily well established at the time resources/reserves are being estimated, but such costs can be defined in cases in which operations are in progress under similar conditions and with similar equipment. Expected revenues, however, are a function of grade estimates and metal prices, and the implications of these two parameters warrant consideration. Some indication of the importance of metal prices in the optimizing of cutoff grade has been indicated (see Section 1.3.2 and Fig. 1.4).

Mine revenues generally depend on the sale of metal concentrates through a smelter contract, which is a short-term, binding, legal agreement between mine and smelter, commonly lasting one to five years. As in most legal agreements, all items of importance are defined and an effort is made to have every likely circumstance considered. The major points covered by smelter contracts are listed in Table 1.8.

Detailed specifications for concentrates can result in the imposition of operational constraints to mining that affect operating profit (e.g., the need for blending of high-Fe and low-Fe sphalerite to maintain an acceptably low Fe content to a Zn concentrate). In some parts of a mineral deposit, deleterious minerals (defined in a smelter contract) may be so abundant as to classify some blocks as waste, even though the average metal content is above cutoff grade.

In estimating costs for a shipment of concentrates, the items listed in Table 1.9 must be quantified,

Table 1.8 Principal topics dealt with in a smelter contract for the purchase of metal concentrates

Commodity market or publication that determines prices to be used
Date and time for period for fixing the price
Currency used for price determination and payment
Method and time of payment
Weights and measures used and any abbreviations
Approximate contents of the concentrates
Estimated production and amount of notice for increase/decrease
Regularity of shipments and identification by lot/car
Types of rail car/vessel acceptable for shipment
Title, risk, and insurance in transit
Weighting and sampling procedures
Limits of error and umpire laboratories
Metals paid for and deductions
Smelting charges for wet concentrates
Metal price escalation/deescalation clause
Penalties for impurities
Added costs for smelter environmental controls
Added costs for labor, fuel, and electricity
Force majeure for both mine and smelter
Means and requirements for communications between parties
Assignment of contract to successive owners of mine and smelter
Country/province where disputes will be settled

Table 1.9 Factors to be quantified in estimating costs related to shipment of concentrates

Road freight
Rail freight
Ocean freight
Loading and wharfage
Shipping representative
Insurance
Treatment
Refining
Marketing
Duty
Sales commission
Other items

and distinction between wet and dry concentrates is essential. In estimating revenues, a reasonable price for all the payable metals must be assumed. Adjustments are made to later revenue estimates to account for differences in assumed and received prices. Deductions from the payable metals must be included, followed by any penalties for impurities.

To estimate the revenue for a given grade of material in situ, mill recoveries must also be estimated for a given head grade. At least two types of model are available for this study. If the mine is in operation, recovery, head grade, and tail grade can be plotted together and any relation shown. Commonly, in practice, a constant tailing grade is assumed, giving an apparent increased recovery when head grades improve; such assumptions are dangerous and can lead to milling losses going unrecognized. Alternatively, tailing grade may be a function of head grade, perhaps

with a complicated relation that must be validated by the graphing of daily grades and recoveries.

Using such information, the net smelter return for rock in the ground can be estimated with simple formulae (Table 1.10) that give revenue per ton of rock after smelting and distribution costs have been accounted for, but prior to deduction of mining and milling costs. Hence, operating profit can be obtained by subtracting operating costs. Commonly, the location and grade of impurities must be known so that blending systems can be designed to keep an impurity level below the penalty level. If this is not possible, the penalty must be included in the formula. Formulae can be applied to a range of metal prices, or the metal price can be included in a more complex relation. An example for a copper–molybdenum operation is given in Table 1.10.

Recent arguments have been presented (Goldie and Tredger, 1991; Goldie, 1996) for the use of net smelter returns (NSR) as a variable of merit in the early evaluation stages of a mineral deposit, particularly in cases in which more than one commodity contributes to the value. Such a procedure is useful because it requires early consideration of many technical matters related to mining and metal recovery, although general assumptions in this regard might be necessary. Clearly, the use of NSR as a value measure for estimation purposes should not preclude detailed consideration of the more fundamental underlying grade values. Nevertheless, NSR has merit

Table 1.10 Sample calculations of recovered grade and dollar value formulae for ore milled for a producing mine (in Canadian dollars)[a]

Costs ($C 1,000) of operation for the time period and cost per ton

Item	Total cost	Cost/ton
Milling	$11,741	$2.156
Plant and administration	3,764	0.691
Ore mining	4,780	0.878
Waste mining	3,485	0.640
Nonoperating	1,967	0.361
Interest on long-term debt	3,325	0.610
Total	$29,062	$5.336

Grade mined and smelter and distribution (S&D) costs[b]

Metal	Copper	Molybdenum
Head grade	0.143	0.031
Tail grade	0.023	0.004
(Head–tail)	0.11987	0.02634
Tons milled	5,446,596	—
Tons metal	6,528.98	1,434.87
S&D cost (C$1,000)	5,095,504	161,337
S&D/ton metal	$780.44	$112.44

Revenue formulae after S&D costs and tailing losses[c]

Copper

Gold addition

@ C$500/troy oz, 0.01234 g/ton milled

$$= 500 \times 0.03215 \times 0.01234 \times \frac{5,446,596}{6,528.98}$$

$$= \text{C\$165.48/ton Cu metal}$$

Silver addition

@ C$14/troy oz, 0.6911 g/ton milled

$$= 14 \times 0.03215 \times 0.6911 \times \frac{5,446,596}{6,528.98}$$

$$= \text{C\$4,259.49/ton Cu metal}$$

Total Cu revenue

$$= 1.00 \times 2,204.62 + 165.48 + 259.49$$

$$= \text{C\$2,629.59/ton Cu metal}$$

Revenue formulae

$$= \text{net copper revenue} - \text{net tailing loss}$$

$$= \% \text{ Cu} \times \frac{(2,659.59-780.44)}{100} - 0.023 \times \frac{(2,629.49-780.44)}{100}$$

$$= \$C \ (\% \text{ Cu}) \times 18.49 - 0.425/\text{ton milled}$$

Molybdenum

Total Mo revenue

$$= 5.00 \ 2,204.62$$

$$= \text{C\$11,023.10/ton Mo metal}$$

Revenue formulae

$$= \text{net molybdenum revenue} - \text{net tailing loss}$$

$$= \% \text{ Mo} \times \frac{(11,023.10-168.66)}{100} - 0.004 \times \frac{(11,023.10-168.66)}{100}$$

$$= \text{C\$} \ (\% \text{ Mo}) \times 108.54 - 0.434/\text{ton milled}$$

Total revenue for both metals

$$= \text{C\$} \ (\% \text{ Cu}) \times 18.49 + (\% \text{ Mo}) \times 108.54 - 0.859$$

(continued)

Table 1.10 (*continued*)

Grades required to cover costs[d]

Item	Unit % Cu	Unit % Mo	Cumulative[e] % Cu	Cumulative[e] % Mo	
Milling	0.071	0.016	0.071	0.016	
Plant and administration	0.016	0.003	0.087	0.019	
Ore mining	0.020	0.006	0.107	0.025	Range of possible cutoff grades
Waste mining	0.015	0.002	0.122	0.027	
Nonoperating	0.009	0.002	0.131	0.029	
Interest on long-term debt	0.014	0.003	0.145	0.032	
Total	0.145	0.032	—	—	

Trial of the revenue formulas using actual grades and tons milled[f]

Head grade, 0.143% Cu and 0.031% Mo; 5,446,596 tons milled

Revenue with C\$1 Cu and C\$5 Mo per pound

Revenue per ton (C\$)
$$= 0.143 \times 18.49 + 0.031 \times 108.54 - 0.859$$
$$= 2.644 + 3.365 - 0.859$$
$$= 5.150$$

Total revenue (C\$)
$$= 5,446,596 \times 5.150$$
$$= 28,050,000$$

[a] Necessary information is taken from the mine operating accounts for a specific period of time, such as the previous six months, and is continually updated to ensure the formulae and results developed are current. Assumptions such as the ratio of metals in the ore and the stripping ratio are made based on the best information available at the time. Tons milled for the time period: 5,446,596.

[b] Note that there is always a very small amount of one metal in the other concentrate. If this small amount is not payable, then it must be included on a weighted average basis in the tailing grade. There are also other payable metals such as gold or silver that may be too small to detect or too costly to assay in the ore sampling procedure. The recovered precious metals per ton of ore are then assumed to be constant values for a production period, but can change depending on depth and location of mining.

[c] Assume the expected metal price, here copper and molybdenum, are estimated at C\$1.00 and C\$5.00 per pound, respectively.

[d] The ratio of copper to molybdenum in the ore mined is assumed to be 4.5:1 Cu:Mo, and changes with mining location and relative metal prices. The stripping ratio is assumed to be 0.73:1 waste:ore, and also changes with mining location, costs, and metal prices.

[e] Working down the column, the cutoff grade could be as low as 0.107 percent Cu and 0.025 percent Mo if it is assumed that the material would be drilled, blasted, and loaded into the truck regardless. The cutoff could also be as high as 0.145 percent Cu and 0.032 percent Mo if all costs to the mine are included.

[f] The formula must balance with the mine operating accounts, and is most sensitive to metal prices and head grades.

because of the inherent economic implications and is particularly useful in cases in which the less desirable method of "equivalent" grades has been used for estimation purposes.

1.7: MINING SOFTWARE – APPLICATIONS

Computers are an essential tool in conducting mineral inventory studies (e.g., Blackwell and Sinclair, 1993; Champigny and Grimley, 1990; Gibbs, 1990) and a substantial amount of software exists that has been designed particularly for use in the mining industry. Applying this mining software requires that data be organized appropriately for efficient creation of computer files, preferably as early as possible in the long, ongoing saga of generating information for exploration and evaluation purposes. Such organization and input is extremely labor intensive, time that is more

Table 1.11 Listing of some of the principal applications of computers in relation to mineral inventory estimation

---

A wide range of data evaluation procedures
Determining data quality and sampling protocols
Three-dimensional visualization of complex geologic features and data distribution
Three-dimensional solid modeling of mineral deposits
Preparation of plans and vertical sections
Contoured plots of grade and other variables
Variography (quantifying grade continuity)
Block modeling of a deposit
Rapid calculation of reserves
Evaluating effects of various mining methods
The ready determination of economic feasibility

---

than made up by the relative speed with which numerous applications can be implemented and updates obtained, and the ability to evaluate different decision scenarios rapidly. Applications include a wide range of topics such as those listed in Table 1.11. To attempt mineral inventory studies without the advantage of computing facilities is to promote grave inefficiencies and, more importantly, to be unable to undertake some procedures that are a routine part of modern mineral inventory practice but that are too complicated to be done manually.

Available software can be classed as public domain and commercial. In general, public domain software is free or extremely inexpensive and includes full source code; of course, it generally comes with no guarantees or technical support. Such software is generally available from universities, government organizations, textbooks, the scientific and technical literature, and various user groups accessible through the Internet. Commercial software, although more costly, generally comes with an up-to-date user's manual and a technical support system from the vendor, and source code is generally proprietary. The software packages are commonly "integrated" in the sense that they are available in modules that offer a wide range of capabilities such as those listed in Table 1.11.

The greatest weakness of much of the available software is that methodologies are not described in sufficient detail to allow a user to understand how decisions and estimates are made by the software. Some questions include the following: How are data selected for contouring or block estimation? What interpolation procedure is used for contouring? Are contour intervals selected arbitrarily, or can they be designated? What mathematical models are "hidden" in software? Users must become more aware that an understanding of the procedures and decision-making rules implemented in software is essential for its correct and effective use.

As Champigny and Grimley (1990) conclude

> Orebodies are dynamic entities prone to significant changes through time as metal prices fluctuate and mining technology evolves. The magnitude of these changes can be efficiently examined through the use of computer-based ore reserve approaches. The application of computers for grade control typically improves the accuracy of estimates of mining units, which ultimately results in higher quantities and/or quality of ore recovered. These benefits . . . are finally being acknowledged. (p. 77)

1.8: PRACTICAL CONSIDERATIONS

1. Each specific mineral inventory estimation is a problem that must be cast in its true mining context. Possible (even alternative) operating scenarios must be designed as a basis for a quantitative evaluation. There is no point to considering a totally unrealistic scenario.

2. It is essential to understand thoroughly the terminology in use in the mineral industry. Consequently, the ramifications and ambiguities of definitions are a necessary basis for embarking on a mineral inventory estimation. A range of traditional use exists for such terms as *ore*, *cutoff grade*, *continuity*, *reserves/resources*, and others, and the implications of using these terms must be appreciated. In particular cases, it is wise to provide specific definitions of terms so that no ambiguities arise.

3. Samples for assay generally constitute about 0.000001 of the volume of a mineral deposit for which a mineral inventory is being estimated.

Consequently, errors are to be expected and should be minimized as much as is feasible. Maintaining high-quality data and estimates requires close attention to sampling and analytical and estimation procedures as well as appropriate integration of a confident geologic model into the estimation process.

4. The sequential nature of resource/reserve estimation must be appreciated. Information is obtained in a staged manner, and new information becomes available intermittently. Plans should be made in advance as to how this sequential gain in information is taken into account and at what intervals entirely new estimates of mineral inventory are made.

5. Mineral inventory estimation is a highly structured undertaking. Reference to the various components of this structure – for example, geologic modeling, documentation of data quality, and data evaluation – imposes a high level of rigor into the exercise of estimation.

6. Strengths and weaknesses of individual procedures for mineral inventory estimation must be understood thoroughly as a basis for determining appropriate methods for each particular estimation project. Choice of an appropriate method of obtaining resource/reserve estimates is closely linked to the geologic characteristics of a deposit, as will become apparent, and the decision to use a particular method must be justified in terms of geology.

7. The procedure used in a mineral inventory estimation must be documented. It is not sufficient to use such terms as *polygonal, method of sections, inverse distance weighting*, and so on without providing more detailed information on procedures used in implementing a general method. Each method involves some arbitrary or subjective decisions that should be summarized.

8. Computers are an essential component of modern mineral inventory estimation. There is a need to understand the advantages and limitations of available software and a need for facility in using the software to provide the necessary information at various stages in estimating a mineral inventory. It is not sufficient simply to provide listings of mineral inventory estimates; it is essential to verify results with various graphic outputs (sections, plans) for comparisons with geologic information and as a means of contrasting various types of estimates.

9. The technical complexity and financial enormity of many active and proposed mining ventures and related mineral inventory estimates cannot be appreciated fully without the detailed understanding that comes from visits to the site.

1.9: SELECTED READING

Annels, A. E., 1991, Mineral deposit evaluation: A practical approach; Chapman and Hall, London.

Blackwell, G. H., 1993, Computerized mine planning for medium-size open-pits; Trans. Inst. Min. Metall, 102(A), pp. A83–A88.

Gibbs, B. L., 1994, Computers: The catalyst for information accessibility, Min. Eng., June, pp. 516–517.

Goldie, R., and P. Tredger, 1991, Net smelter return models and their use in the exploration, evaluation and exploitation of polymetallic deposits; Geosci. Can., 18 (4), pp. 159–171.

King, H. F., D. W. McMahon, and G. J. Bujtor, 1982, A guide to the understanding of ore reserve estimation; Australasian Inst. Min. Metall, Proc. No. 281(Suppl.), March 1–21.

Rossi, M. E., and J. C. Vidakovich (1999). Using meaninful reconciliation information to evaluate predictive models; Soc. Min. Engs., preprint 99–20, pp. 1–8.

1.10: EXERCISES

1. In mine development and in producing mines, it is common practice to stockpile material that is below cutoff grade but above some other arbitrary lower threshold. For example, in a porphyry copper deposit with a cutoff grade of 0.35% Cu, material in the range 0.25–0.35% Cu might be stored for subsequent easy retrieval. Explain why this "low-grade stockpile" material might, at some future date, be fed through the mill.

2. (a) Estimate an optimal cutoff grade for the grade–tonnage scenario summarized in Table 1.2 using 90 percent recovery and a metal price of $1.00 per unit. Note that the tabulation is based on short tons of 2,000 lbs. The calculations should be tabulated as in Table 1.3.

(b) Estimate an optimal cutoff grade for the grade–tonnage scenario summarized in Table 1.2 using mining costs of $0.84/ton and other parameters as in question 2(a). The results should be tabulated as in Table 1.3.

(c) Compare the results of question 2(a) with the results of question 2(b). The combined scenario of questions 2(a) and 2(b) can be compared by estimating a cash flow using the revenue data determined in question 2(a) and the operating costs calculated in question 2(b). Suggestion: These questions are dealt with conveniently using a spreadsheet.

3. Construct a close-spaced regular grid (e.g., 1 × 1 mm^2 cell) on transparent paper (e.g., photocopy appropriate graph paper onto a transparency). Use the grid to estimate in two dimensions the effect that dilution will have in reducing the grade of mined material relative to the expected grade in the hypothetical example of Fig. 1.12. Confine your estimate to the area between the two drill holes. Assume average grades of 1.0% Ni for drill hole No. 1, 1.5% Ni for drill hole No. 2, and 0.2% Ni for the diluting material. Assume further that mining will be constrained exactly between the dashed lines of Fig. 1.12. Estimate relative areas by counting squares with the transparent grid and use relative areas as weights. Using a similar counting technique, estimate the relative amount of ore left unmined between the two drill holes. In place of gridded paper, this problem could be solved using a planimeter or a digitizer and appropriate software.

4. A stope with rectangular outline on a vertical section has been defined on the longitudinal section of a near-vertical gold–quartz vein cutting basaltic rocks. Stope definition is based on 25 vein intersections by exploratory diamond-drill holes.

The stope has a horizontal length of 24 m, a vertical height of 15 m, an average thickness and standard deviation of 2.7 ± 0.4 m, and an average grade of 19.5 g Au/t. Estimate the effect of dilution on grade and tonnage of mined material if the mining method overbreaks the ground by an average of 0.4 m on both sides of the vein and the wallrock averages (a) 0.05 g Au/t, and (b) 5.3 g Au/t. Ignore any complications that arise because of a minimum mining width, and assume that bulk densities of ore and wallrock are the same.

5. Use of the polygonal approach to block estimation (cf. Figs. 1.11 and 1.13) is in conflict with the recognition that the dispersion of average grades decreases as the volume being considered increases (the support effect, see Fig. 1.8). With this difficulty in mind, comment on the two following situations:

(a) Continuous samples from vertical drill holes on a square 50-m grid are used to estimate an array of blocks, each 500 m^3 (i.e., 10 × 10 × 5 m^3).

(b) Blocks to be estimated (500 m^3) each contain a roughly regular array of four blasthole samples.

Constructing plans of the two estimation scenarios is useful.

6. Use the data of Table 1.6 and Fig. 1.9 to estimate the mean grade of the large block of Fig. 1.9 and each of the "quadrant blocks" using $1/d$ and $1/d^2$ weighting schemes. Compare results with the estimate interpolated from contours and comment on the similarities or differences.

7. The data of Table 1.7 and Fig. 1.15 have been used to determine a block estimate using $1/d^2$ as the weighting procedure.

(a) Calculate the estimated block grade using a weighting factor of $1/d$ and again with a weighting factor $1/d^3$.

(b) Compare results with those given in Table 1.5 and the nearest neighbor estimate. Comment on any systematic pattern to the variation in estimated grades as a function of change in the power, x, of the weighting factor ($1/d^x$).

8. Construct graphs of distance, d (abscissa), versus $1/d$, $1/d^2$, $1/d^3$, and $1/d^4$ for $0 < d < 200$. Assuming a case of three samples with distances from a point (block) to be estimated of 10, 50, and 150 units, comment on the relative importance of the data with $d = 10$.

9. A press release by a mineral exploration company contained the following statement regarding a mineral occurrence in a remote part of South America: "An immediate drilling program is planned to expand the existing reserves of 5,132,306 tons of ore grading .07 oz/t gold 6.9 oz/t silver, which is open in all directions." Discuss the content of this statement in detail.

2

Geologic Control of Mineral Inventory Estimation

...computation formed only part, and perhaps not the most important part, of ore reserve estimation;... the estimate in situ *should be seen primarily as a facet of ore geology.* (King et al., 1985, p. 1)

Chapter 2 is concerned with the impact of geology on mineral inventory estimation. For the purposes of this book, geology is considered under an arbitrary classification of topics: geologic mapping in and near a mineral deposit, three-dimensional geometry of mineralized zones, ore deposit models, and continuity and mineralogy of ore and waste. These divisions are arbitrary and the topics overlap considerably. *Continuity* is a term of sufficient import to warrant a separate chapter (Chapter 3). Emphasis throughout this discussion is on the need to document aspects of geology that are important in making decisions related to resource/reserve estimation.

2.1: INTRODUCTION

Understanding the geologic character of a mineral deposit as thoroughly as possible is an essential base on which to build an estimate of mineral inventory. Several recent reports on the classification of resources/reserves contain reference to "exploration information" (e.g., Anonymous, 1994), that is, the source information that contributes to an understanding of the geology in and around a mineral deposit. That information normally is obtained in a structured and sequential fashion, beginning with several working hypotheses about the true geology (Chamberlain, 1965) that become fewer and more refined as the information base increases. A general idea of the nature of exploration information and the systematic manner in which it is obtained during detailed property exploration and evaluation is given in Table 1.1.

The term *geology* encompasses much of what is generally envisaged by the term *earth sciences*, including such traditional specialties as mineralogy, petrology, structural geology, stratigraphy, geochemistry, hydrogeology, economic geology, and so on. Geology affects the estimation procedure in a variety of ways that can be considered under several general and overlapping topics, as follows:

(i) Geologic mapping and general geologic history
(ii) Three-dimensional (geometric) modeling
(iii) Ore deposit (genetic) models
(iv) Mineralogic attributes
(v) Continuity.

The first four topics all contribute to the fifth – continuity. The concept of geologic continuity is commonly confused with an understanding of the continuity of grade; both warrant special attention. The impact of all five categories of geologic information on mineral inventory estimation is summarized here and in Chapter 3.

In general, geologic interpretations are continually evolving components of mineral inventory studies (e.g., Sinclair and Vallée, 1993; Vallée and Cote, 1992). Any arbitrary procedure by the estimator that minimizes or neglects systematic upgrading of geologic concepts as new information is gained should be avoided or viewed with reasonable scepticism. In particular, "hidden assumptions" incorporated in geologic modeling, including automated procedures of geologic interpolation, should be reviewed periodically.

2.2: GEOLOGIC MAPPING

"Geological maps are used in planning future exploration, directing development work and coordinating stoping" (Faddies et al., 1982, p. 43).

Factual geologic information is the base from which a three-dimensional image of a mineral deposit is developed. As a rule, this information is obtained from surface rock exposures, trenches, drill core or cuttings, and underground workings. These sources provide direct observations of rocks and minerals, but represent a very limited proportion of the total volume of a mineral deposit and its surroundings. Even for a well-sampled mineral deposit, the total volume of all samples could be about one one-millionth of the deposit volume. Hence, a substantial *interpretive component* is required in order to develop a three-dimensional model of a mineral deposit and adjacent rocks. This interpretive component involves the interpolation of geologic features between control sites (i.e., extensions of features between known data) and may include some extrapolation (extension outward from known data). The interpretive process is aided by a variety of surveys, particularly geophysical and geochemical surveys that can help localize specific geologic features such as faults or particular rock types, improve confidence in the continuity of "ore," and, in some cases, provide improved grade estimates relative to traditional assaying of samples (e.g., Calvert and Livelybrooks, 1997; Cochrane et al., 1998; Killeen, 1997; McGaughey and Vallée, 1997; Mwenifumbo, 1997; Wong, 1997).

Geologic information is normally recorded on maps and cross sections at a scale appropriate to the aims. Property geology might be mapped at a scale of 1:5,000, whereas mineral-deposit geology might be mapped to a scale of 1:1,000 or even more detailed. The types of information that are recorded and displayed on maps include:

(i) Rock types: Rock composition influences reactivity to mineralizing solutions and controls response to deformation. Rock types (including mineralized ground) are one of the most fundamental pieces of geologic information; their chemical and physical attributes and age relations provide the basic framework for understanding the geologic history of an area (e.g., pre-ore and post-ore dykes in Fig. 2.1).

(ii) Faulting: Faults disrupt and complicate the lithologic record (Figs. 2.1 and 2.2). The ages of faults are important: premineralization faults might be mineralized; postmineralization faults might disrupt a primary deposit and form a boundary across which it is inappropriate to extend grades for block estimation purposes (Fig. 2.3).

(iii) Folding: Folding can provide ground preparation for some types of deposits (e.g., saddle veins) and can disrupt a preexisting mineralized zone extensively to produce a complex geometry. In the case of shear folding of a tabular deposit, mineralization in the fold limbs can be greatly attenuated, whereas a substantial and perhaps augmented thickness can be present in the crests of folds.

(iv) Fracture/vein density and orientation: Sites where fractures have controlled mineralization spatial density and evidence of preferred orientation provide insight into localization of ore and preferred directional controls (Figs. 2.1 and 2.4).

(v) Evidence of primary porosity/permeability: Permeability for mineralizing fluids can be controlled by structure [cf. (iv)] or by lithologic character (e.g., reactive carbonate beds or breccias with substantial interconnected porosity).

(vi) Successive phases of mineralization: Many deposits are clearly the product of more than one phase of mineralization. Sorting out the characteristics of each phase (i.e., understanding the

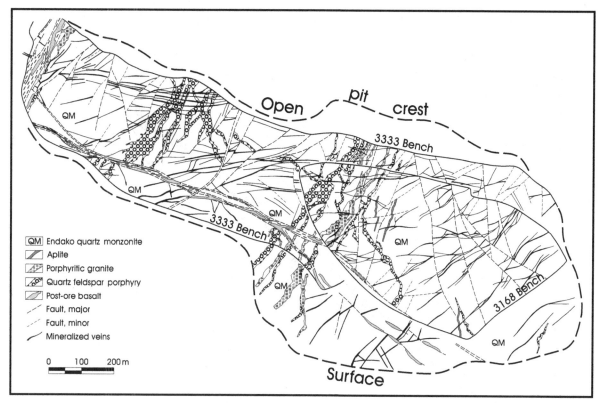

Figure 2.1: Detailed geologic plan of the open pit at Endako porphyry-type molybdenum mine, central British Columbia. Note the different trends of principal veins from place to place in the deposit (i.e., different structural domains with gradational boundaries) and premineral and postmineral dykes. Mineral zoning (not evident on this diagram) is emphasized by a pyrite zone along the south margin of the pit. After Kimura et al. (1976).

paragenesis of the mineralization) and determining the extent to which various phases are superimposed spatially (i.e., through detailed geologic mapping) have important implications to mineral inventory estimation.

For most mineral deposits, much of the detailed geologic information is obtained from logging (i.e., "mapping") the drill core. Consequently, it is wise to record such data in a systematic fashion that is easily adapted to a computer so that the information can be output in a variety of forms (sections, maps, correlation matrices, etc.) to assist in a mineral inventory estimation. Several systems are available commercially and have the advantage of providing sensible codes (e.g., Blanchet and Godwin, 1972) for efficient recording of many geologic attributes, such as miner-

alogy, alteration, lithology, vein density, and so on. In general, a clear, accurate, standardized logging procedure is essential (cf. Atkinson and Erickson, 1984) to promote uniformity of data through what is commonly a long data-gathering period. The general methodology should be adapted easily for rapid entry into computer files and should combine graphic output with descriptive notes; drill-hole graphs (profiles) should allow multiple variables to be displayed and take advantage of such techniques as color coding of different categories of a variable. As geologic information and concepts evolve, circumstances are likely to require the core be relogged. In an evaluation of the Crandon volcanogenic massive sulphide deposit, Rowe and Hite (1984) note that "By the time approximately 150 holes had been completed it was determined that core relogging was necessary..."

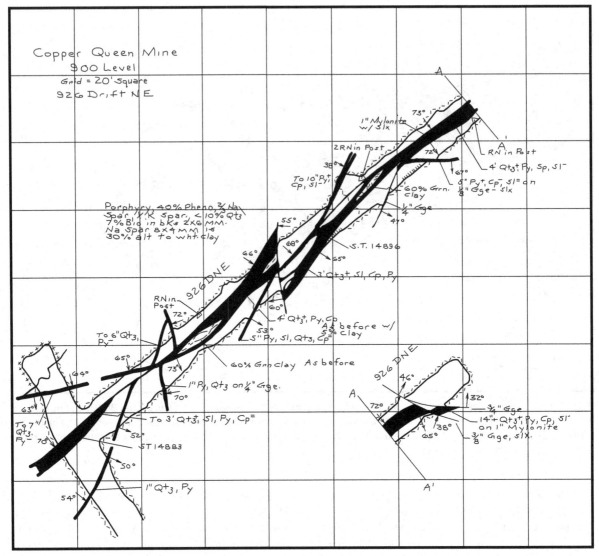

Figure 2.2: Detailed geologic map of underground workings forming part of the Copper Queen Mine, Montana (after Ahrens, 1983). Note the complexity of the vein system and the uncertainty of interpolating the physical form of the vein between any imagined, drill-section spacing.

(p. 14). Drilling generally produces core or rock chips that provide direct observation of material at various depths along the drill hole. A drill hole is a "line" in space, the locus of which can be highly complicated due to deflections of the drilling bit in response to the character of the rocks intersected. Each drill hole should be surveyed to ensure correct positioning of samples and geologic features in space. Lahee (1952) documents "crooked" drill holes for which the end of hole is removed horizontally from the planned position by up to 20 percent of the hole depth. A variety of methods are used for surveying drill-hole orientations at various depths and translating that information into locations in three-dimensional space. Killeen and Elliott (1997) report than modern commercial methods report errors of about 0.1 degree of dip and 1.0 degree of azimuth, but older magnetic compass and gimbal measuring techniques have substantially greater

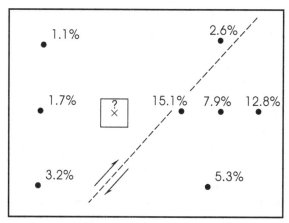

Figure 2.3: Geology as a control on data selection for block estimation. Black dots are sample sites, associated values are hypothetical grades, and the dashed–diagonal line is a fault with major strike–slip offset. Estimates of the block grade made prior to knowledge of the fault might include a value of 15 percent and result in a serious overestimation of the block grade. With knowledge of the fault, the 15 percent sample value can be omitted on geologic grounds (i.e., the fault separates two domains with very different grade characteristics). Modified from Srivastava (1987).

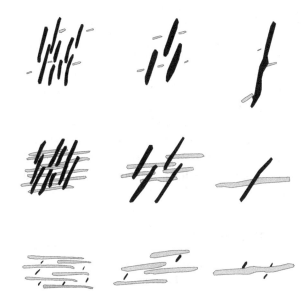

Figure 2.4: Systematic variations in the nature of structural control in "stockwork" mineralization. Two mineralized structural directions are shown: black and shaded. From left to right there is a decrease in the density of structures; from top to bottom there is a gradual change as to which of the two structural directions predominates. Both density of structure and direction of predominant structure are mapable features that can strongly affect grade continuity and geologic interpretation. After Sinclair and Postolski (1999).

errors. Clearly, a high degree of confidence in locating samples in space contributes to the quality of estimates of mineral inventory.

Geologic information that contributes to mine planning (e.g., Grace, 1986) includes the following:

(i) Depth and character of overburden
(ii) Extent (thickness, strike length, depth) of mineralization
(iii) Nature of deposit margins (gradational, sharp, sinuous, straight, jagged)
(iv) Character of ore continuity within separate geologic domains
(v) Drillability of rocks (strength)
(vi) Blasting characteristics
(vii) Pit-slope stability
(viii) Distribution of rock types (especially with regard to internal and external dilution)
(ix) Water inflow.

Important geologic factors affecting the selection of an underground mining method and its most economical execution include the following:

(i) Deposit geometry
(ii) Rock types
(iii) Bedding—thickness, strike, and dip
(iv) Folding and faulting
(v) Geologic contacts
(vi) Fracture, cleavage, and hardness
(vii) Wall characteristics
(viii) Hydrology, aquifers, water quantity, temperature, pH, etc.

Geologic characteristics have an impact on mining dilution and mining recovery. *Recovery* refers to the proportion of the existing resource that is identifiable as being available for extraction. Dilution affects both the grade of material extracted and the amount of material extracted.

It is clear from the foregoing that sound geology is essential to the entire process of exploring, evaluating, developing, and mining a mineral deposit. At any one time, the geologic information available from direct observations represents only a very small proportion of the total rock volume under study, and substantial interpretation is necessary. As increasing amounts of geologic information are obtained, some aspects of an earlier interpretation can change. Hence, factual information should always be distinguished from interpretations, and interpretations should be reviewed periodically in the light of new information and ideas.

2.3: GENERAL GEOLOGY

The geologic framework of a mineral deposit is essential to both mineral exploration and mineral inventory estimation in providing (i) an integration of mineralization into the general geologic evolution of the area and (ii) a three-dimensional (geometric) framework for the mineral deposit. Geologic history is important because it is essential to distinguish pre-, syn-, and post-ore processes and to understand all geologic features that might affect ore continuity – in particular, to appreciate primary aspects of geologic continuity (e.g., veins, strata, shear zone) in contrast with superimposed features (folding, faulting, metamorphism) that commonly disrupt primary mineralized zones. Geologic controls of mineralization can be established as a product of routine and detailed geologic mapping, as can the general character and spatial extent of these controls. A few examples illustrate some simple ways in which geologic knowledge affects mineral inventory estimation.

Consider the simple but realistic example of Fig. 2.3, in which the average grade of a block is to be estimated from local data at two stages of geologic awareness. At an early stage of evaluation, the presence of a fault is unknown and all surrounding data might be incorporated into the block estimate, resulting in a serious estimation error. Later, when the presence of the fault is recognized, the sample contributing excessively to estimation error (the sample located nearest the block but on the opposite side of the fault from the block) can be ignored on geologic grounds in obtaining the block estimate. In the

idealized situation pictured, the difference in the two estimates approaches an order of magnitude.

Structural information about veins can be obtained from routine geologic mapping of a deposit and can be integrated into the structural evolution of the surrounding region. Figure 2.1 is a more complicated example illustrating systematic variations in the orientation of the dominant, molybdenum-bearing veins in the Endako prophyry molybdenum deposit. The directions of physical continuity of ore differ from one part of the deposit to another; in particular, the preferred strike direction is northeasterly in the eastern extremity, whereas in the extreme western end the strike is roughly easterly. Projections of grades for estimation purposes generally can be performed over greater distances along preferred geologic directions versus across such directions. Consequently, the recognition of preferred structural directions is useful geologic information at an early stage of exploration in the design of drilling (especially orientation of drill holes) and sampling programs and can contribute to a high-quality information base on which to build a mineral inventory. The Endako deposit is also characterized by mineralized premineral dykes and barren postmineral dykes.

The geometry of ore and associated structures and rock units has direct implications to mine planning, and thus dramatically affects the problem of ore/waste selection and its impact on mine design. Geologic interpretation based on a thorough understanding of the geologic history of an area is used to provide spatial interpretations of the distribution of various rock units. One "rock unit" is mineralized ground, which, in general, only partly coincides with "ore," as illustrated in Fig. 1.17, where a cutoff grade contour separates ore-grade material from mineralized waste.

Another very different example indicates the practical importance of recognizing whether a deposit has been modified by metamorphism. An originally fine-grained massive sulphide deposit might be modestly affected by regional metamorphism, or extensively recrystallized in a contact metamorphic aureole associated with a later intrusive body, with attendant grain-size increase, such that mineral separation can be done efficiently. Mineralogic mapping that defines metamorphic aureoles, including textural information of the deposit itself, is a sensible approach to this problem. An example is the Faro Camp, Yukon, where several deformed and metamorphosed, shale-hosted Pb–Zn deposits occur in different positions relative to metamorphic grades such that some deposits are more coarsely recrystallized than others.

A thorough understanding of the regional and local geology, including details of structural, metamorphic, igneous, and sedimentologic history, provide the basis for confidence in making the many geologic decisions required in the more detailed physical modeling of a mineral deposit, which are known only from isolated sampling points. Mineral deposits commonly are localized in highly altered areas that have undergone extensive deformation or metamorphism; consequently, recognition of important pre-, syn-, and post-ore features can be complicated and difficult to interpret without the guidance of more widely based geologic information. Moreover, geologic information is fundamental in defining distinctive zones within a larger mineralized volume, each zone being characterized by its own geologic features and perhaps by its own ore continuity. There are many examples in which the presence of several geologic domains has been important to methodology in mineral inventory estimation. The Boss Mountain molybdenum deposit (Fig. 2.5) illustrates three fundamentally different styles of mineralization (i.e., breccia, stringer zone, and high-grade vein), each with different characteristics of ore continuity. Hence, each of the three domains should be estimated independently of the others.

2.4: GENERAL GEOMETRY OF A MINERALIZED/ORE ZONE

For practical purposes, the geometry of a mineralized zone and associated rock units generally is illustrated on a series of cross sections or plans in a systematic fashion. Cross sections generally coincide with drill sections (dense planes of information), but in some cases, cross sections might be interpolated between drill sections. Plans are constructed routinely at various elevations by transferring interpreted information

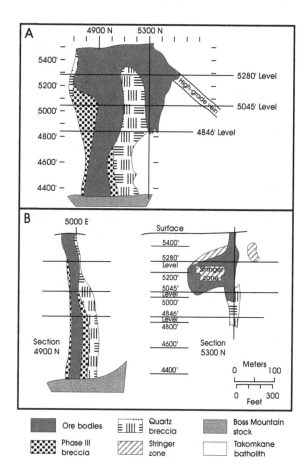

Figure 2.5: Cross sections of the Boss Mountain molybdenite mine, central British Columbia, showing three geologically distinctive domains of mineralization (i.e., breccia, stringer zone, and high-grade vein). (A) A north–south vertical projection. (B) A composite of two east–west vertical projections at 4900 N and 5300 N. Note that two breccia styles are mapped, and that ore-grade mineralization encompasses parts of both breccias. Each domain has different continuity characteristics and must be estimated independently of the others. Redrawn from Soregaroli and Nelson (1976).

from sections to plans at desired elevations and then interpolating between sections. This approach is the norm by which data are collected and displayed, both in routine geologic mapping and for the evaluation of a mineral deposit. Generally, data are more concentrated along sections arranged perpendicular to the principal direction of geologic continuity of the deposit under investigation (for example, an array of drill holes along a section); such sections are commonly relatively widely spaced compared to data spacing within a section. The accessibility of many computer-based graphic display systems also provides the ready availability of three-dimensional views, in some cases from any desired direction. Some caution is urged in using such graphic displays, principally because the highly regular interpolation routines that are a part of the software may lead to smooth interpolations that depart substantially from reality. These highly sophisticated software packages (now available for three-dimensional modeling) are an important component of modern capabilities in data handling and appreciation. However, insight into their built-in interpolation routines is important for the user because they can oversimplify geologic contacts or domain boundaries. Nevertheless, such three-dimensional views are a powerful aid to conceptualizing a mineral deposit for purposes of mineral inventory estimation and mine design.

Of course, even in manually prepared sections and plans it is common practice for the purpose of a mineral inventory estimate to provide a single, very specific interpretation as if it were fact. It is important to remember that interpretations can change significantly even in cases in which relatively short-range interpolations have been made in developing a three-dimensional image of a deposit or mineral zone. Consequently, on plans and sections that depict an interpretation it is useful, wherever possible, to indicate the locations of data on which the interpretation is based. For example, to know the locations of underground workings and drill holes that provide the data used for interpretations provides a factual basis for evaluating one aspect of the "quality" of an interpretation, and indicates where the interpretation is most suspect. An example, the Neves–Corvo copper–tin mine in Portugal (Fig. 2.6; Richards and Sides, 1991) illustrates the evolution in conceptual geometry as additional geologic information is obtained during exploration and development. In this case, as initial underground information (Fig. 2.6b) was added to earlier exploration drill information (Fig. 2.6a), a greater structural complexity to the geometric continuity of the upper part of the Neves–Corvo ore became evident. Along with this better local understanding of geometric form came an improved appreciation of the

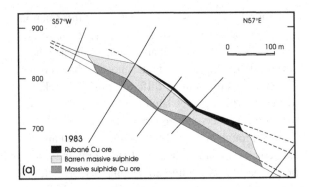

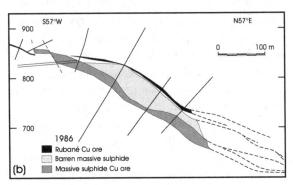

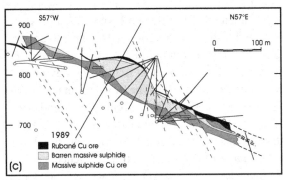

Figure 2.6: Geologic interpretations in 1983 (a) based on surface exploration data, and in 1986 (b) with added underground information, compared with 1989 (c) detailed mapping results based on extensive underground development, Neves–Corvo copper–tin mine, Portugal. Redrawn from Richards and Sides (1991). The conceptual model of a mineral deposit changes in response to additions to the geologic information obtained during exploration and development.

nature of possible complexities in geometry at depths where only widely spaced exploration drill data existed as a basis for interpolation. Subsequently, with abundant underground information (Fig. 2.6c), the ore zones are seen to be highly complex geometrically compared with the initial highly continuous and simple interpretation.

Geometric definition of a mineralized volume is controlled by knowledge of the internal character of the mineralization, a principal goal of detailed deposit evaluation commonly referred to as *delineation*. Extensive sampling is required during delineation, and geologic characteristics provide the principal control in sampling design, including support, number, and disposition of samples. Samples are examined for geologic information, assayed, and perhaps subjected to various kinds of physical measurements or tests. Sampling provides insight into rock character that is important in resource/reserve estimation. Bulk density, for example, is an essential variable for converting volume to tonnes and can vary extensively within and between ore types. Physical characteristics (schistosity, fracture density, etc.) of ore and wallrocks can have a dramatic impact on dilution. Indirect indicators include water loss during drilling (indicating the presence of major structural weaknesses) and rate of penetration of drills (related to physical strength of rock). Geology and detailed sampling thus can be seen as a classic "feedback" situation (cf. Grace, 1986).

2.5: GEOMETRIC ERRORS IN GEOLOGIC MODELING

In mineral inventory estimation, the need for a high level of accuracy in the position of the outer limits of a mineral deposit, and more specifically of the ore margins, is evident because of the control that volume definition exerts on mine planning. Any deficiencies in locations of deposit and ore margins lead to uncertainties in deposit evaluation and the possibility of problems in production planning – problems that may not be recognized until after major commitments and expenditures have been made.

The types of errors incorporated in the process of geologic modeling have been categorized by Sides (1994) as follows:

(i) Inaccuracies associated with original data (e.g., gross errors, such as incorrect drill-hole locations, errors in assumed continuity)

(ii) Sampling and analytical errors (e.g., uncertain-

ties of ore margin locations associated with imprecise grade estimates)

(iii) Errors due to natural variations (e.g., roughness or sinuosity of ore/waste margins)

(iv) Errors in data capture (e.g., mistakes made during input of information to databases)

(v) Computer-derived errors (e.g., uncertainties related to software packages that are not completely debugged or are not ideally adapted to the problem in question).

Gross errors are difficult to analyze systematically, but should be minimized progressively as the geology of the deposit becomes better known. Of course, the potential for surprises in geometric modeling is always present where the object of study is being investigated by widely spaced control points, and substantial interpolation is required. Sampling and analytical errors should be minimized by a well-defined quality control program and the assurance that sampling techniques are appropriate for the material being sampled (cf. Chapter 5). A duplicate sampling program should be designed to quantify the nature of sampling and analytical error. A consideration of "geometric" error in placement of ore/waste margins can begin once the sampling and analytical error is quantified. Errors due to data capture should be recognized and corrected through an independent editing of the data (cf. Chapters 5 and 7). Computer-derived errors are most often the result of interpolated margins being relatively smooth surfaces, as discussed later.

Virtually all geologic models assume smooth interpolation between control points (cf. Houlding, 1991b), particularly when it is early in the exploration history of a deposit and information is in short supply. From a practical point of view vis-à-vis geometric modeling, the reliance on smooth interpolation may be essential, despite the fact that it necessarily leads to errors. Consider again the Neves–Corvo massive sulphide example (Richards and Sides, 1991) illustrated in Fig. 2.6. The differences between the 1983 interpretation and the approximation of reality shown in the 1989 cross section is large. Ore gains roughly offset ore losses on this particular cross section, so the "global" error is not large. However, local errors are large, and their potential is clearly foreseen by the cross section interpreted in 1986, which had the advantage of some underground information in the upper reaches. If anything, a slight increase in global reserves as the information base increases is indicated for the cross section illustrated. The effect of prevalent faulting complexity at upper and lower extremities, of course, can affect recoverable reserves. This example illustrates forcefully that smooth ore/waste contacts, interpolated early in the data accumulation phase of deposit delineation, are unlikely to hold up in reality. Hence, an estimation project of several months duration could be significantly out of date on completion, if substantial changes to the geologic model have emerged from the continuing accumulation of exploration information.

Another case in point is the north limb of the Woodlawn massive sulphide deposit in Australia, where smooth geometric outlines from an early interpretation were proved to be substantially in error during production (King et al., 1985). Figure 2.7 compares the interpreted Woodlawn ore outline (in cross section) based on limited exploration data with the true ore outline mapped after production. Clearly, the early interpretation greatly overestimated the volume of ore by more than 50 percent of the ore actually found and mapped between the two diamond-drill holes.

The two foregoing examples illustrate the uncertain nature of geologic interpolation. Remember that ore resources/reserves normally are determined within a geologically controlled (interpreted) volume. These examples show clearly that smooth interpolation is not necessarily reality, and large local errors are possible. Global estimates of these errors may be possible at an early stage of resource/reserve definition. However, it must be realized that the errors may not be compensating. If a relatively inflexible mining plan is designed to follow a smooth and interpolated ore/waste contact, as occurs commonly in underground mine design, then short-range errors can result in ore being left as waste and waste being included in ore. There are a variety of approaches suggested to deal with this problem (e.g., Sides, 1994; Stone and Dunn, 1994; see also Chapter 16 on dilution). Most important for purposes here is to recognize how geologic information can be gathered and directed toward dealing effectively with the problem (i.e., recognizing the existence of the problem and gaining some insight into the magnitude of errors possible).

Case histories are useful in understanding the nature of errors inherent in interpolation. For individual deposits, an analysis to indicate the cause of discrepancies between estimates and reality is useful and even essential. The Neves–Corvo and Woodlawn examples are sufficient to illustrate the error inherent in standard interpolation practice during geologic modeling of a mineral deposit; the problem is one of understanding ore/waste contacts, their detailed nature, and their short-range variability. Sides (1994) describes a case history for the Graca orebody in which he demonstrates quantitatively the smaller errors in locating ore/waste contacts near control points, in contrast to interpolations that are far removed from control points.

Geologic information about ore/waste margins can provide insight about the magnitude of errors

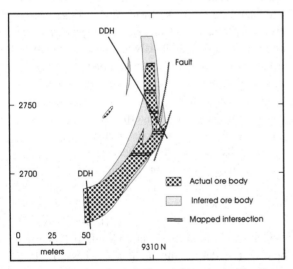

Figure 2.7: Section through the north orebody, Woodlawn massive sulphide deposit, Australia, showing ore inferred from exploration drilling (gray tone plus heavy dots) compared with results from underground mapping during production (heavy dots). Note the remarkable difference in cross-sectional area of ore actually present compared with the much larger, exploration-based estimate. Redrawn from King et al. (1985).

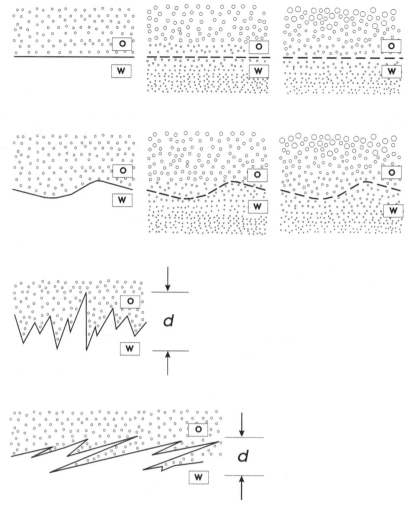

Figure 2.8: Graphic modeling of systematic variations in character of an ore/waste boundary. The boundary changes from sharp (far left) to gradational over increasing distances (middle to far right). The boundary changes from simple (planar) to increasing sinuosity/irregularity from top to bottom. Both boundary characteristics, sharp/gradational and regular/irregular, are functions of scale. Highly complex boundaries can be characterized as having a measurable width, d, that represents a zone of interdigitation of mineralized rock and country rock. After Sinclair and Postolski (1999).

inherent in the construction of simple geometric interpolations. Clearly, during exploration, effort must be directed toward thorough geologic documentation of the nature of ore/waste margins from both surface and underground exposures. Figure 2.8 is an idealized representation of two boundary characteristics, sharp/gradational and regular/sinuous, that are mappable geologic features. Ore/waste contacts should be accessed at several localities in a deposit as early in the data-gathering program as possible; one

locality is unlikely to be representative for an entire deposit. At each locality, the ore/waste contact should be studied and mapped in great detail over a distance greater than the common interpolation distance (e.g., average spacing between drill sections), and the real pattern compared with a smooth interpolation. Figure 2.9 is an idealized ore/waste margin on which several smooth interpolations have been superimposed. In particular, three straight-line interpretations of an ore/waste margin have been drawn, assuming

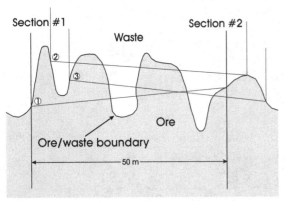

Figure 2.9: Errors of location of irregular boundaries resulting from smooth interpolations between widely spaced control points. In this hypothetical example, drill sections are assumed to be separated by 50 m. Where a boundary can be mapped in detail, a pair of hypothetical intersections 50 m apart can be superimposed on the boundary trace at any location. For example, by starting at one end of the mapped contact, a new and smooth interpolation could be generated at every 5-m separation of a pair of intersections. For each interpolation position, a quantitative, two-dimensional estimate of both "dilution" and "ore lost to waste" can be made. The average of a set of such measurements provides insight into the level of error that arises through relying on smooth interpolations. After Sinclair and Postolski (1999).

the interpolation distance to be the same as the indicated spacing of drill holes. Each of these interpolations provides an opportunity to estimate both the amount of ore lost and the amount of dilution. Such geologically based studies provide an indication of the possible magnitude of error in ore volume (and grade) that can result from the necessarily simplistic, smooth interpolations imposed on early geometric interpretations of mineral deposits. This detailed information can also be used to evaluate the impact of complex ore/waste margins on dilution and loss of ore (see Chapter 16).

The preceding ore loss/dilution calculation contains an implicit assumption that ore/waste boundaries are hard (i.e., sharp and easily recognized). For many deposit types, gradational (soft) ore/waste boundaries are the rule. Where deposit margins are gradational, the data generally contain a significant random component and the best location of the ore/waste

boundary is not obvious. It is common practice to estimate the boundaries by use of a cutoff-grade contour (e.g., Fig. 1.17). Various contouring procedures can place the ore/waste boundary at different positions, and generally it is not clear which method is best. One method developed for porphyry molybdenum deposits (Ranta et al., 1984) is described by Sinclair and Postolski (1998) in the context of porphyry-type deposits but has more general application (Fig. 2.10).

The ore/waste boundary, generally based on a cutoff grade, is first defined where hard information is available and then is extended by interpolation between control points. It may be that geologic character changes smoothly across the gradational zone that marks the ore/waste margin. More commonly, however, the zone is more or less uniform geologically, and is characterized by erratic grade variations from a higher-grade zone (ore) to a lower-grade zone (waste). A common situation for porphyry-type deposits is that samples exist from inclined diamond drill holes (or sampled trenches or underground workings) that pass from ore to waste. An empirical approach

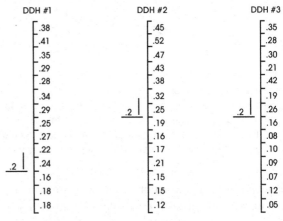

Figure 2.10: Three partial drill-hole sections showing MoS_2 values for contiguous 3-m composites. The example illustrates application of the "9-m rule" (three contiguous composites) to determine the position of a 0.2% MoS_2 grade boundary. Starting in high grade, a lower-grade boundary is fixed at the start of three successive samples (i.e., totaling a length of 9 m) that are below the stipulated grade of 0.2% MoS_2. After Sinclair and Postolski (1999); adapted from Ranta et al. (1984).

to definition of a grade boundary that benefits from success in practice is that described by Ranta et al. (1984) for porphyry molybdenum deposits. Suppose a 0.2 percent MoS_2 contour is desired and samples are for 3-m lengths of drill core. An arbitrary core distance is selected as the basis for decision making (ca. 9 m has proved useful in many porphyry Mo deposits). Then: "a grade boundary is selected at that point where assay values in a drill hole decrease below the minimum grade (0.2%)... and remain continuously below for at least 10 m" (Ranta et al., 1984). The rule can be applied successfully in most cases; some complications occur (e.g., ambiguities, small high-grade pods in low grade, and vice versa) that lead to arbitrary decisions; details of the method with illustrations are provided by Ranta et al. (1999, p. 37).

Exact ore/waste boundaries determined as discussed previously for drill holes and other types of linear samples are the information base from which a continuous boundary is interpolated, normally smoothly, as illustrated in Fig. 2.9.

With the position of a boundary now fixed along a drill hole or other set of linear samples, it is important to characterize the boundary quantitatively and, if possible, define the distance over which the boundary is gradational. There are several methods for approaching this problem. One is to examine the (auto-) correlation of samples on one side of the boundary versus the samples from other side. This can be done by pairing samples that are a uniform distance from the ore/waste boundary, with one pair from one side of the boundary and one pair from the other side. Each drill hole that intersects the boundary provides data for comparing one side of the boundary with the other, a comparison that can be effected on an x–y plot. An example for a large epithermal gold deposit is illustrated in Fig. 2.11, which shows the plotted positions of pairs of 2-m samples taken various distances from the ore/waste boundary. With little difference from one side to the other, the data should scatter about the $y = x$ line and have a relatively high correlation coefficient. A marked difference in average grades on the two sides is evident by values on the x–y plot being substantially removed from the $y = x$ line. It can be useful to monitor variations by a quantitative parame-

Table 2.1 Geochemical contrast for sample pairs various distances from a gradational ore/waste boundary selected at the margin of a large epithermal gold deposit

Distance from boundary (m)		Mean		Geochemical	
	n	x	y	contrast[a]	r
0–2	109	1.12	0.46	2.43	0.043
2–4	107	1.38	0.34	4.06	0.003
4–6	101	1.36	0.33	4.12	0.033
6–8	96	2.10	0.33	6.36	−0.044
8–10	93	1.34	0.34	3.94	−0.029

[a] Data from Fig. 2.11.

ter, such as the correlation coefficient determined for each scatter plot (r in Table 2.1).

A related method (e.g., Sinclair 1991; Sinclair and Postolski, 1999) involves the use of *geochemical contrast of grades*, Cg, determined for various sample support (length) or various distances from the contact. For purposes here, geochemical contrast is defined as

$$C_g = m_o(h)/m_w(h).$$

where $m_o(h)$ is the mean of n_o data (e.g., from n_o drill holes) centered distance h into the ore side of the boundary, and $m_w(h)$ is the mean of n_w data centered distance h into the waste side of the boundary (Fig. 2.10). This procedure can be repeated for sample pairs of increasing distances from the boundary. The procedure normally should be conducted with either raw data (of uniform length) or the composite length used for estimation purposes. For example, consider 4-m composites along drill holes (Fig. 2.10) that cross the ore/waste boundary (0.2% MoS_2) located as described previously. Geochemical contrast is determined for all pairs of 4-m composites that define the boundary; in this case, three pairs have

$$C_g = \frac{\{(0.24 + 0.25 + 0.26)/3\}}{\{(0.16 + 0.19 + 0.16)/3\}} = 1.47.$$

A second geochemical contrast is calculated comparing 4-m composites centered 6 m into ore with 4-m composites centered 6 m into waste; a third

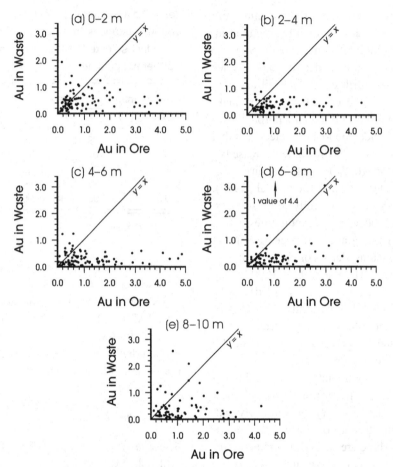

Figure 2.11: Scatter plots of ore grade versus waste grade for sample pairs equidistant from an ore/waste boundary of a large epithermal gold deposit with gradational margins. Each plot is for 2-m samples. For example, plot (c) shows 2-m samples from 4- to 6-m into ore (x), plotted versus 2-m samples from 4- to 6-m into waste (y). Statistics for each diagram are summarized in Table 2.1. Four- to 6-m grades are g Au/t.

geochemical contrast is calculated for all 4-m composites centered 10 m into ore with those centered 10 m into waste; and so on (Table 2.2). In this way, it is possible to track the geochemical contrast as a function of distance of samples from the ore/waste boundary or, for that matter, across any domain boundary. In cases of relatively narrow zones of gradation between domains of high and low grades, respectively, the value of geochemical contrast levels off at a particular distance of boundary to sample. This distance represents half the width of the domain marginal zone (DMZ), over which the boundary is gradational and provides quantitative insight into distances be-

yond which it would be unwise to use data from one side to estimate blocks on the other side. Geochemical contrast for the ore/waste boundary of the epithermal gold deposit (Fig. 2.11) is summarized in Table 2.1. The contrast rises to a rough level of 4 for paired data more than 2 m from the interpreted boundary, indicating a gradational range of 4 m (2 m into ore plus 2 m into waste); consequently, data from one side of this 4-m boundary zone should not be used to estimate blocks on the other side of the zone.

A simple useful approach to examining boundary zones of domains is to examine all available grade profiles that cross the boundary as well as the average

Table 2.2 Sample calculations of geochemical contrast across a gradational boundary

DDH #1	DDH #2	DDH #3	Avg.	C_g (contrast)
0.27	0.38	0.42	0.357	
0.22	0.32	0.19	0.243	
0.24	0.25	0.26	0.250	
0.16	0.19	0.16	0.170	1.47
0.18	0.16	0.08	0.140	1.74
0.18	0.17	0.10	0.150	2.38

Data from Fig. 2.10.

of all these grade profiles. The average profile is especially useful because it helps develop a general approach to block estimation in the immediate vicinity of the boundary. In particular, the character of the average profile helps answer questions such as the following: To what extent, if any, will data from one side of the boundary be used to estimate blocks on the other side? Does the width of a gradational boundary zone between two domains require special limits to the search radius for blocks within the gradational zone? Is a gradational contact zone sufficiently wide relative to block dimensions that the zone should be considered a separate domain for estimation purposes?

Another equally serious problem is that of determining or assuring the physical internal continuity of mineralized ground and ore between control points, which commonly are widely spaced. Errors in assumptions regarding continuity can have an enormous impact on both grade and tonnage estimates, particularly if based on widely spaced data, each of which is assumed to be continuous over half the distance to adjacent data. This essential topic is considered in greater detail in Chapter 3, concerned with continuity.

2.6: ORE DEPOSIT MODELS

2.6.1: General Concepts

Ore deposit models are conceptual views of how ore deposits form combined with an idealized representation of the geometric configuration of various features (alteration, vein type, mineral zoning, etc.) in and around a mineral deposit. Such models are not to be confused with terms such as *block model*, which refers to an arbitrary three-dimensional array of blocks that defines a deposit and to which various attributes or estimates have been assigned (e.g., grade estimates). As Sangster (1995, p. 4) suggests, ore deposit models are "a combination of 'descriptive model' and 'genetic model' . . . inclusion of the latter ensures a finite 'half life' for every deposit model."

Models are useful in organizing ideas and information about a deposit because they represent a standard of comparison for a particular class of deposit. Deposit models generally are developed from an especially important deposit or from the combined information of numerous similar deposits. Consequently, models contain an element of prediction, particularly when certain physical attributes are characteristic of ores of a well-defined deposit type. In some cases, important practical attributes of various deposit types (models) – for example, grade and tonnage – are presented in diagrams such as Fig. 2.12 and provide a rough idea of expectations as to tonnage and grade once a deposit has been recognized as belonging to a particular model or class. Such preconceived ideas about deposit size can lead to overoptimism in both

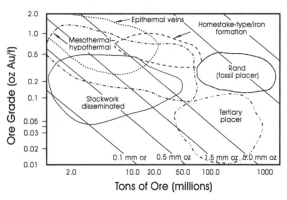

Figure 2.12: Plot of tonnage versus average ore grade for various geologic classes of gold deposits. Such diagrams are generalizations that form implicit and perhaps erroneous expectations as to deposit size once a deposit is classified into an ore deposit model. Redrawn from Babcock (1984).

interpolation and extrapolation of data. They should be avoided in favor of a more deposit-specific approach to decisions related to mineral inventory.

In mineral inventory estimation, models should be considered to be idealized expressions of the generalized nature and spatial distribution of the characteristics of a particular deposit class. In this way, models are used as a means of focussing attention on particular characteristics; the aim should be to determine to what degree a "deposit-to-be-estimated" fits a model, and not to force the particular characteristics of a model to the deposit in question. Note that this is a different philosophy than that of the explorationist, who uses characteristics of generalized models to guide exploration strategy. Once a new deposit has been located, however, the thinking process must change toward the development of a deposit-specific model that forms a guide for interpolation and mineral inventory estimation. In general, it is common for deposits of a particular type to have general characteristics in common, but with detailed differences to be expected among deposits. For this reason, the detailed knowledge of a deposit changes with time as new information is obtained. The concept of *multiple working hypotheses* (Chamberlain, 1965), so necessary in the exploration phase, is a useful philosophical approach to the many assumptions that are necessary in mineral inventory estimation, and in particular to the development of a deposit-specific model.

Ore deposit models incorporate specific information and concepts that are of direct importance to resource/reserve estimation, including the following:

(i) The external form and extent of a mineralized field (e.g., tabular, lenticular, simple vein, gradational alteration halo)

(ii) The nature of ore/waste contacts (e.g., sharp, gradational, sinuous)

(iii) The internal form and local physical continuity of mineralization (e.g., massive, disseminated, stockwork)

(iv) Large-scale continuity of mineralized ground or a mineralized structure (e.g., the use of structure contours to map variability and disruptions in a tabular surface)

(v) Mineral zoning (potentially available early in an exploration program; more likely a mineralogic study will be done during advanced exploration)

(vi) Relation of various deposit attributes to controlling structures or lithologies (which requires detailed geologic mapping, understanding of geologic evolution of the area, and integration of a knowledge of the spatial distribution of geologic features with grade distribution patterns).

The significance of these aspects of ore deposit models to resource/reserve estimation requires that all be documented either as a prelude to or accompanying a report on mineral inventory estimation. It is by such detailed study and consideration that geology is truly integrated confidently into the estimation process. Unfortunately, these topics are rarely well documented in reports relating to mineral inventory estimation, and too commonly are not evaluated sufficiently critically. Thus, even where a substantial effort has gone into the geology, the accumulated information might not be incorporated well into the estimation process.

In general, for mineral inventory purposes, is it essential to appreciate in detail the present status of the ore deposit models that relate most closely to the deposit whose inventory is being estimated. Many deposit models have been developed, and it is impractical to attempt to consider them all in a text such as this; a few models are discussed in the following sections to illustrate their importance to mineral inventory work.

2.6.2: Volcanogenic Massive Sulphide Deposits

Globally, volcanogenic massive sulphide (VMS) deposits are one of the most important sources of precious and base metals. A conceptual model for Kuroko-type VMS deposits is illustrated in Fig. 2.13. Deposits are generally massive, polymetallic, and range in size from less than 1 million to many tens of millions of tons. Those close to the underlying hydrothermal conduit or feeder (i.e., proximal deposits) are commonly lenticular in cross sections parallel to the originally vertical direction of the feeder zone;

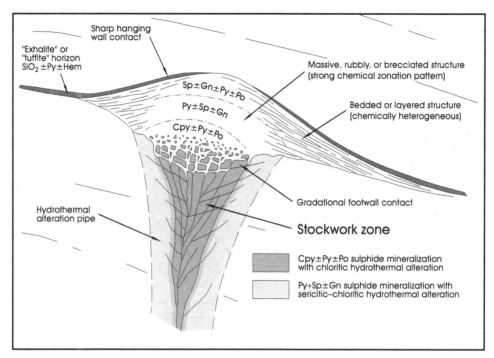

Figure 2.13: A general conceptual model (ore deposit model) for volcanogenic massive sulphide deposits of the Kuroko type (redrawn from Lydon, 1984). The model incorporates significant spatial variations in the character of mineralization (vein, stockwork, massive, bedded), mineral zoning (both vertical and lateral), and the geometric form and spatial position of mineralization types (domains).

perpendicular to this lenticular form, the mass commonly has a preferred elongation. Zones near the vent may be partly to entirely recrystallized, resulting in substantial variation in textures, including grain size, of the sulphides. These lenses are roughly parallel to bedding in the enclosing volcanic pile. The lateral edges of these lenses can be well layered and internally conformable with surrounding stratigraphy. Top layers and lateral margins can be relatively rich in Pb and Zn, and commonly contain significant amounts of barite. The underlying feeder zone is represented by a network of fracture-controlled mineralization, generally rich in Cu, that can vary from an isotropic stockwork to a complex fracture zone with a preferred structural orientation, possibly indicative of an anisotropic character to grade continuity. Later, cross faults are common because of the tectonic setting within which such deposits form. It is not uncommon that shearing is concentrated within these sulphide-rich rocks.

Distal VMS deposits (i.e., deposits that are more remote from the underlying conduit) are generally more tabular in form than proximal deposits, and may be extremely extensive, well-stratified, tabular sheets with a high degree of primary geologic continuity parallel to bedding in the mineralized sheet.

2.6.3: Besshi-Type Cu–Zn Deposits

The Besshi-type model is named for the 33-million-ton deposit of the same name in Late Paleozoic, terrigenous, clastic–carbonate–volcanic sequence in the San Bagawa metamorphic belt of southwestern Japan. Deposits belonging to this model are believed to have formed exhalatively in a submarine basin by chemical precipitation of sulphides from fluids expelled at the sea floor. The Goldstream deposit (Hoy et al., 1984) in southeastern British Columbia illustrates many of the features important in Besshi-type deposits from the perspective of mineral inventory estimation.

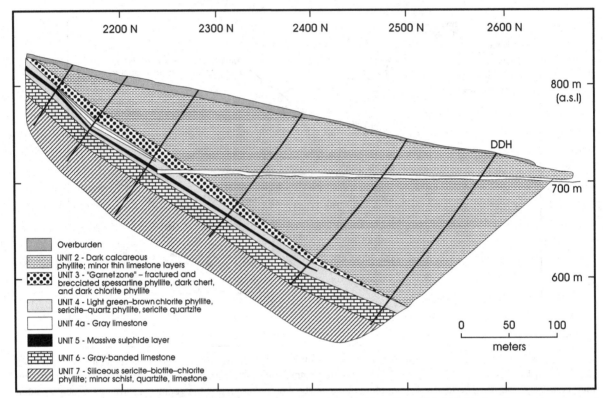

2200 N   2300 N   2400 N   2500 N   2600 N

800 m
(a.s.l)

DDH

700 m

600 m

	Overburden
	UNIT 2 - Dark calcareous phyllite; minor thin limestone layers
	UNIT 3 - "Garnet zone" – fractured and brecciated spessartine phyllite, dark chert, and dark chlorite phyllite
	UNIT 4 - Light green–brown chlorite phyllite, sericite–quartz phyllite, sericite quartzite
	UNIT 4a - Gray limestone
	UNIT 5 - Massive sulphide layer
	UNIT 6 - Gray-banded limestone
	UNIT 7 - Siliceous sericite–biotite–chlorite phyllite; minor schist, quartzite, limestone

0   50   100

meters

Figure 2.14: Cross section of the Goldstream Besshi-type deposit (adapted from Hoy, 1984) illustrating the regular, physical continuity of massive sulphides, parallelism of the sulphide layer to stratigraphy, and relative uniformity of thickness of the sulphide-rich layer. The section extends south from the portal shown on Fig. 2.15. Coordinates are in meters; elevations are meters above sea level.

A cross section of the Goldstream deposit (Fig. 2.14) emphasizes

(i) The sheetlike form
(ii) The parallelism to surrounding stratigraphy
(iii) The continuous nature of the sulphide-rich zone.

Goldstream and other Besshi-type deposits generally occur in metamorphic terrains that have been deformed extensively with the common result that sulphides have been variably mylonitized, granulated, mobilized, and recrystallized. Hence, the internal character of the sulphide sheet can differ significantly from one place to another. Commonly, tectonism has been localized in the sulphide zone, with the result being brittle interlayers among the sulphides that are broken and milled, and fragments of wallrock are included in varying amounts in the sulphides. Metal

recovery from sheared sulphides can be very different relative to coarse-grained ore.

Sulphides are characteristically massive, but disseminated sulphides are not uncommon; either form can be ore grade. Deposit margins are locally sharp and, in some places, shear surfaces, but margins of sulphide with wallrock can be highly interdigitated as a result of mobilization of sulphides during metamorphism and deformation. Metal zoning, particularly vertical zoning, is uncommon. The Goldstream deposit appears to have well-defined zoning, from Cu-rich deposits on the west to Zn-rich deposits in the east (Fig. 2.15), although this must be substantiated because of the relatively sparse and widely spaced data used as a basis for the contouring. Because of the fairly regular sheet-like form, these deposits can be examined usefully with isopach maps (see Fig. 2.15). It is common for metamorphism to obscure evidence

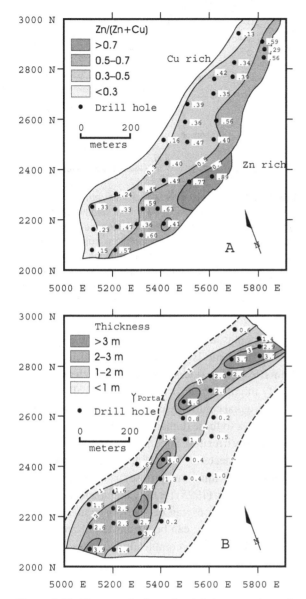

Figure 2.15: Planar projection of a tabular, massive sulphide zone contoured for (A) Cu/(Cu + Zn) ratio, and (B) thickness in meters. These contoured plots are based on exploration data and are smoothed relative to reality. Nevertheless, general trends are apparent that can be important in developing procedures for resource/reserve estimation. Adapted from Hoy (1984). A north–south line through the portal is the line of section for Fig. 2.14. Coordinates are in meters.

of primary alteration assemblages associated with these deposits.

2.6.4: Porphyry-Type Deposits (see also Sinclair and Postolski, 1999)

Porphyry copper and related systems are so variable in character that a variety of models have been formulated to describe the spatial patterns of their various geologic features. For example, the class includes porphyry copper–gold deposits (Sillitoe, 1993), porphyry molybdenum deposits (White et al., 1981), and porphyry copper–molybdenum deposits (Drummond and Godwin, 1976). Numerous other model types are described in the geologic literature. Here, the Lowell–Guilbert (1970) model for porphyry copper–molybdenum deposits is used to illustrate the importance of models to grade estimation in porphyry-type mineralized systems.

The Lowell–Guilbert model pertains to copper and copper–molybdenum porphyry deposits zoned concentrically about a core of igneous rock that is part of an igneous system, including at least some porphyritic units. These mineralizing systems are from several kilometers to hundreds of meters in diameter, and generally are of economic importance for their contents of copper, gold, or molybdenum. Ore minerals are commonly controlled in a stockwork of veinlets, in disseminations, or a combination of the two (Fig. 2.16), and less commonly in breccia pipes. These styles of mineralization may be the basis for defining separate domains for resource estimation purposes. Mineralization may be predominantly in the core intrusion, in adjoining wallrock, or may straddle the contact zone; hence, rock characteristics can vary with the geology. Hydrothermal alteration can be extensive, to the point that original rock material is totally replaced, or relatively much less intense. Many alteration minerals (quartz, sericite, pyrophylite, clay minerals) can have important effects on milling procedures and thus affect metal recoveries or cost of milling.

Both ore minerals and alteration minerals are zoned concentrically about the core intrusion (cf. Fig. 2.16), and overlap of various metals can result in large

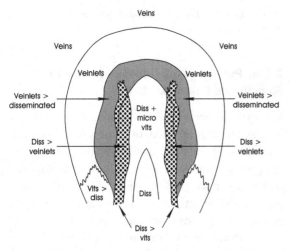

Figure 2.16: An idealized cross section illustrating part of the Lowell–Guilbert (1970) porphyry deposit model. The general disposition of mineralization styles (veins/stockwork and disseminations) are shown relative to the positions of pyrite-rich (gray stippled) and Cu-rich (dotted) zones. The diameter of the system can range from a few hundred to a few thousand meters.

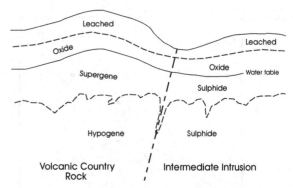

Figure 2.17: Idealized cross section illustrating leached, oxide, and supergene zones derived from weathering of hypogene, porphyry-type mineralization. Because of the downward movement of copper to form the supergene zone, grade continuity in all superimposed zones need bear no similarity to grade continuity in hypogene zones. Similarly, grade continuity in lithologic domains (volcanic versus intrusive rocks) can vary. After Sinclair and Postolski (1998).

volumes of high grade that change gradationally to low grade in a systematic manner (e.g., Fig. 2.19). Vein paragenesis can also be important in controlling high- and low-grade volumes; several stages of veins may contain ore minerals, and the superposition of two or more such stages can lead to high-grade zones that are a function of vein paragenesis. Style of mineralization, paragenesis, and mineral assemblages all contribute to the possibility of several geologically distinctive domains as the product of primary mineralization in a porphyry-type environment, as is well illustrated by the Boss Mountain molybdenum deposit in central British Columbia (e.g., Soregaroli and Nelson, 1976, and Fig. 2.5).

The central intrusion is generally complex, of intermediate composition (granodioritic to quartz monzonitic), and includes related pre-, syn-, and postmineralization dykes. Consequently, dykes can be either mineralized and potentially contribute to ore, or they can be barren and contribute to dilution.

A substantial weathering profile has developed over many porphyry deposits (Fig. 2.17) in which Cu is leached from near-surface rocks and deposited at greater depths, at and below the water table (Chavez, 2000). The result is a sequence of subhorizontal domains (leached, oxide, supergene) in which grade continuity might have been changed drastically relative to the original continuity of hypogene mineralization. Boundaries to these domains can be both gradational and variably irregular. Superimposed faulting can also lead to the necessity of defining separate domains for resource estimation purposes as demonstrated by Krige and Dunn (1995) for the Chuquicamata deposit. Domain boundaries can be sharp or gradational (cf. Fig. 2.8), and there can be a transitional zone many meters thick between completely weathered material and underlying hypogene mineralization.

2.6.5: General Summary

It is not possible to discuss all the ways in which the conceptual model of ore genesis affects mineral inventory. The examples cited in this chapter, however, clearly indicate the importance of models as they affect interpreted geometry, domains of mineralizations with differing characteristic value continuity, and the nature of data available for mineral inventory estimation. Perhaps the most important aspect of a model is that it requires an internal integration of all types of information. In addition, the development of a model meshes observations with science; consequently, confidence in an interpretation increases as more information is brought to bear on the development of a model. This translates into improved confidence in interpretations involving interpolation, extrapolation, physical continuity of domains, nature of domain margins, and so on. Increasingly, geochemical and geophysical survey information is being brought to bear on questions of physical continuity of ore between drill holes (e.g., Cochrane et al., 1998). In addition, downhole analytical methods are under development, and have the advantage that they produce metal estimates representative of a much larger volume than do assayed core (normally half the core) or cutting methods (drill-hole volume). Multielement geochemical analyses supplement assay information and help define a variety of elemental distribution patterns that assist in developing an ore deposit model. Detailed discussions of these methods are beyond the purpose of this text, but an awareness of their relevance is important.

2.7: MINERALOGY

"Assay values alone can never represent the compositional, structural and textural nature of the valuable minerals in a deposit" (Kingston, 1992, p. 49).

A detailed mineralogic study of a mineral deposit provides insight into mineral assemblages; relative abundances of minerals; and spatial variations in the form of mineralization, grain size distributions, nature of mineral intergrowths, host rock variability, and so on. These products of mineralogic examination have important implications to metal and mineral zonal patterns of both ore and gangue, metal recovery, the presence of deleterious minerals, oxide–sulphide distribution, possible by-products/co-products, systematic grain-size variations of ore minerals, and so on (Petruk, 1987). All of these can have significant bearing on mineral inventory estimation because they affect metal recovery and operating profits. Such studies contribute to an understanding of short-range continuity of various types of mineralization.

Routine mineral examination (cf. Sinclair, 1978) should include identification of both gangue and ore minerals, the occurrence and associations of which should be described in a thorough and systematic manner. An example of a fact sheet for recording mineralogic information is shown in Fig. 2.18. Samples and specimens should be from locations more or less evenly distributed throughout a deposit. Proportions of all mineral species should be reported either semiquantitatively or quantitatively. Textural data should include routine information about average grain size, grain size distribution, grain shape, nature of gangue–grain boundaries, and types of intergrowths.

In some cases the quantitative distribution of minerals is best studied using assay data. For example, in many ores lead occurs only in galena, zinc is entirely in sphalerite, and chalcopyrite is the only copper mineral. However, some caution is warranted, as such contours can mask dramatically different forms or modes of occurrence. Examples include chalcopyrite, which may be divided between free grains and minute inclusions in sphalerite; sphalerite, which can range from 1 percent to 18 percent Fe, even in a single deposit; and gold, which may be partly free milling (present as grains that are easily liberated) and partly present in sulphides as microscopic or submicroscopic inclusions. Nevertheless, selected contour plots are useful in defining zonal distributions of metals (and minerals) or, in other cases, in demonstrating spatial correlations among metals (and minerals), as illustrated in Fig. 2.19. The impact of zoning on viability for production can be dramatic. Goldie (1996, p. 40) states, "a very strongly zoned VMS deposit could be worth more than an unzoned deposit, because it might make sense to mine and mill one end of the orebody first, then re-tune the mill and mine the other end."

Significant variations in mineralogy must be considered in developing an estimate of mineral inventory. King et al. (1985) describe a case concerning the Woodlawn massive sulphide deposit, New South Wales, Australia, for which production could not meet estimates because mineralogic features were not taken into account adequately during the feasibility stage. Figure 2.20 shows the general disposition of different

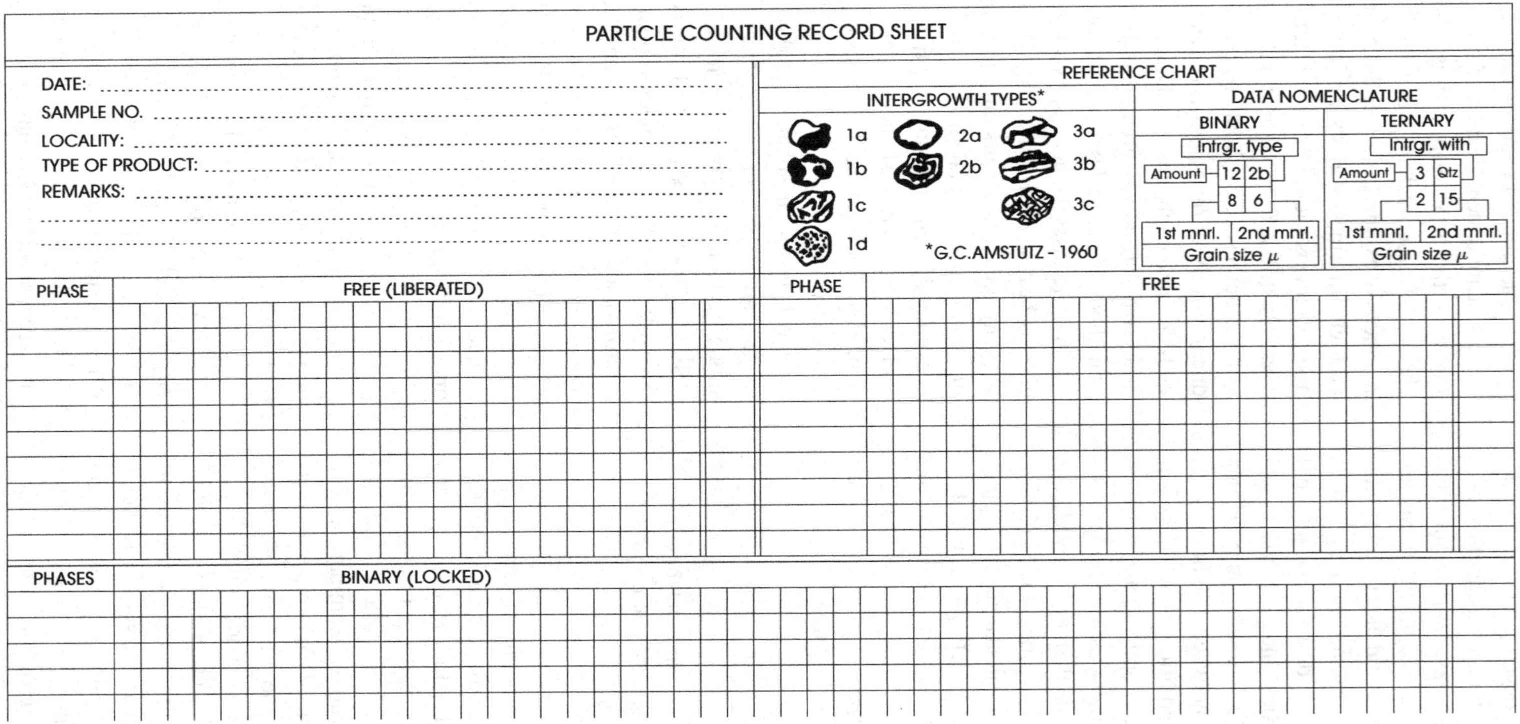

Figure 2.18: Part of a "fact sheet" for recording information obtained during a detailed mineralographic study of a mineral deposit. This particular sheet pertains to crushed materials and the evaluations of particles as liberated (free) or locked (binary, ternary). Amstutz's classification of locked particles is illustrated as a guide. Provided by C. Soux.

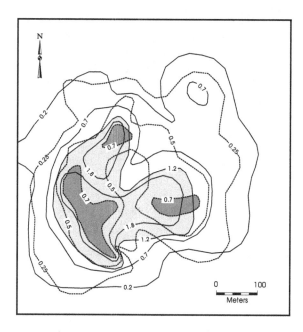

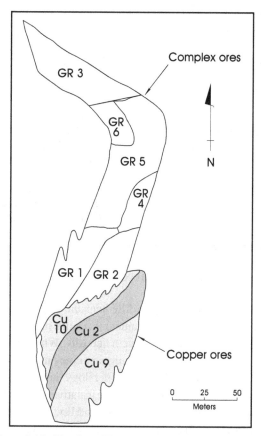

Gold isopleth (ppm) ⌒ Copper isopleth (%)

Figure 2.19: Metal distribution patterns in the Dizon deposit, Phillipines. Solid contours are for Au (g/t); dotted contours are for Cu (%). The 1.2 g/t Au contour delineates a zone characterized by both high-gold and high-copper grades. After Sillitoe (1993).

Figure 2.20: Woodlawn Mine, Australia, 2760 bench, showing distribution of various mineral assemblages as mapped during and after production. Labeled domains are different ore types, based on mineralogic character, and include various combinations of high- to low-talc content, high- to low-pyrite content, coarse- to fine-grained pyrite, and moderate to very high Zn grades. Redrawn from King et al. (1985).

mineralogic domains as they were mapped eventually, and in particular, emphasizes the areas of complex (e.g., talc bearing) and copper-rich ores. Notably, several of these mineralogic types of ore were identified early in the evaluation of the deposit, each with its own suite of grades, gangue minerals, and metallurgic response. Ore types range from high talc (with attendant milling problems) to low talc, from high pyrite to moderate pyrite, and having wide variability in metal (Cu and Zn) grades. A detailed map of the 2760 bench indicates the relative positions of various ore types that, although recognized in a general way when the mineral inventory was being prepared, had not been built into the predictive ore-reserve estimate. This omission contributed to subsequent milling problems, including lower metal recovery than expected and higher than expected contamination levels in both lead and zinc concentrates. These problems could have been reduced had more effort been directed to the engineering aspects of geology, partic-

ularly toward definition of the spatial distribution of ore types. Day-to-day production is not the mine average, but varies with the local mineralogic variations that characterize a deposit.

There are numerous other examples in the literature of the importance of mineralogy in relation to ore reserves. Ahrens (1984) reports that hydrothermal alteration is mapped as a guide to blending ore at Cananea, Mexico. Silicified limestone at an open-pit operation in New Mexico is a concern because inclusion with ore leads to grinding problems. Peterson and McMillan (1992) provide an interesting description of the use of spatial variations in compositions

Table 2.3 Metal recovery versus mineral zone, zinc mine, Broken Hill District

Ore type	Characteristics[a]	Metal recovery (%)		
		Pb	Ag	Zn
Siliceous (fringe)	20–50% SiO$_2$, very fine-grained, py-gn-sp (Fe poor), complex textures	70	40–45	75–80
Pyritic (intermediate)	70–95% sulphides, py-gn-sp (Fe poor), homogeneous texture/mineralogy with local silica; less complex textures than siliceous ore	60–65	40–45	70
Pyrrhotitic	70–80% sulphides, po-py-gn-sp (Fe rich), relatively coarse grained	75	55	85

[a] gn = galena, po = phyrrhotite, py = pyrite, sp = sphalerite.
Source: McMahon et al. (1990).

of members of the tetrahedrite–tennantite solid solution series to define fundamental zoning patterns in both sediment-hosted and vein deposits. Whiting and Sinclair (1990) provide a cautionary note on the interplay of structural domains, mineralogy, and dilution in producing an apparent decrease in native gold abundance with depth at the San Antonio Mine, Manitoba. Kingston (1992) discusses several examples of mineralogic studies of platinum-group minerals as an aid to the evaluation of Pt-bearing deposits. Metal recovery is a function of metal zoning in the Zinc Mine, Broken Hill District, Australia (McMahon et al., 1990), as summarized in Table 2.3. To appreciate the impact of these recovery figures on ore value, consider a representative grade of 3 percent Pb, 6 percent Zn, and 30 g Ag/ton and corresponding metal prices (US$) of 0.25/lb., 0.45/lb., and $5.00/oz. The recovered metal value per ton for each ore type is presented in Table 2.4. The maximum difference in value of recovered metals is more than US$10.00 for pyritic versus pyrrhotitic ores, a difference of more than 15 percent.

Gold ores represent a case of special mineralogic interest for evaluation purposes (e.g., Gasparinni, 1983; Kingston, 1992). Two principal types are readily identifiable through mineralogic investigation – free-milling and refractory, both of which can be represented in a single deposit. Refractory ores, which contain much of the gold as small grains encased in minerals such as pyrite and arsenopyrite, are of particular concern because the small grains are difficult to expose to chemicals (e.g., Na or K cyanide) for dissolution. Hence, high metal recovery can be difficult or impossible. Liberation of gold (Kingston, 1992) can be by grinding (for coarse gold), roasting of the sulphide host, aqueous pressure oxidation (which solubilizes sulphides), and biological leaching (using thiobacillus ferro-oxidans). Additional benefits of mineralogic studies (e.g., Kingston, 1992; Gasparrini, 1983) of gold deposits include the following:

(i) Recognition of cyanicide minerals, such as pyrrhotite, which react with cyanide solution and thus increase the use of chemicals in the concentration process

(ii) Recognition of oxygen-consuming minerals (e.g., orpiment, realgar, stibnite) that leave the pregnant solution deficient in oxygen, thus inhibiting Au dissolution by cyanides

(iii) Identification of Au minerals that are weakly soluble or insoluble in cyanide solution (e.g., Au tellurides, Ag-rich electrum) and hence are not recovered by cyanide treatment

(iv) Recognition of carbonaceous material (e.g., graphite) that absorbs cyanide solution in important quantities.

Table 2.4 Value of recovered metal for various ore types, zinc mine, Broken Hill District, Australia

Ore type	Metal	Recovery (%)	Metal recovery (kg/ton)	Value recovered (US$/ton)	Metal total
Siliceous	Pb	70	21.0	11.59	
	Zn	77	46.2	45.90	
	Ag	43	0.0129	2.07	59.56
Pyritic	Pb	63	18.9	10.43	
	Zn	70	42.0	41.73	
	Ag	43	0.0129	2.07	54.23
Pyrrhotitic	Pb	75	22.5	12.42	
	Zn	85	51.0	50.67	
	Ag	55	0.0165	2.65	65.74

All of these problems lead to higher than necessary operating costs, with the significant effect of increasing the cutoff grade.

An aspect of mineralogy not used to advantage is the determination of the sampling constant in Gy's equation for fundamental sampling error, to be introduced in Chapter 5. Suffice it to say here that the sampling constant is the product of four separate factors, each of which is related to mineralogic features of the mineralization. The names of these factors – mineralogic, liberation, size range, and shape – make it clear that mineralogic characteristics (including textural features) have an overriding control on the value of the sampling constant. A thorough mineralogic study relatively early in the exploration of a deposit can provide the mineralogic information necessary for early optimization of the subsampling protocol to be used for the many samples that form the basis of deposit evaluation (cf. Sketchley, 1998).

2.8: GEOLOGIC DOMAINS

The concept of domains evolves naturally from the variable character of mineralization inherent in mineral deposit models, the common occurrence of mineral zoning, and spatial variations in other physical characteristics. Each domain is characterized by more-or-less uniform geologic characteristics (see Figs. 2.1, 2.3, 2.5, 2.13, 2.16, and 2.17). The boundaries between adjacent domains can be sharp or gradational and either smoothly sinuous or highly irregular. In many cases, domains differ significantly in

grade parameters (mean and range of values) and spatial characteristics (smooth versus erratic variability in space). For practical purposes, domains are generally defined on a scale significantly larger than the size of the selective mining unit (SMU) on which mine planning commonly is based.

Domains are important in mineral inventory estimation because characteristics of one domain can have a very different impact on estimation than do characteristics of another domain. Because the continuity model for one domain can be very different than that for another domain, delineation of domain boundaries is important. An important factor from a mining perspective is that a geologic domain too small to be mined selectively probably is not worthy of formal delineation. Geologic criteria are used routinely to distinguish adjoining zones and provide precise locations of zone boundaries wherever possible. Sharp boundaries generally do not present serious problems when they can be interpolated with confidence over a large volume of a deposit; however, gradational zone margins are more problematic. A common procedure for localizing gradational zone margins involves examination of variations in grade and other characteristics along diamond-drill holes to identify relatively abrupt changes in geologic character, repeated in several neighboring drill holes. For example, high grades might be centered on a breccia zone and might merge somewhat erratically with neighboring lower-grade ground. If the occurrence is repeated in enough nearby drill holes that physical continuity is a reasonable assumption, domain margins are defined by a subjective evaluation of sudden changes in grade values or grade variability along appropriate drill holes. It is common practice to interpolate smooth domain boundaries between control points on drill sections and then to project (interpolate) these outlines from one drill section to another. The matter of defining domain boundaries is similar to that for ore/waste boundaries discussed in Section 2.5. Similarly, characterization of errors in estimating domain boundaries is comparable to error estimation of ore/waste boundaries.

Any arbitrary method of boundary positioning can be subjected to a rigorous evaluation of the effectiveness of the original control points obtained along drill holes (or other linear samples); evaluation of the smooth interpolations is much more subjective. The quantitative testing procedure involves two elements: (1) correlation of paired values from each side of the interpreted margin and (2) the ratio of average values on each side of the interpreted margin (i.e., geochemical contrast, viz. Sinclair, 1991; Sinclair and Postolski, 1999) as outlined in Section 2.5.

2.9: PRACTICAL CONSIDERATIONS

Following is a general consideration of types of geologic and related information that can be useful in developing a confident knowledge of mineral inventory. Not all suggested information types require equal emphasis for all deposits, nor are all suggestions necessarily applicable in the simplistic manner in which they are presented. Nevertheless, the specific suggestions for data gathering are presented in a manner to illustrate their potential importance to decisions regarding mineral inventory estimation – the general tone of the suggestions is what is important; the details can be altered to suit a particular situation.

1. Geologic mapping, including drill-core logging, is the basis of all resource/reserve estimation. Data must be presented on clear maps and cross sections, and fact must be easily distinguishable from interpretation (interpolation and extrapolation). Documentation is essential for all assump-

tions that have an impact on resource/reserve estimation decisions and to provide a basis for audit by others. The use of standard codes for geologic properties (e.g., Blanchet and Godwin, 1972) can increase the efficiency of logging and field mapping, and provides a convenient form for incorporation into computer files.

2. Conduct district geologic mapping of a broad area, including pertinent mineral deposits, on a scale sufficient to develop the geologic history of the area, and the relation of the deposit to that history. Pay particular attention to pre-, syn-, and post-ore processes.

3. Conduct detailed geologic mapping to provide geometric constraints to the mineralized zone. This work includes maps, sections, plans, isopach maps, structure contour maps, Connolly diagrams, stratigraphic correlation, and so on as required to deal with the development of an appropriate ore deposit model for a specific mineral deposit.

4. Detailed characterization of mineralization styles (e.g., stockwork, simple vein, sheeted vein, preferred orientations of veinlets, crackle zone, breccia, disseminated, massive), with emphasis on characteristics such as orientations, mineralogy, structure, lithology, alteration, and so on.

5. Deposit mapping should include a comprehensive mineralogic (applied) study that includes attention to mineral associations, mineral zoning, textural variations (grain size, inclusions), features affecting metal recovery, paragenetic relations, association of precious metals with base-metal sulphides, alteration mineralogy, and so on that might have an impact on mineral inventory (e.g., different stages of veining that do not have the same spatial distribution). A specialized application of mineralogy is the early estimation of the sampling constant in Gy's equation for fundamental sampling error.

6. Detailed mapping of ore/waste contacts where exposed (trenches, stripped areas, underground exploration workings), preferably at several distinct sites throughout the deposit, with at least one such

area per mineralization style. At each locality, this mapping should be done over a distance at least as great as the normal distance of interpolation (e.g., distance between sections) so as to provide a basis for determining the magnitude of error that results by imposing a smooth ore/waste contact between control points. Ore/waste contact mapping should not be limited to drill-core logging because that produces only a point view of the contact and a one-dimensional view of the contact zone. The aim of this category of mapping is to gain a detailed appreciation of the nature of the contact zone over distances that are at the scale of the normal spacing between control points so that a realistic idea can be obtained of errors implicit in smooth interpolations. Where mining follows such smooth surfaces, there is a definite possibility that some ore will be left in the walls and some wallrock will dilute the mined ore material.

7. Detailed assay mapping across ore/waste boundaries in order to document the scale of the gradations. Sample lengths might be much shorter for this mapping than the common sample lengths of the exploration program. Sampling lines across the contact zone should be close spaced relative to the distance between sections and should be guided by the detailed geology. Sample design should be such that composites of roughly comparable length to the standard exploration sample could be formed from the smaller, more detailed samples. Samples should be obtained contiguously along closely spaced sampling lines; individual samples can be of variable length to take geology into account, with the main aim being to understand the relation between grade and individual geologic features.

8. In modern exploration programs, most samples are linear (e.g., drill core, drill cuttings representing a particular interval of drilling, channel samples). Sampling design should include an effort to maintain a more-or-less uniform sample support (common length, mass, orientation) or to arrange samples so that those of one support type can be composited to rough equivalents of another support type. In general, samples should not cross geologic boundaries; where they might, smaller samples should be taken to conform to contacts. It is easy to combine three 1-m sample assays to a 3-m composite, but not possible to subdivide a 3-m sample assay into 1-m assays without resampling and reanalysis. A common support provides sound data for determining an unbiased histogram, as well as improving the estimation of autocorrelation functions. Moreover, the use of a common support contributes to minimizing local estimation errors. In addition to common support, attention should also be directed to considering different spatial densities of sampling (sample spacing) in volumes characterized by different mineralization styles, taking into account the staged nature of exploration and the likelihood that sample density will increase as exploration proceeds.

9. Hard geologic boundaries are those across which there is a sudden and significant change in geologic character (definition depends to some extent on scale). Hard boundaries must be defined during exploration and used as a control in designing a sampling plan. Individual samples should not cross hard geologic boundaries such that data from different geologic environments or different ore types can be evaluated without contamination of one data type with another. Soft boundaries should be characterized by geochemical contrast or correlation studies of data from both sides of the boundaries.

10. Integrating all available geologic information into an ore deposit model is useful because the process requires critical evaluation of all characteristics of a deposit, including continuity of various features. Hence, developing an ore deposit model in critical fashion provides confidence in interpolations and extrapolations of geologic features (e.g., mineral zones, mineralizations styles), as well as the detailed nature of ore/waste and domain boundaries.

11. Geologic interpolations and extrapolations can be made with improved confidence using a variety of

drill holes, geophysical techniques, and multielement lithogeochemical surveys.

2.10: SELECTED READING

Dominy, S. C., A. E. Annels, G. S. Camm, B. W. Cuffley, and I. P. Hodkinson, 1999, Resource evaluation of narrow gold-bearing veins: problems
and methods of grade estimation; Trans. Inst.
Min. Metall. (Sect. A: Min. Industry), v. 108,
pp. 52–69.

Grace, K. A., 1986, The critical role of geology
in reserve determination; *in* Ranta, D. E. (ed.),
Applied mining geology: ore reserve estimation;
American Inst. Min. Metall. Engs., Soc. Min.
Engs., pp. 1–7.

King, H. F., D. W. McMahon, and G. J. Bujor, 1982,
A guide to the understanding of ore reserve esti-
mation; Australasian Inst. Min. Metall., Suppl. to
Proc. No. 281, 21 pp.

McMillan, W. J., T. Hoy, D. G. McIntyre, J. L. Nelson, G. T. Nixon, J. L. Hammack, A. Panteleyev,
G. E. Ray, and I. C. L. Webster, 1991, Ore deposits, tectonics and metallogeny in the Canadian
cordillera; British Columbia Ministry of Energy,
Mines and Petroleum Resources, Paper 1991–4,
276 pp.

Roberts, R. G., and P. A. Sheahan, 1988, Ore deposit models; Geological Association of Canada,
Reprint Series 3, 194 pp.

Sinclair, A. J., 2001, High-quality geology, axiomatic
to high quality resource/reserve estimates; Can.
Inst. Min. Metall. Bull., v. 94, no. 1049, pp. 37–
41.

Sinclair, A. J., and T. A. Postolski, 1999, Geology
– a basis for quality control in resource/reserve
estimation of porphyry-type deposits; Can. Inst.
Min. Metall. Bull., v. 92, no. 1027, pp. 37–
44.

2.11: EXERCISES

1. Select a well-described ore-deposits model from
   one of the many compilations of such models (e.g., McMillan et al., 1991; Roberts and
   Sheahan, 1988; Sheahan and Cherry, 1993). Write
   a three- to five-page account of the model and
   include specific discussion of those features
   that have a direct impact on mineral inventory
   estimation.

2. The proceedings volumes of symposia on the geology of mineral deposits invariably contain cross
   sections of the individual deposits considered. Select one such deposit/paper and provide a critique
   of the cross sections or plans from the perspective
   of their use in separating fact from interpretation
   and their potential use for resource/reserve audit
   purposes.

3

Continuity

Resource/reserve estimation depends first and foremost on a geological model that provides a sound, confident expectation that a well defined volume (deposit/domain) is mineralized throughout. Without this explicit decision regarding geological continuity of a delimited mineralized zone, neither estimates nor classification of mineral inventory is possible. (Sinclair and Blackwell, 2000, p. 34).

In Chapter 3, *continuity* is defined in relation to mineral inventory estimation, and the concept of a mineral deposit consisting of several distinct domains of continuity is presented. Several case histories are discussed to illustrate the distinction between geologic continuity and value continuity, as well as to review some of the methods available for studying continuity. Reference is made to both classic and new approaches to considering continuity.

3.1: INTRODUCTION

Continuity is a topic of international concern in the study of mineral deposits and the classification of mineral inventories. This characteristic is an important parameter in several national resource/reserve classification systems used to describe formally those parts of a mineral deposit that can be regarded as being well-defined assets of mining and exploration companies. Examples of such systems are those of the United States (USGS, 1980), Australia (AIMM, 1988), and Canada (National Policy Statement 2A of the Canadian Security Administrators). These resource/reserve classification schemes describe the near certainty with which the best-defined reserve category should be known (by observation and very limited interpolation) and the decreasing certainty of continuity in other categories of resources/reserves.

During the 1980s, many gold exploration and producing companies placed too little attention toward confirming the physical continuity of mineralization prior to an actual production decision (e.g., Clow, 1991; Knoll, 1989). The resulting errors in estimating metal grades and ore tonnages contributed to the early closing of several mines and the abrupt termination of plans for production at others. More reliable estimates of mineral inventories require better understanding of continuity as a prelude to detailed mineral deposit appraisal.

Two types of continuity are recognized in mineral inventory studies (Sinclair and Vallée, 1994), geologic and value continuity. Definitions are summarized in Table 1.4. The following discussion of continuity is adapted largely from Sinclair and Vallée (1994).

3.2: GEOLOGIC CONTINUITY

Geologic continuity is the physical or geometric occurrence of geologic features that control localization and disposition of mineralization. These controlling

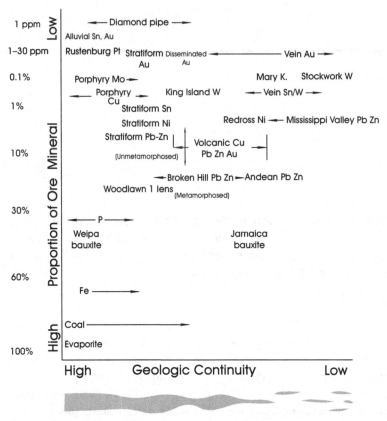

Figure 3.1: Qualitative relation of geologic continuity as a function of ore mineral abundance. The diagram is useful in showing the relative difficulties of obtaining mineral inventory estimations in average deposits of the various classes shown. The concept of geologic continuity is illustrated schematically along the *x* axis. Redrawn from King et al. (1982).

features can be lithologic or structural, primary or secondary, and commonly there is a complex interplay of more than one control. Superimposed metamorphic, structural, or alteration processes can disrupt (or enhance) an originally continuous body. Geologic continuity is a geometric feature and a function of scale; increasing continuity within a mineralized zone can be imagined (cf. King et al., 1982) in the progression from widely dispersed mineral grains through larger blebs and semimassive ore to massive ore (*x* axis in Fig. 3.1). This is a useful if simplistic view because the relative scales of sample size and the size of mineralized blebs also must be taken into account. For example, 10-m blastholes in a porphyry-type deposit are many orders of magnitude larger than the individual blebs of ore minerals. Thus, physical continuity of mineralized ground should be viewed in terms of

the sample size – in this case, the zone of disseminated mineralization rather than the dispersed mineral blebs.

Geologic observations regarding the nature of primary or secondary features is the input from which the physical continuity of a mineral deposit is interpreted. This geologic information is based on some combination of surface observations, drilling, and underground information that provide the basis for observing and recording the main features of the mineral concentration of interest (mode of occurrence and spatial distribution) and the major features controlling mineral distribution: intrusion; volcanic or sedimentary layer; faults or shear zones; and folds, stockwork, and so on. The methods that can be used and their effectiveness depend on the level of information available and on the geologic framework and

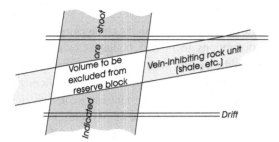

Figure 3.2: Example of the importance of using geologic information to interpret physical continuity of ore. This idealized example shows the vertical projection of a vein. Recognition of the local discontinuity in the vein depends on (i) a knowledge of lithologic control on presence or absence of mineralization, and (ii) detailed geologic mapping and interpretation. Redrawn from Rostad (1986).

deposit type present, but have much in common with techniques of stratigraphic correlation and include theoretic studies, alteration patterns, chemical profiles across mineralized structures, mineral association patterns, and so on, all of which also contribute to the development of an ore deposit model.

Geologic information is used to interpret explicitly, systematically, and in three dimensions (cf. Sides, 1992b) the general geologic environment and general extent and character of mineralized ground. Then follow assumptions (interpolations and extrapolations) about the presence, extent, and limits of a mineralized structure or mass in relation to the sample control sites and the known geology (e.g., Fig. 3.2). These assumptions are based on an understanding of continuity derived from a geologic framework known only within limits. For convenience, deposit types can be grouped into a few basic categories. For example, King et al. (1982) propose a useful geometric scheme as follows: massive and/or disseminated, stratiform (or planar/tabular), vein systems, surficial (residual), and alluvial (placer) deposits. These descriptive categories can be further subdivided if necessary. Direct geologic observations and correlations are supplemented by indirect geophysical evidence to assist in developing a three-dimensional image of the geology in and around a mineral deposit.

In many cases, a particular geologic character persists in much the same manner in all directions within a domain (i.e., a feature is *isotropic*). However, most

geologic attributes are directional or *anisotropic* in nature, and differ in their character as a function of direction in space. Several examples emphasize the importance of this attribute of anisotropy. Within a zone of sheeted veins, it is evident that the physical continuity of a single vein is more extensive within the plane of a vein than across the vein. Similarly, it is common that the regular array of sheeted veins has greater physical continuity parallel to the plane of the vein than across that plane. A syngenetic massive sulphide deposit generally is more extensive parallel to bedding than across bedding. Similarly, mineralization in shear zones is generally more elongate within the plane of the shearing rather than across the shear zone. Anisotropy of shapes of mineralized zones is a common product of the processes that form such zones and reflects underlying anisotropic geologic attributes. This concept of anisotropy is fundamental in the application of geology to obtaining high-quality resource/reserve estimates. Experience suggests that preferred directions of geologic continuity commonly are also preferred directions of grade continuity, as illustrated in Fig. 3.3 for the South Tail zone of the Equity Silver Mine, central British Columbia.

Figure 3.3: Open-pit limits, 1310 level, South Tail zone, Equity silver deposit, central British Columbia. The dashed line separates the deposit into two domains, each characterized largely by stockwork mineralization. In the northern (smaller) domain, the predominant veins strike roughly easterly; in the southern domain, the predominant vein direction is parallel to the length of the open pit. These different directions of strong continuity of veins are illustrated schematically by the ellipses (the axes of which are proportional to semivariogram ranges). After Giroux et al. (1986).

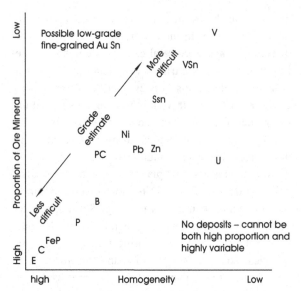

Figure 3.4: Homogeneity of mineralization versus ore mineral abundance. As used here, the term *homogeneity* is akin to the concept of grade continuity. Highly homogeneous ores are relatively easy to estimate with confidence; less homogeneous ores are more difficult to estimate. Redrawn from King et al. (1982). Symbols are as follows: E = evaporite; C = coal; Fe = bedded iron ore; P = phosphate; B = bauxite; Pb Zn = stratiform lead–zinc; Ni = nickel; SSn = stratiform tin; PC = porphyry copper; VSn = tin veins; V = gold, silver veins; U = uranium. The diagram is highly schematic and exceptions exist.

It is important to realize that adopting a deposit model introduces implicit assumptions about both geologic continuity and value continuity, as implied in Figs. 3.1 and 3.4. For example, the well-established physical and grade continuity parallel to bedding contrasts markedly with the highly irregular geometric form and erratic grade distribution characteristic of many skarn deposits. These model-related assumptions, built into early resource/reserve estimates, must be documented explicitly as work progresses. Once deposit delineation has reached a sufficient level of confidence, physical continuity can be studied effectively through the use of many traditional procedures. In tabular deposits – the use of structure contours (e.g., Fig. 3.8) and isopach maps (e.g., Fig. 2.15b) for evaluating trends and physical disruptions to trends – is well established. Similarly, Conolly diagrams, based on contoured distances from an arbitrary plane near

and subparallel to the tabular form of a mineralized zone, are useful for recognizing changes in orientation and disruptions in tabular bodies (Conolly, 1936). Contoured maps of such variables as fracture density, vein density, and grade (one or more elements) as well as mineral or metal zoning maps for successive levels or vertical sections are other useful procedures (e.g., Fig. 1.17). They are particularly useful for evaluating continuity of equi-dimensional deposits and for comparing spatial distributions of various metals. For example, two metals may have been deposited simultaneously with the result that they have a similar spatial distribution (e.g., Fig. 6.12), or they may have been deposited at different paragenetic stages of deposition, in which case there is a possibility that their spatial distributions will differ significantly.

Geologic features that affect physical continuity of a mineralized mass can predate, postdate, or be synchronous with the mineralization process; hence, a detailed geologic history is essential to sorting out all possible complexities that might affect an interpretation of continuity. Preexisting structures can themselves be physically continuous, but this does not guarantee the existence of a continuously mineralized zone. Undetected en echelon structures can cause uncertainty in developing models of physical or grade continuity (e.g., Leitch et al., 1991). The effect of faulting or folding, which potentially disrupts mineralized ground, also must be considered. Clearly, a detailed geologic evaluation, with particular attention to mineralization control and possible subsequent disruption, contributes to the understanding of physical continuity of geologic bodies and is an essential prelude to mineral inventory studies.

Generally, the limiting scale on which one needs to define geologic continuity is the size of the selective mining unit. In the case of value continuity, the required scale of knowledge is substantially less than the dimensions of the selective mining unit. The question of scale clearly is important for samples used in reserve estimation, if for no other reason than the constraints of possible mining methods and the implications to ore/metal recovery. Composites that are large relative to the size of original samples (e.g., 3-m core samples vs. 12-m composites) have a smoothing effect on original grade values; consequently, a mineral distribution pattern that is highly irregular, as based on contouring grades of short samples, might appear much more regularly distributed if based on much larger composites.

3.3: VALUE CONTINUITY

Value continuity is a measure of the spatial character of grades, mineral abundances, vein thicknesses, or some other value or quality (or impurity) measure, throughout a specified domain of interest. As an example of value continuity, grades are said to be *continuous* over distances for which they show a recognizable degree of similarity. Hence, continuity of grade is linked closely with the concept of homogeneity of mineralization (Fig. 3.4). Whereas a geologic attribute is commonly a present or absent feature, value continuity is a question of degree. Mineralization may extend between control points; the problem is to as-

certain how representative the grades of the control points are of the intervening ground. Generally, the structural and/or lithologic zones that localize or control mineralization (i.e., zones of geologic continuity) are the limits within which value continuity is defined. It is one thing to have identified the structure(s) controlling mineralization, but another thing to have reasonable expectation that the structure, or a particular part of the structure is continuously mineralized (and of ore grade) between control points. Grades normally are continuous over much shorter distances than the dimensions of the controlling geologic structure.

In the past, value continuity was examined subjectively by using such traditional techniques as grade profiles (Fig. 3.5) and grade–contour maps/sections (e.g., Fig. 1.17); both are useful techniques and should form part of the data evaluation stage in preparation for a mineral inventory study. Grade profiles along linear samples (e.g., drill holes, trenches) are useful because they illustrate the spatial character of

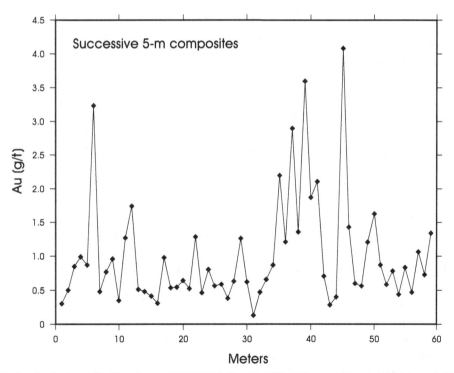

Figure 3.5: Example of a grade profile (5-m composite grades) along a drill hole in an epithermal gold deposit. The drill hole is entirely within mineralized/altered volcanic rocks and illustrates different physical continuity for lower grades versus higher grades. A low-grade population of grades is continuous over greater distance (on average) than is a high-grade population.

contiguous grades (or nearly so) and closely spaced samples over short to intermediate distances. Obviously, it is useful where possible to examine grade profiles for drill holes with different orientations through a deposit. Contoured values (commonly widely spaced control points) reflect an implicit assumption that the variable is continuous between control points. Hence, plots of contoured grade values, while instructive, must be viewed critically in the study of value continuity; there must be geologic reason to believe that mineralization is continuous within the contoured area. For example, grade contours for a bench of a porphyry-type deposit probably incorrectly depicts that part of a mineralized field cut locally by barren dykes.

More recently, value continuity has been studied by the use of autocorrelation functions such as semivariograms and correlograms that quantify a statistical or average continuity in various directions throughout a deposit or a significant domain within a deposit. In general, these models show an increasing average disparity between samples as the distance between samples increases. For many deposits, such measures level off at a sample spacing referred to as the range (i.e., range of influence of a sample). Ranges can be the same in all directions (isotropic continuity) or can vary with direction (anisotropic continuity). Relative ranges can be used to construct ellipses that demonstrate variations in continuity in different directions and from place to place in a deposit (Figs. 3.6 and 3.7).

These quantitative measures of continuity are built on an assumption concerning the physical continuity of a mineralized body. Commonly, this statistical continuity is determined with greatest confidence along the main axis of sampling (e.g., along drill-hole axes). Sampling in the other two dimensions is commonly much more widely spaced (i.e., distance between sections and distance between drill holes along these sections is much greater than sample spacing along drill holes; see Fig. 3.5). For these less well-sampled directions, a conceptual understanding of continuity is very dependent on geologic interpretation. In Figs. 3.3 and 3.7, the long axes of the ellipses are parallel to the principal directions of geologic and value

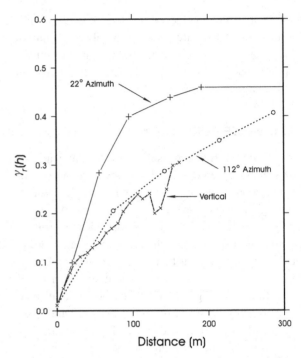

Figure 3.6: Experimental semivariograms (autocorrelation functions) for horizontal and vertical directions for the east domain of the East zone, Huckleberry porphyry copper deposit, central British Columbia (after Postolski, 1998). Note that the ranges (distances at which the experimental semivariograms level off) differ with direction in the deposit (i.e., value continuity is strongly anisotropic).

continuities in the various domains indicated. All ellipse axes are proportional to the ranges of influence as determined from autocorrelation functions (in this case, ranges of semivariograms), which characterize average value continuity as a function of direction. The use of autocorrelation functions as a tool with which to characterize, compare, and contrast value continuity quantitatively from one domain to another is evident. Such ellipses are also useful in a relative sense in depicting changing geologic continuity as a function of direction in space.

Primary factors that affect the estimation of value continuity in a particular geologic environment are: (i) mineral/metal concentrations and (ii) mineral distribution patterns and controls at various scales. Sample size (support) interacts with these primary factors.

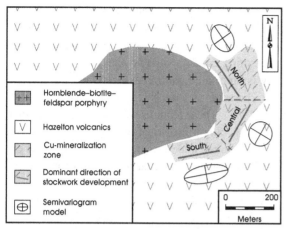

Figure 3.7: Main zone, Huckleberry porphyry Cu–Mo deposit, central British Columbia. The zone is divided into three domains (North, Central, and South), each characterized by a continuity model illustrated schematically for horizontal directions by an ellipse. The radii of the ellipses represent ranges of influence for Cu as a function of direction. A circular pattern indicates isotropic continuity of grades; ellipses indicate anisotropic continuity of grades. See Postolski and Sinclair (1998a).

In certain cases, value continuity is approximately related to the concentrations of the metals/minerals of interest and the geologic deposit model (cf. Fig. 3.4). In particular, the average local variability of grades is directly proportional to average grade. This generality is consistent with the concept of proportional effect, in which the absolute value of an autocorrelation function (e.g., the level of average differences between samples) varies systematically with mean grade, as discussed in Chapter 8.

3.4: CONTINUITY DOMAINS

A geological domain is a spatial entity that represents a well-defined mineralized body. A qualified domain for mineral estimation should contain no, or a minimum amount of, "nonmineralized" materials. The domain boundaries are usually defined on the basis of both assay and geological information. Assays are used to determine a cutoff criterion for the boundary, while geological information, such as faults, may assist to verify or refine the boundaries and to infer the boundaries in the sections where insufficient assay information is available. (Pan et al., 1993, p. 382)

Different parts of a single deposit can be distinctive geologically, and thus can be characterized by different models of physical and statistical continuity (Vallée and Sinclair, 1993). Consequently, for mineral inventory purposes it may be desirable, even necessary, to subdivide a deposit into separate domains, using as a basis the geologic features that control or characterize mineralization. Even a simple vein can give way over a short distance to a zone of horsetail veins. Similarly, where conjugate fractures control mineralization, one fracture direction can predominate in one part of the deposit and the second fracture direction elsewhere in the deposit (e.g., Figs. 2.1, 2.4, and 3.3). In certain cases, a uniform sampling grid size or orientation may not be appropriate for all domains or zones of a deposit. The Kemess South porphyry-type copper–gold deposit, described by Copeland and Rebagliatti (1993), is characterized by five distinct continuity domains with differing geologic characteristics. These authors strongly emphasize the importance of geologic control in optimizing continuity assumptions for mineral inventory purposes. Similarly, each of the five distinctive lithologic domains at the Golden Sunlight gold deposit has its own characteristic autocorrelation model for gold-grade continuity (Sinclair et al., 1983).

In practice, many problems in establishing physical continuity are related to shortcomings of the geologic information base. For example, basic information dealing with the geologic framework and the actual stratigraphy or structure of the rocks hosting a deposit may be missing or very sparse because only limited drill intersections are available. In such a case, the geologic model, the deposit (geometric) model, the derived continuity assumptions, and the interpreted grade and tonnages are all vulnerable to large changes as new information is obtained.

Some of the types of domains that can be anticipated in a porphyry-type deposit are illustrated in Figs. 2.16, 2.17, and 3.7. In Fig. 2.17, distinction is made between leached, supergene, and hypogene zones whose geologic (and ore) character might be expected to differ from one rock type to another. In order to integrate such information into a mineral inventory estimation it is apparent that the basic geologic characteristics must be mapped spatially and examined in conjunction with assay information. Even where the host rock appears uniform in a porphyry environment, different domains might result because of different intensities or directions of predominant structures that control primary mineralization (e.g., Fig. 3.7).

3.5: CONTINUITY IN MINERAL INVENTORY CASE HISTORIES

A multitude of methods have been developed for studying continuity, not all of which can be illustrated here. Three gold-bearing mineral deposits of different geologic types serve to demonstrate some of the useful approaches that can be applied to developing an understanding of both geologic and value continuity as a prelude to undertaking a mineral inventory study (i.e., the Silver Queen epithermal, polymetallic [Zn, Pb, Ag, Au] vein, the Shasta epithermal Au–Ag vein, and the Nickel Plate Au-bearing skarn [cf. Sinclair and Vallée, 1994]).

3.5.1: Silver Queen Deposit

The Silver Queen property in central British Columbia includes polymetallic, epithermal veins that were mined briefly during 1972–1973. Production ceased because of too few headings to provide sufficient mill feed and liberation problems that led to very low metal recoveries (W. Cummings, personal communications, 1990). Production and most exploration were centered on the No. 3 vein system, which strikes northwesterly, extends for about a kilometer of strike length, and dips moderately to the northeast (Leitch et al., 1990). Thickness is variable, commonly in the range of 0.1 to 2.0 m. Two recent independent mineral inventories of the central segment of the No. 3 vein (Nowak, 1991) indicate reserves of about 700,000 tons averaging 0.08 oz Au/t (2.7 g Au/t), 4.8 oz Ag/t (163 g Ag/t), 0.2 percent Cu, 0.8 percent Pb, and 5.4 percent Zn.

Several geologic characteristics of the No. 3 vein system influence the methodology used to evaluate continuity, namely, changes in orientation of veins, crosscutting faults, a variable alteration halo that locally is brecciated and mineralized, and the *en echelon* character of the No. 3 vein system (Leitch et al., 1991). These features were recognized through detailed geologic investigations of exploration drill core and limited underground workings, and their effects on estimation procedures are worth considering briefly.

Most of the 118 exploration, diamond-drill-hole intersections indicate clearly defined vein intervals; a few anomalously thick vein intersections were found to include both a vein interval and adjacent mineralized brecciated ground. Where intersected in existing workings, these brecciated margins were found to have limited lateral extent relative to the more continuous vein. The widely spaced drill data also provide some insight as to the limited physical extent of these breccia bodies, in particular, recognition that they do not extend between any two adjacent drill holes. In contrast, the vein structure and its associated mineralization are evident in all 118 exploration drill holes, thus establishing the general continuity of vein material within the controlling structure. Precious metal grade profiles were found to define vein thickness where marginal breccias occur (Leitch et al., 1991). Thus, in several drill holes with abnormally thick vein intersections from the late 1960s for which logs and core were not available for reexamination, true vein thicknesses were estimated using precious metal profiles; the excess thicknesses were attributed to noncontinuous breccia zones.

Vein continuity was investigated in detail by means of a structure contour map (Leitch et al., 1991) that displays the following features (Fig. 3.8):

(i) The *en echelon* character of parts of the vein system
(ii) A substantial segmenting of the vein due to offsets along cross faults, some with more than 100 ft (31 m) of apparent horizontal movement
(iii) An abrupt large change in vein strike near the south end of the system.

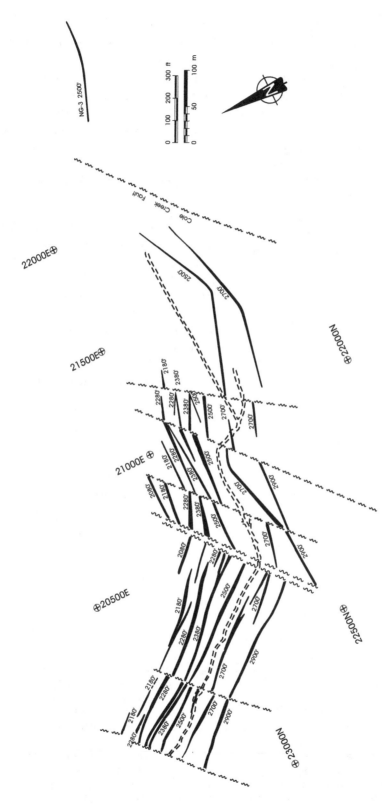

Figure 3.8: Plan of structure contours on part of the No. 3, polymetallic, epithermal, vein system, Silver Queen Mine, central British Columbia (cf. Leitch et al., 1991). Features that affect continuity include the *en echelon* character of veins, curvature of veins, splays from veins, and cross-cutting faults. Redrawn from Sinclair and Vallée (1994).

Examination of underground workings substantiates this view of geologic continuity, as initially developed from exploration drilling.

Detailed investigations by Nowak (1991) demonstrate well-defined autocorrelation for vein thickness and various grade measures (Pb, Zn, Cu, Au, Ag) using both variograms and correlograms (Fig. 3.9). The resulting geostatistical estimates (ordinary kriging) of grades and tonnage are comparable to, but slightly less than, global estimates obtained in an independent polygonal study (Nowak, 1991). This is a pattern that commonly exists in such comparative studies. Of course, the value continuity models assumed for the two estimation methods are very different, and this contributes to the disparity in the estimates. The polygonal method assumes the unlikely situation of uniform, local continuity that is perfectly known for each polygon of influence surrounding each data point (drill hole); the geostatistical approach assumes a statistical continuity to ore grades represented as an average continuity by an autocorrelation function.

3.5.2: JM Zone, Shasta Deposit

The JM structure at the Shasta Mine, northern British Columbia, is a highly altered and silicified zone in basalt of the Toodoggone Formation. Gold and silver values of economic interest are associated with small, epithermal quartz veins that strike northerly with near vertical dips, located within the much broader altered zone. Individual veins extend along strike from a few to 15 m; many occur in clusters across the strike to define zones of economic interest, as shown schematically in Fig. 3.10. Drilling shows the broad, altered zone to range to 100 m in width and commonly

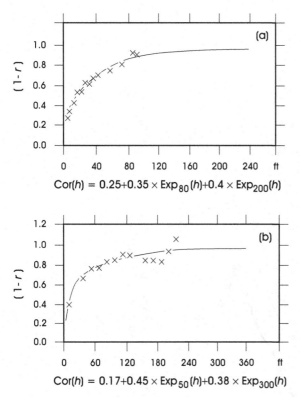

$$Cor(h) = 0.25 + 0.35 \times Exp_{80}(h) + 0.4 \times Exp_{200}(h)$$

$$Cor(h) = 0.17 + 0.45 \times Exp_{50}(h) + 0.38 \times Exp_{300}(h)$$

Figure 3.9: Average value continuity for gold accumulation (a) and vein thickness (b) for the No. 3 vein system, Silver Queen Mine, central British Columbia (see Fig. 3.8) illustrated by an autocorrelation function. In this example, the autocorrelation function is a modified correlogram (i.e., a correlogram for which all values have been subtracted from 1 so that the form is that of a semivariogram). In this case, exponential models have been fitted to the data (see Chapter 9). Data from Nowak (1991).

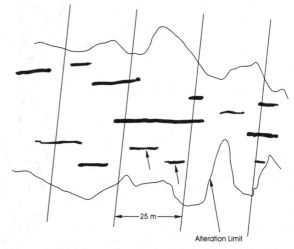

Figure 3.10: Schematic representation of the physical distribution of high-grade values with short-distance continuity, relative to spacing of drill sections, as occurs in the Shasta deposit. Clusters of high grade, northerly striking quartz veins are in a northerly striking altered zone within unaltered volcanic rocks. Note that high-grade zones can occur between drill sections and be unrepresented in data available for estimation; hence, local estimates by projection of known grades are subject to large errors. Redrawn from Sinclair and Vallée (1994).

contains one to three separate vein concentrations across its width. Individual zones of vein concentrations are commonly from less than 1 m to several meters in width (Nowak et al., 1991).

The two principal mappable features whose physical continuities are of interest are as follows:

(i) A variably mineralized, broad alteration zone
(ii) Confined zones of high concentrations of small quartz veins or quartz-infilled breccia.

Surface exposures, exploration drill holes, and underground workings all show that the alteration zone is a continuous, crudely tabular zone. This zone appears to have developed outward from various fracture zones now occupied by quartz veins, concentrated in less well-defined, vertically dipping lenticular masses. Most individual quartz veins appear to have a physical continuity very much less than the 25-m spacing of drill sections; even clusters of veins commonly do not extend between two adjoining sections.

The clusters of quartz veins are associated with the highest precious metal grades; understanding their continuity is fundamental to forecasting the locations, tonnages, and grades of ore zones. The physical occurrence of ore-grade material between relatively widely spaced sampling information from exploration drilling provides an added component of uncertainty.

A quantitative model for grade continuity (semivariogram model) has been constructed from production data and known geology to demonstrate, through simulation, the nature of grade distribution and the correlation (continuity) problem (Nowak et al., 1991). Figure 3.11 is an example of two such conditional simulations done using GSLIB software (Deutsch and Journel, 1998), and clearly demonstrates the intractable problem of estimating local resources where the physical continuity of ore shoots is short compared with the spacing of available data. This example demonstrates how important it is that exploration data provide insight into the detailed nature of local grade continuity so that there is clear appreciation of whether interpolation is possible. When interpolation is not possible, more closely spaced information might be required, or conditional simulation might

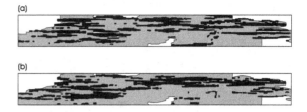

Figure 3.11: Conditional simulations of grades above and below cutoff, JM zone, Shasta Mine, Toodoggone Area, British Columbia. (a, b) Two independent horizontal simulations; (c, d) two independent vertical simulations. Black is ore grade, stippled is altered rock below ore grade. Area of simulation is 500 m × 55 m. White areas are unmineralized or areas for which insufficient data were available to create a simulation.

prove adequate as an estimation procedure. Nevertheless, in such cases of insufficient data, conditional simulation is a potent tool to understand and illustrate the character of grade continuity, and may serve as an adequate estimation procedure in some cases.

In summary, the geologic continuity of the broad altered zones can be established with confidence by a variety of routine geologic mapping procedures that provide the necessary information to construct a three-dimensional geometric model. Exploration data, however, are much too widely spaced (drill sections are about 25 m apart along the structure) to provide a confident interpolation of grades. In such cases, conditional simulation is shown to be a practical means of clarifying and documenting the problem of grade interpolation.

3.5.3: South Pit, Nickel Plate Mine

Mineralization at the South Pit, Nickel Plate Mine (south central British Columbia) occurs as three

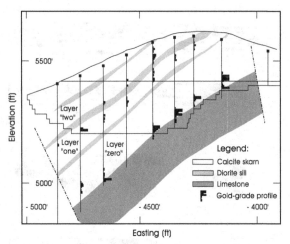

Figure 3.12: Vertical E–W section (1000 N) through the center of the South Pit, Nickel Plate skarn gold deposit, Hedley, British Columbia. Labeled vertical lines are exploration drill holes projected on the plane of the section. Basal limestone and diorite sills are shown as shaded patterns. Relative gold grades in zones of skarn (blank) are indicated by bar-graph profiles on drill holes. Ultimate pit and original topography are shown for reference. Redrawn from Sinclair et al. (1994).

zones of gold-bearing, massive to disseminated pyrrhotite-rich sulphides in a calcite skarn layer striking about N–S and dipping about 30 degrees to the west (Figs. 3.12 and 3.13). Gold values are associated

erratically with sulphides. The geometry of mineralized zones can be approximated from assay plans and sections (Sinclair et al., 1994). Almost all of the 371 surface exploration drill holes used are vertical and intersect one or more of the sulphide-bearing zones indicated in Fig. 3.12. The presence of sulphides in the appropriate stratigraphic positions in large clusters of drill holes demonstrates the geometric continuity of the three sulphide zones in three dimensions. Thicknesses of these zones are variable, and so interpolations between adjacent drill holes are approximate. Nevertheless, the physical continuity of roughly tabular form for each of the three sulphide-bearing zones is established. Physical continuity of sulphide zones is disrupted locally by thick dioritic sills (Figs. 3.12 and 3.13).

A detailed data evaluation included examination of exploration drill-hole profiles (Fig. 3.12 and 3.14) and probability graphs of gold grades (Fig. 6.6). Profiles showed that individual high-grade gold assays were invariably flanked by very much lower grades (i.e., high grades have very limited physical continuity of the order of 3 m or less in a vertical direction along diamond drill holes). No information is available as to the extent of high-grade continuity in the plane of stratification (between drill holes that are separated by about 80 ft). The probability graphs identified

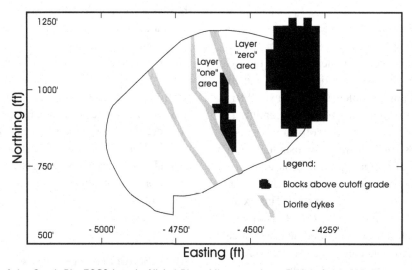

Figure 3.13: Plan of the South Pit, 5280 bench, Nickel Plate Mine, southern British Columbia. Dotted zones are mapped dioritic intrusions that disrupt continuity of ore. Black rectangular areas are blocks estimated to be above cutoff grade (i.e., > 0.03 oz Au/t). Redrawn from Sinclair et al. (1994).

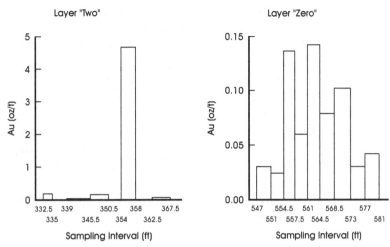

Figure 3.14: Two short-grade profiles from vertical exploration drill data illustrate the different nature of continuity of high grades (left) and low grades (right). In general, extreme grades are isolated from each other by relatively large intervals of low grades, whereas low grades are more highly correlated spatially (i.e., low grades are more similar over much longer distances than are high grades). After Sinclair et al. (1994).

this upper population of gold grades with low continuity as values above about 17 g Au/t (0.5 oz/t); coincidently, this is the cutting factor used at the minesite (Sinclair et al., 1994). Results of this study illustrate the dramatic impact of accounting or not accounting for the different characters of high- and low-grade gold continuity; exploration-based estimates using the inverse squared distance method overestimated grade by about 45 percent, in contrast to a block kriging approach that considered different continuity models for high- and low-grade gold subpopulations. When the differences in continuity are taken into account, resulting reserve estimates based on exploration data closely match production.

3.5.4: Discussion

In a discourse such as this, it is not possible to consider all aspects of continuity in detail. Each mineral deposit is, in many of its attributes, different from any other. The three mineral deposits used as examples, however, present very different geologic characteristics that lead to different approaches to the study of geologic or value continuity. Thus, the examples provide an indication of the diversity of approaches available, the importance of geologic control, and the intellectual flexibility required by mineral inventory estimators in adapting methodology to geology.

For the Silver Queen deposit, the traditional method of structure contours is particularly useful in defining primary and secondary aspects of *en echelon* vein (geologic) continuity. Autocorrelation functions were determined to quantify value continuity measures such as follows:

(i) Vein thickness (thus contributing to vein geometry and tonnage estimates)
(ii) Metal accumulations (grades × thickness), hence, grades (Nowak, 1991).

Both widely spaced drill-hole information (70-m spacing) and more localized, closely spaced information from exploratory underground workings (about 3-m spacing) were used to develop the autocorrelation models. The vein system was treated as a single domain by "unfolding" the two segments of different orientations to a common plane coincident with the central part of the vein. Alternatively, the vein system could have been considered two separate geologic domains, separated where there is a sharp change in orientation.

In the case of JM zone, the relatively sparse exploration drill-core assays were adequate to define geologic continuity of a broadly mineralized/altered zone, but were too widely spaced (25 m) to characterize grade continuity (Nowak et al., 1991). Closely spaced underground information, although localized, permitted the development of an autocorrelation function for grade, providing a basis for undertaking conditional simulations of grade. These simulations clearly demonstrate the absence of ore continuity between sections spaced at 25 m, and assist in defining the type of additional sampling information that would be necessary to document details of value continuity, and serve as a sound basis for mineral inventory estimates and, eventually, production quality control (Nowak et al., 1991).

The crudely tabular shapes of mineralized zones at the Nickel Plate Mine are well demonstrated by two-dimensional views showing grade profiles as bar graphs along exploration drill holes (Fig. 3.9 and Sinclair et al., 1993). Geologic continuity of three mineralized domains (layers) separated by barren ground is evident. Autocorrelation studies done independently in each domain indicate that the same autocorrelation model for gold can be applied to all three layers. Grade data within individual layers demonstrate the common problem of multiple populations – in this case, two, each of which must be examined separately with respect to continuity. How continuity for each data population is estimated or dealt with depends on the data density and spatial disposition of each population. At the Nickel Plate Mine the low-grade gold population is sufficiently abundant to warrant estimation of an autocorrelation function (semivariogram). However, the high-grade population occurs as isolated values or clusters of very few values (i.e., the high-grade population is not concentrated in large enough volumes for a characteristic autocorrelation function to be determined). Instead, high grades were cut to a threshold defined by a probability graph analysis of the total assay data and involving partitioning of the cumulative curve into two subpopulations, high and low grades. This graphic technique has not been widely used in the study of grade continuity, but clearly has potential for such a purpose and warrants wider consideration.

Generally, a substantial amount of data is required to establish a sound autocorrelation model for a mineral deposit. With foresight and planning, such data can be obtained early in the exploration history of a deposit, although the absolute amount of data required cannot be forecast in advance. Once an autocorrelation model has been established, both grade estimates and grade simulations based on such models can be generated. The simulation example for the JM deposit illustrates the advantage of simulations in appreciating grade continuity early in the exploration history of a deposit. At advanced stages of exploration and at the feasibility stage, conditional simulations, being closely tied to increased quantities of grade information, improve our local understanding of grade continuity and provide sound planning and quality control information for both mining and milling.

3.6: PRACTICAL CONSIDERATIONS

All the practical considerations summarized for Chapter 2 apply here. Assumptions regarding continuity must be adapted periodically to a continually improving knowledge of the geologic characteristics of a deposit.

1. The adequacy of mineral inventory estimates depends on a thorough appraisal of two types of continuity: geologic and value.
2. Confidence in both geologic continuity and value continuity requires a thorough understanding of the general and detailed geology of a mineral deposit and the surrounding area. Models of both continuity types must be refined and verified as the geologic information base and the sampling base for a deposit evolve.
3. Models of geologic and value continuity are enhanced by strict quality control procedures for data acquisition, interpretation, and modeling.
4. Evaluating continuity requires systematic geologic and sampling check work, perhaps using a formal audit procedure. Many methods are traditional; increasingly, however, computer-based data analysis software is becoming essential to this work. More attention must be given to the

three-dimensional aspects of sampling grids, particularly in situations in which the range of value continuity is shorter than sample spacing in a particular direction.

5. Several established evaluation procedures that have not been used widely in continuity studies can provide useful insight into an understanding of value continuity. They are: (i) the use of probability graphs to identify thresholds between grade categories (subpopulations) with different continuities (cf. Noble and Ranta, 1982); (ii) the use of autocorrelation studies to quantify value continuity (regardless of whether geostatistical estimates are to be obtained); and (iii) the use of conditional simulation as a means of representing and understanding the nature of value continuity.

6. The physical continuity of an ore-controlling structure and the statistical continuity of ore-grade material within that structure are fundamental attributes of a geologic model that serves as a base for mineral inventory estimation. These are not attributes that can be easily quantified in terms of risk. However, they are attributes that, on average, can be considered to be characteristic of a deposit type, because deposit types generally are characterized by particular attributes.

3.7: SELECTED READING

King, H. F., D. W. McMahon, and G. J. Bujor, 1982, A guide to the understanding of ore reserve estimation; Australasian Inst. Min. Metall., Supplement to Proc. No. 281, 21 pp.

Owens, O., and W. P. Armstrong, 1994, Ore reserves – the 4 C's; Can. Inst. Min. Metall. Bull., v. 87, no. 979, pp. 52–54.

Rowe, R. G., and R. G. Hite, 1982, Applied geology: the foundation for mine design at Exxon Minerals Company's Crandon deposit; in Erickson, A. E. Jr. (ed.), Applied mining geology; AIME, Soc. Min. Engs., pp. 9–27.

Sides, E. J., 1992b, Modelling of geologic discontinuities for reserve estimation purposes at Neves–Corvo, Portugal; in Pflug, R., and Harbaugh, J. W. (eds.), Computer graphics in geology: three-dimensional modeling of geologic structures and simulating geologic processes; Lecture Notes in Earth Sciences, v. 41, pp. 213–228.

Vallée, M., and D. Cote, 1992, The guide to the evaluation of gold deposits: integrating deposit evaluation and reserve inventory practices; Can. Inst. Min. Metall. Bull. v. 85, no. 957, pp. 50–61.

3.8: EXERCISES

1. Contrast the different character of geologic continuity and value continuity in each of the following scenarios:
   (a) A zone of sheeted veins versus a stockwork zone
   (b) Massive Cu–Ni versus network Cu–Ni in ultramafic rocks
   (c) A feeder zone of a volcanogenic massive sulphide deposit versus an upper, stratified sulphide sequence of the same system. The problem can be answered effectively by constructing sketches of the various mineralization styles and superimposing ellipses to represent geologic and value continuity in a relative manner.

2. Table 3.1 provides U_3O_8 data (%) for contiguous samples from eight drill holes (after Rivoirard, 1987). Construct a grade profile for any three consecutive drill holes and comment on the continuity of both low and high grades (see also Fig. 7.9). Assume that individual values are for 2-m samples and that the drill holes are vertical, collared on a flat surface, spaced at 25-m intervals, and numbered consecutively across a mineralized field of flatly dipping sandstone beds.

3. Figure 2.1 is a geologic plan of the surface pit, Endako molybdenum mine, central British Columbia (after Kimura et al., 1976). Assuming this level to be representative of the deposit, comment on geologic and value continuity models for the deposit.

Table 3.1 Sample values of simulated U_3O_8 grades in eight drill holes[a]

No. 1	No. 2	No. 3	No. 4	No. 5	No. 6	No. 7	No. 8
0.79	0.09	0.10	0.62	1.13	0.08	0.12	0.16
0.19	0.09	0.94	0.52	1.32	0.08	0.12	0.16
0.51	0.09	0.10	0.27	2.13	0.00	1.52	0.16
0.56	0.83	0.53	0.35	2.82	0.08	0.62	0.18
1.26	0.16	0.10	0.28	0.62	0.08	0.12	0.42
1.14	0.09	0.10	0.30	2.35	0.08	0.12	0.16
2.47	0.09	0.97	5.46	19.17	0.06	0.12	0.1
5.86	0.82	0.45	25.47	1.81	0.08	0.12	0.45
26.89	1.14	3.16	0.15	9.06	0.08	0.12	0.16
24.07	6.52	5.41	0.15	10.98	0.08	0.12	0.16
20.59	0.24	50.43	0.15	12.05	0.08	0.12	0.16
10.30	0.09	11.17	0.15	3.66	2.10	0.12	0.16
5.31	0.20	0.23	0.88	6.76	0.98	0.12	0.16
57.94	0.09	0.20	0.99	3.37	3.53	0.12	0.16
26.04	0.09	0.33	0.15	0.23	9.63	0.12	0.16
22.34	1.82	0.10	0.56	1.74	20.33	0.12	0.16
11.52	0.09	0.19	0.53	0.21	12.11	0.12	0.16
42.79	0.09	0.22	4.51	0.17	4.17	0.12	0.16
1.50	18.07	0.20	0.25	2.57	1.25	0.12	2.17
9.89	38.72	1.14	0.15	2.68	0.08	0.12	0.23
2.33	27.98	1.04	0.15	0.92	0.69	0.94	0.16
0.67	3.93	0.10	5.00	1.94	0.08	5.60	0.16
1.48	5.81	0.10	4.54	0.17	0.08	0.82	0.16
0.15	0.65	0.10	1.64	0.17	0.19	1.40	0.16
0.42	0.09	0.10	0.15	0.17	0.08	6.77	0.26
0.82	0.09	0.10	0.15	0.17	0.20	18.26	3.36
1.48	0.09	0.10	0.15	0.17	0.30	11.14	1.43
4.72	0.09	0.10	0.15	0.17	0.56	4.82	5.00
6.57	0.09	0.10	0.15	0.17	0.69	3.98	17.88
3.31	0.09	0.10	0.15	0.17	0.08	1.67	1.79
4.13	1.43	0.25	3.04	0.17	0.08	1.42	1.36
11.31	0.32	0.10	9.57	0.17	0.08	0.23	11.84
12.48	0.09	0.10	6.67	0.17	0.08	1.61	1.73
7.68	5.19	0.10	5.95	0.17	0.08	1.58	0.23
12.17	1.74	0.10	0.96	0.17	0.08	1.96	0.53
0.59	0.09	0.10	5.66	0.17	0.08	3.72	0.16
0.15	1.52	0.57	0.58	0.17	0.08	9.16	0.16
1.04	12.20	0.55	0.15	0.17	0.08	3.09	0.16
1.05	2.19	0.10	0.15	0.17	0.08	0.49	0.16
1.73	1.28	0.10	0.15	0.17	0.08	0.12	0.16
1.98	0.21	0.96	0.15	0.17	0.08	0.12	0.16
3.54	0.09	1.08	0.59	0.17	0.71	0.12	0.16

[a] After Rivoirard (1987).

4. Consider a 100-m (northerly) by 30-m horizontal zone of vertical, sheeted veins striking N–S. Individual veins are 2–4 cm wide, can be traced for 20–30 m along the strike, and are spaced at 10–15 cm intervals across strike. Three lines of samples have been taken across strike: 25, 50, and 75 m north of the south boundary. Each line crosses the entire 30-m width of the deposit and is composed of six contiguous 5-m samples.

(a) Comment on the quality of the sampling plan, assuming it to be an early stage evaluation.

(b) The grade estimation problem is two-dimensional at this stage. How would you arrange an array of 5×10 m^2 blocks for estimation? Why?

(c) Categorize the blocks in your array into several groups of relative quality of estimate.

4

Statistical Concepts in Mineral Inventory Estimation: An Overview

Statistics . . . should not be involved in ore reserve estimation until all other factors such as geological continuity and contacts, lost core, representativeness, sampling and assay errors have been identified, examined and assessed. (King et al., 1982, p. 18)

The aim of Chapter 4 is to introduce classic statistical terminology and methods in the context of mineral inventory estimation. It is advantageous for the reader who is unfamiliar with statistical methodologies to have available an introductory text in statistics from which to obtain a more general background and insight into proofs or developments of various parts of the commentary. Emphasis here is on simple and useful statistical procedures. Topics include central tendency, dispersion, covariance, histograms, probability density functions, probability plots, linear correlation, autocorrelation, and linear regression. All are presented in the context of their use in mineral inventory estimation.

4.1: INTRODUCTION

Statistical methods and terminology have been applied to the characterization of ore since about 1945 (cf. Sichel, 1952; Swanson, 1945). Applications to the study of quantitative numeric variables such as metal grades or other deposit characteristics are commonly concerned with central tendency, dispersion of values, the form of probability density functions (histograms), simple correlation, autocorrelation,

relations among groups of variables, and a variety of probabilistic statements related to some of these topics. These traditional statistical approaches to summarizing and understanding data are considered here in their relation to mineral inventory estimation procedures. For deeper insight into the fundamentals of statistics, reference should be made to the myriad of texts available.

Statisticians speak of a *population* or *universe* (i.e., the entire feature under study, e.g., a mineral deposit). This universe or deposit is characterized by variables (e.g., grades) with particular parameters (e.g., mean, standard deviation) and a particular form or shape to the spread of all possible values (data items) about the mean (i.e., the probability density function or the histogram). A general aim of statistical work is to infer the parameters (characteristics) of the universe (deposit) from a subset or sample of possible items (rock sample assays). For example, 345 rock samples analyzed for Cu and Au might be adequate to characterize a particular deposit within an acceptable level of error. Note the two uses of the word *sample*. In statistical usage, the n individual values combine to make a sample (consisting of n items) of the deposit. In mining usage, a sample is commonly a physical quantity of rock material, a representative fraction of which can be analyzed to produce numeric measures of quality (e.g., grades). Samples in mining evaluation projects generally are not randomly located in space, but are more or less regularly positioned; sample patterns vary from fairly regular to highly irregular, two- or three-dimensional spatial arrays of data. Individual samples need not correspond in volume or mass of rock to other samples, although some effort commonly is directed to having a uniform support (sample size/shape) for most of a database.

4.2: CLASSIC STATISTICAL PARAMETERS

4.2.1: Central Tendency

Central tendency, the preferential clustering of values in a data set, is most commonly measured by the arithmetic average (m), determined by summing n values and dividing the sum by n:

$$m = \frac{\sum x_i}{n}.$$

If the n values are a random sample of a more extensive population, the average of the sample is an unbiased estimate of the mean of the population. This mean value can also be considered the expected value of a random draw from the population.

In mineral inventory work, a common problem is to estimate the mean grade from the grades of a limited number of samples of differing size, as in the case of forming composites (Section 6.3.1). As an example, consider two successive drill-core samples that represent very different lengths of core. The mean value of the combined sample is determined as a weighted mean, with weights proportional to volumes or masses of the original samples (e.g., length in the case of split core samples of different length, but of comparable densities). A weighted mean is estimated as

$$m_w = \sum w_i x_i \quad \text{for} \sum w_i = 1$$

where x_i are the values being averaged and w_i are the corresponding weights. The relation $w_i = 1$ is the *nonbias condition*, which is essential to make the weighted mean unbiased. As a simple example of a weighted-mean estimate, consider two copper assays of 1.5 percent and 0.5 percent for drill-core lengths

of 3 m and 1 m, respectively. The average, weighted by lengths, is $(1.5 \times 3/4 + 0.5 \times 1/4) = 1.25\%$, and assumes that the densities of the two samples are identical. Suppose that the densities are 3.3 and 2.7 g/ml respectively. The weighted average, m_w, becomes

$$\begin{aligned}
m_w &= \sum (w_i x_i) = \sum (\ell_i d_i x_i) \Big/ \sum (\ell_i d_i) \\
&= [(1.5 \times 3 \times 3.3) + (0.5 \times 1 \times 2.7)]/ \\
&\quad [(3 \times 3.3) + (1 \times 2.7)] \\
&= (14.85 + 1.35)/(9.9 + 2.7) = 1.29
\end{aligned}$$

where $w_i = \ell_i d_i / \Sigma(\ell_i d_i)$ and ℓ_i and d_i represent the length and density, respectively, of sample i.

In mineral inventory work, weighting assays by sample density might be important, but the procedure is not as widely used as warranted, perhaps because of the additional time and cost of obtaining density data. A hypothetical example from Bevan (1993) demonstrating the potential impact of density weighting is given in Table 4.1 and involves the procedure for combining grades of two equal volumes: one of chalcopyrite and the other of quartz, two minerals with very different densities. The first weighting procedure in Table 4.1 correctly takes relative weights of the two volumes into account. The second weighting procedure ignores the effect of different specimen densities and produces a serious underestimate of the grade (17.3% Cu versus a true grade of 21.25% Cu, an underestimate of about 19%).

It is common practice in the mineral industry to make subgroups (divide data into two or more grade ranges) of duplicate assay data for the purpose of comparing average grades by two laboratories for each grade range. For example, average grades above a cutoff grade for two different laboratories might be compared. In practice, it is incorrect to apply a cutoff to the results of only one lab because this biases the comparisons by including some second-lab values below the cutoff grade. There are several ways to overcome this problem. The first is to apply the threshold to results of both labs. This method should be used for thresholds that coincide with changes in sampling and/or analytical methods by one of the labs. A second method is to apply the threshold to the average of paired data.

Table 4.1 Example of volume weighting versus mass weighting in the estimation of weighted grades

	Chalcopyrite cube (1 cm^3)	Quartz cube (1 cm^3)
Specific gravity	4.25	2.67
Cube weight	4.25 g	2.67 g
Cu grade (%)	34.6	0

Correct grade of combined cubes = [(34.6 + 1 + 4.25) + (0 + 1 + 2.67)]/[4.25 + 2.67] = 21.25% Cu with average specific gravity = 6.92/2 = 3.46

If differences in specific gravity are not considered and a volume weighting is performed to calculate grade, incorrect grade of combined cubes = [(34.6 + 1) + (0 + 1)]/2 = 17.3% Cu

Source: After Bevan (1993).

The mean value, m, of a mixture of two populations is given by the expression

$$m_w = p \cdot m_1 + (1 - p) \cdot m_2 \qquad (4.1)$$

where m is mean, subscripts 1 and 2 refer to populations 1 and 2, and p is the proportion of population 1. Note that the mean of a mixture of two populations is a weighted average that is constrained by the means of the individual populations.

The *median*, another important measure of central tendency (particularly for nonsymmetrically distributed data) is the value corresponding to the middle data item in an ordered data set (ordered from high to low values, or vice versa); that is, 50 percent of the values are higher than the median and 50 percent of the values are lower. For small numbers of items the median is a more stable estimator of central tendency than is the *mean*. Medians are an integral part of the Thompson–Howarth approach to the characterization of sampling and analytical errors as a function of concentration (Section 5.2.1).

Modes are narrow class intervals of data that are more abundant than are data in both adjacent class intervals (i.e., *modes* are local peaks on a histogram). Although a mode can correspond to either mean or median values, the three measures of central tendency are, in general, different (Fig. 4.1); mode, median, and mean are equivalent in the case of a normal distribution. Modes are also important in signaling the possible presence of complex distributions made up of two or more subpopulations (cf. Sinclair, 1976, 1991) and are important for an understanding of outliers (espe-

cially abnormally high values) and their recognition (see Chapter 7).

4.2.2: Dispersion

Dispersion is a measure of the spread of data values. An obvious, but generally impractical, characterization of dispersion is the range of data (i.e., the difference between the minimum and maximum values of a data set). The range is generally unsuitable for defining dispersion because it is highly susceptible to the presence of a single extreme value. The most fundamental measure of dispersion in a data set is the *variance*, s^2, defined as the mean squared difference

$$s^2 = \frac{\sum(x_i - m)^2}{(n - 1)}$$

where x_i is any data value, m is the mean of the data, and n is the number of data items. The term $(n - 1)$,

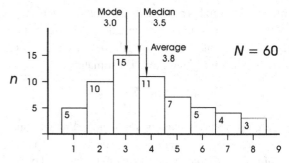

Figure 4.1: Histogram of a hypothetical data set of integer values (abscissa) illustrating that modes, medians, and means can differ for a single data set. Numbers of data within each class are listed in each bar.

referred to as *degrees of freedom*, originates in statistical theory involving the sampling distribution of s^2 (variance of a sample) in contrast to σ^2 (variance of a population). The divisor $(n-1)$ is used so that s^2 is not biased when a small sample $(n < 30)$ is used to characterize an entire population. Note that variance is a "squared" quantity so that no distinction is made between positive and negative differences used in its calculation. The square root of the variance (i.e., the *standard deviation*) is the commonly used practical measure of dispersion because it is in the same units as the variable in question, rather than being (units)^2, as is the case for the variance.

One of the important characteristics of variances is that they are additive when the individual sources of variability are independent. As an example, consider the variability (error) that occurs among analytical data for a set of duplicate field samples; variability arises due to the sampling method used (s_s^2), the subsampling procedures (s_{ss}^2), and the analytical technique (s_a^2), so that the total variance, s_t^2, can be expressed as the sum of variances arising from each of the three sources, as follows:

$$s_t^2 = s_s^2 + s_{ss}^2 + s_a^2. \tag{4.2}$$

In this formula the analytical error variance is estimated from duplicate pulp analyses. With this information the subsampling error variance can be determined from pulp versus reject analyses, and from this the total error and sampling error can be estimated from duplicate sample (e.g., both halves of split drill core) analyses. For these purposes, an internally consistent set of data is required; that is, the reject duplicates should be selected from samples for which there are duplicate half-core data. Likewise, duplicate pulps should be run on some of the same samples. Data not obtained in the forgoing fashion is likely to be inconsistent, meaning that the populations (means and standard deviations) represented by the paired pulps, rejects, and half-cores might be very different, in which case the variances would not be additive. It is particularly useful to have a consistent set of paired data because this leads directly to an appreciation of the magnitude of error arising from each major source of variability, sampling, subsampling, and analysis.

Practical sampling and analytical experiments can be designed to estimate the individual components of variance and perhaps improve the quality of one or more of those sources of variability in order to reduce the overall error.

If a mean value m and related dispersion s have been obtained from n data items, then the sampling dispersion of the mean value (i.e., the standard error of the mean, s_e) is given as

$$s_e = (s^2/n)^{1/2} = s/n^{1/2}. \tag{4.3}$$

In other words, if means were determined for many samples of size n, those means would have a dispersion (standard deviation) estimated by s_e.

In mineral inventory work, there is occasionally confusion of the terms *dispersion variance* and *error variance*. The distinction is clearly evident in the two terms *standard deviation* and *standard error of the mean*, as presented previously. The *standard deviation* (or *variance*) is a characterization of the dispersion of a set of values (e.g., a set of grade values). The *standard error of the mean* (or its squared equivalent) for the same data is an estimate of the average error made in estimating the true mean (of the population) by the mean of the data set. In general, error dispersion is much less than data dispersion.

The weighted variance attributable to a weighted mean is given by the expression

$$s_w^2 = \sum [w_i(x_i - m_w)^2] / \sum w_i.$$

Consider the three values 0.5, 1.0, and 4.0, which have a mean of 1.18 and a standard deviation of 1.89. If the respective weights for these three numbers are 0.2, 0.3, and 0.5, the weighted mean is 2.4 and the weighted standard deviation is 1.14.

The sampling variance for a weighted mean (weights are implied to be unequal) when weights sum to 1, is as follows:

$$s_w^2 = \left\{ \sum [w_i(x_i - m_w)^2] \right\} \sum \Big/ \left[\sum w_i(1 - w_i) \right]. \tag{4.4}$$

In Eq. 4.8, $\sum w_i(1 - w_i)$ is equivalent to $(n-1)$ of Eq. 4.3, and the w_i summed in the numerator has the effect of n in Eq. 4.3 (i.e., the denominator takes into account degrees of freedom and the formula produces results equivalent to Eq. 4.2 if weights are equal).

The variance of a mixture of two populations, w^2, is given by the following expression:

$$\sigma_w^2 = p \cdot \sigma_1^2 + (1 - p)\sigma_2^2 + p(1 - p)(m_1 - m_2)^2$$

where the σ squared terms are variances, the m is means, the subscripts refer to populations 1 and 2, and p is the proportion of population 1 in the mixture. This relation is particularly useful in considering the effect of outliers on data populations (see Chapter 7).

Percentiles (including quantiles) are values below which a stated proportion of values in a data set occurs. Thus, the median is the 50th percentile. In some cases, percentiles are used to describe dispersion; commonly used percentiles are as follows:

P_{10}, P_{90}	values corresponding to 10 and 90 cumulative percent of data, respectively
P_{25}, P_{75}	values corresponding to 25 and 75 cumulative percent of data, respectively. Also referred to as quartiles (Q_{25}, Q_{75})
P_{50}	value corresponding to 50 cumulative percent of data, the median, M_d.

Percentiles correspond closely with the concept of probability, and are particularly useful in dealing with cumulative frequency distributions (cumulative histograms and cumulative probability plots). They provide rapid and simple insight into the symmetry or asymmetry of a distribution of data about the mean value. For example, if data are distributed more or less symmetrically about the mean, then the mean value is approximately equal to the median, and a value such as ($P_{50} - P_{25}$) should be almost equal to ($P_{75} - P_{50}$).

4.2.3: Covariance

Covariance (s_{xy}) is a quantitative measure of the systematic variations of two variables (x and y) and is given by the formula

$$s_{xy} = \sum [(x_i - m_x)(y_i - m_y)]/n$$

where m_x and m_y are the means of the two variables being compared. If high values of x are associated with high values of y and low values of x are associated with low values of y, the covariance is positive; low values of x associated with high values of y (or vice versa) produce a negative covariance. When x and y are statistically independent, the covariance is zero, although the converse is not necessarily true; two variables can have a covariance of zero and yet be dependent.

The covariance is an essential component of the simple linear correlation coefficient discussed in Section 4.6; moreover, it is fundamental to many geostatistical concepts that are introduced in subsequent chapters.

4.2.4: Skewness and Kurtosis

Skewness is an indication of the departure of tails of a distribution from symmetry about the mean. *Positively skewed distributions* have an excess of values extending as a tail toward higher values; *negatively skewed distributions* have a tail extending toward low values (Fig. 4.2). *Kurtosis* is a measure of peakedness (i.e., the relative height of a distribution in a tight range about the mean). Quantitative measures of skewness and kurtosis, available in most statistics texts, are not used in mineral inventory studies. Skewness as a general characteristic, however, is of considerable interest as an indication of whether a distribution is better described as normal or lognormal. In practice, the coefficient of variation (CV) is commonly used for this purpose:

$$CV = s/m.$$

Values of CV less than 0.5 are likely to approach a normal distribution, whereas values greater than 0.5 are skewed and may be described better by a lognormal distribution or a combination of distributions.

4.3: HISTOGRAMS

Histograms are graphs showing frequency of a variable within contiguous value intervals (commonly

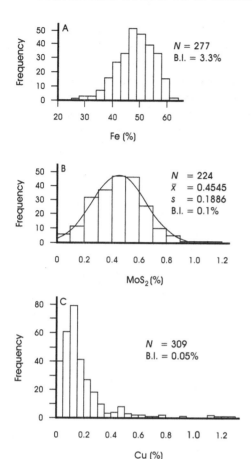

Figure 4.2: Three examples of drill-core assay data presented as histograms: (A) negative skewness of 277 Fe assays, (B) approximate symmetry of 224 MoS_2 values, and (C) positive skewness of 309 Cu values. B.I., bar (class) interval in percent metal; N, number of drill-core assays. A normal probability density function (smooth symmetric curve) has been fitted in (B) with the same mean and standard deviation as the data of the histogram. Data from Becker and Hazen (1961).

a uniform class interval) that extend over the range of the variable. The term is too often applied incorrectly to various types of profiles, particularly bargraph profiles along plots of diamond-drill holes. Histograms are a simple and effective method of displaying many attributes of grade data (Fig. 4.2). The form of the distribution (negatively skewed, symmetric, positively skewed) is readily apparent, as are a qualitative evaluation of dispersion, the extent to

which data cluster centrally, and the presence of one or more modes. Note that these are all features of shape of the histogram (i.e., distribution of data). A clear appreciation of form of data distribution is essential in a consideration of sampling and analytical errors, determination of grade, and tonnage above cutoff grade, and for various statistical tests. Histograms are a common means of displaying such information for mineral inventory studies.

In constructing a histogram (cf. Sinclair, 1976, 1991), the class interval should be uniform, a convenient figure in the range $1/4\,s$ to $1/2\,s$ (Shaw, 1964); the mean value should form a class boundary and frequency should be as a percentage rather than an absolute number (to allow comparison of histograms based on different numbers of items). Each histogram should be accompanied by a list of information that includes number of items, class interval, mean, and standard deviation. Data grouped for a histogram can easily be recast in cumulative form and plotted as a cumulative histogram (e.g., Fig. 4.3).

In the process of obtaining mineral inventory estimates, histograms are commonly used in ways that purport to be representative of a large volume (e.g., a mineral deposit). Because grade data are commonly biased (i.e., clustered) in their spatial distribution (Fig. 4.4a), it is important that some effort be made to examine for and remove the influence of location bias in a histogram of raw data (*naive histogram*). For example, it is not uncommon in exploring mineral deposits to obtain a greater density of data from high-grade zones than from nearby lower-grade zones. In such a case, the available assay data are biased toward high values and produce a histogram that is positively skewed, perhaps even indicating the existence of a separate high-grade population where none exists. Such positively skewed biased distributions also might be mistaken for lognormal, when the true underlying form is normal.

One procedure to correct for location bias in a data set is to weight individual values by quantities proportional to polygonal areas or volumes of influence (i.e., the polygonal method of mineral inventory estimation described in Chapter 1; see also Isaaks and Srivastava, 1989). A method commonly used by geostatisticians

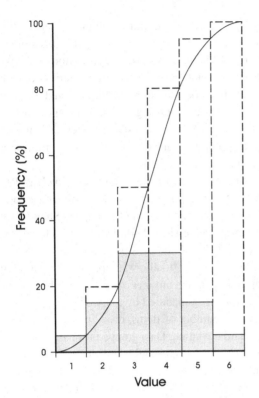

The grid size is clearly important in determining unbiased estimates of the mean, standard deviation, and form of the underlying distribution. If the grid size is so small that each cell contains only one item, the weighted average is the mean of the original data; if the cell size is so large that all data are contained within one cell, the weighted average is again the mean of the original data. Various cell

Figure 4.3: Illustration of the relation between a histogram (shaded), a cumulative histogram (dashed bars), and a cumulative curve (smooth).

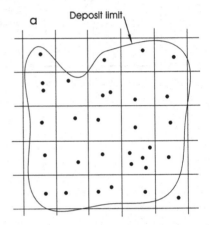

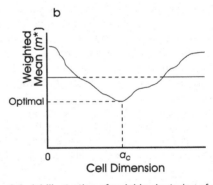

(cf. Journel, 1985) is to superimpose a uniform grid (two dimensional or three dimensional, as required) over the area (volume) of the entire data set in such a way that each cell of the grid contains one or more data items (Fig. 4.4a). Individual data items are weighted by their proportion relative to total data within a cell (i.e., each cell has the same weight regardless of how many data it contains, but the weight of individual data items varies depending on how many occur in a particular cell. In equation form, the weights of all cells are equal (say, 1) and the weight of an individual sample j within cell i is $w_{ij} = 1/n_i$ such that $w_{ij} = 1$ for each cell containing data. In some cases, the weights are scaled so that the total weight equals the total number of data points in the data set ($N = \sum n_i$), but this procedure is not recommended because it implies incorrectly that the resulting histogram is known by N equivalent data.

Figure 4.4: (a) Illustration of variable clustering of data geographically, with an arbitrary, superimposed grid. For declustering to produce an unbiased histogram, each cell receives equal weight and that weight is divided equally among all data within the cell. Hence, the five data in one cell each receive a weight of 1/5th; each datum in a cell containing two data receives a weight of 1/2, and so on. (b) When clustered data are in high-grade areas, the optimum cell size for declustering is found by a plot of weighted-mean grade versus cell size. The cell size producing a minimum-weighted mean is the appropriate cell size with which to generate weights that produce an unbiased histogram. See text for a discussion of complications of this simple pattern.

sizes between these two limits, however, produce a weighted average lower than the raw data mean if samples are overly concentrated in high-grade zones (Fig. 4.4b). Similarly, if there is a concentration of samples in low-grade zones, weighted averages for intermediate cell sizes are greater than the average of the original data. Optimum cell size is that which produces a minimum weighted mean if data are clustered in high-grade zones or a maximum weighted mean if data are concentrated in low-grade zones. Where data are concentrated irregularly in both high- and low-grade zones, the foregoing simple patterns cannot be expected. Instead, it is necessary to determine subjectively an underlying grid size that produces an unbiased histogram.

Consider the case of samples preferentially located in high-grade zones. Weighted averages can be determined for a variety of grid spacings. A plot of these weighted average grades (y axis) versus grid spacing (x axis) define a curve that is concave upward (Fig. 4.4b); the cell size that produces a minimum weighted average provides unbiased estimates of mean, standard deviation, and form of the underlying data distribution. An estimate of the true form of the distribution (normal, lognormal, bimodal, etc.) can be determined by examination of a histogram of appropriately weighted data (see example in Chapter 12). The nature of a data distribution can have an impact on procedures for variography (e.g., use of raw data, log-transforms, relative semivariograms) and calculations involving dispersions for various supports. The presence of more than one mode (subpopulation?) in the distribution indicates the possibility of fundamentally different grade continuities for the various subpopulations, a topic developed further in Chapter 7.

4.4: CONTINUOUS DISTRIBUTIONS

Probability density functions (PDFs) are mathematical models used to describe the probability that random draws from populations defined by the functions meet particular specifications. For example, a randomly selected item from a group of assays de-

scribed by such a function has a probability $p = 0.25$ of being below the first quartile; similarly, there is a probability of 0.1 (i.e., 10%) that a random draw will be higher than the 90th percentile; and so on. Unbiased histograms, plotted with frequency as a proportion, can be viewed as equivalent to a probability density function, albeit discrete rather than continuous, as are many of the commonly used probability density functions. This similarity can be appreciated by picturing a smooth, continuous curve fitted through the tops of class-interval frequencies of a histogram (e.g., Fig. 4.2b).

A majority of variables that are commonly estimated in calculating mineral inventories (grade, thickness, accumulation, etc.) can be described satisfactorily by a few probability density functions, most commonly normal or lognormal models, or mixtures of two or more such models.

4.4.1: Normal Distribution

The *normal* or *Gaussian probability density function* is the common bell-shaped curve, symmetric about the mean value, so prevalent in much statistical literature. A normal distribution is defined by:

$$y = [(2\pi)^{-0.5}s^{-1}]\exp[-(x_i - m)^2/2s^2]$$

where m is the estimate of the arithmetic mean, x_i is any measurement, and s^2 is the estimate of variance of the population. A normal distribution is illustrated in Fig. 4.5. Normal curves can be fitted to an unbiased histogram to demonstrate the likelihood that the variable in question is normally distributed (e.g., Fig. 4.2b). In such cases, the normal curve is assumed to have the same parameters (mean and standard deviation) as does the histogram. A simple procedure for fitting a normal distribution to a histogram is described in the following section.

The normal distribution is widely used to describe discrete data sets. In mineral inventory studies, normal distributions are particularly useful in dealing with various types of errors, particularly errors in analysis and sampling (see Chapter 5).

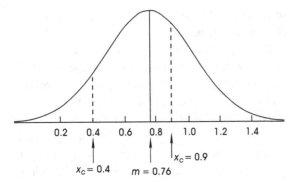

Figure 4.5: Example of a normal (Gaussian) probability density function (PDF). The curve is symmetric about the mean value, $X_m = 0.76$. Spread (dispersion) is measured by the standard deviation, $s = 0.28$. The distribution simulates a Cu-grade distribution for which two arbitrary cutoff grades, 0.4 and 0.9, are shown.

4.4.2: Standard Normal Distribution

All items of an ideal normal distribution can be transformed to the standard normal distribution as follows:

$$z_i = (x_i - m)/s. \tag{4.5}$$

That is, each item is transformed into a number of standard deviations from the mean value. This transformation produces a distribution of standardized normal scores or z values with a mean of zero and a variance of 1, and leads to a formula for the standard normal distribution of

$$y = (2\pi)^{-1/2} \cdot \exp\left(-z_i^2/2\right).$$

The standard normal distribution is shown in Fig. 4.6. Many commonly used statistical tables are based on the standard normal distribution because any normal distribution, regardless of mean and standard deviation, can be related to standardized statistical tables through the transform of Eq. 4.5.

The standard normal distribution is a useful basis with which to consider the concepts of probability and confidence limits. Probabilities are in the range 0–1.0 and may be quoted as percentages. The likelihood that a randomly drawn sample from a normal distribution is less than a specified value x is given by

the proportion of area under the normal curve from minus infinity to x (see Fig. 4.5). This area or probability can be found by transforming x to a corresponding z value (Eq. 4.5) and searching a set of tables of cumulative area from minus infinity to any z score. The difference between such cumulative areas for any two z values gives the probability that a randomly drawn sample lies between the two z values. Note that if the probability of a randomly drawn sample being less than z is given by $P_{>z}$, then the probability that the random draw will be greater than z is given by $P_{>z} = 1 - P_{<z}$. For example, the probability that a sample drawn from a normal population will be greater than the mean value is 0.5; similarly, the probability that the sample will be less than the mean value is 0.5. The probability that a randomly drawn sample is in the range $(m - s)$ to $(m + s)$ is about 0.68 (Fig. 4.6). Proportions of the standard normal distribution above (or below) given z values are tabulated in most statistical texts.

The sampling distribution of the mean values of samples of size n is a normal distribution with a standard deviation given by the standard error (Eq. 4.3). Thus, any sample of size n can be used to define the distribution of means. In other words, the standard error can be used to define confidence limits for the mean value of a sample of size n. This generalization is true for large values of n that closely approach a

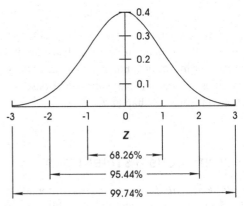

Figure 4.6: The standard normal distribution shows the distribution of standard deviates, or z values, as determined from Eq. 4.1.

normal distribution. For small values of n (< 30), the confidence limits are underestimated using the standard error and must be determined using a t distribution, as defined in many statistical texts.

4.4.3: Approximation Formula for the Normal Distribution

There are practical situations involving normal (and lognormal) distributions for which it is important to know the proportion of the distribution that lies above or below a particular value. For example, in a normal distribution of copper grades, it might be important to determine the proportion of grades above a particular cutoff grade. To solve this problem, it is possible to transform the cutoff grade to the corresponding z value using the transform of Eq. 4.5. That z value can then be located in a set of tables (in most introductory statistical texts) giving $P_{<z}$ the proportion of the area under the curve from minus infinity to z. The required proportion is then $1 - P_{<z}$. In many cases the use of tables is awkward, and it is more convenient to approximate the proportion from one of several formulae that exist for this purpose. One formula recommended by David (1977) for positive values of z is

$$P_{<z} = 0.5[1 + \{1 - \exp(-2z^2/\pi)\}^{1/2}]$$

$$(4.6)$$

or

$$P_{>z} = 1 - P_{<z} \qquad (4.7)$$

where $P_{<z}$ is the proportion of the population below the selected positive z value and $P_{>z}$ is the proportion of the population above the selected positive z value. Note that where z is negative, Eq. 4.6 calculates directly the proportion of the population that is greater than z. These formulae provide estimates of proportions of a normal distributin that are correct to better than 1 in 1,000 for a practical range of z values (i.e., values in the range -2 to $+2$). They can be applied to lognormal populations if data are transformed to logarithms, which are normally distributed.

As an example of the application of Eq. 4.6, consider the cases of samples from a 50,000-ton block

of potential ore for which an unbiased estimate of mean Cu content is $m = 0.76\%$ Cu and grades are normally distributed (Fig 4.5), as the variance is $s^2 = 0.08$, $s = 0.28$. For a cutoff grade of 0.4% Cu, the corresponding z value is $(0.4 - 0.76)/0.28 = -1.286$. With $z = -1.286$ in Eqs. 4.6 and 4.7, 0.903, or 90.3% of the tonnage (i.e., approximately 45,200 tons) is shown to have an average grade above 0.4% Cu. Of course, this is an optimal estimate for ore/waste selection because it is based on the distribution of samples; practical selection units are several orders of magnitude greater in size than are the sample, and the dispersion of mean grades would be substantially smoothed relative to sample grades. In this example, when the cutoff grade is less than the mean grade, a normal distribution of block grades can give a higher proportion of tonnage above cutoff grade than does a sample grade distribution. If the cutoff grade were higher than the mean value, the selection mining unit (SMU) grade distribution would estimate less tonnage above cuoff grade than would sample distribution.

In addition to knowing the proportion of tons above cutoff grade, it is advantageous to know the average grade of the material above (and perhaps below) cutoff grade. For a double truncated normal distribution, the average grade of material between A and B is given by the expression

$$E[x_{A-B}]$$
$$= m + \frac{Z[(A - m)/s] - Z[(B - m)/s]}{\Phi[(B - m)/s] - \Phi[(A - m)/s]} \cdot s$$

$$(4.8)$$

where

A is the value of lower truncation
B is the value of upper truncation
m is the mean of the normal distribution
s is the standard deviation of the normal distribution
$Z[z] = (2)^{-1/2} \exp(-z^2/2)$
$\Phi[z]$ is the proportion of area under the standard normal curve from minus infinity to z.

For a lower truncation of A and no upper truncation

(comparable to the application of a cutoff grade), Eq. 4.8 reduces to

$$E[x_{>A}] = m + \frac{Z[(A-m)/s]}{1 - \Phi[(A-m)/s]} \cdot s$$

$$(4.9)$$

where symbols are as for Eq. 4.8.

As an example of the application of Eq. 4.9, consider the population described two paragraphs previously and shown in Fig. 4.5, for which $A = 0.40\%$ Cu and the parameters of the normal distribution are $m = 0.76$ and $s = 0.28$. Hence, $Z[(0.4 - 0.76)/0.28] = Z[-1.286] = 0.1745$; $\Phi[-1.286]$ can be estimated from Eq. 4.15 or from tables in statistical texts as 0.903. Substitution of these values in Eq. 4.6 gives an expected mean value of Cu above cutoff grade of $E[X_{>0.4}] = 0.814\%$ Cu. For a second example, assume the same normal distribution, but with a cutoff grade of 0.9% Cu (i.e., the cutoff grade is greater than the mean value). Therefore, $Z[(0.9 - 0.76)/0.28] = Z[0.5] = 0.35$, and $\Phi[0.5] = 0.69$. Substitution of these values in Eq. 4.9 gives $E[x > 0.9] = 0.76 + 0.28(0.35/0.31) = 1.075\%$ Cu. Calculations of this sort are useful in developing the concept of grade–tonnage curves that are so prevalent in mineral inventory studies (see Chapter 12).

It is occasionally useful to be able to plot a normal distribution or to fit a normal distribution to a histogram (see Fig. 4.2b). The standard normal distribution can be constructed by solving Eq. 4.10 (cf. Johnson and Kotz, 1970) for various values of z and drawing a continuous curve through the points. In constructing the curve, it is useful to be aware that inflection points occur at $x_m - s$ and $x_m + s$:

$$y = 0.3979(0.6065)^{z \cdot z}.$$

$$(4.10)$$

Values of y for the standard normal curve can be adjusted so that a normal curve can be fitted to any histogram by transforming them (to y'), as in Eq. 4.12:

$$y' = y(n \cdot i/s)$$

$$(4.11)$$

where n is the total frequency of data ($n = 100$ if the histogram is constructed with frequency as a percent rather than an absolute frequency), i is the class interval, and s is the standard deviation of the data.

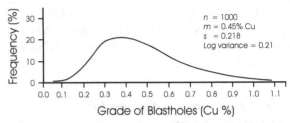

Figure 4.7: An ideal lognormal distribution that approximates the distribution of blasthole assays at Bougainville copper deposit (see David and Toh, 1989). Parameters of the raw data of this distribution are: $m = 0.45\%$ Cu and $s = 0.218$.

4.4.4: Lognormal Distribution

If a variable x is transformed to logarithms [i.e., $t = Ln(x)$] and these log values have a normal distribution, the variable is said to have a *lognormal distribution*. The raw data (untransformed values) of a lognormal distribution are positively skewed (Fig. 4.7), but not all positively skewed distributions are lognormal. It is of interest to note that certain negatively skewed distributions might be treated as lognormally distributed through the transformation

$$t = Ln(C - x)$$

where no value of x can be greater than the arbitrary constant C. The product of two variables that are lognormally distributed is also lognormally distributed.

In some cases, positively skewed data that are not lognormally distributed can be made so by a simple transformation involving the addition of a constant; that is

$$t_i = Ln(x_i + k)$$

is lognormally distributed. Estimation of k is discussed in Section 4.5.1. When such a transformed variable is used for mineral inventory estimation purposes, the initial estimates must be back-transformed (i.e., k must be subtracted from the initial estimates to produce correct estimates). Various values of k generally have little effect on an eventual estimated value of a point or block; however, changes in k can result in changes in logarithmic variance of as much as 50 percent (cf. Clark, 1987). The Sichel t estimator (discussed later) also can be affected substantially by variations in k because the estimator depends on the logarithmic variance. The lognormal distribution has found widespread application in characterizing grade distributions for mineral inventory estimation. This general acceptance should not lead to noncritical acceptance of a lognormal form; detailed evaluation of the probability density function of grade variables is an essential precursor to mineral inventory estimation. Many preliminary lognormal grade models turn out to be significant oversimplifications of reality.

When manipulations of estimates are done with transformed data, it is generally necessary to do a back-transform to provide an estimate in terms of original units. In the case of the lognormal distribution, the back-transform of the mean logarithmic value produces the geometric mean of the original data, which is an underestimate of the true mean of the data. Fortunately, parameters (mean and variance) of the logtransformed distribution (natural logarithms) can be used to estimate the mean and standard deviation of the skewed distribution of raw (untransformed) data using the maximum likelihood estimators of Eqs. 4.13 and 4.14:

$$m = b \cdot e^{v/2} \qquad (4.12)$$

$$s^2 = m^2(e^v - 1) \qquad (4.13)$$

where b is the antilog of the mean log value (i.e., the geometric mean) and v is the variance in natural log units. Equivalent formulas exist for Log 10 units (e.g., Sinclair, 1986). The coefficient of variation can be determined as follows:

$$CV = [e^v - 1]^{1/2}.$$

For small samples ($n < 30$) from lognormal distributions, Sichel (1952) showed that the maximum likelihood estimator of Eq. 4.12 consistently overestimates the arithmetic mean; hence, he introduced the t estimator, as follows:

$$t = e^{\bar{x}} g_n(V)$$

where $\bar{x}$ is the mean log value

$$g_n(V) = 1 + \sum_{r=1}^{\infty} \{[(n-1)^r V^r]/[2^r r!(n-1) \\ \times (n+1)(n+2r-3)]\}$$

for r's that take on successive integer values beginning with 1, V is the variance of logtransformed (natural) values, and n is the sample size.

Sichel's t estimator has found limited use in mineral inventory work outside the South African gold fields because (i) it ignores autocorrelation and (ii) it is complicated and cumbersome to use in practice. Nevertheless, the t estimator is a useful conservative estimator of the arithmetic mean for small data sets when a lognormal distribution can be assumed with confidence. Tables for rapid determination of the t estimator are provided by David (1977).

For a lognormal distribution of grades, as with the normal distribution, it is possible to estimate the proportion of tonnage ($P_{>c}$) above a particular cutoff grade (Eq. 4.14) and the average grade ($g_m > c$) of that proportion above cutoff grade:

$$P_{>c} = 1 - \Phi\{\text{Ln}(x_c/m)/d + d/2\} \qquad (4.14)$$

where d is the standard deviation of logtransformed (base e) data, x_c is the cutoff grade (original units), m is the mean of the distribution (original units), and $\Phi\{z\}$ is the cumulative distribution function of a standard normal variable from $-\text{inf}$ to z. The "recoverable" metal, $R_{>c}$ (i.e., the proportion of total metal that is contained in the tonnage above cutoff grade), is given by

$$R_{>c} = 1 - \Phi\{\text{Ln}(x_c/m)/d - d/2\} \qquad (4.15)$$

where symbols are as described previously. The average grade ($\bar{x}_{>c}$) of that proportion of material above cutoff grade (x_c) is given by

$$\bar{x}_{>c} = m \cdot R_{>c}/P_{>c}. \qquad (4.16)$$

Values of $R_{>c}$ and $P_{>c}$ can be determined from tabulations of the cumulative standard normal variable, or can be estimated using Eqs. 4.14 and 4.15.

Equations 4.14 to 4.16 have many applications in mineral inventory estimation because lognormal distributions or close approximations to lognormal distributions are relatively common for metal grades. For example, the equations can be used to construct grade–tonnage curves for lognormally distributed variables in the same manner that comparable equations for the normal distribution can be used (see Chapter 12).

The cumulative lognormal distribution has been proposed to describe extremely skewed distributions found for diamond grades and, more recently, certain gold deposits (Sichel et al., 1992). However, the distribution has four parameters, the mathematics are complex, and change of support and small sampling theory are yet to be solved.

4.4.5: Binomial Distribution

The binomial distribution (Fig. 4.8), that is, the probability that a characteristic occurs in some items of a data set, is a discrete distribution. For large n, the distribution is closely approximated by a normal distribution, with mean equal to np and variance equal to npq, where n is the total number of items, p is the proportion of items that have a particular attribute, and q is the proportion that do not have the attribute (i.e., $p = 1 - q$). For example, 100 tosses of an unbiased coin might be expected to provide 50 heads and 50 tails (i.e., $p = q = 0.5$). A test of the unbiasedness could consist of 100 tosses, and for $\alpha = 0.05$ (i.e.,

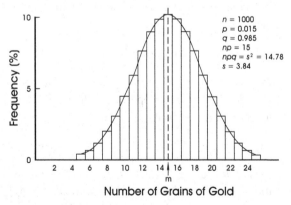

Figure 4.8: Example of an ideal binomial distribution (frequency as a percentage versus number of grains) of the number of heavy mineral grains in alluvium samples, each sample consisting of 1,000 grains. The corresponding normal distribution is shown as the smooth, symmetric curve with a mean of 15 grains. The plot shows, for example, that there is a more than 7 percent chance that a sample of 1,000 grains contain less than 10 grains of heavy minerals when the expected number of grains is 15.

95 percent confidence range) the number of heads should lie in the range $np \pm 2(npq)^{1/2}$, that is, 50 ± 10. If the number of heads is outside this range, it is likely that the coin is biased.

Consider the question of evaluating a set of duplicate analyses for bias. Data in Fig. 5.11b are duplicate Ag grades measured by two different laboratories. If no bias exists, the expectation is that because of random error the data points will be distributed more or less evenly on either side of the line, with slope $= 1.0$ (i.e., both labs will, on average, obtain the same values). The data of Fig. 5.11b show that 19 of 21 values plot on one side of the line. Assume that the binomial distribution can be approximated by a normal distribution with the same parameters as the binomial distribution. For $\alpha = 0.05$, the range of no bias is $np \pm 2(npq)^{1/2} = 10.5 \pm 4.6$, which clearly does not contain the quantity 19. Consequently, the data distribution leads to the conclusion that one lab overestimates relative to the other, although which is the faulty lab remains unknown without additional information.

4.4.6: Poisson Distribution

Experiments yielding numeric values of a random variable x, the number of successes occurring during a given time interval or in a specified region, are often called a *poisson experiment*; some examples are the number of telephone calls received by an office during each hour, the number of days that school is closed due to snow, the number of nuggets in a subsample of a gold ore, and so on. The poisson distribution is fundamental to sampling theory.

A poisson experiment possesses the following properties:

1. The number of successes occurring in one time interval or specified region are independent of those occurring in any other disjoint time interval or region of space.

2. The probability of a single success occurring during a short time interval or in a small region is proportional to the length of the time interval or the size of the region, and does not depend on the

number of successes occurring outside this time interval or region.

3. The probability of more than one success occurring in such a short time interval or falling in such a small region is negligible.

The probability of x successes in a poisson experiment is given by Eq. 4.17:

$$P(x;\mu) = (e^{-\mu}\mu^x)/x! \quad \text{for } x = 0, 1, 2, 3, \ldots$$
$$(4.17)$$

where μ is the average number of successes occurring in the given time interval or specified region, $e = 2.71828$, and x is any integer including zero. Note that for $x = 0$, $x! = 1$.

The mean and the variance of the poisson distribution $P(x;\mu)$ both have the value μ. As Miller and Kahn (1962, p. 374) state, there are many situations in which the mean and variance are equal but that do not fit the poisson distribution. Hence, the close approach to the value unity, of the ratio of the mean to the variance, does not necessarily imply a poisson distribution. Miller and Kahn (1962, p. 378) state: "If the ratio of mean to variance exceeds unity, a rough inference may be drawn that a clustering effect is present. If, on the other hand, the ratio is less than unity, the inference is that there is a regular dispersion of points such that individual points tend to 'repel' each other." The frequency of success (and the variance) is a function of the quadrant size (i.e., the sample size).

The poisson distribution is a limiting form of the binomial distribution when n approaches infinity and np remains constant. Hence, the poisson distribution can be used to approximate the binomial distribution when p is very small and n is very large.

Krumbein and Graybill (1965, p. 110) refer to the use of the poisson distribution for "rare minerals in rocks expressed as number of grains in sub-samples of fixed size." Consequently, the distribution has application to deposits characterized by rare grains of a valuable commodity, as is the case for deposits of gold and diamonds. Consider application of the poisson distribution to samples consisting of rare gold grains in alluvium. Assume that a 500-g sample of sand contains, on average, two grains of gold. Assuming equal grain size (say, 2 mm in diameter) throughout, equivalent to an assay of about 320 g Au/t. The probability that a randomly drawn sample of 500 g contains zero grains of gold is determined from Eq. 4.17 to be 0.27 (i.e., approximately one-quarter of such samples assay zero gold even though the average grade is extremely high). The potentially disastrous effect of the serious lack of representativity of small samples is apparent. An example of the poisson distribution is shown in Fig. 4.9.

A further use of the poisson distribution has been described by Stanley (1998) as a basis for appreciating the quality of assays for any component characterized by the rare-grain effect. For the poisson distribution the variance is the numeric equivalent of the mean

$$\sigma^2 = \mu$$

and the coefficient of variation (CV) expressed as a percentage is

$$CV\% = 100\sigma/\mu.$$

If precision (P) is defined as twice the coefficient of variation, then

$$P = 2CV\% = 200\sigma/\mu = 200/\mu^{1/2} \quad (4.18)$$

where μ is the average number of grains in the sample. Consequently, the calculated precision can be used to determine the effective number of grains of gold in a sample by application of Eq. 4.18. This calculated number of grains is actually the number of "uniform size" grains that produce a poisson sampling error equivalent to the precision of the sample in question.

For a sample size s (in grams) and gold concentration as a proportion (e.g., $c = $ grams per ton divided by 10^6), the total mass of gold (m_t) in the sample can be determined (Eq. 4.19), as well as the mass per "uniform" grain $(m_g$, Eq. 4.20):

$$m_t = s \cdot c \tag{4.19}$$
$$m_g = m_t/\mu. \tag{4.20}$$

It is also possible to calculate the effective grain size as a volume (V_g)

$$V_g = m_g/\rho$$

where ρ is the specific gravity of gold. Knowing the volume, various ideal shapes can be attributed to the grain (e.g., disk, sphere), providing there is independent knowledge of the various diameters for nonspherical cases.

4.5: CUMULATIVE DISTRIBUTIONS

Data grouped into class intervals for the purposes of constructing a histogram can also be considered in terms of cumulative percentages for successive class intervals, cumulated either from low values to high or vice versa. This cumulative information can also be shown as cumulative histograms (Fig. 4.3). As with the histogram, a continuous curve can be used to express the information of a cumulative histogram (Fig. 4.3). Such curves, although easy to understand and in relatively common use, are difficult to interpret because of the ambiguity of evaluating curvature by eye. Thus, other cumulative graphs find more widespread use – in particular, probability graphs.

Table 4.2 Calculated probabilities that a sample with a mean content of two gold grains will contain various numbers of gold grains, assuming a Poisson distribution (see Fig. 4.9)

Expression	Probability	Cumulative probability
$e^{-2}2^0/1$	0.1303	0.1303
$e^{-2}2^1/1$	0.2707	0.4010
$e^{-2}2^2/2$	0.2707	0.6717
$e^{-2}2^3/6$	0.1804	0.8521
$e^{-2}2^4/24$	0.0902	0.9423
$e^{-2}2^5/1,200.0361$		0.9784
$e^{-2}2^6/7,200.0120$		0.9904

4.5.1: Probability Graphs

Probability graphs are a practical, graphic means of evaluating the form of the cumulative distribution of a set of numeric data. Probability paper is constructed so that the ordinate commonly is equal interval or logarithmic as required (i.e., depending on whether the concern is with normal or lognormal distributions);

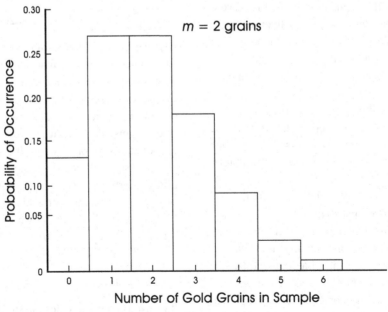

Figure 4.9: Example of a poisson distribution for which the mean sample content is two grains of gold. Each bar of the histogram can be generated by substituting its abcissa in Eq. 4.17. In this example there is about a 13 percent chance that a sample of 1,000 grains contains no gold grains. Calculations are summarized in Table 4.2.

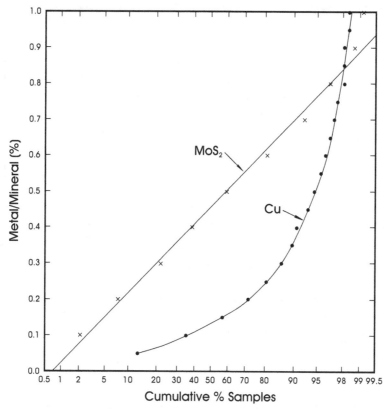

Figure 4.10: Histograms of Fig. 4.2b and 4.2c are plotted as probability graphs. The near-normal distribution of MoS_2 can be approximated closely by a straight line. The strongly positively skewed Cu distribution forms a curved pattern, concave upward. Had the negatively skewed histogram of Fig. 4.2a been plotted, it would have formed a curve concave downward. Note that estimates of the mean and standard deviation of the "normal" distribution can be read from the straight line: $m = 0.453$, $s = (P97.5 - P2.5)/4 = (0.84 - 0.092)/4 = 0.187$. These compare with estimates from the data of $m = 0.455$ and $s = 0.189$.

the abscissa is a variable scale arranged so that a cumulative normal distribution plots as a straight line. Examples are shown on arithmetic probability paper (ordinate scale is equal interval) in Fig. 4.10. In Figure 4.11, the ordinate is a logarithmic scale; cumulative points that form a straight line on this plot have a lognormal distribution. If a data set consists of a mixture of two normal (or lognormal) subpopulations of differing means and partly overlapping ranges, the cumulative data plot on a probability graph as a sigmoidal, curved line (Fig. 4.12). The example of Fig. 4.12 has been constructed manually as a mixture of 20 percent of lognormal population A mixed with 80 percent of lognormal population B. Each point on the curved line can be determined from the equation

$$P_m = f_A \cdot P_A + f_B \cdot P_B$$

where

P_m	is the cumulative percentage of the "mixed" population
P_A	is the cumulative percentage of population A
P_B	is the cumulative percentage of population B
f_A	is the fraction of population A in the mixture, and
f_B	is the fraction of population B in the mixture ($f_B = 1 - f_A$).

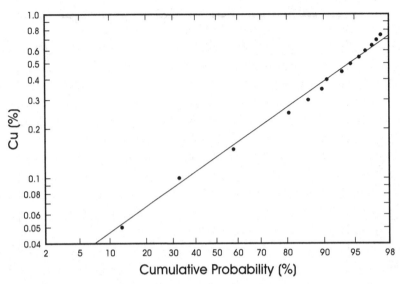

Figure 4.11: The positively skewed Cu histogram of Fig. 4.2c shown as a cumulative plot on a log probability plot (ordinate is a log scale). These data, so strongly concave in Fig. 4.10, are much closer to a straight line when plotted on log probability paper; hence, they are approximated, if crudely, by a straight line representing a lognormal distribution. The mean, estimated from the 50th percentile of the straight line approximation on this diagram, is the geometric mean and substantially underestimates the mean of the raw data.

A sample calculation is shown on the diagram. Mixtures of any two populations can be constructed in a similar fashion to examine the form of a "mixed population" on probability paper. The construction of the mixed population in Fig. 4.12 is proof that a normal distribution that plots as a straight line over the whole range of probabilities (cumulative percentages) does not form a straight line if plotted over a portion of the range; instead, the line is curved, commonly strongly. There are many misconceptions in the mineral deposit literature in which curved probability graphs are approximated by a series of linear segments, each of which is incorrectly stated to represent a normal distribution or a lognormal distribution, as the case may be.

Probability graphs (Sinclair, 1976) are a more useful means of depicting density distributions than are histograms because linear trends are easier to interpret than bell-shaped curves and departures from linear trends are easily recognized. Such graphs are particularly sensitive to departures from normality (or lognormality), are less ambiguous than histograms

for the recognition of multimodal data, and are particularly useful for the selection of optimal thresholds separating the various subpopulations (Sinclair, 1991). Thresholds may bear some relation to cutting factors, as used so commonly in many traditional resource/reserve estimation procedures (cf. Sinclair and Vallée, 1993). Methods for threshold selection, however, commonly involve assumptions as to the density distributions (e.g., normal, lognormal) of the subpopulations.

A common but not necessarily wise practice in the mineral industry is to accept skewed data distributions as either two-parameter or three-parameter lognormal distributions (e.g., Krige, 1960; Raymond and Armstrong, 1988). The three-parameter lognormal distribution is particularly widely used in an effort to transform data to near-normal form through the use of a constant k (the third parameter) added to all raw data values prior to a log transformation. An example is shown in Fig. 4.13. The constant k can be determined by trial and error; a first and perhaps adequate estimate can be determined as follows

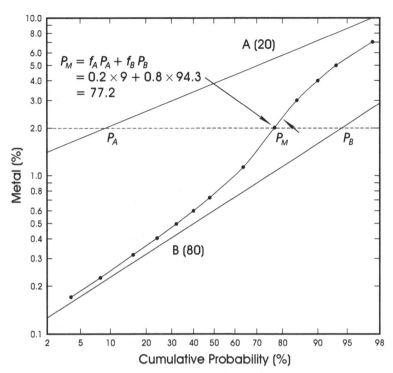

Figure 4.12: Many cumulative distributions have a sygmoidal shape, such as the curve shown here. This curve was generated artificially by combining the two perfect lognormal distributions A and B in the proportions 0.2A and 0.8B. Each of the points defining the mixed cumulative distribution can be generated as illustrated for the point P_M, using the formula shown. Other points are determined by solving the formula for the appropriate values of P_A and P_B, read at a particular ordinate value. Note that any arbitrary distribution can replace A and B, but the calculation method remains the same. Hence, application of this simple formula allows construction of any combination of any two known distributions.

(Krige, 1960):

$$k = \left(M_d^2 - F_1 \cdot F_2\right) / (F_1 + F_2 - 2M_d) \quad (4.21)$$

where M_d is the median (50th percentile) of the data and F_1 and F_2 are data values at two symmetrically disposed cumulative frequencies, say at the 20th and 80th percentiles. The values of M_d, F_1, and F_2 can be read directly from a log probability plot of the raw data. An alternative means of interpreting the surplus of low values is as a separate subpopulation (procedures for dealing with subpopulations using probability plots are described in Chapter 7). One advantage of the concept of subpopulations is that it fits conceptually with a general consideration of multiple subpopulations in assay data.

The cumulative distribution is a characteristic of a mineralized field and can be examined in various local domains to test for uniformity/nonuniformity over a larger field. A smooth curve with a similar shape throughout a deposit (i.e., in various disjoint volumes) suggests uniformity in the character of mineralization (cf. Raymond, 1982).

Probability graphs are relatively well established in mineral inventory work for illustrating cumulative distributions, particularly as a subjective justification of two- and three-parameter lognormal distributions. However, they have not been used extensively in this field for partitioning populations and selecting optimal thresholds, procedures that are useful in data evaluation but may also have application to the question of multiple-continuity populations. Thus, partitioning

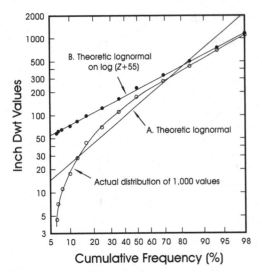

Figure 4.13: Cumulative frequency distribution of 1,000 values (open circles) from the Merriespruit Mine, South Africa, as a logprobability plot. The lognormal distribution with the same mean and standard deviation as the logtransformed data is labeled "A. Theoretical lognormal" and clearly does not describe the distribution (open circles) well. If a constant, $k = 55$, is added to all values and the cumulative curve is replotted (filled circles), a straight line fits the transformed data well. Redrawn from Krige (1960).

procedures are introduced in Chapter 7, with more detailed methodology provided by Sinclair (1974, 1976).

4.6: SIMPLE CORRELATION

Correlation is a measure of similarity between variables or items (e.g., samples). Simple, R-mode correlation is concerned with similarity between pairs of variables; Q-mode correlation deals with the similarity between pairs of samples, perhaps using many variables to make the comparison. Here, attention is restricted to R-mode correlation. When two or more variables are to be examined, a quantitative study of correlation can be useful. The widely familiar, simple, linear correlation coefficient is given by

$$r = s_{xy}/(s_x \cdot s_y) \qquad (4.22)$$

where

r	is a simple, linear correlation coefficient $(-1 < r < 1)$
s_{xy}	or $\mathrm{Cov}(x, y)$ is the covariance of x and y
s_x	is the standard deviation of x
s_y	is the standard deviation of y.

A perfect direct relation between x and y gives $r = 1$; a perfect inverse relation between x and y results in $r = -1$; a value $r = 0$ means that no relation exists between x and y; real data give fractional values and may approach the limits noted (Fig. 4.14). If variables x and y are normally distributed, nonzero values of r can be tested for statistical significance in relation to critical r values obtained from tables in many standard statistics texts (e.g., Krumbein and Graybill, 1965), provided, of course, that the two variables themselves are distributed normally. The presence of outliers and/or nonlinear trends can lead to false values of the correlation coefficient (Fig. 4.15). To offset such problems it is useful to examine $x-y$ plots of all possible pairs of variables in a data set.

Correlation coefficients have a variety of applications in resource/reserve studies. They are necessary in certain formal calculation procedures such as

(i) Error estimation for grades determined from separate estimates for thickness and accumulations (e.g., Sinclair and Deraisme, 1974)
(ii) Construction of the correlogram autocorrelation function
(iii) Establishing models of one variable in terms of another (e.g., instrumental eU_3O_8 versus chemical U_3O_8)
(iv) Examining interrelations among many variables
(v) Dealing with many variables to provide indications of zonal distributions of metals (e.g., Sinclair and Tessari, 1981).

These applications are illustrated elsewhere in this text, particularly in Chapter 6.

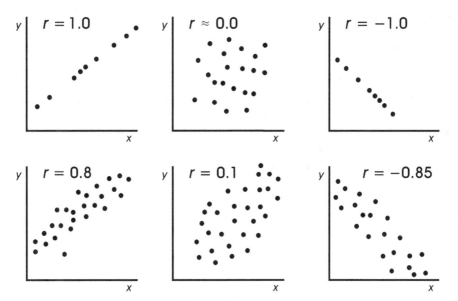

Figure 4.14: Schematic plots to provide a conceptual interpretation of various values of the simple linear correlation coefficient, r. The upper row of diagrams represents ideality and the lower row indicates small departures from ideality. Redrawn from Sinclair (1986).

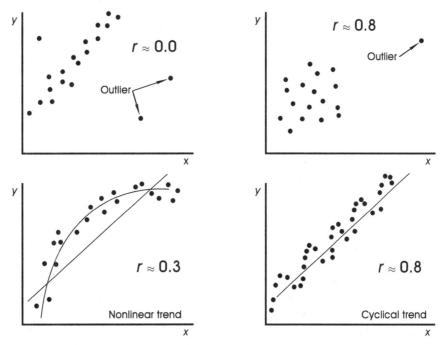

Figure 4.15: Schematic representation of potential complications, specifically, outliers and nonlinear trends, in the interpretation of a simple, linear correlation coefficient. Redrawn from Sinclair (1986).

4.7: AUTOCORRELATION

Autocorrelation involves the correlation of a variable with itself, the paired values being separated in either time or space. Consider a line of equis-paced samples (along a drill hole, trench, drift, etc.). Imagine tabulating the data in successive adjoining pairs, the centroids of which are separated by a distance (lag) h. The first member of each pair is x and the second member is y. Thus, a correlation coefficient can be determined for the tabulated data by applying Eq. 4.18. Variables x and y, being the same, have the same parameters – $x = y$ (or, $x_i = x_{i+h}$) and $s_x = s_y$ (or, $s_{xi} = s_{xi+h}$). Thus, Eq. 4.22 reduces to

$$r = \text{Cov}(x_i, x_{i+h})/s_x^2.$$

The process can be repeated successively for pairs of samples separated by $2h$, $3h$, and so on, and eventually all the calculated correlation coefficients can be plotted versus their respective lags (hs) to produce a correlogram as shown in Fig. 4.16. In many mineral deposits, nearby sample pairs are similar and more distant sample pairs are more different. Consequently, a common form for a correlogram is to have high values of r for short sample separation and lower values of r for increasing sample separation. Each calculated r value can be tested statistically for significant difference from zero. The sample separation at which the correlogram is not significantly different from zero is the *range* (i.e., the average range of influence of a sample). Clearly, autocorrelation is an important attribute of grades in quantifying the average range of influence of a sample in a mineralized field.

It should be apparent that the range (and form) of the correlogram can vary from one direction to another in response to differing geologic character of the substrate being examined (e.g., in an iron-formation unit, the autocorrelation of Fe grades in a direction parallel to bedding need not be the same as the autocorrelation perpendicular to bedding). Autocorrelation is a vector property that can be anisotropic, with the anisotropy controlled by geology. In general, the

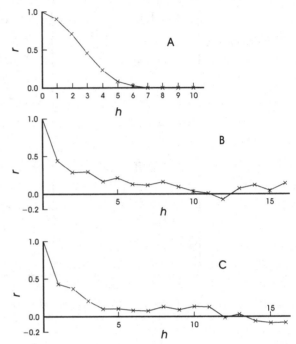

Figure 4.16: Examples of correlograms; that is, correlation coefficient (r) versus sample separation (h). (*A*) A theoretical example; (*B*) correlogram for 118 contiguous Zn percentage values from sphalerite-bearing mineralization, Pulacayo, Bolivia; and (*C*) correlogram for 129 contiguous Ti percentage values from an anorthosite complex, Black Cargo Area, California. Redrawn from Agterberg (1965). The general pattern of all three diagrams indicates that nearby samples are more similar than are more widely spaced samples.

degree of autocorrelation decreases as sample spacing increases in all directions.

In addition to the correlogram, other autocorrelation tools include the *covariogram* and the *variogram* (or *semivariogram*), both of which are essential in geostatistics; hence, a more detailed discussion of autocorrelation is given in Chapter 8.

For a lognormally distributed variable, the (auto)correlation coefficient of the logtransformed data is given by

$$r_{\text{lognormal}} = \{[1 + E^2]r - 1\}/E^2$$

where E is the coefficient of variation of logtransformed data (i.e., $E = s_{\text{lognormal}}/X_{\text{lognormal}}$).

4.8: SIMPLE LINEAR REGRESSION

There are many practical situations in which it is desirable to fit a straight line to a set of paired data. In the previous discussion of correlation it is apparent that a geometric view of the significance of the correlation coefficient could be that it represents a relative measure of how well two variables approach a straight line on an x–y graph. A linear model is expressed as follows:

$$y = b_0 + b_1 x \pm e$$

where x and y (independent and dependent variables, respectively) are the variables being considered, b_1 is the slope of the line, b_0 is the intercept on the y axis, and e is the random dispersion of points about the line (as a standard deviation). A commonly used procedure to produce an optimum linear model in x and y is to minimize the squares of the error e. This criterion is equivalent to solving the following two normal equations:

$$\sum y_i - n\,b_0 - b_1 \sum x_i = 0$$

$$\sum y_i x_i - b_0 \sum x_i - b_1 \sum x_i^2 = 0$$

where all summations are known from a set of n pairs of data. Note that different equations are determined depending on which variable is taken as the dependent variable y. The equation with one variable as the dependent variable can be very different from the equation with the alternate variable taken as y (Fig. 4.17).

The normal equations must be solved to determine values of b_0 and b_1 that minimize the error (dispersion) parallel to the y direction:

$$b_1 = \frac{\sum x_i y_i - (\sum y_i)(\sum x_i)/n}{\sum x_i^2 - (\sum x_i)^2/n} \quad (4.23)$$

$$b_0 = \bar{y} - b_1 \cdot \bar{x}.$$

Scatter about the line (parallel to the y axis), s_d^2, can be determined from Eq. 4.24 and can be used to determine confidence limits on estimated y values:

$$s_d^2 = \left[\sum y_i^2 - b_0 \sum y_i - b_1 \sum x_i y_i \right] / n$$

$$s_d^2 = \sigma_y^2 (1 - r^2). \quad (4.24)$$

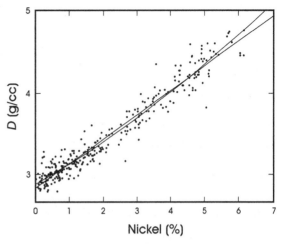

Figure 4.17: Plot of sample density (D g/cc) versus nickel grade (%), Falconbridge Nickel Mines, Sudbury (after Bevan, 1993). A straight line and a quadratic trend have been fitted to the data; both describe the data well. Least squares equations are: $D = 2.839 + 0.297$ Ni and $D = 2.88 + 0.238$ Ni $+ 0.013$ Ni^2. In both cases, all the error has been attributed to D.

Confidence limits on b_0 and b_1 can be determined if necessary (e.g., Krumbein and Graybill, 1965).

This simple linear relation is important historically in the development of practical geostatistics. In an exhaustive study of the relationship between sample assays and block grades for gold deposits of the Witwatersrand, Krige (1951) recognized an empirical, linear relation between estimated panel grades and the average grade of samples within the panel, as follows:

$$y_b = m + b_1(x_b - m) \quad (4.25)$$

where y_b is a particular block grade, m is the mean of both samples and blocks, and x_b is the average of sample grades within the block. This equation is equivalent to Eq. 4.23. Krige's work centered on the fact that such an empirically determined relation was different from the $y = x$ line, the hoped for result. This difference he subsequently referred to as *conditional bias* (i.e., a bias that is conditional on grade). He showed that high-grade estimates, on average, overestimate true grades, and low-grade estimates, on average, underestimate true grades. The coefficients of Eq. 4.25

are (Matheron, 1971)

$$1 = r(\sigma_x/\sigma_y)$$
$$b_1 = r(\sigma_y/\sigma_x)$$
$$b_1 = \sigma_y^2/\sigma_x^2 < 1.$$

A slope less than 1 guarantees that sample grades, on average, overestimate true grades above the mean and, on average, underestimate true grades below the mean. Note that for these relations to hold, estimates are plotted on the x axis and "true" grades are assigned to the y axis. The general result indicated here, the so-called regression effect, applies wherever a slope of one might be expected (hoped for) but for which the variances are different for the two variables being compared.

In many situations in which the relationship between two variables is not linear, it may be possible to approximate a linear relation by transforming one of the variables. Providing x and y are completely determinable from available data, the linear model applies, as in the following examples:

$$y = b_1 \cdot e^{-kz} \pm e$$
$$y = b_1 \cdot z^2 \pm e. \qquad (4.26)$$

Examination of scatter diagrams thus becomes a useful data analysis procedure in recognizing the presence of pairs of correlated variables and pairs of variables that exhibit a linear relation. Some nonlinear relations that can be made linear through transformations, such as those defined by Eq. 4.26 can also be investigated in this manner.

Figure 4.14 demonstrates the inherent linear character of paired variables characterized by a high absolute value of the correlation coefficient. Figure 4.17 is a practical example of the use of least squares relations; both linear and quadratic least squares models are shown, relating density of ore (D) to nickel grade (Ni). These traditional least squares models place all the error in D because Ni is assumed to be known perfectly in order to use it as an estimator of density. In this case, the linear relation is not improved significantly by fitting a quadratic equation to the data. Traditional least squares models are used routinely in

cases such as this, in which one variable is used to estimate another.

4.9: REDUCED MAJOR AXIS REGRESSION

A reduced major axis (RMA) regression is desirable when it is important that errors in both variables be taken into account in establishing the relation between two variables (Sinclair and Bentzen, 1998). The methodology for reduced major axis regression has been described in an earth science context by Agterberg (1974), Till (1974), Miller and Kahn (1962), and Davis (1986). Agterberg (1974) describes RMA as the normal least square of standardized variables when standardization involves the following transforms:

$$z_i = x_i/s_x \quad \text{and} \quad w_i = y_i/s_y$$
$$y = b_0 + b_1 x \pm e.$$

Note that these transformations are analogous to forming standardized z values except that the resulting transformed z_i and w_i values are centered on x/s_x and y/s_y, respectively, rather than zero. In fact, identical least-squares results would be obtained using standardized data (see Kermack and Haldane, 1950). Till (1974) emphasizes the importance of using RMA in comparing paired (duplicate) analytical data.

The general form of the reduced major axis line is given by Eq. 4.23 as follows:

$$y = b_0 + b_1 x \pm e$$

where x and y are paired values, b_0 is the y-axis intercept by the RMA linear model, b_1 is the slope of the model, and e is scatter about the line as a standard deviation. For a set of paired data, b_1 is estimated as

$$b_1 = s_y/s_x$$

where s_x and s_y are the standard deviations of x and y, respectively, and b_0 is estimated from

$$b_0 = \bar{y} - b_1\bar{x}$$

where $\bar{y}$ and $\bar{x}$ are the mean values of y and x, respectively. Usually, we are interested in whether the line passes through the origin because, if not, there

Table 4.3 Parameters of gold analyses for duplicate blast hole samples and three fitted linear models, Silbak Premier gold mine, Stewart, British Columbia

Variable	Mean (g/t)	Standard deviation	Correlation coefficient (AU vs. AUD)
AU	3.429	4.088	0.727
AUD	3.064	3.817	

Linear model	Intercept[a]	Slope[a]	Dispersion about line	Remarks
AUD regress on AU	0.737	0.679	2.62 g/t	All error attributed to AUD
RMA	0.137	0.934	4.13 g/t	Error shared between AU and AUD
AU regress on AUD	1.340	1.284	2.81 g/t	All error attributed to AU

[a] All linear models are of the form AUD = sl + AU + int + e, where sl = slope and int = intercept.

is clearly a fixed bias of some kind. The error on the y-axis intercept, s_0, is given by

$$s_0 = s_y\{([1 - r]/n)(2 + [\bar{x}/s_x]^2[1 + r])\}^{1/2}$$

where r is the correlation coefficient between x and y. The error on the slope is

$$s_{sl} = (s_y/s_x)([1 - r^2]/n)^{1/2}. \qquad (4.27)$$

The dispersion S_{rma} about the reduced major axis is

$$S_{rma} = \{2(1 - r)(s_x^2 + s_y^2)\}^{1/2}$$

where s_x and s_y are the standard deviations of x and y, respectively; x is the independent variable; and y is the dependent variable.

These errors can be taken as normally distributed (Miller and Kahn, 1962) and, for duplicate analytical data, can be used to test whether the intercept error range includes zero (in which case the intercept cannot be distinguished from zero) and the slope error range includes 1 (in which case the slope cannot be distinguished from 1). The dispersion about the RMA line can be used in several practical comparisons, including: (1) the comparison of replicates of several standards by one laboratory with replicates of the same standards by another laboratory; and (2) the comparison of intralaboratory paired analyses. An example is provided in Fig. 4.18, where analyses of duplicate blasthole samples are plotted for a set of quality-monitoring data from the Silbak Premier

gold deposit, Stewart, British Columbia. In this example, the two traditional least-squares models (error attributed entirely to one variable) contrast strongly with the centrally positioned RMA model. Parameters of the variables and the three illustrated linear models are summarized in Table 4.3. In this case, the

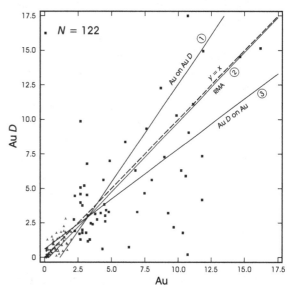

Figure 4.18: Plot of original Au analyses (Au) of blasthole samples versus duplicate blasthole sample analyses (AuD) for the Silbak Premier gold mine, Stewart, British Columbia. Three linear models are shown fitted to the data: (1) all the error attributed to Au, (2) reduced major axis solution, and (3) all the error attributed to AuD. Parameters of the three models are summarized in Table 4.3. Large differences exist among the linear models because of the relatively wide scatter of data.

RMA model is appropriate for interpreting the significance of the data. The intercept has an error, $s = 0.31$, and therefore cannot be distinguished from zero. The slope, although apparently very different from zero, has a large error of $s_0 = 0.058$ reflecting the wide scatter of data; hence, the slope cannot be distinguished from 1. The large dispersion of data can be quantified by Eq. 4.27 as $S_{rma} = \pm 4.13$ g/t, a remarkably large value that leads to great concern about the blasthole sampling procedure.

4.10: PRACTICAL CONSIDERATIONS

1. A rigorous training in basic statistical concepts and procedures is a useful practical background from which to consider the applications of statistics to mineral inventory estimation. Simple statistical concepts must be appreciated in general terms and, more particularly, with regard to specific uses in mineral inventory estimation.
2. Statistical approaches to dealing with data are conveniently accessible for use in a wide range of commercial computer software, including various data management and display programs designed specifically for the mineral industry. The algorithms used in such software packages must be understood.
3. Graphics tools are an increasingly important part of the arsenal contributing to an understanding of data and software that incorporates many means of viewing data in simple conceptual ways are widely available. The ease of use and quality of graphics must not cloud a basic understanding of the principles used and the true nature of the data.
4. No single software package necessarily provides all the forms of output that are desirable when dealing with the extremely large amounts of multivariate data that are commonly available for a mineral inventory study. Practitioners will find it useful to have sufficient familiarity with various software packages so that transfer of data from one to another becomes a trivial matter. Basic programming skills are particularly useful as a time saver in achieving a range of goals (e.g.,

building files of particular subsets of the data, sorting out outliers, testing various data transforms).
5. Statistical procedures must be understood so that they are applied confidently and correctly in practice. An example is the contrast between and interpretation based on classic regression (formula dependent on which variable is selected as the dependent variable), which assumes all the error is in one variable, and regression techniques, such as reduced major axis, which incorporates error in both variables.
6. The use of statistics in mineral inventory estimation should be guided by geologic information, as indicated in Chapters 2, 6, and 7.

4.11: SELECTED READING

Davis, J. C., 1986, Statistics and data analysis in geology; John Wiley and Sons, New York, 646 pp.

Henley, S., and J. W. Avcott, 1992, Some alternatives to geostatistics for mining and exploration; Trans. Inst. Min. Metall., v. 100, pp. A36–A40.

Sinclair, A. J., 1976, Applications of probability graphs in mineral exploration; Assoc. Expl. Geochemists, Spec. Vol. No. 4, 95 pp.

Sinclair, A. J., and A. Bentzen, 1998, Evaluation of errors in paired analytical data by a linear model; in Vallée, M., and A. J. Sinclair (eds.), Quality assurance, continuous quality improvement and standards in mineral resource estimation, Expl. and Min. Geol., v. 7, nos. 1 and 2, pp. 167–174.

Swan, A. R. H., and M. Sandilands, 1995, Introduction to geological data analysis; Blackwell Science, London, 446 pp.

4.12: EXERCISES

1. Given three contiguous drill core samples of lengths 2, 3, and 5 m that assay 3%, 6%, and 12 percent Zn, respectively, compare the arithmetic mean of the assays with the weighted mean, assuming that density is uniform throughout. What is the weighted mean Zn grade if the respective specific gravities are 3.2, 3.7, and 4.1?

Compare the unweighted standard deviation with the weighted standard deviation.

2. (a) Calculate the mean and variance of the 60 integer values of Fig. 4.1. Construct a lognormal distribution with the same mean and variance.

    (b) Assuming perfect lognormality and a cutoff grade of 3.5, determine (1) the proportion of volume above cutoff grade and (2) the average grade of the material above cutoff grade. Compare these estimates with those determined directly from the data.

    (c) Suppose that the spread of original values includes an abnormally high sampling plus analytical variance ($\%^2$) that has been reduced by an absolute amount of 0.55 through the implementation of an improved sampling procedure. Recalculate the proportion of volume above cutoff and the average grade of material above cutoff (assuming lognormality) using the new dispersion. The difference between these estimates and those of part (b) provide an indication of false tonnage and grade that can result from large errors in assay data.

3. A grain count of 320 grains of heavy mineral concentrate yielded 7 grains of cassiterite (SnO_2). Calculate the error of the estimate. What is the likelihood that the true grain count is 4 or less?

4. Calculate the probability of 0, 1, 2, 3, ... nuggets of gold in a sample of 2,000 grains known to contain, on average, 2 nuggets per 5,000 grains?

5. A 0.5-kg sample of a placer sand has an average grain size of 2 mm, and grain counts show that the average scheelite content is 3 grains. What is the proportion of random samples that on average contain 0, 1, 2, ... grains of scheelite? What is the expected WO_3 content of samples with a weight of 3 kg, assuming that only quartz and scheelite are present?

6. A sample of approximately 2,500 grains yields 0 grains of gold in a region where other samples contained some gold grains. What is the probability that the 0-grain result is simply an expected sampling variation of a material that contains an average of 0.5 grains of gold per 2,500 grains? What would the probability be of 0 gold grains if the sample size were doubled?

7. The following tabulation contains class interval and cumulative percentages for 120 Au analyses from a polymetallic, massive sulphide deposit:

Cutoff Grade (g/t)	% Data $\geq$ Cutoff
0.0	100
1.0	99
2.0	81
2.5	60
3.0	35
4.0	8
4.5	3
5.0	1
5.5	0

    (a) Construct the histogram and fit by a normal or lognormal probability density function.

    (b) Calculate the mean and standard deviation of the data. Note that this can be done using frequencies (specifically, Freq/100) as weights.

    (c) Plot the data on both arithmetic and logarithmic probability paper. Estimate the parameters in each case and compare with the results of (b).

    (d) Assuming that the histogram is unbiased, use a normal or lognormal approximation (whichever is best) to calculate the proportion of tonnage above a cutoff grade of 2.25 g/t Au.

8. The gold data (inch-pennyweight) of Fig. 4.19 (cumulated in Table 4.4) are shown in a simple and widely used form of presentation for planning purposes in tabular deposits (Royle, 1972). Plot the data on arithmetic and log probability paper and interpret. Interpretation should include comment

Table 4.4 Tabulation of assay data (in dwt) shown in Fig. 4.19

Value (in dwt)	Lower drift	Number in upper drift	Raise	Total number	Cumulative number	Cumulative percent
Trace	3	5	5	13	13	9.42
<10	1	2	0	3	16	11.59
10–20	1	5	4	10	26	18.84
20–30	4	3	0	7	33	23.91
30–40	0	3	0	3	36	26.09
40–50	0	1	0	1	37	26.81
50–60	5	1	1	7	44	31.88
60–70	0	2	1	3	47	34.06
70–80	1	0	0	1	48	34.78
80–90	3	1	1	5	53	38.41
90–100	0	1	0	1	54	39.13
100–200	7	3	0	10	64	46.38
200–300	1	7	1	9	73	52.90
300–400	2	4	1	7	80	57.97
400–500	3	6	3	12	92	66.67
500–600	2	2	0	4	96	69.57
600–700	3	1	0	4	100	72.46
700–800	4	2	2	8	108	78.26
800–900	1	0	2	3	111	80.43
900–1,000	4	1	1	6	117	84.78
1,000–1,100	3	0	2	5	122	88.41
1,100–1,200	0	0	1	1	123	89.13
1,200–1,400	3	1	3	7	130	94.20
1,400–1,600	0	0	0	0	130	94.20
1,600–1,800	3	2	0	5	135	97.83
1,800–2,000	1	0	0	1	136	98.55
2,000–3,000	0	2	0	2	138	100.0

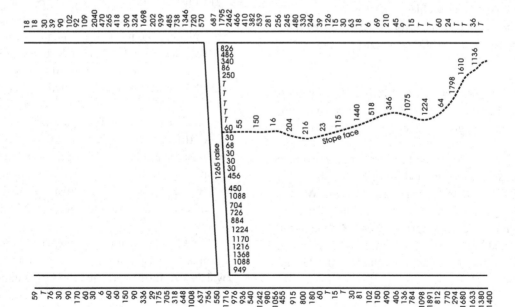

Figure 4.19: Data for underground workings of a gold-bearing quartz vein. Values are in dwt at regular intervals along underground workings and along a stope face. Redrawn from Royle (1972).

on the form of the distribution, the possibility of the presence of two or more subpopulations, and, if possible, a rough estimate of thresholds between subpopulations.

9. The following cumulative data (U_3O_8) represent 1,081 analyses for a uranium deposit (Isaaks, 1984). Parameters are: $m = 4.87\%$, $s = 7.84$, $CV = 1.61$. Plot the data on both arithmetic and log probability paper. Comment on the probable number of subpopulations and the problem(s) that arises in determining this number.

Grade (%U_3O_8) of Data	Cumulative %
0.2	22.0
0.3	34.0
0.5	46.0
1.0	54.0
1.5	60.0
2.5	69.0
5.0	75.0
82.0	90.0
20.0	94.7

5

Data and Data Quality

Sampling theory cannot replace experience and common sense. Used in concert with these qualities, however, it (sampling theory) can yield the most information about the population being sampled with the least cost and effort. (Kratochvil and Taylor, 1981, p. 938)

This chapter provides a brief introduction to the types of data normally encountered in mineral inventory estimation, a consideration of errors in data, designing data gathering effectively, and monitoring the quality of data as they are obtained. Topics proceed to a general consideration of data arrays and their impact on reserve/resource estimation, the importance of sampling experiments in optimizing sampling procedures and discussions of subsampling and analytical quality control. Gy's sampling equation and its implications are considered at length. A general approach to grade control in a porphyry-type, open-pit environment is summarized.

5.1: INTRODUCTION

Mineral inventories are estimated from quantitative measurements (assays) derived from representative samples of rock material; these data are extended in some manner to much larger volumes/masses as part of the estimation process. Volumes for which grades are estimated commonly are up to 1 million times the total sample volume on which the estimates are based; obviously, raw data must be of high quality. Samples must be distinguished from specimens; specimens of ore or waste are selected by geologists to exemplify particular features of mineralogy, texture, structure, or geologic relations and are unlikely to be representative of ore grades in much larger volumes. "Grab samples" of ores can be highly biased with respect to grade; both grab samples and geologic specimens are "point samples" as described by Vallée (1992, p. 48), and should be avoided for purposes of mineral inventory estimates. Point samples are useful for such purposes as characterizing types of mineralization and demonstrating physical continuity of mineralization.

Sample values are used to make estimates of mean grades of blocks of variable size, such that blocks can be classified as ore or waste; clearly, high-quality data are desirable as a basis for this decision. The quality of estimates depends on the number of samples used, the mass of individual samples, the orientation of individual samples, the spatial distribution of samples relative to the block to be estimated, sampling procedures, sample reduction methods, and, of course, the value continuity of the variable (grade) in question.

Deliberate errors are unacceptable and special security measures may be necessary to guard against their occurrence. Illegal activities can include salting (purposeful contamination of samples or outcrops), importing mineralized material (e.g., drill core) from another property, and falsification of supposedly

Table 5.1 Types of samples used traditionally in mineral inventory estimation

Name	Brief description
Point	A localized specimen taken to illustrate a particular geologic/mineralogic feature; a prospector's "grab sample" (a small, localized chip). Generally not representative of grade, and thus cannot be projected from site of origin. Can be indicative of continuity of mineralization. Commonly small in size, perhaps 0.1–0.2 kg.
Linear	Has one very long dimension relative to others. Includes drill core and cuttings, channel samples (highly regular cuts from a rock surface), or a series of adjacent or nearly so chip samples. Individual linear samples are generally 0.5 to several kg. This is the type of sample that forms the main database of most mineral inventory studies.
Panel	Generally a regular array of chips taken over a more or less planar surface such as a surface exposure of ore on the face/wall of underground openings. Such samples commonly range from 1–5 kg.
Broken ground	Large amounts of broken ground can result from trenching and the driving of underground openings. This material can be sampled regularly as it is produced or may be moved to one or more localities with variable and unknown mixing, possible segregation of heavy components, losses during transit, and dilution during sequential moves. Representative sampling is a difficult problem generally requiring large samples. Used to optimize ore-dressing procedures.
Bulk	Very large samples (hundreds to thousands of tons) generally taken to verify grade estimates from smaller samples or to optimize ore-dressing procedures prior to mill design and construction. Small-scale mining procedures are generally required to obtain these samples.

Source: From Vallée (1992, 1998a).

historic documents (Wunderlich, 1990). Verification of technical literature, independent sampling, regular control of sampling and analytical procedures, special security precautions, and the use of reliable staff are means of minimizing deliberate errors (cf, Rogers, 1998). As Agricola (1556) stated: "A prudent owner, before he buys shares, ought to go to the mine and carefully examine the nature of the vein, for it is very important that he should be on his guard lest fraudulent sellers of shares should deceive him" (Hoover and Hoover, 1950, p. 22).

5.2: NUMERIC DATA FOR MINERAL INVENTORY ESTIMATION

5.2.1: Types of Samples

The principal categories of samples useful in mineral inventory estimates are summarized in Table 5.1. Linear samples are most common, particularly samples of drill core and/or cuttings, supplemented in some cases by channel and chip samples. *Channel samples* are those in which all solid material is collected from a long prism of rock of rectangular cross section; *chip samples* are those in which rock chips are collected at one or more points in a regular pattern, along a line in the case of a linear chip sample. Channel sampling is labor intensive, and both channel- and chip-sampling techniques require access to rock faces. Channel and chip samples are suited particularly to outcrops, trenches, and underground workings, and generally are taken with the aid of handheld drills. In the case of strongly preferred orientations (e.g., bedding, shearing), channels should be oriented across the layering; chip samples are more appropriate when directional heterogeneities are less pronounced. The principal problem encountered with both channel and chip sampling is preferential breaking of soft minerals. Soft ore minerals may be overrepresented in a sample and thus may impose a high bias on the resulting grade; conversely, soft-gangue minerals can be overrepresented and may lead to underestimation of grade. The problem can be minimized by taking large samples or, where possible, taking separate samples from soft and hard zones and weighting corresponding grades by proportions to obtain a weighted-grade estimate.

Drill-based sampling methods are now used routinely, particularly for evaluation of large mineral deposits where abundant data are required from what would otherwise be inaccessible parts of a deposit. Rotary, percussion, auger, and diamond drills are used (with variants) depending on specific circumstances. Either solid rock core or fragmented or finely ground cuttings are brought to surface by drilling and sampled for assay. Both core and cuttings provide insight into the nature of inaccessible parts of a deposit and are most informative and representative where they are at a large angle to the direction (plane) of greatest continuity of mineralization. Core is commonly split, one-half being retained for geologic information and the other half providing material for assay. Core splitting can be done with a mechanical splitter or with a diamond saw; use of a saw is labor intensive but may be warranted when it is important to reduce sampling variability, as in the case of many gold deposits. In rare cases, the entire core is used either for assay or to create composite samples for mill tests. Such procedures should be avoided if at all possible because "half core" should be saved and stored safely, as it is an essential reference material with which to develop new concepts of both geologic and grade continuity as knowledge of a deposit evolves (cf. Vallée, 1992). Photographing split core in core boxes is a widely used procedure for preserving evidence of the character of the core and is particularly essential when all of the core is consumed for either assaying or testing milling procedures.

Many types of drilling do not produce core. Instead, they provide broken and finely ground material that is flushed to the surface (by air or water) and is either sampled systematically by machine as it arrives at the surface or accumulates in a pile that must be subsampled. The amount of cuttings from a single drill hole can be enormous and the sampling problem is not trivial. Generally, drill cuttings must be reduced in mass substantially by a sampling procedure such as riffling to produce a sample of manageable size for subsampling and assay.

Drill samples are subject to a variety of sources of error (e.g., Long, 1998). For example, some types of ground (e.g., highly altered rock) are difficult, if not impossible, to core, and core loss may be substantial.

Such core loss can result from the presence of soft minerals in a hard surrounding (e.g., galena in quartz) or from superimposed fracturing and/or alteration (e.g., a shear zone). When drill cuttings are sampled, soft material high in a drill hole can collapse and contaminate cuttings lower in a hole. Water is an inherent part of many drilling procedures and its low relative density can result in material, such as gold, high in a drill hole (e.g., above the water table) being transported to lower depths in the hole.

Panel samples are normally taken underground, intermittently as a working face advances (i.e., as an adit or drift is advanced round by round by drilling and blasting). Generally, the sample is a combination of chips taken from a close-spaced, two-dimensional grid over the surface, and thus commonly represents a surface area of about 2×2 m. Such samples ordinarily are used as a check on samples of smaller or different support (size, shape, orientation). As with any chip-sampling procedure, panel samples can be biased if soft material is included in nonrepresentative amounts in the ultimate sample. When possible, hard and soft materials might be chip sampled and analyzed separately; a weighted average for the panel can be based on the proportions of hard and soft materials and their respective bulk densities.

It should be noted that a substantial amount of assay information for some variables (e.g., uranium, berylium) can be obtained directly by instrument readings from a rock face so physical samples might not be taken. Hand-held instruments on rough rock faces can produce large errors because calibration generally does not take surface roughness into account. There is increasing use of borehole geophysical methods to obtain assay information. Passive techniques (e.g., gamma ray spectrometry) record natural emissions from rocks, and active techniques (e.g., spectral gamma–gamma and X-ray fluorescence) have an electronic or radioactive source that bombards the rocks so that characteristic emissions can be recorded (Killeen, 1997). Active techniques are applicable to a wide range of elements. All instrumental methods must be calibrated with great care.

5.2.2: Concerns Regarding Data Quality

The aim of sampling is to provide value measures (e.g., assays of metal grades) that are the fundamental information to be used in making resource/reserve estimates. Individual value measures are subject to variability from three discrete sources: (i) real geologic variations, (ii) sampling errors, and (iii) measurement (analytical) errors. The effects of real geologic variability can be dealt with through the use of autocorrelation functions and statistical methods closely tied to the concept of "range of influence" of a sample (see Chapters 8 and 9). To a limited extent (generally negligible), this source of variability can be decreased by increasing sample size.

There are several ways in which an approach to sampling and data gathering can be arranged to maximize the quality of the data and the resulting mineral inventory estimates, whatever the method of estimation. Some of the more important factors are as follows:

(i) To reduce and know the magnitudes of errors in the database (assays, thickness, sample coordinates), in particular, to improve and monitor the quality of assays and the representivity of samples

(ii) To develop a sampling plan and procedure that are appropriate to the task

(iii) To ensure that analytical methods and procedures have adequate precision and accuracy

(iv) To integrate geology into the data accumulation and evaluation phases so that data are obtained and used in optimal fashion with respect to the geologic character of the material being sampled.

Minimizing errors in sampling and assaying reduces the nugget effect (random variability) and the sill of the experimental semivariogram (see Chapter 9) and thus improves the quality of model fitting to the experimental semivariogram. An improved autocorrelation (e.g., semivariogram) model is an improved quality measure of average continuity of grades and, if geostatistical methods of estimation are to be used, reduces the estimation error. Using a single, aver-

age semivariogram model for an entire deposit, rather than models adapted to the various geologic domains present, can lead to an inappropriate application of geostatistics with abnormally large and unrecognized errors in a high proportion of the estimates.

Sampling methods, sample preparation procedures, and assaying procedures are too commonly accepted routinely without rigorous checking during information gathering for delineating and detailing a specific mineral deposit (e.g., Pitard, 1989, 1994). Requirements for sample parameters such as sample mass, particle size, and number of particles per sample differ as a function of the character of mineralization. Sampling procedures that are not optimal introduce an additional, possibly large, component of variance to the geologic/mineralogic variations that are described by autocorrelation functions such as semivariograms or correlograms. Systematic tests to help optimize data-gathering procedures are used far too rarely in practice, despite the low cost of such measures and the increased error that is implicit in the added variance. Sampling methods should not be accepted without testing (Sinclair and Vallée, 1993), as methods suitable for one deposit type might be totally inappropriate for another. Often, too little attention is paid to reviewing the implications of the various parameters and procedures associated with standardized sampling methods such as diamond drilling (e.g., changing core sizes, variable core recovery, variable sampling methods).

The same scrutiny used in quality control of sampling procedures should be applied to assaying practices and results (Vallée, 1992). Reducing the assay variability related to procedural and instrumental deficiencies also improves delineation of ore limits and the efficiency of grade-quality control during production. Design of optimum subsampling procedures can be guided by Gy's sampling equation (Gy, 1979; Radlowski and Sinclair, 1993) to appreciate inherent errors and to minimize the nugget effect (Pitard, 1994). Reports of assay results should contain information about sample preparation and analytical methods, including information about detection limits.

5.2.3: Location of Samples

Inherent in the use of sample grades to develop a mineral inventory is the implicit understanding that sample locations in space are well known. Unfortunately, this is not always the case. Sample sites and drill-hole collars must be surveyed for accurate positioning in three dimensions. Drill holes wander, depending on their orientation and the physical characteristics of rocks they intersect. Even small deviations from a planned orientation at the collar of a hole can result in substantial departure of the end-of-hole from the planned/projected position. A drill hole must be surveyed at intermittent locations along its trace and this information is then combined with the surface information (collar coordinates) to determine realistic three-dimensional coordinates for samples along the "true" drill-hole trace. Most commercial, down-hole, surveying devices routinely measure angle of plunge to better than 1 degree and the angle of azimuth to better than 0.1 degree (Killeen and Elliot, 1997). The problem is illustrated in Fig. 5.1, in which small, progressive changes in the plunge of a drill hole (1–3 degrees over 50 m) result in the end-of-hole being shifted about 30 m from its projected position. Lahee (1952) discusses much larger deviations in which drill hole ends are displaced by 20 percent or more of the hole length. Traditional down-hole surveying methods can be affected by variable magnetic character of the surrounding rock. Gyroscopes can be used inside the drill casing where magnetic properties are a problem. New techniques such as ring-laser gyros do not depend on magnetic properties and have almost no moving parts.

5.3: ERROR CLASSIFICATION AND TERMINOLOGY

5.3.1: Definitions

The accuracy of a sampling and analytical procedure is a measure of how closely the true value of the sample is approached by reported analyses. Generally, true values are not perfectly known and tra-

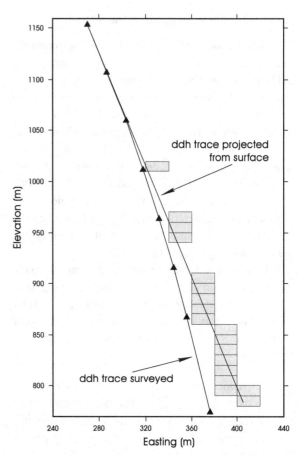

Figure 5.1: An example of a correct drill-hole trace based on survey information, relative to the expected (projected) trace if the hole did not wander. Triangles are survey positions along the drill hole, where hole plunge and azimuth were measured. Dotted pattern indicates eighteen 10 × 20 m blocks intersected by simple projection of the collar orientation, but not intersected by the correct drill-hole position. In this example, small direction changes of a degree or so, at 50-m intervals down the hole, result in an overall 9-degree deviation in orientation (and about 30-m displacement) near the hole bottom.

dition has led to a general procedure that provides a close approach to true values of samples. Fundamental to the method is the availability of a compositional range of standards that are widely analyzed (internationally in some cases) by a variety of proved analytical methods and whose mean values are well known. These mean values are accepted as the true values and the standards are used to calibrate local, secondary standards (natural and/or synthetic materials) that serve as routine calibration controls in most commercial and minesite laboratories. Replicate analyses of internal standards are commonly reported by commercial laboratories along with assay results reported to a client; the consistency of such data is a measure of the reproducibility of the analytical procedure, and the nearness of the average of replicates to the accepted standard value is also a measure of accuracy. It is important not to use a false measure of accuracy. If the aim of an assay is to determine soluble oxide copper accuracy for such a variable, it should not be compared to total Cu content.

Departures from true values are called *bias*, a matter that often receives too little attention, "and yet is normally the most significant source of error in economic terms" (Burn, 1981). Bias can be considered ideally as either proportional to concentration or fixed (constant regardless of concentration).

Reproducibility (precision) is a measure of the inherent variability of a specified sampling and analytical procedure and need bear little similarity to accuracy (Fig. 5.2). For example, a particular acid extraction for copper may not attack certain Cu-bearing minerals, and thus may underestimate copper even though repeated analyses give very consistent results.

Replicate analyses of a sample are not all identical; a variety of errors are involved in any measurement procedures and these accumulate in the reported analytical values. In general, when only random errors are present, the error can be summarized as an average variance (s^2) or as a standard deviation (s) relative to the mean ($\bar{x}$). Commonly, the relative error as a percentage is used ($100s/x$); the quantity s/x is also known as the coefficient of variation (CV). Precision, a quantitative measure of reproducibility, commonly is quoted in terms of two standard deviations as a percentage of the mean:

$$Pr = 200s/\bar{x}$$

but can be generalized as a function of concentration as

$$Pr_c = 200s_c/x_c.$$

The error, s_c, commonly varies systematically as a function of concentration (see Section 5.3.2).

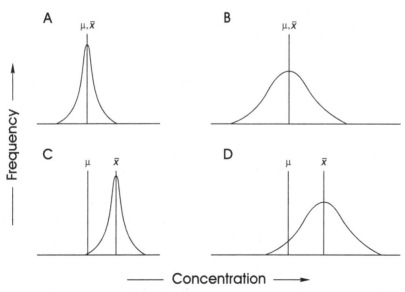

Figure 5.2: Types of errors in measurement data: random and systematic errors, each with narrow (good-quality) and wide (poor-quality) precision. μ is the true population value that is being estimated by samples (histogram); $\bar{x}$ is the mean value of the sample.

Detection limits of analytical techniques should be reported; a useful definition of *detection limit* is that concentration at which $Pr_c = 100$, or equivalent, the concentration at which $2s_c = x_c$ (Thompson and Howarth, 1976). Note that precision can be very different for different metals, even though the metals might be measured by the same analytical method.

The recognition of bias (systematic departures from the truth) is more difficult and requires a quality control system. A program involving periodic analyses of standards can serve as an early indicator of bias and is an essential component of a quality control program. A more extensive consideration of bias is provided in Section 5.8.

A thorough understanding of data quality is clearly important in estimating mineral inventory. Less obvious, however, is the impact of data quality on the extensive reconciliation work that is invariably done during production. It is common practice in operating mines to compare estimated metal with mill heads over practical operational time units (e.g., weekly or monthly production). This procedure serves as a complex indicator of the quality of estimation and is part of a broader mass balance procedure that indicates the extent to which production agrees with expectation (estimation). Assay data that are of low quality, for whatever reason, make these comparative studies (reconciliations) ambiguous or uninformative.

Errors in sampling and analytical methods generally can be treated by statistical methods, providing reasonable design has gone into generating the appropriate information. Random errors, as a rule, are considered to be normally distributed, with several exceptions (Thompson and Howarth, 1976), as follows:

(i) The analyte is concentrated in a very small proportion of the particles constituting extremely heterogeneous samples (e.g., cassiterite or gold grains in a sediment).

(ii) Precision is poor and calibration is nonlinear (e.g., calibration can be logarithmic near the detection limit).

(iii) Concentrations are within an order of magnitude of the digital resolution of a measuring instrument.

(iv) Concentration levels are near the detection limit and "subzero" readings are set to zero or recorded as "less than" (produces a censored distribution).

(v) Outliers are present in the data (considered in Chapter 7).

5.3.2: Relation of Error to Concentration

Experience has shown that errors in the numeric values of assays commonly are a function of metal concentration. As a rule, this relation is poorly documented and is rarely quantified in cases of mineral inventory estimation. The problem has been investigated extensively by Thompson and Howarth (1973, 1976, 1978) and Howarth and Thompson (1976) in relation to geochemical data sets in which values can extend over several orders of magnitude. They demonstrate that it is inappropriate to assign the same average (absolute or relative) error to both low- and high-valued items in a data set that spans a large compositional range (an order of magnitude or more). Their work suggests that a linear model provides an acceptable estimation of error as a function of composition for analytical data of common geochemical media (samples of soils, sands, water, rocks, etc.); assays that serve as a base for mineral inventory are just a special case of analytical data for rock media and are appropriate for study using the Thompson–Howarth method. A cautionary note is that the Thompson–Howarth method is restricted to a quantification of random error as a linear function of metal concentration, bias is ignored, and some other method of testing for bias is essential (see Section 5.8). Moreover, the linear model is assumed to apply to the entire compositional range being investigated. Such an assumption could be inappropriate in cases in which data are derived from two or more completely different styles of mineralization. The method is meant to be applied to situations in which duplicate samples are distributed through the various analytical batches that generate a complete data set and cannot be identified by the laboratory conducting the analyses. Hence, the duplicate samples are representative of the conditions that prevail throughout the generation of the entire data set. The Thompson–Howarth method is not meant for paired data generated by different laboratories, analysts, or sample types.

The Thompson–Howarth procedure considers paired data to be a basis for monitoring errors, although it can be adapted easily to use a variable number of replicate analyses. Similarly, the method as published for geochemical data recommends a minimum of 50 pairs of analyses; for more precise assay data, the authors have found as few as 30 duplicates to be adequate. The general method is as follows:

(i) For a set of paired (duplicate) data, determine the mean concentration of each pair $[(x_1 + x_2)/2]$ and the corresponding absolute difference in concentrations (i.e., $|x_1 - x_2|$).

(ii) Arrange paired data in order of increasing concentration, using means of pairs.

(iii) Divide the full data set into successive ordered groups of 11 for geochemical data. (The authors have found that as few as seven pairs per group is adequate.)

(iv) For each group, find the group mean value (concentration) and the median value of pair differences.

(v) Plot the median difference of paired data in each group versus the corresponding mean value for the group (Fig. 5.3).

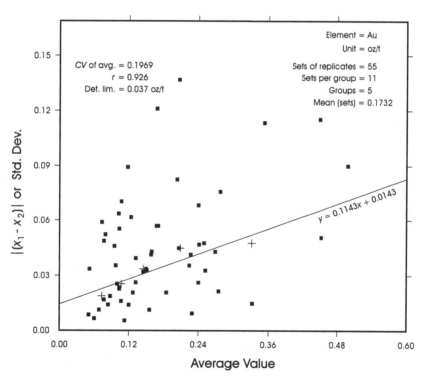

Figure 5.3: Sampling and analytical errors of gold analyses as a function of composition, based on 55 duplicate samples from the J&L massive sulphide deposit, southeastern British Columbia. Ordinate is standard deviation of replicate analyses (or absolute difference of sample pairs); abscissa is average concentration of replicate analyses. Filled squares are individual samples that have been replicated; + signs are median differences (or standard deviations) versus averages of successive groups of 11 sets of sample replicates. A linear model has been fitted to the + signs to describe error as a function of composition. See text for details. CV = coefficient of variation; r = correlation coefficient; Det. Lim. = detection limit as calculated from the data (i.e., concentration at which concentration equals twice the standard deviation).

(vi) Obtain a line through the plotted mean–median points (by eye or by regression). This line has the form:

$$y = b + kx.$$

If b and k are multiplied by 1.048 (because medians have been used as estimates of standard deviations), the equation becomes

$$s_c = s_o + m \cdot c$$

where c is the concentration, m is the slope, and s_o is the y-axis intercept. A slightly modified version of the Thompson–Howarth method of error analysis is available in the program ZERROR (publisher's website), although it is evident from this sequence of steps that the analysis is done easily in a spreadsheet. ZERROR accepts duplicate and/or replicate analyses. Thus, instead of plotting absolute difference of pairs versus mean value, a plot of standard deviation versus mean value is provided. The two are essentially equivalent for paired data over short concentration ranges that result from the ordered grouping of pairs as described for the Thompson–Howarth procedure. Use of the program should be preceded by a general examination of paired data for bias, as outlined in Section 5.8, because the method is not sensitive to the recognition of bias. Figure 5.3 is an example of the output of ZERROR.

The Thompson–Howarth method was designed for duplicate geochemical data from a single laboratory, data that are commonly of relatively low quality in comparison with assay data that are the basis for resource/reserve estimation. In addition, there is an implicit assumption that the two sets of analyses being compared are subject to the same errors, on average. This obviously need not be true where duplicate results by two different labs are being compared – in which case the resulting error model is intermediate between the models for each lab individually. Consequently, despite its increasing use for data evaluation vis-à-vis mineral inventory estimates, the method is not entirely suitable for broader applications involving both systematic and random errors or data from multiple laboratories. An alternative and more general approach to monitoring data quality is outlined in Section 5.8.

5.3.3: Bias Resulting from Truncated Distributions

Truncation of a grade distribution, as in the application of a cutoff grade used to separate ore from waste, necessarily leads to a bias in resulting estimates of recoverable metal, even though high-quality, unbiased assay data are used to make the estimates. The problem arises because block classification as ore or waste is based on estimates, which, no matter what their quality, contain some error. Thus, some values above a cutoff (truncation) grade are estimated as being below cutoff grade and vice versa. In a simplistic example, Springett (1989, p. 287) illustrates the problem:

[C]onsider the trivial but informative example of a gold deposit with a constant grade of 1.7 g/t (0.05 oz per st) completely homogeneously distributed – thus any sample or truck load that is taken from the deposit contains exactly 1.7 g/t (0.05 oz per st) of gold. The operator is unaware of this uniform grade distribution and will carry out selection by means of blasthole samples. Assume a sampling error that is normally distributed with a mean of zero and standard deviation of 0.34 g/t (0.01 oz per st) that is incurred at both the mill and the mine. Obviously, if perfect, error-free selection was possible, then if the cutoff grade was at any value equal to or below 1.7 g/t (0.05 oz per st) the entire deposit would be delivered to the mill, and if the cutoff grade were at any value greater than 1.7 g/t (0.05 oz per st) none of the deposit would be mined. However, given the assumed error distribution described above, the apparent distribution of blasthole sample grades will then be normally distributed with a mean of 1.7 g/t (0.05 oz per st) and a standard deviation of 0.34 g/t (0.01 oz per st). If selection is carried out by the polygonal method, then two curves can be developed showing for a range of cutoff grades: the average grade reported by the mine (and) the average grade reported by the mill.

The foregoing example, illustrated in Figs. 5.4 and 5.5 in units of oz/t, assumes a constant grade of

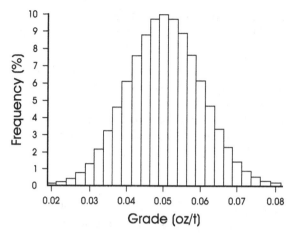

Figure 5.4: Normal estimated grade distribution with mean of 0.05 oz/t and standard deviation of 0.01 oz/t (i.e., 20 percent error as one standard deviation). Note that the true uniform grade is 0.05 oz/t and that dispersion of estimates arises entirely because of error of estimation. Redrawn from Springett (1989).

0.05 oz/t with an error (standard deviation) of 0.01 oz/t to give the distribution of estimated block grades of Fig. 5.4. For various cutoff grades, the expected average grade of material mined (estimated using Eq. 4.18) is much higher than is actually reported as recovered at the mill (Fig. 5.5). It is apparent that even in this simplistic case, a systematic high bias for estimates of contained metal is introduced by truncation (selection relative to a cutoff grade) despite the unbiased character of the sampling error.

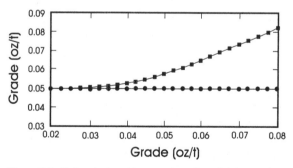

Figure 5.5: Estimated and true grades of production versus cutoff grade for the normal distribution of Fig. 5.4. Filled squares are average grade of blocks estimated (due to random error) as having grade above the true grade. Filled circles indicate true grade of blocks selected as ore. Note that as the cutoff grade increases, the extent of overestimation of grade increases. Redrawn from Springett (1989).

A more realistic detailed example of the impact of various levels of block estimation error on estimated and true copper grades at the Bougainville porphyry copper deposit is documented in Chapter 19. In general, if the cutoff grade is greater than the mean grade of the distribution, the effect of sampling plus analytical error is to increase estimated tonnage and decrease estimated average grade relative to reality. For cutoff grades less than the average grade, the effect is to decrease estimated tonnage and increase estimated average grade relative to reality. In the latter case, the increase in grade is generally slight, perhaps imperceptible.

5.4: SAMPLING PATTERNS

5.4.1: Terminology and Concerns

Traditional sampling patterns are random, random stratified, regular, and irregular (Fig. 5.6). Random

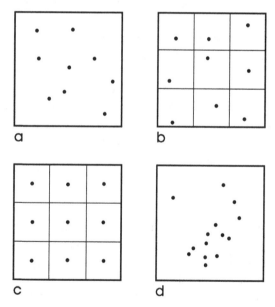

Figure 5.6: Examples of sampling patterns: (a) random sample locations within field V; (b) stratified random samples, with each of n samples located randomly within a small block, v, within field V; (c) regular sampling pattern, with all samples located at nodes of a regular grid with cells summing to V; and (d) clustered (biased) sampling pattern, that is, the sampling is not representative of field V.

sampling rarely is used in mineral deposit appraisals, principally because of the influence of geology on grade distribution. Instead, grid sampling, stratified sampling, or a close approach to these sampling strategies is used to optimize sampling patterns within a framework determined by geologic structures and domains. An important element of this approach generally consists of orienting the most dense sampling direction perpendicular to the plane of greatest geologic continuity, as estimated from the geologic model in vogue at the time of sampling. Such a strategy can lead to several pitfalls because of the assumptions involved; thus, continual review and refinement of the sampling strategy is required as geologic and assay knowledge improves. All regularly spaced sampling grids are subject to the possibility of strong bias because of periodicity in the sampling field. This problem is particularly serious for widely spaced exploration data.

A general problem in all cases of estimation is whether sufficient data are available for the required estimate; obtaining additional data may be desirable, even necessary. As indicated previously, sampling grids commonly provide concentrations of data along more-or-less regularly spaced cross sections that serve as a basis for geologic modeling. When the sampling procedure is drilling, the layout of drill-hole grids directly affects the understanding of both geologic and value continuity, which, because of relatively widely spaced data, can be particularly uncertain in directions perpendicular to the drill sections. Wherever possible, the orientation and plunge of a drill hole is selected roughly perpendicular to the perceived planes of strongest continuity, subject to limitations of surface topography and geologic structure. In consequence, sample points or intersections commonly are far more widely spaced in the horizontal plane; thus, the detailed determination of continuity that is obtained with relative ease and assuredness along the drill-hole axis is not readily obtained in the other two dimensions. In such cases, physical or value discontinuities that are shorter than the drill-hole grid cell may be missed, particularly where actual rock exposures are sparse or absent. For example, experimental semivariograms for the Huckleberry porphyry Cu deposit (Main zone) are illustrated in Fig. 3.6 by ex-

perimental semivariograms in three orthogonal directions. These semivariograms were determined from vertical exploration–drill-hole assay data; down-hole composite samples are spaced about 8 m apart, whereas in the horizontal plane samples are mostly at least six to eight times as far apart. Clearly, the autocorrelation (semivariogram) model is not well known over short distances in two of the three principal grid directions, a situation that could be rectified by gathering appropriate and closely spaced data during exploration.

A relatively widely spaced sampling pattern may be appropriate for deposit delineation and global resource estimates when geologic information is abundant, when a geologic model can be established with confidence, and when value continuity can be shown to lack significant discontinuities. More closely spaced control data are required for local estimation, particularly when the block size for estimation purposes is much smaller than the drill-hole spacing at an early stage of exploration.

These limitations of commonly used exploration drilling patterns are shortcomings for widely used deposit estimation procedures because they directly affect the adequacy of the database for mineral inventory estimation. Such inadequacies cannot be compensated for by mathematic- or computer-based data treatment procedures; instead, they lead to assumptions and more room for unquantifiable error.

Sampling patterns evolve as the deposit evaluation process progresses through various stages. The first sampling pattern (grid orientation, grid spacings, and sample sizes) generally is chosen to delimit long-range geologic continuity and provide a general indication of grade distribution. Early grids commonly are more or less square or rectangular. A substantial literature exists contrasting the efficiencies of square, triangular, and hexagonal grids (e.g., Yfantis et al., 1987), but these highly idealized grids are generally not worth the effort during exploration and evaluation of a mineral deposit because of the overriding control of geology and grade distribution on the need for detailed information. Whatever the initial relatively wide-spaced grid sampling, this work should be supplemented by some local, closely spaced samples in all major directions to provide initial insight into a quantitative, three-dimensional model for short-range, as well as long-range, grade continuity. Short-range information at a relatively early stage of deposit evaluation can be obtained from drill holes with distinctly different orientations and from sampling of surface trenches and areas of surface stripping. At later evaluation stages, additional samples are required for geologic purposes and for local estimation, particularly as deposit definition work approaches the requirements of the mining feasibility stage; some examples of this are denser sampling networks, stricter sampling parameters to ensure high-quality data, and bulk-type samples. Closely spaced samples (e.g., "assay mapping") are required along all the principal dimensions of a deposit so that short-range continuity, possibly anisotropic, can be quantified. Such "infill sampling" may continue the rectangular form common in early sampling plans, although alternative patterns (e.g., triangular and square two-dimensional arrays) are recommended by some authors (e.g., Annels, 1991). This short-range continuity is an essential piece of knowledge on which to base interpolations both at the exploration stage and later, and is too often minimized in many mineral inventory studies.

5.4.2: Sample Representativity

The concept of *sample representativity* refers to minimal bias and acceptable random error, but in common usage, because the term is ambiguous it is used in at least two specific contexts. First, it is used in the sense that a particular type of sample is appropriate for a particular purpose. For example, small-radius drill holes may not provide adequate core samples because of core loss or because the variable in question has a high nugget effect, whereas a larger-diameter core may provide better core recovery and decrease the nugget effect, and thus be more appropriate. Hence, we talk of the representativity of a particular sampling method for a stated purpose. A given sampling protocol appropriate for one deposit may not give adequate (representative) results for another deposit. Vertical drill holes, for example, might not produce a representative sampling where the style of mineral occurrence is as near-vertical veinlets.

A second, perhaps wider use of the term *sample representativity* relates to how well a given assay or set of assay values represents a mass being estimated. In this sense, the question is related not to the sampling method, but to the size of individual samples, the number of individual samples, the sampling pattern (spatial distribution of samples), and to how adequate these factors are in representing the mass to be estimated. A relative measure of the representativity of samples for estimation purposes is difficult to determine because it is based on subjective geologic interpretation as well as implications from assay data themselves. Given that geologic continuity is assured and that an appropriate autocorrelation model can be determined, the relative representativity of a particular sampling pattern and number of samples is given by the global estimation error. Of course, various sampling patterns can lead to essentially equivalent estimation variances, and some of these patterns might contain large local estimation errors.

The aim of sampling is that samples (and their grades) be representative of the larger mass to be estimated. In other words, to qualify as representative, a sample grade must be close to the grade of the block it is meant to represent. Of course, individual samples can vary widely in grade, even over very short distances; consequently, an increase of the mass of sample used to estimate a volume improves the representativity by decreasing the possible grade fluctuations. This increase in mass can be accomplished by increasing the size of individual samples (e.g., increase diameter or length of drill core) or by increasing the spatial density of samples so that more samples are available for estimation purposes within a fixed volume. One global estimate of representativity is an autocorrelation function for samples of a particular support – for example, the semivariogram (see Chapter 9).

Sampling of segregated (stratified) material is a problem commonly encountered in mineral inventory estimation and reconciliation studies. In many situations, broken material with a wide range of sizes is to be sampled. Finely ground material and any heavy mineral particles (perhaps rich in soft or brittle minerals) commonly move physically through interstices to accumulate at the bottom, as in a "muck car," for example. There is obvious difficulty in obtaining a representative sample by a shovel or other kind of cutting system. It is rare that such samples can be taken in a way that guarantees absence of bias unless the entire car of material is the sample. Another common sampling environment is the conical pile of drill cuttings that accumulate at the top of blastholes; these can be strongly stratified in a reverse sense relative to the true stratification in the drill hole. Moreover, piles of cuttings can be deposited asymmetrically because of air turbulence from the drill. Clearly, blasthole cuttings require careful sampling procedures, involving sample supports that cross the stratification and are also representative in the other two dimensions.

A special problem of sample representativity involves the common case of sampling or subsampling broken or granular material to be analyzed for a rare component such as a heavy mineral (e.g., gold, diamonds). As a generalization, the problem can be mitigated to some degree by increasing the sample (subsample) mass. Such samples, each incorporating less than five rare grains, can be considered with a Poisson distribution (Eq. 4.31). Consider a beach sand that contains two grains of gold ($\mu = 2$) per 30,000 total grains (30,000 grains of quartz–feldspar sand is approximately 400 g). Equation 4.31 can be solved for various values of x to give the probability of occurrence of samples containing x grains as follows: the chance of taking a 30,000-grain sample containing three grains of gold is $P(3, 2) = (e^{-2}2^3)/(3.2.1) = 0.180$. Other discrete probabilites can be determined in a similar manner and are listed in Table 5.2. Results such as these, although idealized, are useful because they represent expectations with careful sampling. Note that only about one-quarter of the 30,000-grain samples taken report the correct gold content; a significant number of samples (about 1 in 7) contain no gold; one-quarter report 50 percent of the true average; and the remainder report substantial overestimates. Bold probabilites in Table 5.2 are expected (average) results.

To examine the effect of an increase in sample size, assume that the sample size is doubled. Thus, the number of contained gold particles doubles, so $\mu = 4$. Equation 4.18 can be solved for this situation to produce the results listed in Table 5.2. Note that

Table 5.2 Probabilities that 30,000- and 60,000-grain samples, with respective average contents of two and four gold grains, will contain various numbers of gold particles

Probability		Grains in sample
30,000 grains	60,000 grains	(no.)
$P(0, 2) = 0.135$	$P(0, 4) = 0.018$	0
$P(1, 2) = 0.270$	$P(1, 4) = 0.073$	1
$P(2, 2) = 0.270$	$P(2, 4) = 0.147$	2
$P(3, 2) = 0.180$	$P(3, 4) = 0.195$	3
$P(4, 2) = 0.090$	**$P(4, 4) = 0.195$**	4
$P(5, 2) = 0.036$	$P(5, 4) = 0.156$	5
$P(6, 2) = 0.012$	$P(6, 4) = 0.104$	6
$P(7, 2) = 0.0034$	$P(7, 4) = 0.060$	7
$P(8, 2) = 0.0009$	$P(8, 4) = 0.030$	8

the range $\mu \pm 0.5\mu$ now contains 82 percent of the possible samples, whereas with the smaller sample this range contained 72 percent of possible samples. The increase in sample size has "squeezed" both tails of the distribution toward the true mean value.

Sample representativity also can be evaluated by comparing the measured metal contents of successive parts (unit volumes) of a bulk sample with corresponding estimates of those units based on available sample/assay information (John and Thalenhorst, 1991). A one-to-one relation between estimated value of unit volumes and true metal contents is not to be expected because of conditional bias (Section 4.8).

5.5: SAMPLING EXPERIMENTS

5.5.1: Introduction to the Concept

Sampling procedures such as those referred to briefly in the introduction to this chapter are too commonly accepted routinely without rigorous checking of their viability. Sampling methods found acceptable for one deposit type might be totally unsuitable for another deposit. Sample sizes can differ significantly among sample types – for example, a segment of a AX diamond drill core 15 ft long has a mass of about 0.004 short tons, whereas cuttings from a 12-in diameter blasthole in a mine with a 15-m bench height can be between three and four short tons. Systematic tests, required to help optimize sampling procedures, need not be costly. Too often, we forget to review the implications of the various parameters and procedures associated with a standardized sampling method such as diamond drilling. In some cases, a uniform sample grid size or orientation may not be appropriate for all domains or zones of the deposit.

A common source of ambiguity in interpreting and utilizing quantitative data is that several sampling methods have been used during the evaluation of a deposit. For example, evaluation work can involve linear samples (diamond drilling, rotary drill or channel sampling), surface (two-dimensional) samples (panels), and large-volume samples (trench samples, mining round samples, and bulk samples). In places, some of these sample types may represent different areas of the deposit with little or no overlap; hence, systematic data to make comparative studies and check for representativity (relative random error, possibility of bias) of sampling methods may be lacking. Even when two or more sample types are interdispersed in the same domain there can be difficulties of interpretation to the point of serious uncertainty if the level of accuracy of the various sample types is not known. Establishing the relations among different sample types requires a systematic approach to data gathering; in particular

(i) Areas of overlap of the various sampling methods that are large enough, with numbers of samples sufficient for recognizing correlation, for example, diamond drilling an area of a bulk sample, or directing drill holes parallel to channel samples

(ii) Systematic statistical and geostatistical controls of each sample set by itself and relative to other data sets involved.

5.5.2: Comparing Sampling Procedures at Equity Silver Mine

In mine production, quality control of sampling often can be improved. Of course, accumulated experience in a producing mine offers an initial measure of quality control, but too few operations try to improve the quality of routine sampling procedures beyond this starting point. As an example of the potential impact of sampling method on estimates (and thus on productivity), consider the case history of Equity Silver Mine, central British Columbia, where mine personnel conducted a small but revealing test of several methods of sampling production blasthole cuttings. The general problem of sampling blasthole cuttings is widely recognized, and a selection of practical sampling approaches is presented by Annels (1991).

The blasthole sampling procedure used at the Equity Silver Mine, a *tube (pipe) sampling method* somewhat modified from that described by Ewanchuck (1968), involves shoving a 3-in diameter tube (pipe) into four spots symmetrically located in the cuttings pile (Fig. 5.7), upending the tube so that material remains in it, and transferring the material in the tube to a sample bag. A second method, *channel sampling* (with a 4-in-wide shovel) along four symmetrically distributed radii of the cuttings pile to produce four separate samples and a composite channel sample, was to be tested. Finally, the total bulk of remaining cuttings was taken as a sample to provide a means of mass balance for the entire blasthole cuttings pile. Because all the cuttings from a particular drill hole were used as sample material, it is possible to produce a weighted average grade (weighted by sample mass) – that is, a best estimate of the true value of the cuttings pile. Results of individual sampling methods can be compared with this best estimate (Table 5.3 and Fig. 5.8). Calculations indicate that the use of channel sampling at Equity would have produced a dramatic improvement in assay quality, and thus a significant improvement in ore/waste classification and a corresponding increase in profit. The decreased sampling error by channel sampling is particularly evident in x–y plots of analytical data for tube and channel sampling versus the best estimate (Fig. 5.8).

Some of the important results of this sampling experiment, based on a cutoff grade of 50 g Ag/t and the assumption that the best weighted value is correct, include the following:

(i) Tube sampling of 42 cuttings piles misclassified 7 samples: 4 ore samples as waste and 3 waste samples as ore.

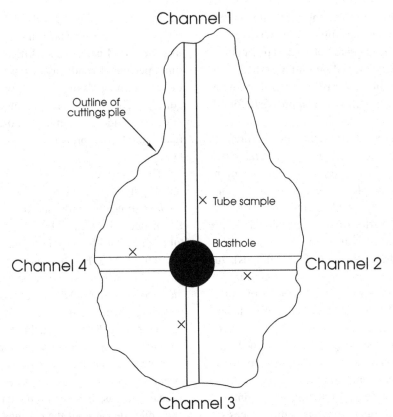

Figure 5.7: Idealized plan view of a pile of cuttings surrounding a blasthole. *X*s mark locations of four scoops (using a tube) combined to produce a tube sample. Four channel positions are shown relative to the elongated form of most piles of cuttings. (when the drill travels to the next hole, the bit drags through the cuttings, showing the direction of drill travel). Redrawn from Giroux et al. (1986).

(ii) Channel sampling of 42 cuttings piles misclassified a single sample (and block) of ore as waste.

(iii) Of the 18 samples identified as ore by tube sampling, metal content was overestimated by about 9 percent (Table 5.4).

(iv) Of the 18 samples identified as ore by channel sampling (not the same as the 18 identified by tube sampling), metal content was overestimated by about 5.2 percent (Table 5.4).

(v) There are hidden losses to each sampling method resulting from misclassification of ore as waste (i.e., lost operating profit from ore classed as waste). In the case of tube sampling, three ore samples were lost because they are incorrectly classed as waste. For channel sampling, the corresponding loss was one sample.

Additional well-documented examples of the design of practical experiments to evaluate sampling procedures and sample reduction protocols include the Colosseum gold mine (Davis et al., 1989), the Mount Hope Molybdenum Prospect (Schwarz et al., 1984), and the Ortiz gold mine (Springett, 1984).

5.5.3: Sampling Large Lots of Particulate Material

The sampling of two- and three-dimensional piles of broken material (e.g., crushed ore/waste, low-grade stockpiles, tailings, dump material) presents serious problems of potential bias and, generally, economic

Table 5.3 Assay data, blasthole sampling experiment, Equity Silver Mine (see text)

	Tube		Channel		Bulk		Best	
	Cu	Ag	Cu	Ag	Cu	Ag	Cu (%)	Ag (g/t)
1	0.14	5	0.25	18	0.26	22	0.246	20.0
2	0.06	20	0.05	22	0.04	15	0.043	17.9
3	0.02	11	0.02	6	0.01	4	0.014	4.98
4	0.08	123	0.07	103	0.07	101	0.068	102.4
5	0.03	36	0.06	70	0.04	54	0.046	59.9
6	0.69	274	0.97	393	0.87	389	0.879	387.5
7	0.69	148	0.6	163	0.52	151	0.540	155.7
8	0.12	10	0.1	8	0.16	20	0.132	15.0
9	0.06	76	0.06	62	0.05	46	0.053	53.1
10	0.11	85	0.15	142	0.14	117	0.139	126.0
11	0.03	28	0.03	32	0.04	47	0.035	40.5
12	0.02	27	0.03	27	0.04	37	0.035	32.8
13	0.28	108	0.07	55	0.03	50	0.051	53.5
14	0.07	76	0.09	98	0.09	86	0.087	90.5
15	0.09	126	0.09	160	0.12	178	0.105	169.5
16	0.06	42	0.07	35	0.06	48	0.062	42.7
17	0.47	11	0.44	131	0.38	11	0.394	58.5
18	0.56	325	0.43	217	0.45	226	0.433	225.1
19	0	0	0.02	5	0.03	7	0.025	6.0
20	0.14	78	0.09	45	0.08	40	0.083	43.0
21	0.33	199	0.24	123	0.3	147	0.271	138.9
22	0.07	31	0.07	28	0.05	23	0.057	25.2
23	0.04	19	0.04	18	0.03	14	0.033	15.7
24	0.04	11	0.03	10	0.03	10	0.029	10.0
25	0.04	29	0.02	18	0.03	21	0.026	20.0
26	0.42	220	0.39	233	0.43	268	0.403	252.8
27	0.18	52	0.11	21	0.09	16	0.097	19.0
28	0.02	301	0.76	175	0.79	181	0.737	181.9
29	0.24	43	0.27	89	0.21	74	0.227	79.1
30	0.02	10	0.02	9	0.02	11	0.019	10.2
31	0.19	15	0.15	26	0.16	24	0.15	24.5
32	0.24	4	0.19	4	0.23	3	0.209	3.4
33	0.15	141	0.25	240	0.17	167	0.194	195.2
34	0.06	51	0.03	22	0.03	21	0.03	22.2
35	0.02	16	0.06	57	0.07	66	0.063	61.1
36	0.03	22	0.01	10	0.02	13	0.016	12.1
37	0.07	6	0.03	12	0.01	7	0.019	9.0
38	0.05	28	0.08	45	0.11	78	0.094	63.6
39	0.1	55	0.03	17	0.05	26	0.043	23.2
40	0.2	68	0.19	68	0.16	54	0.168	59.9
41	0.04	9	0.06	18	0.04	8	0.046	12.0
42	0	2	0.01	2	0.01	2	0.009	2
MEAN	0.149	70.0	0.160	72.3	0.155	68.6	0.153	70.1
ST. DEV.	0.1776	83.67	0.2096	83.94	0.1997	83.10	0.1956	82.30
CV	1.190	1.20	1.31	1.16	1.29	1.21	1.28	1.17

Table 5.4 Comparison of overestimation of blasthole grades by two sampling techniques, Equity Silver Mine

Sampling method	Ore samples (no.)	Average estimate	Grade true	Overestimation (%)
Tube	18	139.2	127.7	9.01
Channel	18	143.2	136.1	5.22
Best	19		132.3	

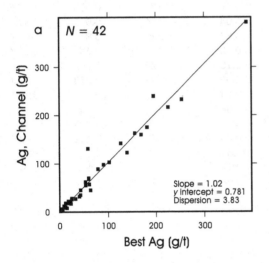

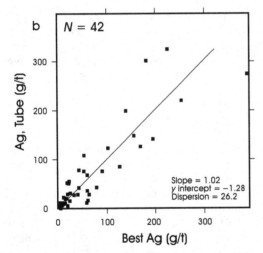

Figure 5.8: (a) Average channel-sample Ag assay (g/t) versus weighted average of blasthole cuttings pile (best Ag). Note the substantial reduction in spread of plotted data for channel samples in comparison with original tube data. (b) Original tube-sample Ag assay (g/t) versus weighted average for all blasthole cuttings (best Ag) for 42 sample pairs analyzed for Ag, Equity Silver Mine, Ltd. An RMA line (parameters shown) is essentially equivalent to the $y = x$ line. Redrawn from Giroux et al. (1986).

constraints limit the amount of sampling that can be carried out. Gy (1979, p. 357) states, "Practically . . . the sampling of three-dimensional particulate objects is a generally unsolvable problem and repeats essentially the same statement with regard to two-dimensional piles (i.e., sheetlike or tabular piles). The best general solution is to deal with what Gy (ibid) calls one- or zero-dimensional modifications of two- and three-dimensional piles. A one-dimensional moving stream of fragmented material is generally a solvable sampling problem; the material can be sampled by taking uniform volumes at regular intervals from the moving stream (e.g., a conveyor belt).

A lot that is naturally divided into a large number of equal units of practically uniform weight (Gy, 1979, p. 359) is said to be zero dimensional if primary sampling consists of a selection of a certain number of these units (Gy, 1979, p. 359). Units are such uniform volumes as bags, barrels, and truckloads. When handling large tonnages under the form of zero-dimensional objects in a routine way, the most accurate and the cheapest of all solutions consists in selecting for instance one unit out of 10 or 20 (primary sample) according to a systematic or stratified scheme, in discharging the increment-units into a surge bin (Gy, 1979, p. 361). Clearly, sampling of zero- or one-dimensional lots (masses) involves moving the entire lot.

5.6: IMPROVING SAMPLE REDUCTION PROCEDURES

It is surprising that low-skilled and low-paid workers are often involved in the critical process of sampling and preparation of samples for assay, particularly in light of the millions of dollars that might be dependent upon the results. (Kennedy and Wade, 1972, p. 71)

Sample reduction (subsampling protocol) is the totality of procedures that are used to extract a much smaller representative amount for actual analysis from a large sample volume. A mineralized sample commonly is a mixture of relatively large but variably sized fragments of solid material that must be reduced in both particle size and weight to a small amount of finely ground material that is analyzed to determine metal content. This overall procedure, known as a *subsampling protocol*, involves a series of steps of alternating particle size reduction and mass reduction that can be demonstrated by a simple, purely illustrative example. Suppose that a sample consisting of 1 m of half-core weighs 2,700 g and consists of fragments up to 10 cm in length. The sample might be crushed so that the maximum particle diameter is 0.5 cm. Then the crushed material is homogenized and a portion is taken – say, one-quarter of the original sample, perhaps by riffling. This smaller portion ($2{,}700 \times 1/4 = 675$ g) is then further crushed or ground to a much smaller particle size, and again the material is homogenized and a fraction is taken, perhaps one-quarter of the material ($675 \times 1/4 = 168$ g). The remaining three-quarters of material (about $675 - 168 = 507$ g) at this stage might be saved as a reject and the one-quarter taken is further ground to provide a pulp, part of which is analyzed. Assuming no loss of material during size reduction, the amount of material forming the pulp is $\frac{1}{16}$ of the original sample, or $2{,}700/16 = 168$ g, of which perhaps 30 g is measured out for actual analysis. The reject and unused pulp commonly are retained for a specified period of time, perhaps one or two years, in case they are required for quality control purposes.

It is wise to have a comprehensive understanding of the mineralogy of ore and gangue minerals as a basis for developing a sample reduction scheme. An example showing many steps in the reduction of a bulk sample (20 tons) to material for assay is illustrated in Fig. 5.9. The procedure must ensure adequate homogenization at each stage of mass reduction in order to

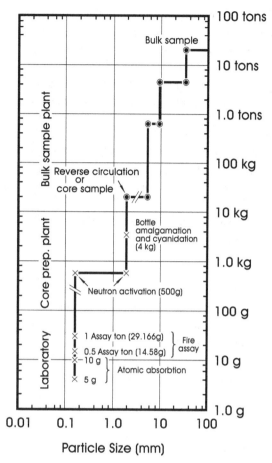

Figure 5.9: Sample reduction diagram for a bulk sample taken from the Ortiz gold deposit. Note the steplike character that reflects alternate reductions in sample mass and particle diameter in order to reduce a large sample volume (ca. 20 tons) to a volume (grams) that can be analyzed. Redrawn from Springett (1983).

eliminate bias and minimize random errors. Sample reduction schemes used in operating mines commonly can be designed to provide high-quality subsamples, and therefore high-quality assays. The normal result of an inadequate sample reduction system is a large random error (sampling plus analytical error) in assays; of course, biases and procedural errors are also possible. These large errors contribute to a high nugget effect and possible masking of the underlying structure (ranges) of autocorrelation. At the very least, they lead to larger than necessary errors in block estimation and in defining ore limits; thus, they contribute to ore/waste misclassification problems, loss of metal, and difficulties with mass balance reconciliations.

Gy (1979) addressed this problem at length in developing a sampling equation that has been widely accepted in the mineral industry (e.g., Springett, 1984), with significant modification in the case of some gold deposits (Francois-Bongarcon, 1991). A practical means of applying Gy's ideas is through the use of a sampling (or sample reduction) diagram on which a safety line, based on Gy's fundamental sampling error equation, is plotted as a guide to acceptable sample reduction procedures (Fig. 5.10). When little is known of the detailed characteristics of an ore, Gy's general safety line can be used as a guide; when properties of an ore are reasonably well understood, a safety line tailored to a specific ore can be determined.

On a sample reduction diagram, the sample path is plotted as a series of connected straight lines representing alternating particle size reduction stages (crushing, grinding, pulverizing) and mass reduction stages (subsampling). In such sample reduction

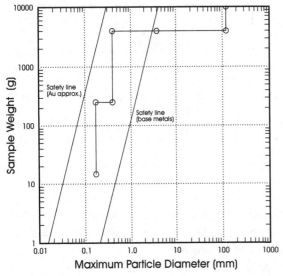

Figure 5.10: Sample reduction diagram for blasthole samples, Nickel Plate Gold Mine. Two safety lines are shown: the base metals line is after Gy (1979), and the Au safety line is an approximate, arbitrary line that emphasizes the relative difficulty of generating gold versus base metal analyses with low error. After Radlowski and Sinclair (1993).

schemes, it is important that the initial particle size reduction stage be sufficient to move the sample position to the left of a safety line (representing a predetermined maximum allowable error) and that subsequent stages not cross to the right of the safety line. When sufficient information is available, a safety line can be determined for each ore (see later); otherwise, a generalized safety line can be used. Further details of the procedure are provided by Gy (1979), Pitard (1989, 1994), and Francois-Bongarcon (1991). The example in Fig. 5.10 is for rotary-drill cuttings and exploration diamond-drill core for the Nickel Plate Mine, Hedley, British Columbia (Radlowski and Sinclair, 1993).

Gy's (1979) simplified sampling equation is as follows:

$$(1/M_S - 1/M_L) = Cd^x/s^2$$

where

M_S(grams)	is the weight of the sample
M_L(grams)	is the weight of the lot being sampled
C(g/cm^3)	is the sampling constant
d(cm)	is the maximum particle diameter (mesh size retaining upper 5 percent)
s (fraction)	is the relative fundamental sampling error
x	is the exponent (commonly $= 3$).

The equation links the variables s, M_S, M_L, and d, and can be solved for one if the others are fixed. Commonly, M_L is very large so that $1/M_L$ is negligible and only two variables must be fixed to solve for a third. The exponent (x) and sampling constant (C) must be estimated prior to undertaking calculations; determination of C requires substantial insight into the character of the material being sampled. The exponent x is commonly taken as 3.0 (Gy, 1979); for some gold deposits, it is 1.5 (Francois-Bongarcon, 1991).

The sampling constant (C) can vary over a wide range, from 10^{-4} to 10^{+4} g/cm^3, and should be determined carefully for each particular type of fragmental material. For very preliminary evaluations, estimation of C can be avoided by reference to Gy's standard safety line for base-metal sampling (i.e., $C = 0.08 \text{g/cm}^3$). In practice, the estimation of C is difficult, but is critical to the successful application of Gy's equation (see Sketchley, 1998; Francois-Bongarcon, 1998). A crude approach is to determine an "effective" C empirically by solving the equation in controlled situations in which all other variables are known, for example, when paired data can be used to determine the relative error. A much more detailed, time-consuming, and costly approach involving a "heterogeneity" study (Gy, 1979; Pitard, 1989) is more accurate and may be warranted. Alternatively, the sampling constant can be estimated with some ambiguity, as described later.

The sampling constant is given by the equation

$$C = m \cdot L \cdot f \cdot g$$

where

m [g/cm^3]	is the mineralogic composition factor
L (0.001–1)	is the liberation factor
f (0.2–0.5)	is the particle-shape factor
g (0.25–1)	is the size-range factor.

As Francois-Bongarcon (1991) pointed out, Gy's sampling constant, C, is not truly constant, but depends on the value of L, as indicated in the following.

5.6.1: The Mineralogic Composition Factor (m)

The mineralogic composition factor

$$m = \frac{1-a}{a}[(1-a)\rho_c + a\rho_g]$$

where

a	is the critical component (mineral) as a weight fraction of the whole

(e.g., in a zinc ore for which sample reduction is being evaluated to obtain acceptable quality Zn assays, the critical component is sphalerite; however, for gold in the same ore, the critical component is the mineral hosting the gold)

ρ_c [g/cm^3]	is the density of the critical component
ρ_g [g/cm^3]	is the density of the gangue component.

5.6.2: The Liberation Factor

The liberation factor ($L = 0$–1.0) is

$$L = [d_L/d]^b$$

unless d exceeds d_L, in which case $L = 1$, where d_L [cm] is the liberation diameter of the critical component (difficult to estimate); and, d [cm] is the actual diameter of the largest particle and b (exponent) is determined experimentally (Francois-Bongarcon, 1998).

5.6.3: The Particle Shape Factor

The particle shape factor ($f = 0.2$–0.5) is

$$0.2 \leq f \leq 0.5$$

($f = 0.5$ assumed for most applications; $f = 0.2$ for gold ores).

5.6.4: The Size Range Factor

The size range factor ($g = 0$–1.0) is

$$g \text{ is a function of } d/d_s$$

where d [cm] is the upper size limit (5 percent oversize) and d_s [cm] is the lower size limit (5 percent undersize).

Empirical estimates used in practical applications (Gy, 1979) are

Large size	$d/d_s > 4$	$g = 0.25$
Medium size	$4 > d/d_s > 2$	$g = 0.50$
Small size	$2 > d/d_s > 1$	$g = 0.75$
Uniform size	$d/d_s = 1$	$g = 1.00$.

5.6.5: Applications of Gy's Equation

Gy's formula is especially useful in designing appropriate subsampling procedures and in the initial evaluation of an existing subsampling protocol. However, the formula calculates an ideal fundamental sampling error that is only approached in practice, and the formula should not be used to replace the use of replicate data to determine and monitor the actual error inherent in a particular sampling protocol.

5.6.6: Direct Solution of Gy's Equation (Simplified Form)

Relative fundamental sampling error(s) can be calculated if the following are given:

(i) Exponent (exp) and sampling constant (C), estimated as discussed previously
(ii) Sample weight (M_S in grams)
(iii) Maximum particle diameter (d in cm).

For example, such information is available from the flow sheet describing sample reduction procedures in assay laboratories. Note that in many cases, M_L is so large relative to M_S that $1/M_L$ is negligible.

Minimum sample weight required, M_S, can be calculated if the following are given:

(i) Exponent (exp) and sampling constant (C), estimated as discussed previously
(ii) Maximum particle diameter (d)
(iii) A known or targeted relative sampling error.

Maximum particle size tolerated (d_o) can be calculated if the following are given:

(i) Exponent (exp) and sampling constant (C), estimated as discussed previously
(ii) Sample weight (M_S)
(iii) Known or targeted relative sampling error.

5.6.7: User's Safety Line

Gy (1979) introduced the concept of a safety line on a sample reduction diagram as a means of quality control. Provided that mass reduction and particle size reduction procedures keep a subsample reduction path to the left of a safety line, then the subsampling system involves a fundamental error at least as low as that used to define the safety line. An equation for a safety line can be determined from Gy's equation, as follows:

$$M_d = k \cdot d^x$$

where k is equivalent to (C/s^2) and x is an exponent, normally taken as 3. The function M_d versus d appears as a straight line on a log–log graph of particle size versus sample mass. C can be estimated as discussed in Section 5.6, and a safety line can then be drawn for a desired limiting error (s). Examples of safety lines for base metal and gold deposits are shown on Fig. 5.10. The base metal safety line is from Gy (1979) and is based on $k = 125,000$. If cost is more important than precision, lower k to 60,000; if precision is more important, increase k to 250,000.

5.7: ASSAY QUALITY CONTROL PROCEDURES

The problem of analyzing Au in copper rich ores is not new or unusual. Any heavy metal in large quantities can hinder the analysis of Au by the traditional Fire Assay Lead Collection procedure. However, an experienced Fire Assay Chemist can usually get around this problem by either lowering the effective sample weight (increase flux to sample ratio), or by pre-leaching the sample. (R. Calow, personal communication, 1998)

5.7.1: Introduction

The use of Gy's equation for fundamental error control, using a safety line on a sampling diagram, represents a somewhat idealized expectation that is approached in practice but may not be attained. A quality control program is necessary in order to know and to monitor variations in data quality. For this purpose, the concept of an *analytical system* (Thompson and Howarth, 1976) is useful. An analytical system is composed of the following:

(i) The set of samples with the analyte in a specific matrix
(ii) The exactly defined analytical procedure
(iii) The particular instruments used.

A corollory of this concept is that the samples for a system should be drawn from a homogeneous type, otherwise one may be attempting to measure a meaningless average precision between two distinct systems. (Thompson and Howarth, 1976, p. 692)

Laboratories usually maintain a system of checking for subsampling and analytical errors. This normally involves periodic reanalysis of internal and international standards. These data should be reported routinely to clients to allow them to evaluate it.

In general, laboratory users should invoke their own systematic approach to quality control involving the taking of duplicate samples (perhaps 1 in every 20 samples should be analyzed in duplicate), the use of project standards, systematic interlaboratory checks, use of different methods for particular analytical situations (e.g., metallic assays for high-grade gold ores), and reassaying of unexpectedly high (or low?) values. Such procedures provide information pertinent to the recognition of bias and the magnitude of random errors. It is important that the results of all these quality control measures be provided to those conducting mineral inventory estimates in a concise and informative way. For all this concern, however, errors arising in the laboratory are generally small compared to errors that arise from sampling procedures.

5.7.2: Using the Correct Analyst and Analytical Methods

Only scientifically tested and generally accepted analytical procedures are appropriate ways with which to generate assay data for use in mineral inventory estimation. "Black-box" methods for which there is no obvious scientific explanation are inappropriate, and results based on such methods are unacceptable.

Bacon et al. (1989) provide a useful discussion of analytical procedures, particularly as they apply to precious metal deposits, and conclude that traditional fire assaying, done by an accredited professional assayer, provides the best approach to precious metal assaying. Only a few jurisdictions maintain a rigorous, formal testing procedure for registering assayers. Hence, it is important to know the reputation of an assayer as well as inquiring into the assayer's qualifications and the general validity of the analytical methods used. The assay laboratory itself warrants an inspection – labs should be clean with an appropriate ventilation system (not too strong, not too weak) in the preparation room and a well-organized reject storage facility. A well-conceived quality control system should be in place.

For many metals, analytical procedures are well established. Problems can arise in special cases (e.g., if instrumental analysis is used to measure uranium content, there is an assumption of equilibrium in the decay scheme of uranium). In certain deposits, this equilibrium is attained only imperfectly and caution is required. When chloride dissolution methods are used for coarse-grained native silver, the silver can become coated with silver chloride, which inhibits dissolution; in such cases, it is important to maximize the surface area of the native silver. Coarse native gold (grains larger than about 100 μ in diameter) represents a common and difficult problem that requires special attention, perhaps involving "metallic" assays (see later).

Even when an acceptable analytical method is used, it is important to maintain internal checks on the quality (including accuracy) of data being provided by a laboratory. A routine system of submission of standards should be followed, as well as having samples checked by another reputable laboratory. An example is the significant bias that can enter fire assay data because of peculiar sample compositions not suited to the fluxes in routine use by a laboratory. Such problems can be identified by a systematic quality control program of duplicate sampling and analyses. Figure 5.11 is a comparison of assay results by two laboratories for Cu and Ag values for tailing, ore, and concentrate samples from the Equity Silver Mine (Giroux et al., 1986). Clearly, both laboratories, on average, produced similar results for Cu, with

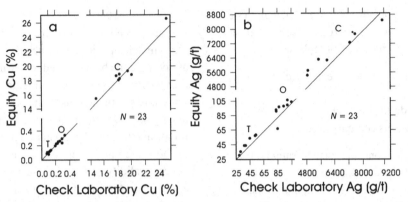

Figure 5.11: Plots of duplicate analyses for 23 samples by two different laboratories for (a) copper and (b) silver at Equity Silver Mine. The 45° lines are equal value reference lines. Samples are tailing (T), ore (O), and concentrates (C). The two labs produce good agreement for Cu values, but slightly different results for Ag.

relatively slight scatter (low random error); in the case of Ag, however, even qualitative inspection reveals a small bias, which can be demonstrated statistically in several ways. Without further insight, it is not possible to know which laboratory is systematically in error (i.e., does one systematically overestimate, does the other systematically underestimate, or are Ag assays from both incorrect?).

The grinding procedure/equipment used in sample preparation can contribute to contamination either from the grinding equipment itself or from smeared particles retained from previous samples (cf. Hickson and Juras, 1986). Cross contamination can be a serious problem, for example, if samples bear native gold. For such materials, cross contamination can be prevented by grinding "blank" material between samples and then cleaning the equipment with nylon brushes and compressed air.

Some problems of quality are specific to specific elements. Coarse gold, for example, may require that samples be examined to identify and collect coarse gold for a "metal" analysis that is then weighted with the corresponding "rock" analysis to give a final assay result. The procedure could involve

(i) Examination of sample descriptions (e.g., drill-hole logs) for reference to visible gold
(ii) Passing the sample through a shaking table or use of a vanning shovel for visual assessment

(iii) Reexamination and reassay of samples in which free gold was not recognized but for which the assay value is abnormally high.

Two graphic approaches are useful for assistance in monitoring and quantifying accuracy of laboratory assay results; routine analyses of standards can be examined regularly on binary plots of assay versus time or assay versus known value. Time-dependent plots provide a check on systematic variations that can result from new staff, equipment problems, variations in quality of chemicals, and variations in the physical operating conditions in the chemical laboratory. Biases and abnormalities show up clearly (Fig. 5.12). Value-dependent plots are particularly useful in documenting the presence or absence of bias and for comparing results by various analytical techniques or from different laboratories. Figure 5.12a, a time-dependent plot for two molybdenum standards, reveals systematic variations over time, as well as a general tendency for the analytical method to underestimate both samples relative to the known values (arrows on the y axis). Value-dependent plots (e.g., Fig. 5.13) show the means and standard errors plotted versus the differences from the corresponding known values. In the case illustrated, replicate analyses of five molybdenum standards using a two-acid dissolution method are seen to underestimate consistently the true values of standards determined by a four-acid

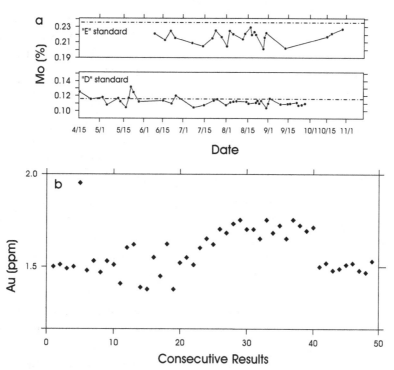

Figure 5.12: (a) Time plots of analytical data for two Mo standards, E and D (redrawn from Schwarz et al., 1984). The expected (known?) values with which these replicates should be compared are 0.2347 ± 0.0184 for E, and 0.1160 ± 0.0038 for D. It is common practice to show the true value and confidence limits of the standard as lines parallel to the x axis. Analyses of both standards are biased on the low side. (b) Idealized time plot to illustrate interpretation of patterns (modified from Smith, 1999). In all cases, assume an expected value of 1.5 ppm Au. Results 0–10 are accurate and precise except for one outlier; results 11–20 are accurate but not precise; results 21–30 show a systematic trend that warrants investigation; results 31–40 are precise but not accurate; results 41–50 are both precise and accurate.

dissolution technique; the bias ranges from almost zero at low values to almost 10 percent in the case of the highest grade standard.

The possibility of fraud involved in the production of assay results must be considered. A reasonable level of security is essential, beginning at the time samples are taken, during transport to the assay laboratory and during treatment in the laboratory. There are numerous examples in which laxity in these matters has led to fraudulent practices that took far too long to uncover and resulted in public announcements of totally erroneous mineral inventory estimates. Frequent independent audits of sampling and analytical procedures are the best safeguard against fraudulent practice.

Once data of acceptable quality are obtained, there remain many opportunities for human errors to creep in accidentally; for example, incorrect plotting, reversal of figures in transcribing data, and order of magnitude errors. Many such errors can be identified by examining sorted tabulations of data for outliers, noting extreme values on postings as plans or sections, comparing similar plots done independently, and examining a variety of common graphical outputs (histograms, probability plots, scatter diagrams) obtained routinely in the normal course of data evaluation.

5.7.3: Salting and Its Recognition

Salting is the surreptitious introduction of material into samples (McKinstry, 1948, p. 67). Salting of possible ore material is an unfortunate occurrence sometimes encountered in the evaluation of some gold deposits, and its presence or absence must be assured

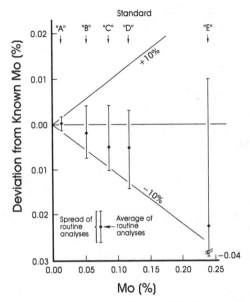

Figure 5.13: Plot showing deviation of five Mo standards from their expected values. Diamonds are the averages for each standard; vertical lines show the ±2s range. Proportional 10 percent error lines are shown. Redrawn from Schwarz et al. (1984).

during the process of data verification. Unfortunately, salting can take many forms (e.g., McKinstry, 1948), from contaminating the rock in-place to surreptitious introduction of extraneous metal at some stage of the sampling protocol or during procedures at the analytical laboratory. Where core drilling provides the great bulk of perhaps tens of thousands of samples, the impracticality of salting the ore in-place on a significant scale is apparent. However, the possibility of salting at some stage of the sampling protocol is always a concern. In the case of gold deposits, the record shows that historically, salting has been achieved by adding material to samples, including the addition of gold shavings, placer gold, or a solution such as gold chloride. Such salting might occur when the sample is first taken, at some intermediate stage in the sampling protocol, or in the analytical laboratory.

When salting occurs, its identification is essential to deposit evaluation and procedures must be implemented to either identify or ensure the absence of salting. Such procedures are part of a general data verification program. Checking sampling procedure and results is therefore essential as a means of iden-

tifying the presence or absence of salting. There are a number of approaches that can be used to corroborate analytical data for this purpose, including the following:

(i) Sampling protocols must be examined in intricate detail to identify those stages most susceptible to salting
(ii) Duplicate sampling from accessible sites (e.g., remaining drill-core halves)
(iii) Comparison of production with estimates
(iv) Microscope examination of rejects and pulps
(v) Twinned drill holes
(vi) Bulk sampling from surface sites
(vii) Exploration drives, including bulk sampling.

In general, the purpose of these undertakings is to verify the quality of reported grades or provide evidence of salting. Consider each of these approaches in turn.

The examination of sampling protocols involves a critical review of the established chain of custody (e.g., Kratochvil and Taylor, 1981) and target hardening procedures (e.g., Rogers, 1998). A *chain of custody* is the detailed procedure that each sample follows such that "the integrity of samples from source to measurement is ensured" (Kratochvil and Taylor, 1981, p. 928). The chain of custody normally is documented as part of the sampling protocol with emphasis on the security incorporated into the protocol. *Target hardening procedures* are those specific procedures designed to overcome perceived weaknesses in a chain of custody. Weaknesses in the security measures that form part of a chain of custody are generally evident to geologists and mining engineers experienced in the evaluation of mineral deposits.

Duplicate sampling from accessible sites is a time-honored procedure for identifying or offsetting salting (e.g., McKinstry, 1948). This could take the form of analyzing the retained or second half-cores for comparison with existing analyses for the corresponding first halves. Standard techniques can then be used to compare original and duplicate sampling results to test for the presence of bias. If a significant difference was noted between two independently obtained sets of analytical results, normal procedure would be to undertake check analyses and, in certain cases, to involve an umpire laboratory. This approach is the best way to guarantee the presence or absence of significant salting in a large data set because it need not be restricted to a small part of the sampling–subsampling–analytical protocol.

"Assay returns which give a very much higher average than records of past production may or may not be correct but in any case they call for an explanation" (McKinstry, 1948, p. 69). Reconciliation of mine production data with estimates, particularly estimates based on diamond-drill data, is difficult in the best of cases because

(i) There are large uncertainties in estimating dilution during mining.
(ii) Identification of ore blocks during mining uses a different, more comprehensive database than is used for making block estimates (e.g., blasthole data vs. diamond-drill-hole data).
(iii) Individual blocks of ore cannot be monitored through the mill, hence their true grade cannot be measured.
(iv) Ore from more than one working face is put through the mill over a short time period, the block estimation method is biased, and so on.

Consequently, reconciliations of production with estimates is an ambiguous means of identifying problems with the underlying database unless very detailed evaluations of other factors can be incorporated into the overall reconciliation.

In some cases, when the introduction of gold to samples is as particulate gold and has taken place prior to the separation of the pulps and rejects, the pulps and rejects can be examined visually and microscopically for the presence of gold shavings or placer gold grains. Such examinations would not provide insight into salting by a gold-bearing solution, nor would they give direct evidence of salting that occurred at a later stage of the subsampling procedure or during analysis.

Twinned drill holes, bulk sampling, and underground drives are all relatively expensive and time-consuming methods of checking the validity of data. Any or all of these grade verification procedures can form part of a due diligence. It should be noted that these three methods are directed toward verifying the general representativity of samples rather than identifying individual samples that are unrepresentative of the true grade. Consequently, they are used mostly when data validity remains a problem after other methods have been used. Of these three approaches, the use of twinned drill holes is by far the most widely used method to verify the general magnitude of grades.

5.8: A PROCEDURE FOR EVALUATING PAIRED QUALITY CONTROL DATA

5.8.1: Introduction

A program of duplicate sampling and assaying is a routine, quality control/quality assurance undertaking in a sampling program designed to accumulate information on which to base a mineral inventory. Such a program should include (cf. Thompson and Howarth, 1976) the following:

(i) Large numbers of duplicates that can be used to provide stable statistics
(ii) Duplicates that span the entire range of values of material being analyzed
(iii) Assurance that the materials selected for duplicate analysis are representative of the bulk of material being analyzed
(iv) Quality control samples that are not recognizable by the analyst.

The evaluation of duplicate sampling data requires thoroughness and regularity in examining paired data, both graphically and statistically. It is essential to test systematically for bias and accuracy and differences in these errors as a function of composition. Histograms and scatter plots are two familiar and simple graphics tools that can be used to aid the data evaluation process.

5.8.2: Estimation of Global Bias in Duplicate Data

Global or average bias in duplicate data can be evaluated by examining differences between paired data values (e.g., original analysis minus repeat analysis). These differences can be viewed in a histogram to appreciate the symmetry or asymmetry of the distribution of the differences, and statistics can be used to conduct a formal statistical test for the existence of bias. When bias is indicated, such tests in themselves do not indicate which of the original or replicate data are incorrect; additional information is necessary to reach a decision as to whether either or both contain bias.

Histograms of a variable are a simple graphics tool that, for a single variable, illustrate the range of values, the range of the most abundant values, and the general disposition of values relative to the mean (i.e., the shape of the distribution). *Histograms of differences* (and the corresponding means and standard deviations) provide insight into the presence and level of global bias between the two components of a set of paired data. In addition, they demonstrate symmetry or asymmetry of the distribution, thus indicating possible incompatibility of paired data for specific ranges of values. The character of a histogram can be quantified by three common statistics: the mean, m (measure of central tendency); the standard deviation, s (a measure of the extent of the spread of values about the mean); and the coefficient of variation (CV), a measure of asymmetry of the distribution.

5.8.3: Practical Procedure for Evaluating Global Bias

1. For paired data, a difference, d, is determined for each pair of values (e.g., $d = Au_2 - Au_1$). It is practical to subtract the original analysis from any repeat analysis; in this way, the original analysis is the standard of reference and negative differences mean that the duplicate underestimates relative to the original, and positive differences means that the duplicate overestimates relative to the original.

2. The histogram of differences is examined for outliers. Any outliers present are removed and a new histogram is used to characterize the relation between paired data for the range in question. The

origin of outlier values requires close scrutiny – they can arise due to inherent sampling error or operational error. When due to an inherent sampling error, they may well relate to abnormal mineralogic/textural form, as is commonly the case for gold.

3. Statistical parameters are calculated and summarized with particular emphasis on mean and standard deviation of the differences and the form of the distribution (histogram). If no bias exists between the duplicate analyses, the mean difference, m, should be close to zero. The mean and standard deviation can be used to test whether or not m is significantly different from zero (paired t-test). In general, if the range $m \pm 2(s/n^{1/2})$ includes zero, then the mean is equivalent to zero and no global bias can be demonstrated by this test (ca. 95 percent confidence level). If this range does not include zero, then bias exists and an estimate of the average or global bias is given by the mean difference. The presence of outliers destroys the validity of this test.

4. Asymmetry of a histogram of differences arises because one of the pair members suffers abnormal error in relation to the corresponding member, generally for relatively high values. Which member is in error is uncertain and an understanding of the source of error requires additional information; this problem can be investigated further by examining the binary plot of the paired data.

5. In certain cases, it is useful to use histograms to compare the quality of several laboratories. If statistical parameters are to be used for this purpose, it is important that n be large, outliers be removed, and the duplicate data reflect an equivalent range of values in all laboratories being compared. Comparisons include: (i) extent of global bias, if any; and (ii) relative spread of differences. In comparing cases involving different units, the comparison of spread of differences must be made using a relative variable, such as the average error as a proportion of the mean grade of duplicates used.

Table 5.5 Statistical summaries for various duplicate data sets, Silbak Premier Mine, British Columbia

Deposit	Type of duplicate	Units	n	Mean diff.	s	$s/n^{1/2}$	Remarks
Silbak Premier, Au	Pulp/reject	oz/t	127	0.0028	0.0127	0.00113	Mine lab vs. Min-En, 83
			126	0.0025	0.0123	0.00110	Less one outlier
	Pulp/reject	oz/t	395	0.0334	.422		Mine lab vs. Min-En, 86
			393	0.0044	.0592	0.00298	Minus two outliers
			350	0.0028	.0223	0.00119	Assays up to 0.35 oz/t
	Split core, $\frac{1}{2}$ vs. $\frac{1}{4}$	oz/t	147	0.0197	0.143	0.0118	Analyses by mine lab

6. For many assay distributions there is an abnormally high proportion of assays of very low values, and a corresponding sparsity of higher values. The high value of n that emerges from this can lead to a bias being recognized through the use of a paired t-test. This does not imply that a bias is demonstrated for the entire range of data.

5.8.4: Examples of the Use of Histograms and Related Statistics

Several replicate data sets obtained from producing or past-producing gold mines are used to demonstrate the use of histograms and simple statistics described above in characterizing quality of data (Tables 5.5 and 5.6). Some of the results of Table 5.5 are illustrated in Figs. 5.14 and 5.15, respectively. In Table 5.5, for $n = 126$, one outlier has been removed so that a fair paired t-test can be made. Such a test shows a small global bias between results of the two labs (i.e., the mean $\pm$ twice the standard error does not include zero). For $n = 393$, two outliers have been removed and a paired t-test cannot identify bias. The data of Table 5.6 are partly for different types of paired samples, and include values greater than 1 oz/t; thus, the results are not directly comparable with those in Table 5.5. Of the examples summarized in Table 5.6, only the Shasta data indicate identifiable bias.

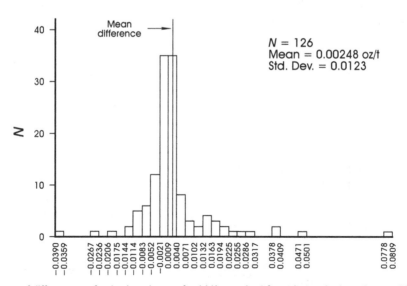

Figure 5.14: Histogram of differences of paired analyses of gold ($Au_1 - Au_2$) for pulps and rejects by two different laboratories in 1983. The mean difference (vertical line through central class interval) is off-center mainly because of a single value on the far right that destroys the symmetry of the histogram. Statistics are summarized in Table 5.5. Silbak Premier gold mine, northern British Columbia.

Table 5.6 Statistical summaries for duplicate data sets from Mascot, Shasta, and Silver Stack gold mines

Deposit	Type of duplicate	Units	n	Mean diff.	s	$s/n^{1/2}$	Remarks
Mascot, Au	One pulp	oz/t	221	0.0017	0.0313	0.00211	Kamloops lab vs. mine lab, assays to 1.5 oz/t
Shasta, Au	One pulp	oz/t	83	0.396	1.68		Mine lab repeats, assays up to 30 oz/t
			81	0.160	0.493	0.0548	Assays up to 11 oz/t
Silver Stack, Au	Pulp vs. reject 1	oz/t	40	−0.0372	0.222	0.0351	Assay ton by mine lab includes outliers
	Pulp vs. reject 2	oz/t	40	−0.0200	0.249	0.0394	One assay ton by mine lab includes outliers

5.8.5: A Conceptual Model for Description of Error in Paired Data

Ideally, with paired analyses of either the same samples or duplicate samples, one hopes that the paired results will be identical or nearly so (i.e., on an x–y graph, the data plot on the line $y = x$). Of course, some random error is always present, so the best that can be hoped for is that the paired data will scatter somewhat about the line $y = x$, producing a cloud of points bisected by the line $y = x$ (e.g., Fig. 5.11a). This pattern is not always met in practice because the amount of random error in one of the sets of data might be very different than the amount of random error in the second set (perhaps from a different laboratory). This means that the dispersions of the two data sets could be different even though both sets of analyses represent the same samples. The obvious effect of these differences is that the general trend of the cloud of plotted points is not well described by the line $y = x$ because of excessive scatter.

A second type of error can be present, that is, systematic differences between the duplicate analyses. In this case, the two sets of data might plot preferentially on one side of the line $y = x$. In fact, there are two extreme kinds of bias that can be present in analytical data: proportional bias and fixed bias. In the case of *proportional bias*, each pair of analyses in

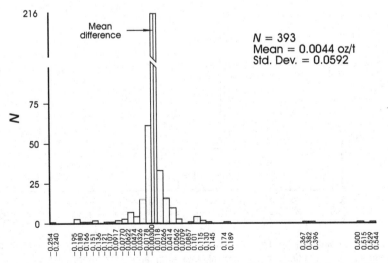

Figure 5.15: Histogram of differences of paired analyses of gold ($Au_1 - Au_2$) for pulps and rejects by two different laboratories in 1986. Statistics are summarized in Table 5.5. Silbak Premier gold mine, northern British Columbia.

a set of paired data has the same or nearly the same ratio as any other pair. The result is shown ideally in Fig. 5.16b. *Fixed bias* arises when one analysis is equal to the corresponding paired analysis plus a constant, as shown in Fig. 5.16c. Of course, both types of bias can be present in a set of paired data (Fig. 5.16d), and random error is always present to some extent (Fig. 5.16a–d), producing scatter about the general trend of plotted values. In the general case, all types of error might be present in a set of paired data, random error plus fixed bias plus proportional bias, as shown in Fig. 5.16d. These combinations of patterns can be represented by a linear regression model with the form

$$y = b_0 + b_1 x \pm e$$

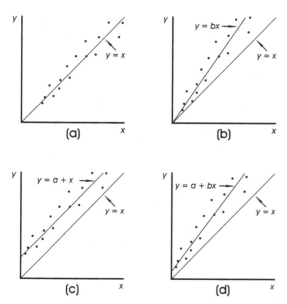

Figure 5.16: A simple linear model to describe errors in paired data as a function of composition. Scales for x and y are equal interval. (a) Dispersion of paired data about the $y = x$ line results from random error in both x and y. (b) Proportional bias plus random error produces a linear trend through the origin with slope different from 1.0. (c) Random error plus a fixed bias produces a line with slope $= 1.0$ and a nonzero y intercept. (d) A general model incorporates random error, fixed bias, and proportional bias to produce a linear array of plotted data with slope different from 1.0 and a nonzero y intercept. After Sinclair and Bentzen (1998).

where

b_0	is the y intercept, an estimate of the fixed bias
b_1	is the slope, a function of proportional bias
e	is the average random error.

The model for errors in paired analyses is even more complicated in nature than shown in Fig. 5.16d because the character of error in assay data can change as a function of composition. The Thompson–Howarth linear model for random error in analyses from a single laboratory is described in Section 5.3.2. In the more general situation, this simple picture is complicated by the possible presence of bias and the likelihood that different styles of mineralization (perhaps different grade ranges) are subject to quite different errors. Some of the patterns encountered in practice are shown in idealized form in Fig. 5.17.

5.8.6: Quantitative Modeling of Error

5.8.6.1: Introduction

In fitting models that quantify a systematic relation between two variables, such as a set of duplicate analyses, it is common to (i) determine or assume the form of the mathematical relation between the two variables, and then (ii) adopt a method to calculate the model parameters specific to a data set. A wide range of mathematical models are available, and even with a single model type (e.g., linear) there are various choices to be made in the calculation procedures available. For example, one might adopt a linear model to describe a relationship between duplicate analyses. However, a number of very different linear models could arise depending on the many different calculation methods and their implicit and explicit assumptions. An incorrect choice of calculation method can lead to an inappropriate model from which incorrect statistical inference can result. Paired analytical data are not immune from this problem.

Regression techniques incorporating a linear model are commonly used for the comparison of one set of analyses (ys) with another (xs), as described by Till (1974), Ripley and Thompson (1987), Sinclair

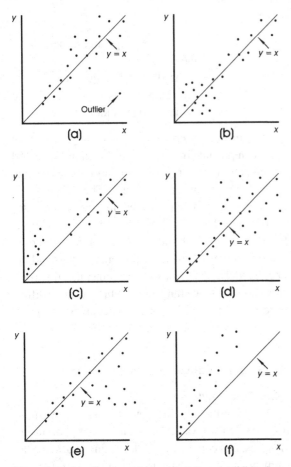

Figure 5.17: Idealized examples of patterns exhibited on scatter plots of paired quality control data incorporating sampling and analytical errors. (a) Random error plus outlier. (b) Two random errors as a function of concentration, perhaps resulting from differing analytical methods. (c) Random error plus proportional bias at low values, only random error at high values, perhaps resulting from errors in calibrating standard. (d) Difference in random error as a function of concentration, perhaps arising from disseminated versus nugget styles of mineralization. (e) Difference in random error as a function of concentration plus a bias in the high-valued data group, possibly resulting from segregation during sampling or subsampling due to differences in density. (f) Proportional bias such as might arise by incorrect calibration of a standard that was then diluted to form standards of lower concentrations. After Sinclair and Bentzen (1998).

and Bentzen (1998), and many others. The principal justification for use of a linear model is the expectation that, with no bias, a set of duplicate analyses will be equivalent except for a component of random error. Hence, the data are expected to cluster along and about the line $y = x$ on a graph of y versus x. If the random difference is small, the spread away from the $y = x$ line is small; if the random differences are large, the spread is large. When there is a significant bias between the two sets, some, even most, of the plotted values will not be centered on the $y = x$ line. Instead, the data may be centered, all or in part, on another line of the form of Eq. 5.1. b_0 and b_1 are called *parameters of the linear model* and for a specific model they are constants. Once b_0 and b_1 are known, the linear equation can be solved by substituting any value of x and calculating the corresponding value of y.

If y values are distributed normally about the line for any value of x, and m and b are estimated by a procedure known as *least squares*, then m and b are also normally distributed (Miller and Kahn, 1962). The advantage of these two parameters being normally distributed is that they can be used to make statistical tests, specifically, whether bias is recognizable in the data. In fact, because of the central limit theorem, these statistical tests can be made even if the underlying distributions are not normal, providing the amount of data is large. As the amount of data increases ($n \geq 40$) and the data distribution becomes more symmetric, the mean values of these parameters tend toward a normal distribution, regardless of the nature of the data distribution.

If statistical tests form part of the evaluation of the significance of a linear model, it is important that the paired data cover the range of concentrations expected; otherwise, conclusions are not generally applicable.

5.8.6.2: Assumptions Inherent in a Linear Model Determined by Least Squares

1. One of the variables, y, is normally distributed about the trend that defines the relation with the second variable, x.

2. The distribution of y values has the same spread regardless of the value of x.

3. The form of the model commonly is assumed to be a straight line.

Consider each assumption separately.

1. y is normally distributed for every value of x

This assumption is not necessarily met other than coincidently. The most obvious reason that data fail this assumption is that data are not represented uniformly throughout the range of paired data values combined with the fact that error can vary significantly with range. Consequently, the distribution of y values can be nonnormal and should be tested. Moreover, as discussed below, assumption 2 rarely prevails.

2. y has the same spread, regardless of the value of x

This assumption is rarely met in a large set of duplicate gold analyses. One reason is that more than one style of mineralization is represented in the data. For example, it is evident that a nugget style (e.g., visible gold) of mineralization will have very different precision than lower-grade, more evenly dispersed mineralization (e.g, fine-grained, disseminated, nonvisible gold). Apart from these geological considerations there is the well-established fact that the average error increases as the concentration increases (e.g., Francois-Bongarcon, 1998; Thompson and Howarth, 1973). Consequently, it might be necessary to subdivide the paired data set into two or more subsets, each of which is more closely in agreement with the assumption.

3. Linear model

There are various sources that can complicate the assumption of a single, simple linear model. For example, sampling or analytical procedures can be more appropriate for one range of data and less appropriate for another range, as in the case of low-grade disseminated mineralization versus high-grade, nugget-type mineralization. Consequently, the nature of errors can be very different for the two styles of mineralization.

Moreover, errors in a single laboratory can be very different from errors in another laboratory for part of the data; for example, visible gold-bearing samples might have an error that is very different than finely dispersed gold. Such a situation could lead to two quite different linear models for different and possibly overlapping grade ranges. When the two grade ranges overlap substantially, the two linear models could appear to form a single, curved relationship. Complexity in the model should be expected and looked for, although generally there is no basis on which to select other than linear models.

5.8.6.3: A Practical Linear Model

In general, the linear model is applied to a comparison of duplicate analyses (paired analytical data) of samples. Commonly, an early set of analyses is being compared with a later set of analyses and the two are expected to give more or less equivalent values on average, unless analytical or sampling problems exist for one or both sets of data. For this reason, the relation between the two sets of analyses is expected to be a straight line with some scatter of data about the line because of ever-present random error. The paired data might be generated in a variety of ways, including the following:

1. Repeat analyses of pulps (or rejects) by the same laboratory
2. Repeat analyses using a second analytical method by the same or another laboratory
3. Repeat analyses by two analytical methods by one or two laboratories
4. Analyses of two sets of samples representing the same phenomenon.

In each of these cases there is a general expectation that the two sets of analyses will be identical, on average, providing that no bias exists in any of the analyses. Reality commonly does not attain this ideal situation. Both sampling and analytical procedures can lead to very different error patterns for different subsets of the total data. Consequently, the subjective process of subsetting the data might be necessary. There are two additional reasons why subsetting of data might be necessary: (i) the presence of outliers in the data, and (ii) the presence of influential samples in the data.

Outliers are those values that differ very greatly from the vast majority of data. As a rule, outliers are fairly straight forward to recognize, although in some cases their recognition is subjective. Influential values represent a small proportion of data that do not class as outliers but have a very strong influence on the particular model calculated for a data set. As an example, consider a data set consisting of 100 paired gold analyses, 95 of which are about evenly scattered in the range of 0–4 g Au/t. The remaining 5 values spread between 4 and 7 g/t. The 5 high values might dominate the linear model to the point that the model is not representative of the other 95 values. Clearly, in such a case it is wise to remove the 5 influential values and calculate a model for the 95 values. The 5 values might be described adequately by the model based on 95 values; if not, they must be compared separately.

5.8.6.4: *Choice of an Estimation Method*

There are four general approaches that have been used to fit linear models to paired analytical data:

1. Weighted least squares
2. Principal axis
3. Major axis
4. Reduced major axis.

In all four cases, a general method known as *least squares* is used to determine the linear model and the result is referred to as a *best-fit* model. In each case, the term *best fit* means that a particular error criterion is minimized relative to the linear model that is determined. Not all of these criteria are appropriate for comparing replicate assay data.

The most widely available method of fitting a line to a set of paired data, *traditional least squares*, is an example of an inappropriate least-squares procedure that, in some cases, has been incorrectly applied to the description of paired assay data. The reason that traditional least squares is inappropriate for such data is that the method assumes that one of the variables (x) is perfectly known and places all the error in the second variable (y). In reality, there are errors in both of the variables being compared, and this must be taken

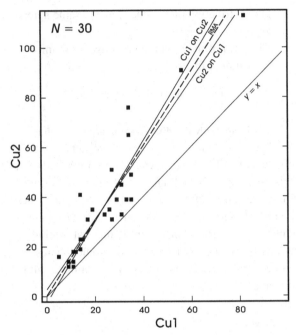

Figure 5.18: A set of 30 duplicate copper analyses (X-ray fluorescence) for metal-bearing muds. Three regression lines are shown, regression of Cu1 on Cu2, regression of Cu2 on Cu1, and reduced major-axis regression (RMA). Cu2 data were obtained at a different time from Cu1 data, and the power settings on the X-ray generator were very different, resulting in a strong proportional bias.

into account in defining the linear model. The problem with traditional least squares is well illustrated by an example, gold assays of duplicate samples of blastholes from the Silbak Premier gold mine, illustrated in Fig. 4.18 (from Sinclair and Bentzen, 1998). Two different lines are obtained, depending on which variable is taken as y, the dependent variable. If we were to incorrectly accept these lines and test one variable statistically against the other, we would arrive at two opposing conclusions for the two lines, both conclusions being incorrect. For example, if AUD is taken as y, we would conclude that bias exists and that AUD underestimates AU by about 27 percent if AU is taken as y, we would conclude that bias exists and that AU underestimates AUD by about 29 percent. These two results are dramatically in conflict and clearly show that the traditional least-squares method is generally inappropriate as a means of defining a best-fit linear model for paired assay data.

A least-squares procedure is required that produces a fair representation of the underlying trend in a set of paired data. This is achieved best by using a method that takes into account the different errors that exist in the two sets of data being compared (e.g., Mark and Church, 1974). Because errors are rarely known in detail in the comparison of many kinds of variables, including assay data, several practical approaches have been offered to producing least-squares models so that relationships between paired variables are determined fairly.

Weighted least-squares procedures (e.g., Ripley and Thompson, 1987) can be highly subjective because of the manner by which the weights are determined. In some cases they provide linear models that lie outside the limits defined by the two traditional least-squares procedures. Consequently, weighted-least-squares methods are not generally acceptable to give an unbiased treatment of paired assay data.

The major-axis solution is based on minimizing the squared perpendicular distances from each point to the line. This is equivalent to minimizing simultaneously in both the x and y directions. This procedure is affected by differences in scale between the two variables being compared.

The reduced major axis (RMA) linear model combines a standardization of the two variables (i.e., divide each value by the standard deviation of the data) and a major-axis least-squares solution to determine the linear model. This procedure avoids any concern of difference in scale of the two variables (e.g., when large biases exist between the paired variables). Dent (1937) showed that for paired variables, the maximum likelihood estimator of the ratio of errors when the errors are unknown is $(s_y/s_x)^2$, which is equivalent to an RMA line through the data. In general, errors are unknown for paired analytical data.

An RMA regression is desirable when it is important that errors in both variables be taken into account in establishing the relation between two variables (Sinclair and Bentzen, 1998). The methodology for RMA regression has been described in an earth science context by Agterberg (1974), Till (1974), Miller and Kahn (1962), and Davis (1986). Till (1974) emphasizes the importance of using RMA in comparing paired (duplicate) analytical data. Detailed equations

for determining slope, intercept, and associated errors of an RMA linear model are outlined in Chapter 4. In general, the errors can be taken as normally distributed (cf. Miller and Kahn, 1962), and can be used to test whether the intercept error range includes zero (in which case the intercept cannot be distinguished from zero) and the slope error range includes 1 (in which case the slope cannot be distinguished from 1). The dispersion about the RMA line can be used in several practical comparisons, including (i) the comparison of replicates of several standards by one laboratory with replicates of the same standards by another laboratory, and (ii) the comparison of intralaboratory paired analyses. An example is provided in Fig. 4.18, in which a single linear model (RMA) has been fitted to the data. However, two subsets of data are shown on the figure with different symbols, and each subset has its own characteristic dispersion pattern relatived to the linear model. In such cases, linear models can be determined independently for each subset of data, thus providing more insight into the character of errors in the data set.

Commonly, the application of RMA analysis is straightforward, as is the case for Fig. 5.18 involving two sets of copper analyses (i.e., duplicate values on the same samples by the same analytical method [XRF] but at different times [and, as discovered after the analyses were obtained, with different power settings for the X-ray generator]). The result is a large proportional bias and a substantial difference between RMA and traditional least-squares solutions for the linear models. Scatter plots and fitted RMA models for two sets of data summarized in Table 5.7 are shown in Figs. 5.19 and 5.20. These diagrams show comparisons of check assays on reject samples at an operating mine for two different years, 1983 and 1986. The RMA statistics in both cases indicate that the y intercept cannot be distinguished from zero and that the slopes are, at most, imperceptibly different from 1. The correlation coefficients (0.962 and 0.991, respectively) do not represent an adequate measure of the relative quality of the two sets of data because with equal reproducibility, the correlation coefficients depend to some extent on the range of the assay data. In this case, the ranges of the two diagrams differ by more than an order of magnitude. A better way to

Table 5.7 Parameters of reduced major axis models for several types of paired gold data for the Silbak Premier mine

Comparison type	Units	n	r	sl[a]	int[a]	s_d[b]
Pulp/reject two labs	oz/t	126	0.962	0.947 (0.023)	−0.001 (0.001)	0.0047
Pulp/reject two labs	oz/t	395	0.989	0.752 (0.006)	0.034 (0.008)	
		393	0.991	1.019 (0.007)	−0.008 (0.003)	0.011
		350	0.955	0.987 (0.016)	−0.002 (0.001)	

[a] Figure in brackets is error as one standard deviation.
[b] s_d is the dispersion about the reduced-major-axis line.

compare relative quality is to use the dispersion about the line (Eq. 4.27). For these two diagrams, the dispersions are 0.017 for 1983 data and 0.083 for 1986 data (i.e., the 1986 comparison is substantially worse than that of 1983).

In compairing duplicate analyses it is common practice in the mineral industry to determine the average of absolute differences of paired duplicate assays, whether those duplicates be pulps, rejects, or samples. Even where the average difference is close to zero (i.e., positive and negative differences closely compenaste) it is useful to know the average abso-

lute difference between paired values. Where there is a substantial number of pairs (> 40), the distribution of real differences generally approaches a normal distribution. In such a case, the average absolute difference can be estimated as $0.798 s_d$ where s_d is the standard deviation of differences. The standard deviation of differences can be determined directly or, where bias is not significant, can be estimated from the dispersion about the reduced major-axis line as follows:

$$s_d^2 = S_{rma}^2/2 = \left(S_x^2 + S_y^2\right)/2$$

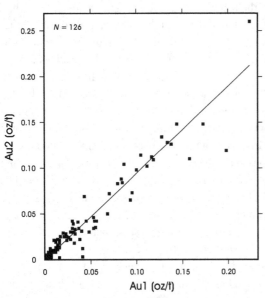

Figure 5.19: RMA line fitted to duplicate Au analyses, Silbak Premier gold mine, northern British Columbia, 1983 data. Au1 are the mine analyses of split drill core; Au2 are reject analyses by an independent laboratory.

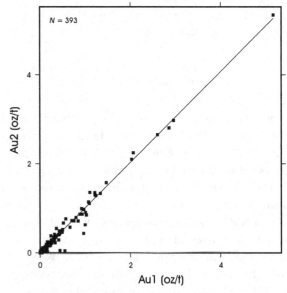

Figure 5.20: RMA line fitted to duplicate Au analyses, Silbak Premier gold mine, northern British Columbia, 1986 data. Au1 are the mine analyses of split drill core; Au2 are reject analyses by an independent laboratory. Note the contrast in scale compared with Fig. 5.19.

If x and y are the same lab the resulting value of s_d^2 is an estimate of the precision of the laboratory. If x and y are different labs, the resulting value of s_d^2 is the average precision of the two laboratories.

When differences are shown to exist between two variables, the procedure for determining a functional relation (as outlined previously) does not in itself indicate which of the two data sets being compared is correct or if either is correct. RMA establishes a fair measure of the difference (relative bias) between two sets of data. In order to determine the relative correctness of one or the other, it is essential to have additional information, for example, (i) measurements of standards included with one or both sets of data and analyzed under the same conditions, and (ii) when information is not available on acceptable standards, it may be possible to use replicate analyses of the same samples by one or more additional laboratories to provide insight into which of the two sets of data is more likely to be correct.

5.9: IMPROVING THE UNDERSTANDING OF VALUE CONTINUITY

Autocorrelation functions (Section 4.7) are required for some procedures in mineral inventory estimation or simply to quantify value continuity. In Section 5.4.1, the point was made that sampling patterns early in the exploration of a deposit can be designed to improve confidence in a three-dimensional model of autocorrelation. Diamond-drill sampling schemes provide continuity information preferentially along drill holes, and spacing between holes may not be adequate to measure continuity satisfactorily in the other two dimensions. Sampling patterns and the continuity estimated from them must be reviewed and new sampling designed to meet such a deficiency.

Sampling should be closely tied to geology and coded systematically so that data can be easily categorized into domains if the need arises. For example, individual samples should not cross major lithologic boundaries, veins and wallrock should be sampled separately to ascertain the detailed control of metal distribution, and so on. Sampling grid dimensions may have to be adjusted depending on the variability of geologic and mineralization parameters. Even at early stages of exploration, local, close-spaced samples should be taken along lines in various orientations to evaluate local continuity and to be integrate into the complementary, detailed geologic information. Journel and Huijbregts (1978) recommend that within a larger sampling field, local crosses of closely spaced data be collected to provide some insight into local (short-range) continuity. Closely spaced information also can be obtained in either appropriately chosen areas of stripping and trenching or exploratory underground working.

In addition, quality of data has a significant impact on experimentally determined autocorrelation functions and the ease with which experimental data can be fitted with a model. In particular, large random errors lead to a high nugget effect and erratic variations in experimental data points, in some cases of sufficient magnitude that interpreting a model is dubious. Figure 5.21 illustrates two semivariograms for the Ortiz gold deposit, New Mexico (Springett, 1983), one based on production blastholes (notorious for large errors) and the other based on confirmatory crosscut samples. Note that the production blasthole data produces a much larger intercept (nugget effect) than the confirmatory samples (0.17 versus 0.02). In

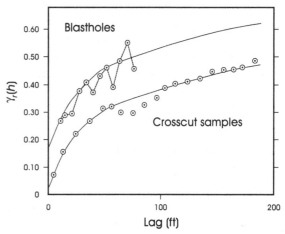

Figure 5.21: Different semivariogram (autocorrelation) models for different sample types, Ortiz Gold Mine, New Mexico. Circles are experimental data; smooth curves are fitted models. Note greater fluctuations and higher nugget effect (*y* intercept) for the blasthole samples versus the crosscut (drill round = ca. 20 tons) samples. Redrawn from Springett (1983).

addition, the experimental points forming the blast-hole semivariogram are relatively erratic.

Following this strategy reduces one of the common problems encountered in applying geostatistics at a prefeasibility stage of exploration, that is, the common scarcity of closely spaced data with which to define both the nugget effect and the short-range grade continuity in a confident manner. Of course, if more than one domain has been recognized, control of short-range continuity is required for each separate domain. In many practical cases a deposit can be divided into several domains, each of which is characterized by its own distinctive semivariogram model. This arises because differences in genesis, lithology, or structure produced differences in the local character of mineralization. The concept of domains is implicit in a discussion by Srivastava (1987) of conditional probability as control for grade estimation; geologic data are emphasized as an essential source of conditioning information.

5.10: A GENERALIZED APPROACH TO OPEN-PIT-MINE GRADE CONTROL

The potential for improved profits in large, open-pit operations from adequate and well-designed sampling methods is large, probably larger than any other productivity improvement approach. For example, in a 10-million-ton per year copper mine, better selection that leads to an improvement in average grade from 0.4 percent Cu to 0.404 percent Cu increases gross annual revenue by more than $1 million (1995 prices in US$).

5.10.1: Initial Investigations

Prior to the development of a grade control program, samples are taken (generally in the form of diamond-drill core) and a test pit may have been excavated and sampled. Sampling of broken material from a test pit is relatively straightforward if all material is passed through a pilot mill, but can be difficult and biased otherwise. In the sampling of both drill core and broken material, potential for bias exists when there are clusters of small rich veinlets, preferred orientations of veinlets, strong differences in hardness

of gangue and ore minerals is a feature of the ore, and so on. The distribution of grade should be examined and data must be checked for bias, preferably with the aid of a computer-based graphic display system. When data abundance permits, samples of different supports should not be mixed at this stage if their dispersions (variances) are significantly different. A probability graph of grades or composite grades (of common support) serves to identify the form of the histogram and recognize the possibility of multiple populations in the data, perhaps leading to the recognition of separate domains of mineralization. When such domains are recognized, each should be investigated independently with regard to characteristics of original grades or grades of composites.

5.10.2: Development of a Sampling Program

Initial production planning in open pits normally is based on exploration diamond-drill-hole data supplemented by smaller amounts of data from other sources (e.g., surface trenches, limited underground workings). Raymond and Armstrong (1988) provide an informative account of an inherent bias in diamond drilling, demonstrated first by comparing diamond-drill-core grades with average grades of 521 100-ton rounds obtained from developing a decline in the Valley copper porphyry deposit. Two diamond-drill holes were drilled in advance of the decline and were sampled in increments corresponding to decline rounds. Grades shown in Fig. 5.22 for a portion of

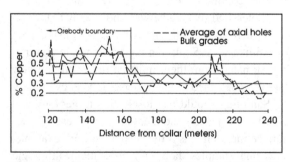

Figure 5.22: Comparison of sampling results by two methods along a decline in the Valley Copper deposit. Dashed sawtooth curve indicates grades for axial, diamond-drill core obtained in advance of the drive; solid sawtooth curve is result of bulk sampling obtained as the decline was driven. Redrawn from Raymond and Armstrong (1988).

the decline; reveal a consistent underestimation of Cu grade by diamond-drill hole, estimated at 0.044% ± 0.013% (95% confidence). With the advantage of 17 months of test production data, Raymond and Armstrong (1988) go on to demonstrate a roughly equivalent bias for 114 diamond-drill-hole bench composites that correspond with the volume of ore produced during the test period. Unfortunately, no explanation is offered as to the cause of these biases. At the Brenda Mo–Cu deposit, a similar problem arose because soft material was preferentially washed out of the drill core by the drilling water.

Sampling programs in open-pit operations generally are constrained by details of the production system; within this framework there is variable room for flexibility. Blasthole patterns, for example, are designed to break ground efficiently; hence, blasthole sample spacing is not a parameter that can be changed easily, if at all. However, the detailed procedure used to sample blasthole cuttings is an important variable and lack of attention to quality of sampling and assaying can lead to significant bias or large random errors (cf. Raymond and Armstrong, 1988).

Sampling generally is conducted by the blasthole driller or helper; a common procedure involves the use of a piece of pipe or tube that is shoved into the cuttings pile at a predetermined number (generally 4 to 12) of roughly evenly distributed sites to obtain a scoop of cuttings at each site. The sample is the mixture of all such scoops. Random sampling errors can be large if sample volume is too small (cf. Giroux and Sinclair, 1986). Analytical error is commonly small compared with sampling error; for example, Johnson and Blackwell (1986) show that at the Brenda porphyry Cu–Mo mine the analytical error is less than one-tenth the variability introduced by sampling. Because the cuttings pile is stratified, with the last material drilled on top, very different grades can exist throughout the pile. If all strata in the pile are not sampled, bias can result. The cuttings from a blasthole 60 ft deep and $12\frac{1}{4}$ in in diameter totals about four tons. Twelve systematically positioned scoops from this material provide a sample that is reduced in a riffle at the sample site to perhaps 2 lb in the case of a copper deposit and 5 to 10 lb in the case of a gold deposit.

Alternative methods of obtaining grade information can depend on a variety of equipment. Equipment available for sampling includes diamond and reverse circulation drills, mechanical sampling machines for blasthole cuttings, hand drills for channel sampling, X-ray and other drill-hole probes, and various surface scanners. Specialized equipment is costly and may require operation by technical personnel; such equipment can be justified only if selectivity is improved sufficiently compared with other procedures. Where production drilling equipment is used to provide cuttings, sample bias can result if values concentrate in certain size fractions and these fractions are partly lost as dust or washed out by drilling fluid. To test for this possibility, several samples should be separated into size fractions for individual assay. Many operations use electric/hydraulic rotary drills that require a minimal amount of water for dust supression, and the drill cuttings available are representative of all size fractions. This is not the case for drilling in wet ground. Water forced out of the hole washes better-grade fines (ground sulphides) away from the cuttings pile, reducing the grade of the remaining material.

The increasing demand of environmental control on drills can provide additional problems vis-à-vis sampling. In many cases, blasthole drills are designed to leave the coarse cuttings near the hole collar and to store the fines in a dust collector, which automatically dumps fines (the consistency of flour) nearby when the collector is full, not necessarily at the completion of a hole. The density of the fines is low and the volume large; proportionate sampling of the coarse cuttings and fines is necessary when grade is not uniformly distributed among size fractions.

5.10.3: Sampling Personnel and Sample Record

Samples can be collected by personnel hired especially for the purpose, but this can result in no samples being taken on weekends in an effort to avoid shift differential payments. The logical person to take the sample is the driller or drill oiler, when available. Poor sampling practice can result from poor supervision and management. Impressing the driller with the significance and importance of the sample and the effect of poor sampling on grade control and profitability is a useful approach to quality control.

Samples must be tagged and numbered unambiguously using numbered metal tags. Pairs of tags can be supplied to the driller who keeps a log of the blast number and hole number from a survey plan, the matching tag number, the drill number, and name of the drill oiler. One tag is attached to a hole marker for later reference by the hole dewatering and explosive loading personnel. The other tag is included with the sample. The mine survey crew maintains the tag system and records all necessary information; the operations shift supervisors are responsible for timely delivery of samples to the assay laboratory. Long-term success in sampling and grade control is only possible with good supervision and supportive management.

5.10.4: Implementation of Grade Control

The result of a grade control program is the clear demarcation in the pit of the separation between ore and waste. When possible, ore and waste should be blasted separately. In many cases several grades of stockpile material must be distinguished, perhaps for leaching or processing near the end of mine life. Plans showing the blasts should be color coded to specific material; similar colored flagging and string can be used in the pit. Despite the concept of a selection mining unit (SMU), the ore/waste contact cannot be so simplistically defined in practice. For example, a single waste block surrounded by ore will almost certainly be included with ore for logistical reasons. "Dog-leg" ore/waste contacts are confusing to production personnel and should be replaced by smoother contours to avoid unnecessary loss of quality of grade control.

Personal computers are now in routine use at most mine sites. Networking of computers for assay laboratory, grade control, geology, and engineering is now common and provides highly efficient data transfer. Timely receipt of assay results and grade interpolations is essential to efficient, open-pit operations, particularly when two- or three-row, free-face blasting of less than 100 blastholes is the practice. Grade interpolation is increasingly being done by one or another form of kriging. Most common is ordinary kriging; indicator and multiple indicator kriging are useful in special cases, and conditional simulations are widely used, especially in more erratic deposits. The results should be displayed on a plan of appropriate scale, currently generally computer generated, and as a grade–tonnage curve for each blast. At most operations the cutoff grade is changed as required to respond to changes in metal prices, availability of ore, and to minimize deviations from long-term plans. At large, modern, open-pit operations, drills are monitored using global positioning systems (GPS) and the driller includes a tag indicating the coordinates of the hole as well as the blast and bench with the sample, eliminating the need for computer-aided matching of sample tags and locations.

5.10.5: Mineral Inventory: Mine–Mill Grade Comparisons

Direct comparison of mine and mill grades is meaningless in the short term of a few days for conventional mills or several weeks in leaching operations. The conventional mill separates and stockpiles coarse material, which is mixed and milled days later, but immediately processes the fines, which are generally of better grade. Daily comparisons of mine and mill head grades on time plots can be helpful in indicating that particular low- or high-grade material has been processed, but only in qualitative form. Over a one-month period (possibly longer), the mine and mill estimates should balance, but it is the mill figure that is accountable. The mine figure is used for comparative purposes as an indication of the success of the grade control system.

Mines operate on the basis of short-term grade control, which depletes the mineral inventory. At regular intervals of several months to a year, the volume mined must be removed from inventory and the contained grades and tonnages compared with actual production. If the figures do not balance, the various components of the reconciliation must be examined with the possibility that the mineral inventory methodology and the grade control procedures must be reassessed. Bear in mind, however, that many factors including technologic changes permitting a change in mill throughput and significant changes in cutoff grade due to changes in metal prices can result in substantial differences between estimates and eventual output.

5.11: SUMMARY

Data acquisition procedures may not be of adequate quality to properly represent a mineral deposit or, in particular cases, to properly define the geologic and value continuities on which mineral inventory estimates depend. For high-quality, resource/reserve estimation, high quality of the database can be maintained by systematic critical evaluation of

(i) Sample acquisition (including three-dimensional aspects of deposit sampling patterns)
(ii) Sample reduction and assaying procedures
(iii) Geologic data acquisition, interpretation, and modeling.

Sampling patterns and methods, sample reduction schemes, and assaying procedures must be designed and tested for the local geologic substrate. A regular program of duplicate and check sampling and analyses serves as a continuing measure of quality control. Attention to these matters reduces the nugget effect and improves the quality, and therefore the ease of interpretation, of experimental semivariograms, as well as improving the quality of estimation.

Detailed geologic mapping and three-dimensional deposit reconstruction is essential and must be integrated fully into models of continuity of value measures (e.g., grade, thickness) based on well-designed sampling patterns. Many deposits can be divided into two or more geologic domains (lithologic, stuctural, genetic), each of which is characterized by its own statistical distribution of grades and grade continuity model. In such cases, the indifferent application of a single "average" continuity model (semivariogram) to determine mineral inventory can lead to substantial "hidden" errors unrelated to errors calculated during block estimation.

Improved sampling and geologic procedures contribute directly to improved metal recovery and income. Data quality can be set in a broader perspective such as total quality or continuous improvement management methods that are being adopted by increasing numbers of mining companies. The essential element of these philosophies, whatever the name, is a continuous process of review and change that is oriented to an overall improvement in profitibility. For deposit/reserve estimation, the systematic scrutiny inherent in these methods applies to all stages, from data acquisition and editing to mine planning and production control. To be effective, such a review process must be carried out in a systematic manner.

5.12: PRACTICAL CONSIDERATIONS

1. Basic assay data can become extraordinarily abundant as exploration, development, and production proceed. A computer-based storage and retrieval system for data is essential and should be implemented early in the exploration of a deposit.
2. Program design should attempt to provide as much data of uniform support as is practical.
3. The design of a sampling program should include consideration of a planned systematic increase in the spatial density of data as evaluation progresses.
4. Sampling methods in use elsewhere should not be adopted blindly at a deposit without considerations of alternative methods or the possibility of improving results by modification of the existing method. Sampling experiments can be designed to meet this need.
5. A sampling protocol should be designed for mineralized material of each well-defined type, using Gy's sampling equation and sampling diagram with an appropriate safety line.
6. Data should be requested from the analytical laboratory coded as to analytical batches, including the various quality control samples (blanks and standards) that have been analyzed as part of each batch. As data are received, all control sample values should be monitored for possible contamination and problems with analytical quality.

7. An analytical method should be suitable for the analytical goal. Even for an individual method, there is room for choice (e.g., in gold analyses by fire assay, concern may arise as to whether one assay ton is sufficient material on which to base an assay).

8. A quality control system involving duplicate sampling should be implemented at the outset of a sampling and assaying project. Details of the system vary depending on circumstances, but should include, as appropriate, duplicate samples (perhaps 1 in every 30 to 60 samples to monitor total sampling and analytical variances), duplicate pulps (perhaps 1 in every 20 samples to monitor subsampling and analytical errors of the analytical laboratory), and duplicate rejects (perhaps every 30 to 40 samples to monitor error in the first stage of sampling/subsampling). Some of the duplicate pulps and rejects should be analyzed by reputable independent laboratories as a check against bias. A few samples from a project, representative of the important analytical range of values, should be analyzed at intervals as project "standards" to monitor possible bias in the laboratory. All of these duplicate analyses should be reported and coded in such a way that they can be monitored graphically, regularly, and with ease. These quality control procedures are in addition to those implemented by the analytical laboratory and should not be identifiable by the laboratory.

5.13: SELECTED READING

Burn, R. G., 1981, Data reliability in ore reserve assessments; Mining Mag., October, pp. 289–299.

Francois-Bongarcon, D., 1998, Extensions to the demonstration of Gy's formula; in Vallée, M., and A. J. Sinclair (eds.), Quality assurance, continuous quality improvement and standards in mineral resource estimation. Expl. and Min. Geol., v. 7, nos. 1 and 2, pp. 149–154.

Long, S. D., 1998, Practical quality control procedures in mineral inventory estimation; in Vallée, M., and A. J. Sinclair (eds.), Quality assurance, continuous quality improvement and standards in mineral resource estimation, Expl. and Min. Geol., v. 7, nos. 1 and 2, pp. 117–128.

Schwarz, Jr., F. P., S. M. Weber, and A. J. Erickson, Jr., 1984, Quality control of sample preparation at the Mount Hope molybdenum prospect, Eureka County, Nevada; Soc. Min. Engs. of AIME, Proc. Symp. on "Applied Mining Geology," pp. 175–187.

Thompson, M., and R. J. Howarth, 1976, Duplicate analysis in geochemical practice: Part I. Theoretical approach and estimation of analytical reproducibility; Analyst, v. 101, pp. 690–698.

Vallée, M., M. Dagbert, and D. Cote, 1993, Quality control requirements for more reliable mineral deposit and reserve estimates; Can. Inst. Min. Metall. Bull., v. 86, no. 969, pp. 65–75.

Vallée, M., and A. J. Sinclair (eds.), 1998, Quality assurance, continuous quality improvement and standards in mineral resource estimation; Expl. and Min. Geol., v. 7, nos. 1 and 2, 180 pp.

5.14: EXERCISES

1. The following sampling protocol for a gold ore (after Davis et al., 1989) can be plotted graphically, either manually or using GYSAMPLE software (which can be downloaded through publisher's website). A 3- to 5-kg sample of blasthole cuttings was dried and crushed to 95 precent −2 mm. A 1-kg split is pulverized to 100 μ, and 500 g is taken for assay. Of the 500 g split, 50 g were used for fire assay. Construct and evaluate the sampling diagram. Compare your interpretation with the conclusion of Davis et al. (obtained by independent experimentation) that "assay repeatability in the Colosseum ores would be enhanced by a particle size reduction before the initial sample split" (1989, p. 829).

2. Plot and comment on the adequacy of the following subsampling protocol for the Mt. Hope molybenum prospect (Schwarz et al., 1984): each 3-m segment of drill core is split and one-half (approximately 10,000 g) is bagged; this sample

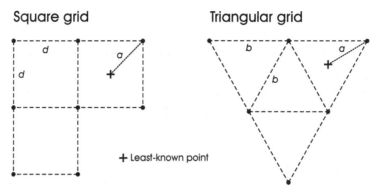

Figure 5.23: Square and triangular grid patterns (cf. Annels, 1991) arranged such that the central point in each case is the same distance from all data in the first surrounding aureole.

is crushed to 2 mesh (jaw crusher) and further crushed to 10 mesh (cone crusher). The cone-crushed product is reduced to about 1/8th volume (about 1,200 g) with a riffle splitter. The split is pulverized to 80 mesh, rolled, and further split to produce two 100-g subsamples. One of the 100-g subsamples is further pulverized to 200 mesh and a 25-g split is taken for commercial assay. Metric equivalents of mesh sizes are given in Appendix 2.

3. The data file J&L.eas (which can be downloaded through the publisher's website) contains 55 quadruplicate samples, each set of quadruplicate samples representing coarsely crushed material from one drill round (ca. 30 short tons). Each set of quadruplicate samples were cut as material passed into a sampling tower, and all four subsamples were passed through the same subsampling protocol and analyzed in the same way for five elements: Au, Ag, As, Zn, and Pb. Evaluate the quality of the assay data using the program ERRORZ, available through the publisher's website.

4. Annels (1991, p. 61) discusses the comparative efficiencies of square and offset (triangular)

grids for drilling patterns and concludes in part that an "offset grid is ... more efficient in that drillholes are further apart ... 29.9% fewer holes to cover the same area with the same degree of confidence." Evaluate the comment "with the same degree of confidence" by discussing the estimation of a central point in each of the basic arrays using a single aureole of data. Assume the patterns and dimensions of Fig. 5.23.

5. The data set sludge.eas contains 84 half-core assays for gold (oz/t) as well as corresponding weighted core/sludge assays. Compare the two sets of data using regression analysis. Comment on the results in light of the limitations/advantages of sludge assays.

6. Compare the Cu analyses in Table 5.3 by each of the two sampling methods (tube and channel) with the best estimate. Recall that the Ag data indicates a significant improvement in metal recovery by the channel sampling results compared with the tube sampling results (see Section 5.5.2). Do the Cu data support the conclusion based on Ag analyses? Data in file equitybh.eas.

6

Exploratory Data Evaluation

Possibilities that had escaped the notice of previous operators may become obvious when historical information is analyzed systematically. (McKinstry, 1948, p. 436)

Chapter 6 introduces the need for a thorough data analysis as a prelude to a mineral inventory study. Orderly preparation of data and close attention to data editing are essential precursors to data evaluation. Composites are introduced and discussed in some detail because they commonly form the basis of mineral inventory estimation. Finally, a highly structured approach to data evaluation is recommended, and is illustrated in terms of univariate, bivariate, and multivariate procedures. Generally, a range of computer software is necessary for thorough and efficient data analysis; however, use of computers should not preclude a fundamental understanding of the methods implicit in their use!

6.1: INTRODUCTION

Data evaluation forms an essential part of every mineral inventory estimate and involves a thorough organization and understanding of the data that are the basis of a resource/reserve estimate. The ultimate purpose of exploratory data evaluation in mineral inventory work is to improve the quality of estimation; specific aims include the following:

(i) Error recognition

(ii) To provide a comprehensive knowledge of the statistical and spatial characteristics of all variables of interest for resource/reserve estimation

(iii) To document and understand the interrelations among the variables of interest

(iv) To recognize any systematic spatial variation of variables such as grade and thickness of mineralized zones

(v) To recognize and define distinctive geologic domains that must be evaluated independently for mineral inventory estimation

(vi) To identify and understand outliers

(vii) To evaluate similarity/dissimilarity of various types of raw data, especially samples of different supports.

These aims are not mutually exclusive, but each can have its own impact on resource/reserve estimation. Error, of course, must be minimized. Quantification of various types of errors has been discussed in detail in Chapters 2 and 5. Here, concern is with gross human and mechanical errors, such as incorrect coordinates and errors in transcribing data. Individual variables can have characteristics that lead to different decisions for each variable during the course of estimation. Similarly, interrelations of variables might contribute to the ease (or difficulty) of estimating several variables. Identifiable trends can lead to the definition of two or more domains, each of which might

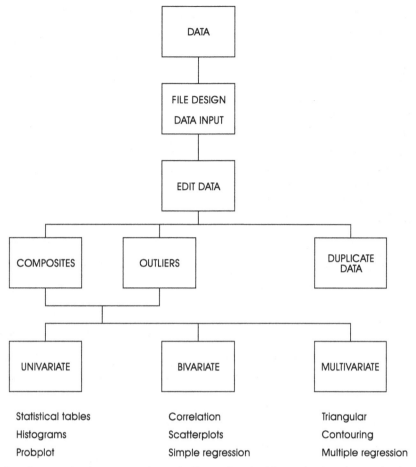

Figure 6.1: Simple flow diagram showing a general organization and content for exploratory data evaluation related to mineral inventory estimation.

be estimated independently of the others. "Outlier" grades, a persistent problem, require detailed evaluation because they have an impact on reserves much out of proportion to their abundance; this topic is reserved for Chapter 7.

Data organization and evaluation can involve as much as 50 percent of the time necessary to conduct a mineral inventory estimation. The principal aspects of a data evaluation system designed to meet the previously mentioned aims include:

(i) File design and data input
(ii) Data editing
(iii) Quantification of data quality
(iv) Grouping of data by geologic domain, sample support, and so on

(v) Univariate statistical analysis (e.g., histograms, probability graphs)
(vi) Bivariate statistical analysis (e.g., correlation, linear regression)
(vii) Spatial patterns and trends.

An orderly procedure to the evaluation is essential (Fig. 6.1) and it is useful if such a system progresses from simple techniques to more complicated techniques, as required. This evaluation is facilitated by a well-organized structure to the data; therefore, an early emphasis on file design and data editing is recommended. Some specific aims of data evaluation are to define the probability density functions of important variables and obtain insight into their spatial characteristics. Quality control procedures are a sufficiently important aspect of data treatment to warrant a separate discussion.

6.2: FILE DESIGN AND DATA INPUT

In many cases, large quantities of data are involved in mineral inventory estimation, and ease of storage, retrieval, and handling are essential; disorganized storage and retrieval procedures can drastically affect the efficiency and cost of obtaining a mineral inventory estimate. Assay data, however essential, must be accompanied by appropriate geologic (rock type, character of mineralization, alteration, etc.) and location (three-dimensional coordinates) information. Specific geologic characteristics to be recorded are a function of the geologic type of deposit being evaluated. Information can be maintained within a single database management system or in a variety of files linked by an identity number, commonly sample number, drill hole number, and the like.

It is convenient to begin systematic collection and organization of data early in the exploration/evaluation history of a mineral deposit. Even at the detailed exploration stage, enormous quantities of data are generated, and failure to develop a computer-based system of data storage and retrieval can lead to errors and loss of some information at later stages of evaluation.

Today there is little excuse for not using computer-based filing systems for the handling of assay and related information (cf. Blackwell and Sinclair, 1992). Numerous commercial database management systems are available, some of them designed especially for the mineral industry, with modules designed explicitly for mineral inventory estimation by a variety of commonly used procedures. Of course, each of these systems has particular limitations, and it is wise to examine their capabilities thoroughly in advance of purchase or use so as to ensure they meet the needs of the project. Although these systems have become relatively complex, most are accompanied by user's manuals, and their purchase typically provides access to technical advice through a troubleshooting phone contact.

One aspect of data organization not fully appreciated by many mineral inventory practitioners is the

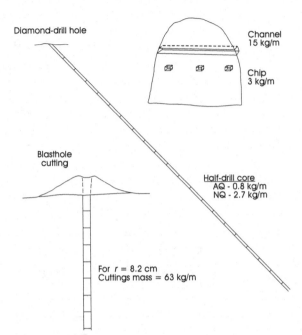

Figure 6.2: The concept of sample support. Size, shape, and orientation all contribute to the support of a sample. Samples of different orientations, different dimensions, and different masses thus have different supports, and therefore can have different dispersions (standard deviations). Representative masses of several common types of samples are illustrated.

importance of data support (see Chapter 1). *Support* is the size, shape, and orientation of samples (Figure 6.2). In many practical situations, the variability of assays of one support differs substantially from the variability of assays of a second support. It is easy to imagine that a set of small chip samples along an underground drift has much more variability than a set of very large samples (each totaling all the material from a drill round in the drift) taken from the same length of underground opening. This fact is one of the principal justifications for using composites of uniform support when samples of smaller, perhaps variable, supports have been taken in the first instance.

Two examples illustrate the importance of ore character on support; hence, the importance of ore classification as a basis for grouping samples. In a bedded deposit such as many iron formation and shale-hosted Pb–Zn deposits, contiguous linear samples (e.g., drill core or channel samples) taken parallel to bedding are more similar on average than are contiguous samples taken across bedding. Similarly, in a sheeted vein zone, samples taken perpendicular to the veins have an extremely high variability if sample length is less than the average spacing between veins (because some samples can be zero grade) in contrast to samples whose lengths are greater than the vein spacing (all such samples must contain some vein material). These two examples make the case for the importance of support in an intuitively obvious way; the first case emphasizes the importance of orientation of samples, and the second demonstrates the importance of lengths of samples.

6.3: DATA EDITING

It is essential that data be checked thoroughly once a database has been constructed and, subsequently, as new data are added. A variety of procedures can be used, but most important is a systematic approach so that significant errors are spotted early and can be corrected or avoided. With small data sets, it may be adequate to check sample locations and values against original documents, including geologic maps and cross sections, surveyed sample location maps, and tabulations of data as received from the analyst. When large data sets are involved, such procedures are not efficient and a variety of computer-based, graphic-output procedures are commonly used to test visually for outliers and other problems.

(i) Sample locations can be output on plans or sections for comparison with original documents. Errors in locations of drill holes and hole deviation can impinge significantly on resource/reserve estimation and, unfortunately, are common.

(ii) Geologic coding of samples can be checked by output of maps (symbols for different sample categories) of appropriate scale for comparison with master maps and cross sections containing the original information.

(iii) Outlier samples commonly can be recognized using a variety of simple graphic output that is an inherent part of data analysis procedures, including histograms, probability plots, scatter-grams, and triangular diagrams. Isolated highs (and lows) on a contoured diagram of a variable may also indicate the possibility of an outlier that requires explanation. Some of these techniques are introduced elsewhere in this chapter.

The graphic and statistical techniques for outlier recognition do not necessarily imply that the data so recognized are incorrect; the methods simply identify the samples that might represent errors and for which an explanation is important. In addition to these procedures, a formal evaluation of data quality is essential. Outliers are discussed in detail in Chapter 7.

6.3.1: Composites

Raw data for a mineral inventory estimation generally are obtained from a variety of supports (Figure 6.2); hence, they are commonly combined in such a way as to produce composites of approximately uniform support. Composites are combinations (mixtures) of either samples or analytical data. In some cases, several individual samples or representative parts of individual samples are combined physically in order to form a single representative sample for purposes of certain kinds of testing. Material for mill or bench tests of various ore types is commonly obtained in this manner. For example, early in the evaluation history of a deposit, it is common practice to combine core from a group of diamond-drill holes in order to conduct initial milling tests. In some cases, physical mixing of samples is used as a cost-saving procedure with regard to analytical costs.

For mineral inventory purposes, however, compositing generally involves the combining of existing data values (i.e., compositing is a numerical procedure that involves the calculation of weighted average grades over larger volumes than the original samples). Commonly, such compositing is linear in nature, involving the calculation of weighted averages of contiguous samples over a uniform length greater than a single sample length. A substantial smoothing effect (reduction in dispersion of grades) results as illustrated in Fig. 6.3 because compositing is equivalent to an increase in support.

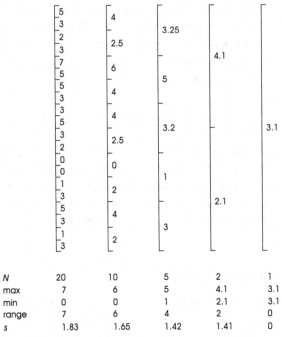

N	20	10	5	2	1
max	7	6	5	4.1	3.1
min	0	0	1	2.1	3.1
range	7	6	4	2	0
s	1.83	1.65	1.42	1.41	0

Figure 6.3: Idealized example of the formation of composite values from smaller subsamples along a linear sample (e.g., a drill hole). Note the decrease in dispersion of values as the composites become longer. This is an example of a more general geostatistical equation that relates the semivariogram and dispersion variances for any two sample supports (see Section 8.9).

Data are composited to standard lengths as a basis for mineral inventory studies to

(i) reduce the number of samples. If there are many thousands of samples, numbers of values can be substantially reduced (two- to four-fold), thus speeding data handling and increasing efficiency of various computer programs involving data searches (cf. Raymond, 1982).

(ii) bring data to a common support. For example, in their geostatistical study of the Golden Sunlight gold deposit Sinclair et al. (1983) combine drill core samples of various lengths to a common length of approximately 10 ft. Raymond (1982) combined 1.5- and 2.5-m sample assays at the Mt. Isa copper deposit to produce 10-m composite grades for use in obtaining mineral inventory estimates.

(iii) reduce the effect of isolated high-grade samples and thus reduce the erratic character of experimental semivariograms and attendant difficulty in semivariogram modeling (Sinclair et al., 1983). This procedure also somewhat reduces the problem of overestimating high-grade blocks that can result from isolated high-grade samples.

(iv) produce bench composites (i.e., composites whose lengths extend from the top of a bench to the base). Such composites are particularly useful when two-dimensional estimation procedures are used for benches. The production of bench composites from raw data, which themselves are not bound by bench floors, normally involves a component of artificial smoothing of the data. As an example, consider a 5-m long sample, only 1 m of which occurs in the bench interval of a composite. The 5-m sample grade is assigned to the 1-m portion and weighted into the bench composite. The repetition of this procedure introduces a slight artificial smoothing to bench composites, in addition to the inherent smoothing involved in the compositing process (see previously and Fig. 6.3). To avoid this artificial smoothing, composites should not be restricted to a bench interval; in which case, block estimates for a bench should be made by three-dimensional rather than two-dimensional estimation procedures. Note that the summation of grade times length and length values should be the same before and after compositing.

(v) reduce the likelihood of assigning peculiar weights by the kriging procedure (see Chapter 13).

Rendu (1986) notes that several matters need to be considered to ensure the most effective compositing results, as follows:

(i) definition of geologic domains within which compositing is to be done.

(ii) categorization of boundaries between geologic domains as sharp or gradational (fuzzy). Composites should not cross sharp boundaries, but can extend into gradational contact zones.

(iii) the choice of composite length for large deposits

is controlled by bench height or the smallest dimension of the selective mining unit. For tabular deposits, composite length might be limited by the distance between footwall and hanging wall and whether they are sharp or gradational.

(iv) When the forgoing decisions have been made, the following information should be calculated and recorded for each composite: coordinates of the top and bottom of each composite, coordinates of the center of each composite (because composites are generally considered to be points in mineral inventory estimation), weighted average grades, length of composite (because of boundary limitations, they are not all the desired length), and important geologic characteristics.

Detailed attention to the compositing procedure is essential because composites generally replace raw data as the basis for mineral inventory estimation. As a rule, composites from a given geologic domain are used in making mineral inventory estimates of that domain only. A question arises as to how to deal with segments of lost core; a common operating procedure is to form a composite only if information is available for at least half of the composite length. A similar procedure can be used for the variable lengths at the end of a drill hole. Where exploration drill holes are stopped in high-grade ground, a good practice is to form a composite from the bottom of the hole toward the collar. This procedure provides a sounder database within the high-grade zone.

In general, if both the data and the composites are unbiased, the average grades of all composites should be centred on the same mean value as determined for the smaller components (raw data) from which the composites are constructed. Minor departures of the two means are to be expected because small amounts of data might be ignored in the compositing procedure. In some cases, however, composites are filtered to produce the database for resource/reserve estimation, and the data and composite mean values can be very different. Lavigne (1991) summarizes a preliminary mineral inventory study of the Lac Knife graphite deposit in which he shows that selected 4-m composites have a significantly lower Cg (carbon as graphite)

grade than do the original 1-m samples (14.91 percent vs. 17.10 percent). The reason for this discrepancy is that a filtering of the composites was done for estimation purposes; only composites above 4 percent Cg were accepted in the database. Numerous "isolated" values greater than 4 percent were incorporated into composites with average grades below this threshold and thus were lost for estimation purposes when the filter was applied.

In certain cases, especially for tabular bodies crossed by lines of contiguous samples (e.g., drill holes crossing roughly tabular bodies, for which the margins are not clearly defined geologically), it might be desireable to produce a composite of nonuniform length whose length is constrained by a minimum length (e.g., minimum mining width), a maximum length, and might include a mining locus (e.g., the center of a vein). A program to achieve some or all of these goals is discussed by Diering (1992), who describes a procedure that optimizes the composite by maximizing either monetary benefit or tonnage within constraints such as those listed previously. The procedure uses a *seed value*, or a location along a linear sample, that is constrained to be included in the eventual composite. The method has useful application when distances between adjacent composites are short and there is a reasonable expectation of physical continuity between neighboring, optimized composites. However, in many practical situations involving irregular, gradational variations in grade, it would be dangerous to assume a physical continuity between widely spaced optimized composites (e.g., in the case of widely spaced drilling through variably mineralized, shear, or alteration zones).

6.4: UNIVARIATE PROCEDURES FOR DATA EVALUATION

Statistical parameters are useful in summarizing data and as a basis for comparisons of various data subgroups. Means and dispersions can be used for this purpose and a variety of graphic or classic statistical approaches (hypothesis tests) can be used to make such comparisons. Statistical tests might include χ^2 tests (perhaps to test formally whether a histogram can be assumed to represent a normal distribution), t-tests (to compare mean values of two data sets representing different geologic domains), and F-tests (to compare variances obtained by two laboratories for analyses of the same standards). Procedures for these and other tests are provided in many introductory statistical texts.

6.4.1: Histograms

Histograms are a simple and familiar method of displaying information about numeric variables. Box plots are a simpler but less informative alternative. Three histograms of ore grades are shown in Figure 4.2, illustrating negatively skewed, symmetric, and positively skewed data. It is evident that histograms are useful in determining the form of a distribution, spread of values and range of greatest concentration of values, and the presence of multiple subpopulations. Unfortunately, this information can be clouded unless histograms are prepared carefully. From a practical point of view, histogram shape may suggest certain approaches to the estimation procedure (e.g., Raymond, 1982); for example, a lognormal sample distribution suggests that a proportional effect must be taken into account in the development of a semivariogram model.

6.4.2: Raw (Naive) versus Unbiased Histograms

Exploration data commonly are concentrated in zones of relatively high grade; therefore, the histogram of raw grades is biased (see Section 4.3). It is necessary to remove the effects of clustering in order to produce an unbiased histogram that (i) removes misconceptions of mean grade, (ii) shows the true form of the distribution, and (iii) serves as a basis for constructing grade-tonnage curves. *Declustering* refers to methods used to minimize the effects of such biased spatial distribution.

The principal problem in declustering is to decide on the cell size to use in the determination of weights. This problem is solved, in some cases, by determining the unbiased mean of declustered data for a range

of cell sizes and accepting the cell size that produces a minimum (or maximum) unbiased mean value (see Section 4.3). An example for the Virginia zone, Copper Mountain porphyry district, is shown in Fig. 6.4. Figure 6.4c is a plot of the unbiased mean versus various sizes of blocks used for declustering. A substantial difference exists in means for the two histograms (0.075 vs. 0.098 g/t Au). The unbiased histogram is now available for use in establishing the true form of the data distribution and, in combination with the volume-variance relation, to develop a global resource estimate for the zone (see Chapter 11).

6.4.3: Continuous Distributions

Continuous distributions are commonly used to describe data, and it may be useful to fit such a distribution (e.g., normal, lognormal) to a histogram. The fitting procedure is relatively simple and is described in Chapter 4 (Eq. 4.12) and in many introductory statistical texts; an example is shown in Fig. 4.2b.

Variables of interest in mineral inventory commonly do not fit a normal distribution, but in some cases, approximate a lognormal distribution or mixtures of lognormally distributed subpopulations. Some skewed distributions can be forced to fit a lognormal distribution by the addition of a constant to each raw data item (cf. Krige, 1960). This improvement in lognormal fit arises where there is a surplus of low values in the original data relative to an ideal lognormal distribution. This surplus can be real or can be a result of nonrepresentativity of the data. As a general practice, unbiased histograms (or cumulative histograms) should be used as a basis for characterizing the form of a distribution. In particular, the additive constant of a three-parameter lognormal distribution should be estimated using the nonbiased data. This can be done easily by plotting the nonbiased cumulative data as a probability plot. Then the additive constant k can be estimated from Eq. 4.22.

In many cases, a simple transformation such as an additive constant is inappropriate because it masks the presence of more than one lognormally distributed subpopulation, each of which may have its own geologic and spatial characteristics, and thus may

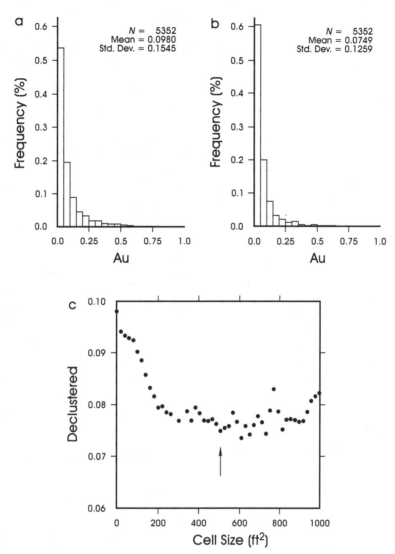

Figure 6.4: (a) Histogram of 5,352 Au assays (ppm) of 10-ft lengths of diamond-drill core, Virginia porphyry zone, Princeton, British Columbia. (b) Unbiased histogram of 5,352 Au values obtained by the cell declustering method. (c) Plot of weighted-mean Au grade versus block size used to produce the weighted mean. Any block size greater than about 200 ft^2 produces a reasonably unbiased histogram. The arrow shows the size used to obtain the histogram of (b).

require separate treatment for resource/reserve estimation purposes.

6.4.4: Probability Graphs

Background information concerning graphic manipulations using probability graphs is given by Sinclair (1974, 1976, 1991). A cumulative normal (or lognormal) distribution plots as a series of points (either individual values or the cumulative data of a cumulative histogram) that define a straight line on equal interval (or log) probability graph paper (e.g., Zn; Fig. 6.5). The literature is full of misinterpretations or misrepresentations of this statement; too commonly, several straight-line segments of a single plot are each interpreted incorrectly to represent normal (or lognormal)

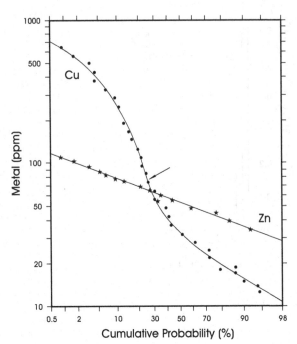

Figure 6.5: Probability graphs (cumulative curves) of metal abundances in residual soils over the Daisy Creek stratabound copper prospect, western Montana (Sinclair, 1991), cumulated from high to low values. The straight line is a lognormal population approximating the Zn data (stars), and the curved line approximating the Cu data (filled circles) has the form of a bimodal distribution (i.e., a mixture of two lognormal subpopulations). The arrow points to an inflection in the Cu curve at the 25th percentile.

distributions. A normal (or lognormal) distribution plotted over a fraction of the probability range on the appropriate probability paper is strongly curved. A case in point is the widespread interpretation of a log probability plot that can be approximated by three "linear" segments to represent three lognormal populations; Sinclair (1974, 1976) shows that such curved patterns are the natural outcome of mixing two lognormal populations (e.g., Cu; Fig. 6.5). Here, several examples suffice to demonstrate the several uses of probability graphs in mineral inventory estimation.

6.4.5: Form of a Distribution

As an example of the use of a probability graph in analyzing the form of a distribution, consider the example

of Fig. 4.13 for 1,000 gold values from a South African gold mine. Cumulative classes of raw data (e.g., the equivalent of a cumulative histogram) are plotted as open circles to which a smooth, concave-downward curve has been fitted. The continuous straight line is a perfect lognormal population with the same parameters (mean and standard deviation) as the raw data. Clearly, the raw data depart substantially from lognormality. However, with the determination of a constant ($K = 55$) using Eq. 4.22, and the addition of this constant to all raw values, the plotted cumulative data (black dots) very closely define a straight line, and the transformed variable closely approaches a lognormal distribution.

Commonly, cumulative plots have the form of Cu in Fig. 6.5 and can be interpreted as mixtures of two lognormal subpopulations. Proportions of the two subpopulations are given by an inflection point on the cumulative curve (cf. Sinclair, 1976). In the case of Cu in Fig. 6.5, the inflection point is at the 25th percentile, indicating 25 percent of an upper lognormal subpopulation and 75 percent of a lower subpopulation. This information is essential in partitioning the curve into its two components, as described in Chapter 7.

6.4.6: Multiple Populations

The presence of multiple populations in a data set to be used for mineral inventory estimation is generally a function of the geology of the deposit. When present, multiple populations must be examined in detail because of the likelihood that each geologic population also represents a distinct continuity environment, a possibility that is too commonly ignored in practice. The amount of data in a particular subpopulation is not necessarily a controlling factor in defining that distinctive populations exist, a topic discussed at greater length in Chapter 7.

Consider the case of Au assays for the Nickel Plate skarn deposit (South Pit) as shown in Fig. 6.6 for one of three sulphide-bearing layers (Sinclair et al., 1994). In this case, the upper subpopulation represents a small proportion of the total data and individual items occur as isolated values in a field of lower-grade mineralization (i.e., the upper population data

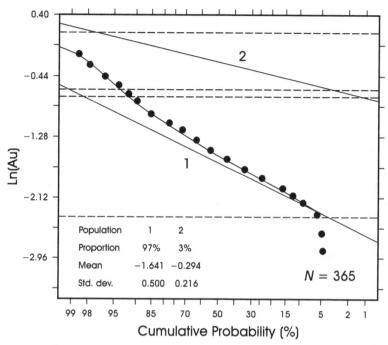

Figure 6.6: Probability graph (cumulative curve) of 365 gold assays, middle sulphide-bearing layer, Bulldog Pit, Nickel Plate Mine (Sinclair and Vallée, 1994). Black dots are cumulative raw data fitted by a smooth (bimodal) curve. Straight lines are the two partitioned subpopulations. Dashed horizontal lines are at the 2.5 and 97.5 percentiles of the partitioned populations; the two central dashed lines collapse to a single value equivalent to a threshold of 0.5 oz/t (−0.7 expressed as a log value) separating the two partitioned subpopulations. The partitioning procedure is described in Chapter 7 and by Sinclair (1976).

are distributed irregularly throughout the mineralized layer). Two lognormal populations, perhaps with differing continuity, are indicated. Therefore, the two populations must be evaluated separately in terms of their contributions to tonnage and grade in a final mineral inventory estimate.

6.5: BIVARIATE PROCEDURES FOR DATA EVALUATION

6.5.1: Correlation

Simple, linear-correlation coefficients range from −1 to +1, depending on whether the two variables whose similarity is being examined are well or poorly correlated (Fig. 4.14). High absolute values of correlation coefficients indicate a close approach to a linear trend of the two variables on a binary plot; a value near zero indicates the absence of such a linear trend. An

example (Fig. 6.7) is provided by a plot of Au versus Ag ($r = 0.81$) from a data set for 10-ft diamond-drill-core samples from a part of the Aurora epithermal Au deposit, Nevada (Poliquin, 1994). For all practical purposes, this linear trend passes through the origin, indicating a remarkably uniform ratio of Au/Ag. With smoothing of data (in block estimation for example) the correlation would become even better and the grade of by-product Ag can be determined directly from the measured Au content and the known Au/Ag ratio. Such a strong relation in small samples is even more pronounced in much larger blocks, and indicates that estimated Au values for blocks can be used to estimate the corresponding Ag abundances.

When variables have more or less symmetric (normal) distributions, either naturally or by transformation, a correlation coefficient can be tested for "significant difference from zero" at a selected level of statistical significance (see Krumbein and Graybill,

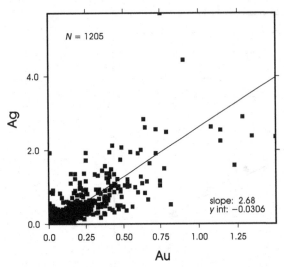

Figure 6.7: Scatter diagram of Ag (ordinate) versus Au (abscissa) for assay data from the Humboldt vein, Aurora gold deposit, Nevada. Both variables in oz/st. The fitted straight line has an equation: Ag = 2.68 (Au) − 0.03. Thus, gold assays provide an estimate for accompanying silver grade. Redrawn from Poliquin (1994).

1965). In this way, attention can be focussed on the significant correlation coefficients.

In addition to concerns about the effect of outliers and nonlinear trends, there are other concerns that affect the interpretation of calculated values of the correlation coefficient; in particular, the incorporation of ratios and percentages among the variables considered. In the case of percentage data, the data are *closed* – that is, they must sum to 100 percent – as opposed to *open* data, which have no such constraint (e.g., Krumbein and Graybill, 1965). The difficulty with percentage data can be appreciated with a simple system of three variables; as one variable increases in relative amount, the sum of the other two must decrease. Therefore, correlations are forced by the arbitrary procedures of reporting information as percentages. The problem decreases as the number of components increases and if components of relatively low percentages are considered. Similarly, ratios contain built-in correlations when they are compared with one of the components of the ratio (e.g., Miller and Kahn, 1962). When possible, ratios and percentage

data should be avoided in correlation studies with very few variables involved.

When there are many variables, their interrelations can be investigated in a matrix of all possible correlation coeficients and in various types of correlation diagrams. An example of a correlation matrix is given in Table 6.1 (see also Fig. 6.9) for 33 *run-of-mine* chip samples (each sample is a continuous series of chips across a polymetallic vein) spaced about 2- to 3-m apart along a drift that exposes both metal-bearing and barren vein material (No. 18 vein, Keno Hill Camp, Yukon). The order of the variables has been rearranged so groups of highly correlated variables are readily apparent. Absolute values of *r* greater than .339 are significant at the 0.05 level of significance (critical value of *r* from tables in Krumbein and Graybill, 1965). Grouping of significantly correlated variables (see Table 6.1 and Fig. 6.8) facilitates interpretation of their mineralogic implications (see Sinclair and Tessari, 1981; Tessari and Sinclair, 1980). Group I is principally galena and a variety of closely associated, less-abundant, silver-bearing minerals, including tetrahedrite and mathildite; Group II is sphalerite; Group III is carbonate; Group IV is pyrite and siderite; and Group VI is arsenopyrite. Interpretation of Group V is uncertain, but it appears to be related to wallrock components, and thus may represent sample dilution. An additional advantage of grouping data so that highly correlated variables are adjacent to each other is that within- and between-group correlations become apparent and give an indication of possible zoning patterns, as discussed in Section 6.5.2.

Correlation coefficients can be used in some cases to develop ideas relating to relative spatial distribution patterns of several variables (e.g., metal zoning). The Keno Hill data in Table 6.1 is a case in point (Sinclair and Tessari, 1981; Tessari and Sinclair, 1980). High positive correlation coefficients indicate that the paired variables are more or less superimposed spatially; high negative correlation coefficients indicate that paired variables are more-or-less disjoint. Intermediate values of correlation coefficients are ambiguous with respect to spatial distribution.

Table 6.1 Correlation matrix, Keno No. 18 vein

	Ag	Pb	Bi	Sb	Cu	Zn	Cd	Hg	Ca	Sr	Mg	Fe	Mn	Ni	F	B	V	Ba
PB	0.99																	
Bi	0.55	0.56																
Sb	0.76	0.77	0.33															
Cu	0.44	0.32	0.43	0.42														
Zn	−0.01	−0.07	0.03	0.32	0.54													
Cd	0.17	0.09	0.07	0.32	0.58	0.87												
Hg	0.09	0.03	0.07	0.51	0.53	0.81	0.67											
Ca	−0.09	−0.08	−0.05	−0.08	−0.13	0.05	−0.10	−0.11										
Sr	−0.11	−0.10	0.01	−0.06	−0.11	0.14	−0.02	−0.06	0.67									
Mg	0.03	0.03	−0.09	0.08	−0.03	0.06	0.02	−0.10	0.24	0.62								
Fe	−0.05	−0.10	−0.12	−0.06	0.20	0.27	0.37	0.10	−0.06	0.22	0.46							
Mn	0.09	0.07	−0.03	−0.03	0.16	0.25	0.33	−0.09	−0.08	−0.08	0.25	0.48						
Ni	−0.21	−0.24	−0.10	−0.22	0.06	0.19	−0.00	−0.05	0.25	0.36	0.32	0.45	0.43					
F	−0.18	−0.19	−0.16	−0.19	−0.06	0.16	−0.01	0.06	0.01	0.40	0.03	0.26	−0.02	0.44				
B	−0.27	−0.27	−0.16	−0.36	−0.15	0.06	−0.01	−0.10	0.01	0.23	−0.16	0.13	0.17	0.41	0.69			
V	0.02	−0.03	0.00	−0.07	0.20	0.24	0.10	0.27	−0.12	0.22	−0.06	0.21	0.08	0.28	0.70	0.44		
Ba	−0.23	−0.21	−0.13	−0.28	−0.21	0.06	−0.08	−0.04	0.03	0.36	−0.12	0.08	0.04	0.40	0.85	0.75	0.56	
As	−0.08	−0.08	−0.07	0.07	−0.07	−0.02	−0.01	−0.00	0.20	0.50	0.72	0.45	−0.16	0.16	−0.02	−0.25	−0.15	−0.22

Correlation matrix based on $\log_{10}$ transformed data

After Sinclair and Tessari (1981)

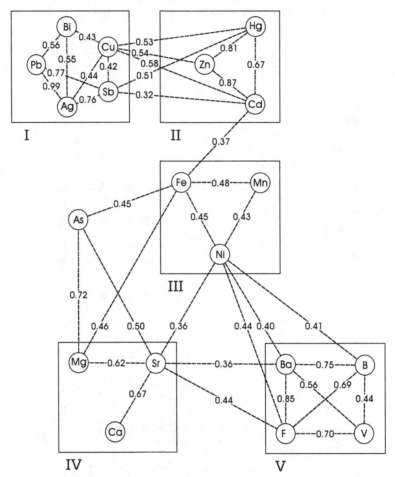

Figure 6.8: Correlation diagram for the data of Table 6.1 for assays from the Keno No. 18 vein, Keno Hill Camp, Yukon Territory (redrawn from Sinclair and Tessari, 1981). Variables are represented by circles; intracorrelated variables are grouped in numbered squares that relate to mineralogy (see text).

6.5.2: Graphic Display of Correlation Coefficients

Graphic displays of correlation information also can be informative. Correlation diagrams (e.g., Krumbein and Graybill, 1965) assign a circle to each variable; circles of highly correlated variables are clustered and members of each pair of highly correlated variables are joined by a line indicating the correlation coefficient. Such diagrams are constructed most conveniently by beginning a cluster of highly correlated variables with the highest correlation coefficient and successively adding the next lowest values to the emerging diagram. Correlation coefficients for the No. 18 vein (Keno Hill, Yukon) give the correlation matrix of Table 6.1, shown as a correlation diagram in Fig. 6.8. Another widely used means of displaying correlations in a multivariate data set that also aids in grouping variables into intracorrelated groups is the dendrograph (McCammon, 1968). The data of Table 6.1 and Fig. 6.8 are shown as a dendrograph in Fig. 6.9.

These graphic techniques combined with an orderly evaluation of a correlation matrix provide insight into groups of correlated elements and are useful in developing attributes of an ore deposit model.

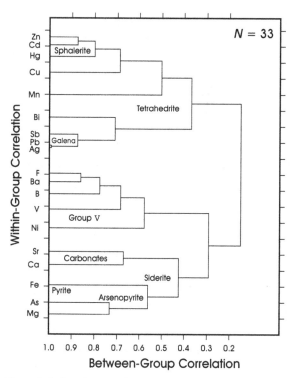

Figure 6.9: Dendrogram for the correlation coefficients of Table 6.1, Keno No. 18 vein, Keno Hill Camp, Yukon Territory. Note slight differences in groupings of elements by this method, in contrast with the correlation diagram of Fig. 6.8.

However, the two graphic methods will not necessarily end up with exactly the same groupings of variables, as is evident in a comparison of Figs. 6.8 and 6.9. This arises because of the averaging of correlation coefficients that is necessary in the construction of the dendrograph. In a large multielement data set, such differences are minimal and generally do not affect the overall interpretation.

6.5.3: Scatter Diagrams and Regression Analysis

Scatter (binary) diagrams are one of the simplest, and yet most useful, graphic procedures for evaluating data. They serve as a rapid check for outliers, either using duplicate analytical data or examining correlated variables; they are excellent for identifying cluster-ing of data and systematic variations between pairs of variables, regardless of whether the relations are linear, quadratic, exponential, or otherwise.

When many variables are under study, binary plots of all possible pairs of a multivariate data set are a simple and effective means of evaluating outliers, recognizing strongly correlated pairs of variables, examining paired data in error analysis, and recognizing the possiblity of more than one bivariate subpopulation, each with its characteristic correlation. It is particularly useful in mineral inventory studies to examine binary plots of by-products and co-products versus the principal metal(s) under investigation. An example for the Humboldt epithermal gold vein (Aurora Mine), Mineral County, Nevada (Fig. 6.7), permits direct estimation of Ag grades from Au assay by the following relation:

$$Ag(oz/t) = 2.68 \times Au(oz/t) - 0.03.$$

In this case, the y intercept is so low as to be insignificant, and the Au/Ag ratio alone can be used to estimate the Ag content without analyzing all samples for Ag.

Linear trends are also indicated by high values of correlation coefficients, but such coefficients can be affected by the presence of nonlinear trends as well as the presence of outliers. Consequently, examination of scatter diagrams is a safer means of identifying outliers. Rapid procedures for viewing all possible scatter diagrams are essential for very large sets of multivariate data; many sophisticated software packages provide this facility to one degree or another. The P-RES program (Bentzen and Sinclair, 1993) offers such a capability.

Generally, it is not safe to assume a linear trend between two variables without a close examination of the scatter plot. An example in Fig. 6.10 illustrates a quadratic model of sulphur grade versus density for massive copper–nickel ores of the Sudbury District. In the case of nickel versus density (Fig. 4.18), it is questionable if a quadratic trend adds any information of consequence to the linear model. However, for sulphur versus density it is clear that a linear trend would be inappropriate; instead, the relation is well described by a quadratic trend.

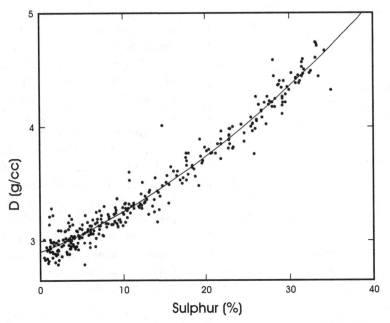

Figure 6.10: Scatter plot of sample density (D) versus sulphur content (percent) for massive Cu–Ni samples of the Sudbury Mining Camp (redrawn from Bevan, 1993). A well-defined curvilinear pattern is approximated by a quadratic trend with equation $D(g/cc) = 2.99 + 0.0283S + 0.000675S^2$.

6.6: SPATIAL CHARACTER OF DATA

6.6.1: Introduction

Knowledge of the spatial character of data is implicit in some mineral inventory estimation procedures. Here, concern is with preliminary methods for early recognition of systematic spatial distribution patterns (e.g., metal zoning), systematic trends and distinctive geologic domains recognized by their characteristic statistical distributions of metals. The principal approaches to evaluating the spatial character of data at an early stage of data evaluation include (1) contouring variables in plan and section; (2) the use of symbol maps to examine for preferred spatial distributions of specific abundance levels of a metal; (3) moving average maps; and (4) grade profiles. Postings of data (plans or sections with values shown) are a useful beginning, particularly when data sites are also coded by symbols that correspond to various geologic attributes (e.g., rock type, alteration intensity).

Moving window statistics are a useful means of examining spatial variations (e.g., Isaaks and Srivastava, 1989). Problems can arise in determining the appropriate size of window for which statistics are determined, overlap of windows (imposing smoothing), window shape (rectangular, circular), number of data points per window to give adequate estimates of parameters, and so on. In general, the local mean and standard deviations are mapped for all windows and spatial variations in the two parameters are examined. Clustering of either parameter is an indication of either a trend or the presence of different geologic domains. A plot of the window means and standard deviations also can be used for recognizing the presence of a systematic relation (e.g., proportional effect) that might be important in a subsequent study of autocorrelation.

6.6.2: Contoured Plans and Profiles

Where variables are gradational in nature and there is reason to be confident that this gradational character exists between control points, useful information can be obtained from contour plots of both plans and

a

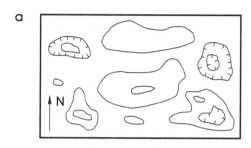

b

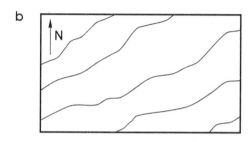

c

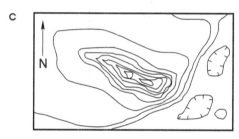

Figure 6.11: Idealized patterns of contoured maps. (a) Isotropy indicated by contours with no pronounced preferred orientation. (b) Planar trend defined by regularly spaced, straight contours. (c) Separate domains indicated by abrupt change in style of contours. Modified from Journel (1987).

sections. In particular, contours indicate trends, directions of preferred elongation (isotropy versus anisotropy to the spatial distribution), and indications of the need for more than one domain (Fig. 6.11). Contouring is most commonly done with computer software and the detailed procedures can be hidden from the user; it is important to have a clear understanding of the contouring criteria contained within a given software package. In the general case, data are irregularly distributed spatially. Most contouring routines use some kind of interpolation criterion with

which to construct a regular grid of values that can be contoured relatively easily (e.g., by simple linear interpolation). The grid spacing has a pronounced effect on smoothing, and the spacing appropriate for one part of a data field may not be appropriate for another. The grid spacing should be roughly equivalent to the average data spacing; as the grid spacing gets progressively larger, the amount of smoothing increases and, conversely, with smaller grid spacing, a level of detail is implied that does not exist in the data. Each node of the grid is estimated by a local data set, commonly selected by octant search, with some limit to the amount of data per octant and a minimum number of values required. Once local data are defined, an interpolation algorithm must be implemented to determine the estimate for the node, for example, an inverse distance weighting procedure such as defined in Section 1.5.4. A variety of smoothing procedures can also be imposed on the final contoured product for esthetic purposes. For interpretation purposes, it is useful to have a contoured plot show a posting of control values because the smoothing involved in contouring commonly results in raw data not being honored by the contours. If an inverse weighting algorithm is used for interpolation, it is important to recognize the following:

(i) The choice of power is of an arbitrary nature (e.g., $1/d$, $1/d^2$, $1/d^3$).

(ii) The minimum number of data required for node estimation controls the extent beyond the data for which node estimates are determined.

(iii) High powers of d in the denominator of the weighting algorithm (e.g., $1/d^3$, $1/d^4$) create less smoothing than do low values. Very high values of the exponent approximate a nearest-neighbor interpolator.

Contoured plans or profiles are commonly used in both developing and operating mines as a means of mine planning and grade control, respectively. Such applications should be avoided or viewed with extreme caution where control data are widely spaced. Figure 6.12 compares high Cu and Au values for one bench interval of the Virginia porphyry copper deposit; the two patterns are very similar, a reflection of

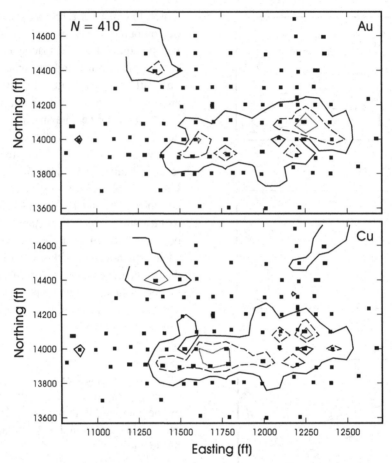

Figure 6.12: Contoured values of Cu (lower) and Au (upper) assays for a 20-ft bench, Virginia porphyry Cu–Au deposit, southern British Columbia (Postolski, unpublished data, 1995). Squares are sample locations from vertical drill holes; rectangles indicate samples from inclined drill holes. Contour values for both metals are: solid contour = mean value; dashed contour = mean + 0.75 std. dev; dotted contour = mean + 1.5 std. dev. The similar contour patterns for the two variables indicate they are strongly correlated.

the high correlation coefficient ($r = 0.61$) between the two variables. The implication of this information is that the grades of Cu and Au will vary sympathetically during production.

6.7: MULTIVARIATE DATA ANALYSIS

Exploration and evaluation programs generate vast amounts of multivariate data with which to attempt to model the geologic–geochemical environment that comprises a mineral deposit. Most of the foregoing data evaluation procedures involve either single variables or simple graphical approaches to examining two or three variables. Modern data-gathering tech-

niques, including new cheap analytical procedures, commonly lead to the quantification of many tens of variables for individual sample sites. This inundation of data has led to attempts at using a wide range of multivariate data-interpretation procedures in an effort to: (i) recognize fundamental groupings of variables that behave similarly; (ii) reduce the time involved in data handling and interpretation to a manageable level; and, at the same time, (iii) retain any information important to the geology and grade-distribution pattern of the deposit.

Multivariate techniques used in the context of mineral inventory studies include applications of multiple regression, cluster analysis, discriminant analysis, and factor analysis. These methods have not been widely adopted in exploratory data analysis as a prelude to mineral inventory estimation and it is not our purpose to review or discuss these mathematically sophisticated procedures in detail. The interested reader is referred to a variety of readily available texts, including Koch and Link (1970) and Miller and Kahn (1962). These procedures are mathematically complex. In some cases, use of these procedures imparts a certain artificial tone to the analysis because of the following:

(i) Some of the variables considered (e.g., length of major fractures, distance to a particular geologic feature) are of a peculiar nature.
(ii) Data transforms are needed in order to bring some variables to a semblance of a symmetric distribution.
(iii) The assumptions involved may be highly idealistic (e.g., normality of axes in n-dimensional space may produce axes that are uninterpretable, i.e., complex variables that consist of weighted contributions of several variables).
(iv) Many complex variables that result from multivariate analyses commonly correspond with a basic geologic grouping that is known in advance.

Here, two simple examples, triangular diagrams and multiple regression, are used to demonstrate some of the potential applications of data-analysis procedures involving more than two variables.

6.7.1: Triangular Diagrams

Triangular graphs are used routinely to display relative compositions of samples in terms of three variables. Indications of absolute abundance levels generally are lost, whereas information on relative abundances is retained. In cases in which metal abundances differ by several orders of magnitude, it is common practice to multiply one or two of the elements by an appropriate factor so that plotted points spread over much of the triangular field. This procedure results in a drastic distortion of the ratio scales that are implicit in such diagrams. For example, a line drawn

from vertex A to the midpoint of the opposite side BC of a triangular diagram normally represents a $B{:}C$ ratio of 1. However, if B has been multiplied by 10 and C by 10,000, this centrally located line represents a true ratio of $B/C = 0.001$. Finally, it is important to bear in mind that relative errors in ratios are larger than errors of the individual elements that comprise them. For example, a relative error of 10 percent in each of the members of a ratio leads to a relative error of about 14 percent in the ratio (assuming independent variables and random error).

Triangular diagrams are useful in recognizing clustering of data in both two and three dimensions. An example (Fig. 6.13) taken from a multivariate data set for the Snip Gold Mine in northern British Columbia (cf. Rhys and Godwin, 1992) shows a well-defined pattern. Linear trends that extend from a vertex of a triangular diagram indicate a constant ratio of the two elements forming the other two vertices. Clustering of data about a point indicates that the relative proportions of all three variables are uniform. Separate concentrations of data on a triangular diagram may aid in the recognition of multiple domains for purposes of mineral inventory estimation. Figure 6.13 is a Au–Ag–Pb triangular diagram for the Snip mesothermal gold deposit (Rhys and Godwin, 1992),

Figure 6.13: Triangular plot of Au (g/mt), Ag (g/mt), and Pb (percent), Snip mesothermal gold deposit, northern British Columbia. Gold and silver are strongly correlated for very high Pb content but are essentially uncorrelated otherwise.

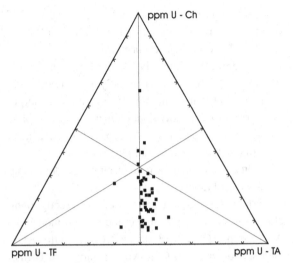

Figure 6.14: Triangular plot of chemical analyses of uranium (ppmU-Ch) versus two borehole radiometric logging estimates for 46 verification samples of a pitchblende-bearing quartz vein. On average, the two drill-hole logging techniques grossly overestimate relative to the chemical results. Moreover, the TA logging results consistently overestimate relative to the TF logging estimates.

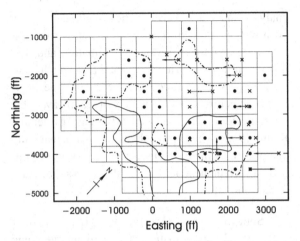

Figure 6.15: Exploration grid (400 × 400 ft[1]) over the Casino porphyry deposit, Yukon Territory, Canada. Dots are drill holes used to construct a multiple regression model for Cu abundance; Xs are drill holes used to verify the model. Contours are for hypogene Cu grades estimated by the model at grid centers. Contours of estimated values are 0.4 percent Cu (solid) and 0.2 percent Cu (dash–dot). Modified from Godwin and Sinclair (1979).

for which two such metal associations are evident. For relatively high Pb contents, there is a concentration of data along a line with an Ag/Au ratio (both variables in ppm) of about 0.4. When Pb is less than about 90 percent of the three elements, the ratio Ag/(Au + Pb) is roughly constant, at less than 0.1. Increasing Au values are thus seen to coincide with decreasing Pb content and a decreasing Au/Ag ratio.

Figure 6.14 compares three methods of uranium analysis for 46 samples. The plot clearly demonstrates that, on average, both radiometric methods (TA and TF) overestimate substantially relative to chemical analyses. Moreover, the TA method, on average, overestimates by about 5 percent relative to the TF method.

6.7.2: Multiple Regression

Godwin and Sinclair (1979) describe a detailed multiple regression study of 20 of the more than 40 exploration-generated variables available for each of the 125 cells of an exploration grid (each cell is 400 × 400 ft²) over the Casino Cu–Mo porphyry deposit, Yukon Territory, Canada (Fig. 6.15). The variables

used, derived from geologic, geophysical, and geochemical surveys over the property (Table 6.2), were transformed to produce variables with similar relative values and dispersions (i.e., to approach homoscedasticity). A variety of multiple regression models were then obtained using a training set of information for 35 well-informed cells with centrally located, vertical diamond-drill holes. For example, one model relates the average Cu grade of hypogene mineralization (averaged over drill-core lengths that generally exceeded 200 ft) to a subset of the transformed variable of Table 6.2. The model, based on 35 control drill holes, was verified by estimating grades in an additional 18 drill holes (Fig. 6.15); grades estimated using the model were remarkably close to reality. Then, the model was used to estimate potential average Cu grades for all 125 cells; the results were assigned to the corresponding cell center and were contoured to provide a guide for future exploration drilling (Fig. 6.15). A complex model such as this must be recognized as a working hypothesis. To attach too much faith would be foolish; the initial quality of many of the variables is relatively poor, and the data transforms introduce some difficulty in interpretation. The procedures,

Table 6.2 Selected variables for casino porphyry Cu–Mo grid, used to establish a multivariate model for the local value of hypogene Cu grade

Dependent variable	
	Hypogene grade, Cu (%)
Independent variables	
Rock geochemical	Cu, cell average (ppm)
	Mo, cell average (ppm)
	Pb, cell average (ppm)
	Zn, cell average (ppm)
Geophysical	Ground magnetic, avgerage cell value (gamma)
	Ground magnetic, range in cell (gamma)
	Resistivity, average cell value (ohm-m)
	Resistivity, range in cell (ohm-m)
	Distance, ground mag high to cell center (ft)
	Distance, main airborne mag low to cell center (ft)
	Distance, secondary airborne mag low to cell center (ft)
	Airborne mag value at cell center (gamma)
	Airborne mag range in cell (gamma)
Lithologic	Area of cell underlain by quartz monzontie (%)
	Area of cell underlain by Patton dacite (%)
	Areas of cell underlain by breccia (%)
Alteration	Area of cell underlain by phyllic alteration (%)
	Area of cell underlain by potassic alteration (%)
	Area of cell underlain by visible hematite or magnetite
	Area of cell underlain by rocks with visible tourmaline and magnetite or hematite

however, provide an objective and quantitative means of establishing relative importance (weights) of geologic variables (the coefficients of the multiple regression model) in contrast to the highly subjective and nonreproducible weights that are commonly used. Such methods cannot be used directly for reserve estimation, but they can contribute to decisions regarding value continuity and limits to domains.

6.8: PRACTICAL CONSIDERATIONS

1. Expect that a substantial portion of work directed toward a mineral inventory estimation will be attributed to data organization and evaluation, perhaps 25 to 50 percent over the duration of the project.

2. A computer-based data-management system is essential and should be incorporated into any mineral evaluation program at the start of the program. A variety of software then can be applied to data evaluation with efficiency.

3. When raw data are to a common support, they may provide a satisfactory base from which to proceed in a mineral inventory study. When composites are constructed, close attention must be paid to choice of composite length and the rules established for determining composite values.

4. Data (perhaps composites) should be evaluated with an orderly or structured approach that emphasizes simple methods first, and increasingly complicated methods only as the study progresses and results warrant.

5. Specific studies should be directed toward (i) characterizing probability density functions, (ii) documenting interrelations of important variables, (iii) recognizing separate data domains and evaluating data separately from each domain, (iv) documenting spatial distribution patterns of important variables (including localization of high grades), and (v) understanding outliers. It is important to understand the aims of each stage in the evaluation process.

6. Interpretation in the course of applying data-evaluation procedures is facilitated by a consideration of geology.

6.9: SELECTED READING

Davis, J. C., 1986, Statistics and data analysis in geology (2nd ed.); John Wiley and Sons, New York, 646 pp.

Sinclair, A. J., 1976, Applications of probability graphs in mineral exploration; Assoc. Exp. Geochem., Spec. v. 4, 95 pp.

Swan, A. R. H., and M. Sandilands, 1995,

Introduction to geologic data analysis; Blackwell Science Ltd., London, 446 pp.

6.10: EXERCISES

1. Calculate the values of 10-m composites for each of the following two drill holes:

| ddh # 1 | | Grade | ddh # 2 | | Grade |
From (m)	To (m)	(% Cu)	From (m)	To (m)	(g Au/mt)
27	28.5	.74	157	159	1.3
28.5	31.5	.60	159	161	3.3
31.5	41.3	.11	161	163	2.2
41.3	43	.44	163	165	4.0
43	45	.16	165	167	4.9
45	47	.51	167	169	1.7
47	49.5	.46	169	171	5.9
49.5	53	.23	171	173	1.5
53 = end of hole			173 = end of hole		

ddh = diamond-drill hole.

2. Simulate a single drill hole consisting of 30 successive samples of 1 m, to each of which an assay value has been attributed. For this purpose, it is adequate to select values using random number tables. Construct 2-m, 5-m, and 10-m composites from this simulated data set. Comment on the means and standard deviations for the various data/composite lengths.

3. Develop a general algorithm for the formation of bench composites from vertical, exploration, and diamond-drill-core assays. Allow for variable length of core assays, variable starting elevation for a bench, variable bench height, and lost core.

4. Plot the data of Table 4.4 on both arithmetic and logarithmic probability paper. Interpret the results.

7

Outliers

How to deal with erratic high samples is one of the knottiest problems in ore estimation. (McKinstry, 1948, p. 49).

This chapter is concerned with the difficult problem of recognizing and classifying outliers and determining if or how to incorporate them into a mineral inventory estimation. A conceptual model of multiple populations is introduced, and probability graphs are recommended as a useful tool for identifying thresholds separating populations. These thresholds provide a means of isolating outlier populations and examining their spatial characteristics.

7.1: INTRODUCTION

An *outlier* is an observation that appears to be inconsistent with the vast majority of data values. In a mineral inventory context, concern is directed especially toward outliers that are high relative to most data. The following are among the problems outliers can cause:

(i) Introduction of substantial variability in estimates of various statistical parameters including the mean, variance, and covariance. Hence, they have an impact on autocorrelation measures such as the experimental semivariogram relative to what would be obtained otherwise.

(ii) An outlier value that is approximately central to a block being estimated can result in an abnormally high mean value being assigned to the block, thus leading to a problem of overestimation of both high-grade tonnage and the grade of that tonnage.

(iii) In block estimation procedures such as kriging, if an outlier value coincides with a negative weight, the resulting kriged estimate can be seriously incorrect, and may even be negative in extreme cases.

These difficulties have been recognized for some time, and a variety of approaches have evolved for dealing with outliers in the course of a mineral inventory estimation, including the following:

(i) All outlier values receive special treatment, which can include a variety of alternatives, such as reanalyzing (if possible) or cutting (decreasing) to some predetermined upper limit based on experience (arbitrary) or using an empirical cutting method tied to the histogram (95th percentile of data) or statistical parameters (mean plus two standard deviations), that experience has shown to be acceptable (see also Parrish, 1997).

(ii) Treatment of outliers in many geostatistical studies involves omission of the outliers during estimation of the semivariogram, but use of the outliers in subsequent ordinary kriging.

(iii) In some cases, so-called outliers represent a separate geologic population in the data that may coincide with an identifiable physical domain, which, for estimation purposes, can be considered independently of the principal domain.

In general, extraordinarily high values should be viewed with scepticism during mineral deposit evaluation. These high values may arise because of errors or may reflect distinct geologic subenvironments or domains within a mineral deposit. Effort must be directed to examining these high values and their geologic context as soon as is feasible after identification in order to distinguish errors from "real" values and investigate the characteristics of the real values and how they will be integrated into mineral inventory estimates.

Treatment of real outliers in resource/reserve estimation is a perplexing problem to which there is no generally accepted solution at the present time. The specific concern is that very high values not be assigned too much weight, or they contribute to an apparent tonnage of high-grade ore that does not exist. Experience over many decades has shown that a small proportion (of the order of 1 percent or less) of gold grades that are very high (perhaps one to two orders of magnitude higher than the average of remaining samples) can lead to serious overestimation of average grade (and possibly tonnage) above the cutoff grade. This is particularly so if the very high grade values are treated in the same manner as are lower-grade values during resource/reserve estimation. Commonly, outlier populations are geologically distinctive and have very limited physical (geologic) continuity relative to lower-grade values; therefore, to assume that high grades can be extended into neighboring rock the same distance as low-grade samples are extended could lead to a significant overestimation of resource/reserves.

7.2: CUTTING (CAPPING) OUTLIER VALUES

[C]utting is a little like democracy – it's a lousy system but it works. (Clow, 1991, p. 34).

7.2.1: The Ordinary Case

Cutting or capping of outlier values is a widespread practice in the mining industry. In brief, *cutting* lowers grades to a critical threshold so that the gold lost by reducing the outlier grades is regained by extending the cut grades (still high grades) in the same manner as low grades are extended for making block estimates. The *cutting threshold* is determined by trial and error. The philosophy of the practice can be illustrated by a simple, two-dimensional example (Fig. 7.1) on which a block is to be estimated by a central drill hole that includes five samples centered within the block. These five values are composed of four values that average 2 g Au/t and one value with a grade of 70 g Au/t. The average of the five samples is

$$g_1 = [(4 \times 2) + (1 \times 70)]/5 = 15.6 \text{ g Au/t}$$

and this value represents one possible estimate (polygonal) of the block. However, suppose we know from a detailed geologic study of the deposit that outlier gold values higher than 30 g/t have a recognizable geologic form with an average physical continuity of 2 m in the two principal directions. Given the various measurements for the two-dimensional block of

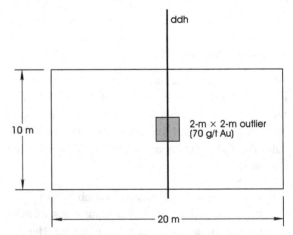

Figure 7.1: Diagram illustrating a small, two-dimensional outlier (i.e., short distance of physical continuity) assaying 70 g Au/t in a field of much lower grades circa 2 g/t. The physical dimensions of the outlier (2 m × 2 m) are small compared with a block to be estimated (20 m × 10 m).

Fig. 7.1, it is possible to produce another block estimate that is weighted according to the likely areas of the two grade values. In particular, the 70 g/t values would have a weight of $(2 \times 2)/(20 \times 10) = 0.02$, whereas the 2-m average would have a weight of $[(20 \times 10) - 2 \times 2)]/(20 \times 10) = 0.98$. These weights produce a true block grade, g_2, as follows:

$$g_2 = (0.02 \times 70) + (0.98 \times 2) = 3.36 \text{ g Au/t.}$$

Consider now the arbitrary decision to cut outlier grades to some predetermined value; for the sake of example, assume that value to be 35 g/t. In this case, the polygonal estimate of the block grade becomes

$$g_3 = [(4 \times 2) + (1 \times 35)]/5 = 8.6 \text{ g Au/t.}$$

The arbitrary decision to cut outliers to 35 g/t has offset some of the overestimation problem, but in this idealized example, the cutting limit (factor) still leads to overestimation. If the value of the block were known, then it would be possible to work backward and find an appropriate value to which the outlier should be cut in order to give the block a correct estimate. For example, on the assumption that g_2 as shown in a previous equation is the best estimate of the block, the following relation can be used to determine the correct cutting factor (F) in this idealized and simple case:

$$3.36 = [(4 \times 2) + (1 \times F)]/5$$

giving

$$F = 8.8 \text{ g Au/t.}$$

In other words, all outlier values should be cut to a value of 8.8 g/t, a value that can be incorporated into the specific block-estimation procedure (polygonal in this case) as if it had the same continuity as the lower grades.

Of course, in practice, the true grade of individual blocks is not known. Instead, average production is compared with estimates based on various cutting factors until a value of F is obtained that gives a reasonable comparison of estimates with production. This emphasizes the principal difficulty with the practice of cutting: choice of a cutting factor is little more

than a guess prior to actual production. An additional problem is that determination of a cutting factor can be based on only a limited part of a deposit; what is optimal for one part might not be appropriate elsewhere in the deposit. Despite these limitations, even when reference to production is not possible, cutting of high grades remains a common practice because of the seriousness of the potential overestimation.

Cutting (capping) of grades is generally undesirable because its arbitrary nature can lead to large uncertainties in grade and tonnage estimates. The procedure can be avoided in a variety of ways, as becomes apparent here and in subsequent chapters.

7.2.2: Outliers and Negative Weights

Some geostatistical procedures of resource/reserve estimation can result in negative weights being assigned to one or more of a group of data selected to estimate a block grade (see Chapter 10). Consider the scenario of five separate grades being used to estimate a block, with four grades averaging 2 g Au/t and the fifth grade having a value of 100 g Au/t and located within the block to be estimated. For the sake of discussion, the block is assumed to be $10 \times 20 \times 20$ m (4,000 m^3) and the outlier is assumed to be $2 \times 2 \times 1$ m (4 m^3). Hence, the true grade of the block is $2(4,000 - 4)/4,000 + 100 \times 4/4,000 = 1.998 + 0.1 = 2.098$ g Au/t.

Now, assume several different weighting factors for the outlier in calculating the block grade.

Outlier weight $= +0.01$
  Block grade $= (0.99 \times 2.0) + (0.01 \times 100)$
            $= 1.98 + 1.00 = 2.98$
42 percent overestimation

Outlier weight $= -0.01$
  Block grade $= (1.01 \times 2.0) - (0.01 \times 100)$
            $= 2.02 - 1.00 = 1.02$
42 percent underestimation

Outlier weight $= -0.03$
  Block grade $= (1.03 \times 2.0) - (0.03 \times 100)$
            $= 2.06 - 3.00 = -0.94$
Impossible.

The lesson of this simple calculation is that even very small weights attached to outliers can lead to significant biases in block grade estimation: positive weights commonly lead to overestimation and negative weights to underestimation. Clearly, in resource/reserve estimation, outliers cannot be treated in the same way as the more abundant range of values.

7.3: A CONCEPTUAL MODEL FOR OUTLIERS

It is useful to examine outliers as part of a more general model of data distribution in which the distribution (histogram or probability density function) is viewed as consisting potentially of multiple populations, each with its own geologic (including grade) characteristics. In particular, well-defined subpopulations commonly can be shown to represent different geologic domains (cf. McKinstry, 1948) characterized by different geologic attributes. Such a data interpretation model is shown schematically in Fig. 7.2 (after Sinclair, 1991), in which outliers are seen to

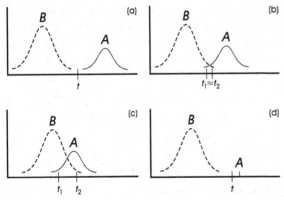

Figure 7.2: Conceptual model for two populations, including the general concept of outliers (Sinclair, 1991). (a) Two subpopulations with no effective overlap of values separated by a single threshold. (b) Two subpopulations with slight overlap of values with two thresholds that for practical purposes can be expressed as an average threshold. (c) Substantial overlap of two subpopulations in which a single threshold is not an efficient separator. (d) The traditional concept of an outlier in mineral inventory use as a few high values (high subpopulation) relative to the bulk of the data (low subpopulation).

be a subpopulation represented by a few values in a multimodal data set (bimodal in the illustration).

This conceptual model attaches no origin to the subpopulations. An outlier subpopulation can be the result of errors entirely, or can consist of real values that differ markedly from the bulk of the data. It is one of the objects of a thorough data evaluation to identify all outliers and classify each of them as errors or as members of a recognizable geologic subpopulation. This aim is assisted immeasurably if outlier values are recognized early during the exploration phase of a property and real (geologic) outliers are examined in detail and characterized geologically.

A thorough discussion of multiple populations is provided by Zhu and Journel (1991), who develop a formal mathematical model for mixed populations and discuss at length the impact that combining populations can have on estimation of parameters using a mixed distribution. Zhu and Journel (1991, p. 670) state, "Traditional models which ignore the mixture (of two populations) are shown to be misleading and even totally inadequate in the presence of extreme conditioning data" (p. 670) and "Mixture of populations might be seen as the norm rather than the exception in real-world practice" (p. 670).

7.4: IDENTIFICATION OF OUTLIERS

Because outliers represent either errors or real anomalies in the character of the material analyzed, data-checking procedures must be designed to recognize errors early and at all stages of data handling. Sound sampling and subsampling procedures minimize a wide range of material handling errors that can mask true values. Similarly, a rigorous systematic approach to obtaining data and monitoring data quality minimizes errors. Verification of data is essential to identify various mechanical errors that can arise in data handling and recording.

7.4.1: Graphic Identification of Outliers

A variety of standard plotting procedures, including histograms, probability plots, and scatter diagrams, (x vs. y) are useful in pinpointing abnormally high

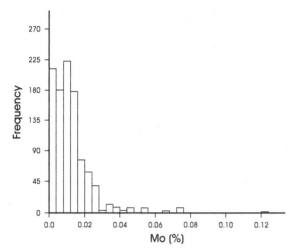

Figure 7.3: Histogram of 1,035 Mo assays, 1972 drilling, main zone, Huckleberry porphyry copper deposit, central British Columbia (after Craig, 1994). Note the difficulty in selecting a threshold with confidence that identifies outlier values that are not simply a part of the upper tail of a lognormal distribution.

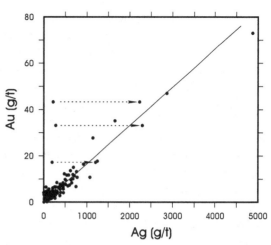

Figure 7.4: A scatter plot of Au versus Ag assays for 335 samples, Shasta gold deposit, northern British Columbia. Three outliers, shown in their incorrect locations, were corrected from assay sheets. The outliers were easily recognized because they departed markedly from the linear trend of Au versus Ag, characteristic of most of the data.

values that can be checked against original data (e.g., assay sheets, drill logs). Reference to Fig. 7.2 indicates the possibility of using histograms for outlier recognition; an example and the problem with their use is illustrated in Fig. 7.3. Probability graphs are generally far more useful than histograms for outlier recognition (e.g., Fig. 6.6).

Figure 7.4 is a plot of Ag versus Au assays for 335 samples from the Shasta epithermal gold deposit (Nowak et al., 1991) illustrating the advantage of binary plots in outlier recognition. Initially, three of the Ag values were recorded incorrectly, one order of magnitude too low. These three values were identified on a plot comparable to Fig. 7.4 and the original assay sheets were rechecked and provided the correct values with which to correct the database. In this case, outliers were recognized because of departures from a very strong correlation between Au and Ag in the Shasta deposit. These outliers would not have been recognizable with univariate graphic schemes such as histograms or probability plots because the few incorrect values are within the range of other values.

Other empirical graphic approaches are also useful for the identification of outliers, including semivariogram clouds and *h* scattergrams (cf. Journel, 1987),

that have become routine tools of geostatisticians and are discussed in Chapter 9. Generally speaking, outlier recognition methods based on semivariograms should not be necessary. A structured data-evaluation program as recommended in Chapter 6 should characterize data distributions and isolate outliers as separate populations long before the stage of semivariogram modeling.

7.4.2: Automated Outlier Identification

A variety of rules can be implemented by computer to check for likely outliers, rules that commonly make use of the standard deviation of a data set or the dispersion of data about a trend. For example, data that are more than three standard deviations from the mean value might be classed as outliers. In some cases, the mean plus two standard deviations might be used as a threshold to separate outliers from ordinary data. Such rules have the advantage that they can be automated for checking very large databases and easily lead to recognition of a small amount of data that can be checked/verified. However, such rules are not perfect, as shown in Fig. 7.4 (error outliers are too low to be recognized by high thresholds), and because they are arbitrary. Eventually, in the course of recognizing outliers, it is wise to examine density distributions for grade as well as referring to sample location plots.

7.5: MULTIPLE GEOLOGIC POPULATIONS

The geologic category of outliers includes those geologically significant high values surrounded by a clearly lower population. This commonly encountered situation generally reflects two distinctive styles of mineralization: (i) disseminated versus (ii) veinlets, in the case of Au in some precious metal deposits. The possible need to recognize separate domains, within each of which estimation is done independently, has long been recognized by geostatisticians (e.g., Journel, 1985). However, common practice has been to include assays for both types of mineralization as the same regionalized variable, without serious thought being given to the fact that each might have a totally different characteristic autocorrelation (value continuity). Consider the overestimation problem that could arise if the greater continuity of widely disseminated Au were attributed to a single outlier Au value that reflected a local vein, centimeters in thickness and of abnormally high grade. Extending this local high-grade Au value over a large volume would lead to a local overestimation of both high-grade tonnage and the average grade of that high-grade tonnage. Unfortunately, despite awareness of the problem, it continues to be treated subjectively in practice, contributing to an all too common overestimation of recoverable grade and tonnage in many precious metal deposits. Different grade subpopulations might be characterized by very different autocorrelation characteristics; consequently, subpopulations must be identified, examined as to value continuity, and, if necessary, considered separately in the evaluation process (e.g., Champigny and Sinclair, 1984).

As an example of the effect of mixtures of populations, consider the simple example presented by Pan (1995) as shown in Fig. 7.5. In this example, a comparison is made of a block estimate with all data more or less the same order of magnitude (#1 = 0.06) versus a similar geometric array in which the value at #1

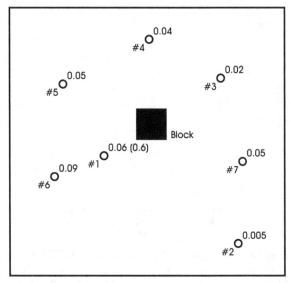

Figure 7.5: An example of the effect of multiple populations on block estimation. Ordinary kriging weights for the samples are: #1 = 0.3, #2 = 0.02, #3 = 0.22, #4 = 0.12, #5 = 0.15, #6 = 0.07, and #7 = 0.12; they provide a block estimate of 0.047 oz/t (cf. Pan, 1995). If sample #1 is replaced with a value of 0.6, the block estimate becomes 0.209, a more than 300 percent increase.

has been replaced by a much higher value (one order of magnitude higher). The presence of two subpopulations in the data seems likely, and there is no reason to expect that they both have the same continuity. An example of this problem is described by Sinclair et al. (1993) for the Nickel Plate skarn Au deposit, southern British Columbia. In this case, two grade populations can be identified on a probability plot of the assay data (Fig. 6.6) and an optimum threshold selected (cf. Sinclair, 1976). Manual inspection of drill-hole grade profiles verifies the presence of two grade-related types of continuity.

7.6: PROBABILITY PLOTS

This simple graphic technique introduced in Chapters 4 and 6 has found wide application in the recognition and characterization of multiple populations in applied geochemistry (Sinclair, 1974, 1976, 1991) and has been used on occasion to deal with multiple subpopulations in mineral inventory studies (e.g., Sinclair et al., 1993). Examples are discussed in detail by Champigny and Sinclair (1984) for the Cinola Au deposit, British Columbia, and by Noble and Ranta (1984) for porphyry–molybdenum deposits. The probability plots for gold data from the Sunnyside, Colorado, and Jerritt Canyon and Sleeper mines in Nevada (Parker, 1991) appear to be comparable examples. Consideration of multiple populations such as indicated in the forgoing examples requires an understanding of partitioning of probability graphs (i.e., a procedure to extract individual subpopulations from the cumulative curve of a mixture of subpopulations).

7.6.1: Partitioning Procedure

There is a lengthy mining literature showing that metal grades are rarely normally distributed, but more commonly approach a lognormal distribution. If this is true for geologically homogeneous ores – that is, ores characterized by a single form of mineralization – then the presence of two distinct types of mineralization might be expected to give rise to a mixture of two lognormally distributed subpopulations, one for each mineralization type. An example is Cu values for the Eagle vein, where Sinclair (1976) demonstrates substantively different lognormal populations for hypogene sulphide assays and grades of leached and oxidized material Fig. 7.6. Such mixtures produce curved patterns on probability graph paper, as shown in Figs. 7.6 to 7.8. In practice, similar patterns are commonly interpreted to consist of two overlapping subpopulations; the central part of the curve contains an inflection point (change in direction of curvature) indicating the possible presence of two lognormal subpopulations. In theory, there is one more subpopulation than there are inflection points; however, sampling and analytical error generally limit the recognition of more than three or four subpopulations in a data set.

Partitioning is the term applied to procedures to separate the cumulative curve of a mixture of subpopulations into the component subpopulations. In practice, partitioning is normally done using a computer program (e.g., Stanley, 1987). A manual procedure for

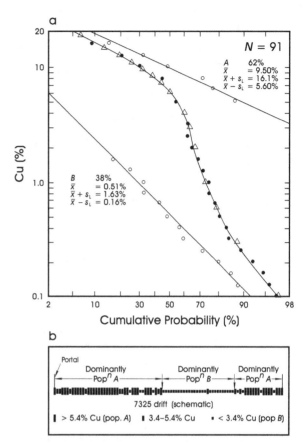

Figure 7.6: (a) Probability plot of Cu assays for 91 samples taken along the 7325 level, Eagle copper vein, at approximately 6-ft intervals. Redrawn from Sinclair (1976). Raw data are shown as dots to which a smooth curve has been hand-fitted. This smooth curve has been partitioned into the two component populations A and B; circles are the partitioning points and the straight lines (fitted by eye) are the estimated lognormal subpopulations that fit the partitioning points. Triangles are calculated perfect mixtures of the ideal subpopulations, A and B; the model describes the raw data very satisfactorily. Most of A subpopulation is higher than 5.4 percent Cu; most of B subpopulation is less than 3.4 percent Cu; very few values lie between these two thresholds. (b) The 91 values coded as to subpopulation (i.e., relative to the two thresholds) showing preferential spatial locations of the subpopulations.

this partitioning is summarized from Sinclair (1976), and the procedure described is for the simple case of two subpopulations. More complicated cases are discussed by Sinclair (1976). The following steps should

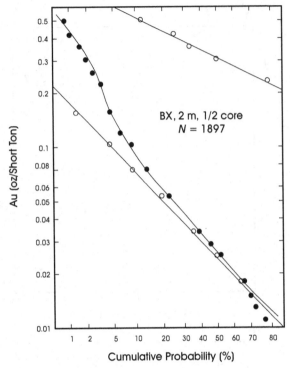

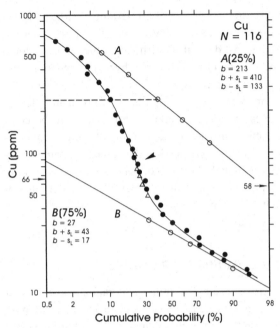

Figure 7.7: Probability plot for 1897 2-m samples of BX half-core, Cinola deposit, Queen Charlotte Islands, British Columbia (Champigny and Sinclair, 1984). Dots are raw data; smooth curve is a hand-fitted bimodal model; circles are construction points involved in partitioning the two log-normal subpopulations (straight lines). A threshold of 0.22 oz Au/st separates the upper subpopulation (3 percent) from the lower subpopulation (97 percent).

Figure 7.8: Example of the partitioning method for a bi-modal probability plot. Dots are cumulative data and the sygmoidal curve is a hand-drawn model fitting the data, with an inflexion point shown the arrowhead. Circles are construction points that define the partitioned subpopulations, A and B. An example is the dashed line (drawn through a dot representing combined data) that is partitioned to give the circle on the solid line representing population A. The values 66 and 58 represent a range within which a single threshold can be selected in order to distinguish the A and B subpopulations efficiently. See text for details.

be read in reference to Figs. 7.6 to 7.10:

(i) Sort data items in order of decreasing value.

(ii) Determine cumulative percentages of the items individually or in groups of uniform class interval.

(iii) Plot the cumulative values on probability paper (see black dots on Figs. 7.6 to 7.10).

(iv) Draw a smooth curve or line through the plotted points to describe the trend of the plotted data. In the case of a straight line, the mean can be read at the 50th percentile and the standard deviation can be estimated, for example, as $(P84 - P16)/2$ or $/(P97.5 - P2.5)/4$.

(v) In the case of a curved trend that can be interpreted as a bimodal distribution (e.g., Figs. 7.6 to 7.10), note the inflection point (change of di-

rection of curvature) in the smooth model hand-fitted to the data.

(vi) Every point P_m on this smooth model has a contribution from each of the two component subpopulations (e.g., the dashed horizontal line in the middle of Fig. 7.8) and can be described by the equation

$$P_m = f_a P_a + f_b P_b$$

where P_m is the probability of the mixture, f_a is the inflection point as a proportion of the data, P_a is the probability (cumulative percent) of subpopulation A, $f_b = (1 - f_a)$, and P_b is the probability of subpopulation B. This equation contains two unknowns, P_a and P_b.

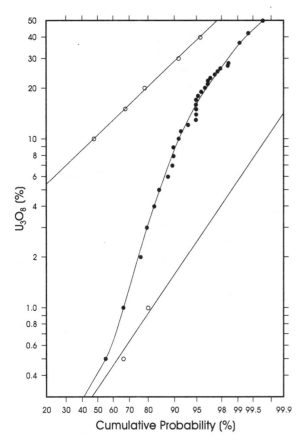

Figure 7.9: Probability graph for 336 simulated U_3O_8 values for eight diamond-drill holes (data from Rivoirard, 1987). The raw data (dots) are approximated by a smooth curve (bimodal model) that has been partitioned into upper ($A = 17$ percent) and lower ($B = 83$ percent) lognormal subpopulations, separated by a threshold of about two. Circles are partitioning values used to derive the two ideal subpopulations. Note that values have been cumulated from low to high, a variation common in technical literature.

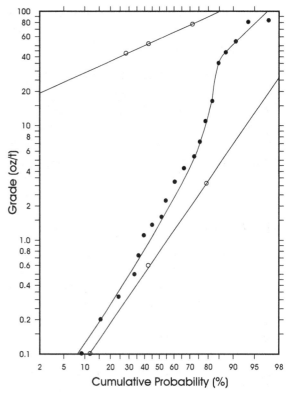

Figure 7.10: Lognormal probability plot for gold accumulation in an epithermal gold deposit, the Sunnyside Mine, Colorado (modified from Parker, 1991). Dots are cumulative data values that have been fitted by a smooth, hand-drawn, bimodal curve. The bimodal curve has been partitioned into the high-grade and low-grade subpopulations that are shown as straight lines through construction points (circles). A threshold of 20 separates the two subpopulations. This model appears to simplify reality; the lower subpopulation is a simplification of what appears to be several subpopulations. Note that values have been cumulated from low to high, a variation common in technical literature.

(vii) If we call the upper subpopulation A and apply the formula to several points on the curve, P_b is effectively zero, and the equation can be solved for several P_m–P_a pairs.

(viii) Thus, several points on the A subpopulation have been determined (shown as open circles in Fig. 7.6) and a straight line can be drawn through them to provide an estimate of the entire subpopulation A.

(ix) Because P_a is now known for all values of P_m, the equation can be solved for several values of P_b, and the B subpopulation can be approximated by a straight line through a plot of the estimated values.

(x) The equation can now be used to calculate a range of P_m values using the estimated subpopulations A and B to verify that the extrapolated lines are satisfactory subpopulations with which to model the mixture (the curved line).

(xi) Thresholds can now be estimated at the upper 2.5th percentile of the B subpopulation and the lower 97.5th percentile of the A subpopulation.

For practical purposes, these two thresholds commonly collapse to a single value (the average).

It is important to understand the partitioning procedure, but the manual method is tedious and time-consuming. Personal computer software is readily available to assist the partitioning process, for example, PROBPLOT (Stanley, 1987) and P-RES (Bentzen and Sinclair, 1993) which can be downloaded through the publisher's website.

7.7: EXAMPLES

Published literature on the topic of outliers has concentrated on the problem of isolated values. In their geostatistical study of the Cinola (Spegona) gold deposit, Queen Charlotte Islands, British Columbia, Champigny and Sinclair (1984) recognized the occurrence of a small percentage of high values (Fig. 7.7) that seemed to be distributed erratically over the large mineralized field. They recognized the very different continuities of the two subpopulations and estimated the impact of each separately on the block model ($100 \times 100 \times 8$ m^3) adopted for their study.

Rivoirard (1987) describes a case history of variography in an uranium deposit. Extreme variations in sill levels of experimental semivariograms are the product of a strongly postively skewed data distribution of U$_3$O$_8$ grades. His study is a classic description of the problem that arises in using a single absolute semivariogram model for values spanning more than two orders of magnitude (mean $= 1.1$, $s = 3.6$, $CV = s/m = 3.3$) in which two subpopulations make up the data. Rivoirard demonstrates that log transformed data provide more stable experimental semivariograms on which to base a structural model, although there is no underlying reason why the two subpopulations should have the same underlying grade-continuity model. Cressie's (1985) work suggests that a relative semivariogram would provide a similar benefit and is more acceptable because it avoids logtransforming the data and subsequent uncertainty with

the "back-transformed" kriging estimates. An alternate interpretation can be considered in the light of the general model for outliers presented here – consider Rivoirard's simulated data from the perspective of possible multiple populations, particularly in view of his descriptions that "the histogram is very skew" (p. 2) and "the distribution cannot be considered lognormal" (p. 2).

A probability plot of the published data ($n = 336$) for eight drill holes shown in Fig. 7.9 suggests the presence of multiple subpopulations – specifically, two lognormal subpopulations. On this assumption, the plot has been partitioned using the procedures described in Section 7.6.1. High- and low-grade populations are shown on the graph, and a threshold of about two can be selected according to the method of Sinclair (1976). This threshold can be used to "isolate" or identify values of the high subpopulation along individual drill holes (Appendix in Rivoirard, 1987), and these identified values can be examined for their spatial characteristics (e.g., average length of groups of contiguous "high" samples). In this case, the average high-grade structure is 3.5 sample lengths (std. dev. $= 2.7$), that is, $3.5 \times 1.5 = 5.25$ m, a composite length that is much less than the range of the various semivariograms illustrated by Rivoirard (1987). This procedure allows location of boundaries between high- and low-grade domains. Hence, each domain can be examined independently for its continuity model. If comparable sampling exists in different orientations through the two subpopulations, they can be characterized in other directions in space. With sufficient data, this can lead to the development of an average three-dimensional model (ellipsoid) representing the high-grade subpopulation. Moreover, domains can be defined in three dimensions; therefore, three-dimensional continuity models for each of the domains can be investigated independently.

Parker (1991) considers the need for a new approach to outlier recognition and uses probability plots as a means of identifying the presence of multiple subpopulations. He discusses three examples involving data for precious-metal deposits

(Sunnyside, Colorado; Jerritt Canyon, Nevada; and Sleeper, Nevada), each of which can be interpreted to consist of two lognormal grade subpopulations. Data for the Sunnyside Mine are reproduced in Fig. 7.10, where they have been approximated by a smooth, hand-drawn curve that has been partitioned into two lognormal subpopulations as described previously. A threshold of 20 separates the two subpopulations efficiently and would serve as a basis for examining the spatial distribution of the two subpopulations. The two other examples discussed by Parker (1991) can be treated in an identical manner.

Of course, multiple subpopulations are not restricted to data containing small proportions of a high-valued subpopulation. A more general example of the application of probability graphs to an understanding of multiple populations and their implications to resource/reserve estimation is provided by a set of 91 samples taken across the Eagle copper vein at approximately 6-ft intervals along an exploratory drift. The probability graph for these data is shown in Fig. 7.6a (from Sinclair, 1976). Ignoring the low end of the distribution, it is possible to interpret the remainder as a mixture of two lognormal populations that overlap slightly. Moreover, effective thresholds can be selected to identify most values within each of the two populations; thresholds are 3.4 percent Cu and 5.4 percent Cu, and the interval between is the main range of overlap. Individual samples are coded as to population (high or low grade) or as being in the range of overlap, on Fig. 7.6b; a preferred spatial distribution is apparent. The high population defines those parts of the vein where fresh copper sulphides occur; the low population occurs where the vein has been highly fractured, oxidized, and partly leached of copper. The few samples in the range of overlap are distributed erratically within both of the identifiable populations. In this case, the probability graph has been useful in establishing fundamentally different geologic domains that are important in establishing a mineral inventory. Sinclair and Deraisme (1974) undertook such a study of the Eagle vein and ignored the oxidized and leached part of the vein when establishing resources.

7.8: STRUCTURED APPROACH TO MULTIPLE POPULATIONS

When multiple populations (here bimodality is assumed) are identified a useful systematic approach to their evaluation is as follows:

(i) Using methods of Section 7.6.1, partition the subpopulations and define a threshold that distinguishes efficiently between the high and low subpopulations.

(ii) Examine grade profiles along drill holes, trenches, and so on in various directions to develop a clear three-dimensional image of the structure of the outlier subpopulation. In particular, it is important to define the dimensions of outlier clusters.

(iii) From Step (ii), an average image or geometry of the outlier subpopulation can be developed three dimensionally. In a simple but uncommon case, outliers are grouped together in a well-defined volume. More commonly, individual values or small clusters of outliers occur in various locations and in seemingly irregular patterns within a much larger volume.

(iv) The distribution of outlier clusters in three dimensions must be examined relative to the total deposit volume in order to define, if possible, a more confined volume (the outlier zone) within which outliers occur and where they must be considered with respect to mineral inventory estimation (Fig. 7.11). This outlier zone can be a single zone of tightly grouped outliers or a zone containing many dispersed outliers or clusters of outliers within a field of lower-valued data.

(v) Within the outlier zone, outliers can be considered in a variety of ways as outlined in some of the forgoing examples. In general, however, it is essential that both outliers and the more abundant typical data be examined separately for continuity. As a consequence of different continuities, outlier and typical mineralization probably require independent contributions to mineral inventory estimation.

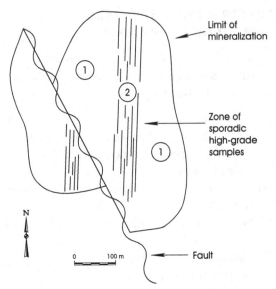

Limit of
mineralization

Zone of
sporadic
high-grade
samples

Fault

N

0   100 m

Figure 7.11: Hypothetical plan illustrating the concept of a large, mineralized zone of low gold grades (sum or areas 1 and 2) on which a late, vein stage of mineralization with preferred orientation has been superimposed as a restricted zone (area 2). The late veins are characterized by sporadic geologic outlier values. Different grade continuity models are to be expected in two such domains. Similarly, different resource/reserve estimation procedures might be required for the two distinctive phases of mineralization.

7.9: INCORPORATION OF OUTLIERS INTO RESOURCE/RESERVE ESTIMATES

A description of cutting (capping) in Section 7.2 indicates how outliers have been dealt with traditionally (i.e., by reducing their value and then incorporating the reduced value in the normal procedure for resource/reserve estimation). The practical difficulty with this approach is that it is arbitrary and uncertain, and can lead to substantial errors. An alternative procedure is illustrated in Fig. 7.1 when geologic investigations have led to an appreciation of the general physical form of outliers so that reasonable weights can be attached to them during block estimation. This procedure is recommended for improving local block estimates, but must be preceded by thorough geologic characterization of outliers, preferably beginning during exploration of the deposit. Too often, dealing with outliers is left to engineers, far removed in space and time from the geologists who mapped the deposit. Consequently, these engineers are then faced

with arbitrary decisions and assumptions about outliers, instead of basing their treatment on geologic knowledge.

Geostatisticians have developed several procedures for dealing with grade populations that extend over one to three orders of magnitude and contain outliers, including restricted kriging and multiple-indicator kriging. Both of these estimation methods have been demonstrated to be useful in a number of cases, and both warrant consideration. Further discussion of these topics is found in subsequent chapters.

This discussion of the identification of outliers has been emphasized as part of a thorough data evaluation. Clearly, the presence of outliers has a dramatic impact on procedures to be used in mineral inventory estimation. Although it is not the aim of this book to develop details of methodology in mineral inventory procedures themselves, it is important to realize that the abundance and spatial distribution of outliers have a serious impact on estimation methodology. Obviously, outliers that can be recognized as errors should be omitted from the estimation process. Rare, isolated values that are real present a problem in estimation because it is unlikely that they have the same continuity characteristics as the bulk of the data; therefore, they must be treated differently. Rare outliers might be ignored during estimation; more abundant, clustered outliers may define a sufficiently large domain so that a value continuity model can be defined with confidence. Intermediate situations demand innovation and common sense in their treatment. Indicator kriging is perhaps the most widely used method of dealing with the outlier (and multiple domain) problem from the point of view of estimation. The procedure is not particularly effective if the outlier domain is an abnormally small percentage of the mineral deposit and outlier values occur as widely scattered individuals, apparently randomly positioned.

7.10: PRACTICAL CONSIDERATIONS

1. Outliers are an important focus for data evaluation. They should be evaluated with an orderly or structured approach as they are encountered in a data gathering (exploration) program, and errors must be distinguished from geologic

outliers. Data should be vetted by a variety of computer-based and graphic techniques during and after the data-gathering process to produce a high-quality data set for resource/reserve estimation.

2. Geologic outliers must be characterized by mineralization type and spatial distribution, each an important feature in determining how outlier values will be incorporated into a mineral inventory estimation program. It is particularly important to realize that firsthand descriptive, spatial, and perhaps genetic understanding of outliers is easiest when assay sheets, core, samples, outcrops, and so on are immediately and easily available to geologists who have a firsthand familiarity with the deposit, rather than much later, when such professionals might be absent from or only indirectly associated with a mineral inventory estimation team.

3. During exploration, effort must be directed to the problem of characterizing outlier populations spatially (i.e., in three dimensions). Such characterization should include an understanding of the dimensions of bodies that give rise to outlier values and the distribution of these bodies in space. It is generally too late for such an undertaking when a feasibility study is in progress.

4. Many standard graphic methods of analyzing data have the advantage that outliers can be recognized easily. Binary plots and probability graphs are two particularly useful procedures; a thorough familiarity with the use of probability plots in recognizing and partitioning multiple populations provides a practical basis for evaluating and classifying outliers.

7.11: SELECTED READING

Sinclair, A. J., 1991, A fundamental approach to threshold estimation in exploration geochemistry: probability graphs revisited; Jour. Geochem. Expl., v. 41, pp. 1–22.

Sinclair, A. J., 1976, Applications of probability graphs in mineral exploration; Assoc. Exp. Geochem. Spec. v. 4, 95 pp.

7.12: EXERCISES

1. Construct an idealized scenario in which a $20 \times 20 \times 10 \text{ m}^3$ block in a porphyry copper deposit is to be estimated by five blasthole samples with values $0.25, 0.35, 0.3, 1.90,$ and 0.3. Assume that high values in general have been found by geologic investigations to represent 1-m-wide vertical veins that cut the block in a direction parallel to one of the sides and have an extent in the plane of the vein, equivalent to about half the cross-sectional area of the block side. (a) What is the true grade of the block if the high value is outside the block and the other values have equal weight? (b) What is the true grade of the block if the high-valued structure is entirely within the block and the other four samples have equal weight? (c) What is the estimated grade of the block if the high value has a weight of -0.04 and other samples have equal weight?

2. The following cumulative data represent 1,081 uranium grades greater than 0.1 percent U, for an unconformity-type uranium deposit in northern Saskatchewan (data from Isaaks, 1984) with a mean of 4.87 percent U and a standard deviation of 7.84. Plot the data on log probability paper and interpret.

U (wt%)	Cum. freq. (%)
0.2	22
0.3	34
0.5	46
1.0	54
1.5	60
2.5	69
5.0	75
10.0	82
15.0	90
20.0	94

3. The following cumulative gold data (grade vs. cumulative percent) are for an epithermal gold deposit (cf. Stone and Dunn, 1994). Plot the data on log probability paper and interpret.

Au (g/t)	Cum. (%)	Au (g/t)	Cum. (%)	Au (g/t)	Cum. (%)
0.1	30.8	2.1	94.3	4.1	98.0
0.2	47.0	2.2	94.9	4.2	98.2
0.3	59.6	2.3	94.9	4.3	98.5
0.4	68.5	2.4	95.1	4.4	98.8
0.5	74.4	2.5	95.4	4.5	99.0
0.6	77.4	2.6	96.0	4.6	99.0
0.7	80.2	2.7	96.2	4.7	99.0
0.8	82.0	2.8	96.2	4.8	99.0
0.9	84.8	2.9	96.2	4.9	99.0
1.0	86.9	3.0	96.2	5.0	99.5
1.1	89.0	3.1	96.9	.	
1.2	89.5	3.2	96.9	.	
1.3	90.4	3.3	97.2	.	
1.4	90.9	3.4	97.2	.	
1.5	91.4	3.5	97.5	.	
1.6	91.8	3.6	97.5	6.8	100.0
1.7	92.1	3.7	97.7		
1.8	92.7	3.8	97.7		
1.9	92.7	3.9	98.0		
2.0	93.7	4.0	98.0		

8

An Introduction to Geostatistics

Geostatistics is of real potential if it is reconciled with the geology of the deposit. (King et al., 1982, p. 18)

This chapter provides a general introduction to geostatistics from the conceptual point of view. The underlying idea of error in the estimation of the mean grade of a point or block, based on an array of samples, is considered. This error can be thought of as resulting from the extension of sample grades to a block (or point) and is referred to as an *estimation error*. *Estimation variance* (error) is distinguished from *dispersion variance* (spread of data). *Auxiliary functions*, expressions for estimation variance for particular geometric arrays of data relative to a point/block being estimated, are introduced.

8.1: INTRODUCTION

Mineral inventory estimation has, until the advent of geostatistics in the late 1950s, been an empirical, applied science in which the estimator's experience and judgment have been of fundamental importance. During the first half of the twentieth century, a variety of subjective methodologies were developed that, if applied cautiously, produced a reasonable reconciliation of estimates with production, particularly for tabular deposits, investigated and mined through underground workings. The trend to larger-scale, near-surface deposits, investigated largely by drilling and surface stripping, with limited underground

examination, led to the necessity of more "interpretation" as to the occurrence of ore, and in developing mineral inventory estimates. Because of expansion in the use of drilling rather than much more costly underground workings to explore and evaluate mineral deposits, the subject of mineral inventory estimation became much more interpretive. This increasing reliance on interpretation coincided with the advent of computers in the estimation process, leading to the widescale implementation of automated, mineral inventory estimation procedures. Unfortunately, new, automated procedures that could be applied rapidly also contributed to errors from misapplication. As a consequence, empirical (subjective) methods of mineral inventory estimation were prone to serious errors, and an underlying theory of resource/reserve estimation became even more desirable than in the past.

Consider a simple rectangular block to be estimated by a centrally located sample value. It is apparent that were the sample reassayed, a different value might result for reasons of sampling and/or analytical errors. It is also clear that had the sample been displaced only slightly from its position, a substantially different assay value might have been obtained as a result of real, short-range, geologic variations in grade. This simple example illustrates the inherent error that exists where the value of one sample is attributed to a point or extended to a block (i.e., when the value is used as an estimator of the point or block). The error in this particular case can be expressed as a variance

$$e^2 = (Z^* - Z)^2$$

where Z is the true value of the block and Z^* is the estimated value. Of course, the same concept of error applies however many data are used for determining Z^* and whatever their locations in space relative to the block being estimated. When many blocks are estimated, the average extension/estimation error is given by

$$e^2 = \sum (Z_i^* - Z_i)^2 / n.$$

This extension error cannot be determined explicitly without additional information, because the true block grades Z_i are unknown.

There are many examples of the excessive errors that can result from uncertain or inadequate methodology in reserve estimation; some recent cases are documented by King et al. (1982), Knoll (1989), and others. Even when traditional methods have produced more or less acceptable estimates, they have suffered from inefficient use of the available data and little quantitative appreciation of the errors involved in estimation. The general concept of estimation error is fundamental to an appreciation of mineral inventory estimates by any method. It is evident that a block grade cannot be determined exactly from a limited number of contained/nearby sample grades; some level of error is involved and a quantification of that error is highly desirable.

Geostatistics, originally developed with reference to mineral resource/reserve estimation, is defined as "applications of the theory of regionalized variables" (Matheron, 1971, p. 5). It is a statistical approach to estimation that uses the presence of some degree of spatial dependence for grade variables in mineral deposits. *Spatial dependence* means that two nearby samples are likely to be similar in grade (not identical, but relatively close in value); in a high-grade zone, they are both likely to be high grade, whereas in a low-grade zone they are both likely to be low grade. Clearly, spatial dependence implies that any two nearby grades (within the range of dependence) cannot be considered two random draws from a grade distribution representing the entire deposit.

Furthermore, a natural outcome of nearby samples being correlated is that samples closest to a block being estimated should be given more weight than samples further away.

The concepts of geostatistics can be applied much more widely than the early applications to mining, but here attention is limited to uses in the field of mineral inventory estimation. Matheron (1971) developed the basis for geostatistics in the mineral industry during the 1950s and 1960s, although other attempts had been made to apply statistical concepts to resource/reserve estimation (e.g., Krige, 1951; Sichel, 1952; Swanson, 1945). Most of the early geostatistical literature appeared in French and was not widely read outside the French-speaking world, thus delaying a broad acceptance of geostatistical methodology until the 1970s. Since then, the theoretical base and applications have expanded widely, although the subject is not without its detractors (see Chapter 17).

Geostatistical theory is important because it provides a basis for optimizing estimates according to well-accepted criteria (e.g., least squares) and produces a quality measure (error variance) of those estimates. Traditional estimation methods, including polygonal, inverse-distance weighting, and other techniques, are empirical, and even if based on experience in other deposits, major problems can arise in their application to a new deposit; quantitative indication of the quality of empirical estimates must await reconciliation during production. Commonly, such reconciliations require large volumes of ore and cannot be obtained conveniently for small blocks.

Of course, geostatistical estimates are not necessarily demonstrably better than estimates made by a more subjective method. All mineral inventory studies involve assumptions and approximations, whether implicit or explicit. Any estimation procedure can produce incorrect results because of poor application of the procedure, inappropriateness of the procedure, or a change in the geologic model resulting from new information obtained as exploration of a deposit progresses.

Large quantities of data are necessary for making detailed local estimates with a high level of confidence; relatively fewer data are needed for global estimation. It is difficult to quantify "large" and "fewer" in this context because they are functions of the variability that characterizes each individual deposit. In general, forethought to data collection starting at the exploration stage can optimize (i.e., minimize) the data needed for a geostatistical resource/reserve estimate. Early planning with regard to data distribution is particularly important in contributing to a confident, three-dimensional characterization of grade continuity of the variable(s) under study. Data on a regular grid are not essential, but a high proportion of quasi-regular data is useful.

8.2: SOME BENEFITS OF A GEOSTATISTICAL APPROACH TO MINERAL INVENTORY ESTIMATION

All mineral inventory estimation methods make use of concepts such as (1) *zone of influence* of a sample, (2) influence of data geometry on estimation (including the effect of clustered data), and (3) the use of different types of samples with different support for the estimation process. Geostatistics provides a quantitative approach to all of these topics, in contrast to subjective decisions that are necessary in empirical approaches to mineral inventory estimation. Examples are shown in Fig. 8.1.

The autocorrelation function of a grade variable quantifies the concept of *radius of influence* of a sample (cf. Readdy, 1986). Figure 8.1c demonstrates that assays for pairs of nearby samples have a small, squared difference, and that the squared difference increases as the sample spacing increases up to a sample separation "a" (commonly referred to as the *range*). For sample spacings greater than this range, the mean-squared difference of paired data is essentially constant. The range a has been equated to the radius of influence of a sample (e.g., Matheron, 1971).

Grade data commonly exhibit both a random component and a structured component. An autocorrelation function (e.g., Fig. 8.1c) is a practical means of quantifying these two components. For example, the intercept at the origin of Fig. 8.1c, C_0, reflects the random or so-called nugget component of a variable,

whereas the range a quantifies the average dimension of the structured component.

Figure 8.1a illustrates two different data arrays relative to a block to be estimated. One array involves extrapolation; the other involves interpolation. In both cases, the three data have approximately the same weight by most automatic interpolation methods applied to a homogeneous geologic environment. Intuition tells us that the estimate based on interpolation is better than the estimate based on extrapolation; the geostatistical estimation procedure of kriging permits the average errors in the two cases to be quantified. Similarly, kriging takes into account the relative quality of estimates when there are differences in the number of data (Fig. 8.1d).

The problem of conditional bias, introduced in Section 4.8, is the situation in which limited data, on average, overestimate high grades and underestimate low grades. Kriging, although not necessarily correcting this problem in its entirety, does minimize conditional bias.

Perhaps most importantly from a geologic point of view, the autocorrelation function (Fig. 8.1c) can be determined in various directions, and anisotropies (i.e., different ranges in different directions) of the spatial structure of grades can be quantified. Such anisotropies commonly have a geologic explanation and can be integrated easily into the kriging estimation procedure.

More detailed insight into the forgoing advantages of a geostatistical approach to mineral inventory estimation appear in subsequent chapters. It is important at this stage, however, to have a conceptual appreciation of the use of geostatistical procedures.

8.3: RANDOM FUNCTION

A *random function* is a probabalistic description of the spatial distribution of a variable. It is a useful means by which to consider geostatistics because it incorporates the concepts of both random and structured components to the spatial variability of a variable, such as grade. Moreover, random functions can be applied to estimation problems of the type encountered in mineral inventory estimation. In describing the general

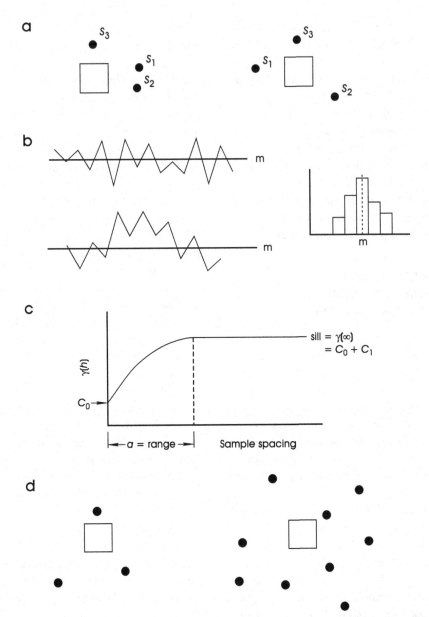

Figure 8.1: Idealized examples illustrating the importance of geostatistics. (a) Two blocks, each to be estimated by three samples; estimation by extrapolation (left) and estimation by interpolation (right) are distinguished in geostatistical estimation. (b) Two grade profiles shown relative to the mean value m have the same histogram shown schematically. The upper profile is distributed randomly about m, whereas the lower profile incorporates a structured data arrangement (i.e., a general trend) in space with a relatively small component of random variability about that structure. Geostatistical estimation takes advantage of the structured nature of data. (c) A semivariogram [$\gamma(h)$ vs. lag] illustrating the average similarity of nearby sample pairs [low values of $\gamma(h)$ at low values of lag], the systematic increase in $\gamma(h)$ as sample spacing increases, and the leveling of variability beyond a sample spacing a. C_0 is a random component of variability constant for all sample spacings; the average range of influence of a sample is lag (i.e., sample spacing) equals a. Note that $\gamma(h)$ is half the mean squared difference in grades that are separated by distance h. (d) Two blocks, one of which is to be estimated by relatively few data (left) and the other by more abundant data (right). In addition to providing block estimates, geostatistics quantifies the error of estimation. This error reflects both the number and array of data relative to the block being estimated.

concept of a random function, it is convenient to do so by comparison with a random variable.

A *random variable* is one that takes a certain range of values, the relative frequencies of which can be described by a probability density function (cf. histogram). One sample randomly drawn from the population is a single realization of the variable. Two randomly drawn samples have uncorrelated grades and in this sense can be contrasted with two grades drawn from a more structured, regionalized variable. In the case of a regionalized variable, it is a useful concept to consider all the available sample data as being one realization (sample) of the random function that characterizes the data three dimensionally. This realization incorporates both the randomness and the structural features present in the data and, in fact, the random function can be defined by a quantification of this randomness and structure using an autocorrelation function, such as the semivariogram illustrated in Fig. 8.1c. In the case of a two-dimensional field of ore samples, a complex mathematical surface can be imagined to describe the topography of the grade values. This complex surface is one realization of the random function and can be characterized by its autocorrelation features.

8.4: STATIONARITY

What does *stationarity* mean? It means, simply, that the mean and variance of values do not depend on location. Now how could an ore deposit possibly be thus when it proceeds from margins without mineralization or below economic cutoff, hopefully to a central, richly endowed region? (Philip and Watson, 1986, p. 96).

The concept of stationarity can be difficult to grasp but for purposes here bears a close relation to the term *homogeneity* used by geologists to characterize domains of similar geologic characteristics such as styles of mineralization (see Fig. 3.4). A field (domain) of data is said to be stationary if the same population is being sampled everywhere in that field; implicitly, no trend exists in the data. Thus, for every location x_i in domain D, the expected value of

$Z(x_i)$ is m, the mean value of the domain. Second-order stationarity requires that the mean is stationary and the covariance between sample points exists and is given by Eq. 8.3. For resource/reserve estimation, these extreme forms of stationarity are not required. In practice, quasi-stationarity generally is a sufficient condition [i.e., $Z(x)$ is locally stationary]. Moreover, this local stationarity need only apply to the differences between pairs of data and can be visualized as a series of overlapping small (local) volumes, each of which is stationary and which together define the total volume/field. The concept of quasi-stationarity (i.e., local stationarity) is consistent with the concept of high-grade and low-grade zones being spatially separate in a mineral deposit.

8.5: GEOSTATISTICAL CONCEPTS AND TERMINOLOGY

Geostatistics is couched in mathematical terminology, particularly associated with expectations and random functions. A fairly widely used set of symbols has evolved that is adhered to here.

The expected value of a regionalized variable, $Z(x)$ (read "Z at location x"), is the mean, denoted here by m:

$$E\{Z(x)\} = m.$$

Variance of a regionalized variable is given by

$$\text{Var}\{Z(x)\} = E\{[Z(x) - m(x)]^2\}. \qquad (8.1)$$

Covariance of a regionalized variable is given by

$$C(x_1, x_2) = E\{[Z(x_1) - m(x_1)][Z(x_2) \\ - m(x_2)]\}. \qquad (8.2)$$

This covariance expression is equivalent to Eq. 4.10, where x equals x_1 and y equals x_2, to give $C(x_1, x_2)$ equals $C(x, y)$ equals s_{xy}. In Eq. 8.2, however, x_1 and x_2 represent two spatially distinct positions of the same variable. Hence, $m(x_1)$ equals $m(x_2)$, and for an autocorrelated variable and a stationary field of data, Eq. 8.2 reduces to

$$C(x_1, x_2) = C(h) = E\{Z(x_1) \cdot Z(x_2)\} - m^2 \\ (8.3)$$

where h is the vectoral distance from x_1 to x_2. Moreover, where h is zero, x_1 and x_2 are identical, and Eq. 8.3 becomes equivalent to the variance (Eq. 8.1).

8.6: THE VARIOGRAM/SEMIVARIOGRAM

The variogram, $2\gamma(x_1, x_2)$, the fundamental tool of geostatistics, is defined as

$$2\gamma(x_1, x_2) = \text{Var}\{Z(x_1) - Z(x_2)\}$$
$$= E\{[Z(x_1) - Z(x_2)]^2\} = 2\gamma(h)$$
$$(8.4)$$

where h is the vector from x_1 to x_2. Half of the variogram becomes the semivariogram (gamma h):

$$\gamma(x_1, x_2) = \gamma(h) = \text{semivariogram}$$
$$= (E\{[Z(x_1) - Z(x_2)]^2\})/2. \quad (8.5)$$

As originally defined, the term *semivariogram* refers to half the variogram. Recent widespread use has evolved to the point that the term *variogram* is now widely used for what is really the semivariogram. Here we retain the original meaning of the terms and use *semivariogram* throughout the text.

The semivariogram can be replaced by covariances. The relation between the two is given by

$$\gamma(h) = C(0) - C(h) \quad (8.6)$$

where $C(0)$ is the covariance for a lag of zero (equivalent to the variance). Equation 8.8 can be demonstrated by substituting each of the terms in the right side of Eq. 8.7 by corresponding expectations in Eqs. 8.1 and 8.4 to reproduce Eq. 8.5. Because $C(0)$ is a constant parameter (i.e., the variance) for a given data set, the close interdependence of $\gamma(h)$ and $C(h)$ becomes apparent. Thus, in geostatistics many calculations can be done using equations in terms of semivariogram values or equivalent equations in terms of covariances.

Figure 8.1c is a useful conceptual image of a semivariogram showing a systematic variation in $\gamma(h)$ as a function of lag, the distance between two sites. Note (1) the so-called nugget effect, C_0 (i.e., the intercept for zero lag), characterizes the random component of the data; (2) the systematically increasing values of

$\gamma(h)$, until lag a (range); and (3) $\gamma(\alpha)$ is the sill of the semivariogram [i.e., the constant value of $\gamma(h)$ for lags greater than the range]. Note that $\gamma(\alpha)$ equals $C_0 + C_1$.

Quantifying the similarity patterns (spatial characteristics) of grade or other regionalized variables pertaining to a mineral deposit normally is preceded by a critical examination of the geology of the deposit and a thorough data analysis. These studies provide the information base on which a knowledge of both geologic and value continuity is obtained. A detailed structural study of a regionalized variable is aimed at quantifying the spatial characteristics of the variable in a statistical sense. Such a quantification also defines average value continuity and provides the basis for a wide range of geostatistical calculations concerned with mineral inventory estimation. The semivariogram is the principal measure of similarity (i.e., autocorrelation) used for geostatistical purposes in this text, although, as indicated previously, other autocorrelation functions (e.g., covariogram, correlogram) would be equally as good. Autocorrelation studies in geostatistics are often referred to as *variography* because of this traditional emphasis on the variogram (or the semivariogram).

8.7: ESTIMATION VARIANCE/EXTENSION VARIANCE

It is now possible to examine further the concept of estimation variance referred to in Section 8.1. In particular, an understanding of the semivariogram leads directly to a further appreciation of the concept of extension variance. The semivariogram is half the mean squared difference between two values, x_i and x_{i+h}, separated by a distance h. If one of those values (x_i) is known and the other (x_{i+h}) is unknown, the known value can be assigned to the unknown site and the average error associated with this extension of grade is, by definition, the $2\gamma(h)$ value for the distance between the two points. Hence, the semivariogram is a quantification of error when an uninformed location is estimated by an informed location.

A more generalized concept of the estimation variance is given by once again considering the expression for error

$$\sigma_e^2 = E\{(Z_i - Z_i^*)^2\}$$

where Z_i represents true values of the body being estimated and Z_i^* represents the estimates. Journel and Huijbregts (1978) show that this expression can be expanded and expressed in terms of average semi-variogram values as follows:

$$\sigma_e^2 = 2\bar{\gamma}(V, v) - \bar{\gamma}(V, V) - \bar{\gamma}(v, v) \qquad (8.7)$$

where

V and v	are large and small volumes, respectively (e.g., SMU and sample)
$\bar{\gamma}(V, v)$	is the average semivariogram value when the two ends of vector h describe all possible positions in V and v, respectively
$\bar{\gamma}(V, V)$	is the average semivariogram value when the two ends of vector h describe all possible positions in V
$\bar{\gamma}(v, v)$	is the average semivariogram value when the two ends of vector h describe all possible positions in v.

The estimation variance is a function of several semivariogram values and, hence, is seen to depend on (i) the distance between V and v, (ii) the sizes of V and v, (iii) the quantity and spatial arrangement of information in v, and (iv) the semivariogram model. The fundamental nature of the semivariogram is evident and quantificaiton of the semivariogram model leads directly to quantification of estimation variance. *Extension variance* is a term that also has been applied to this situation; extending the value of sample v to a larger volume V results in an extension error. A practical example is the use of a centrally located blasthole assay to represent the grade of the surrounding block.

A simple example of the application of Eq. 8.7 is illustrated by Fig. 8.2. In this case, the aim is to produce two estimates; first, determine the error to be attributed to a point estimate at the block center using

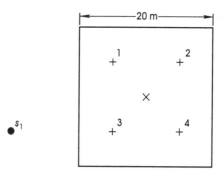

Figure 8.2: Estimation of a point (x) and a two-dimensional block by two samples s_1 and s_2. Distances can be substituted in a model for the semivariogram to calculate various $\gamma(h)$ values so that the estimation variance can be calculated for both the point and block estimates. Details are given in the text, and $\gamma(h)$ values are summarized in Table 8.1. In this case, the block is discretized by only four points in order to estimate the average gamma value within the block (i.e., the F function). The semivariogram model adopted is a linear model, $\gamma(h)$ equals $0.1h$.

the point samples s_1 and s_2 and the semivariogram model $\gamma(h)$ equals $0.1h$. Then, with the same data and semivariogram model, determine the error for an estimation of the block mean. For the purpose of these calculations, the block is approximated (discretized) by the four points $+^1 \cdots +^4$. All the necessary gamma values are summarized in Table 8.1. Substitution in Eq. 8.9 for the point estimation problem gives

$$e^2 = 2 \times 2.47 - 2.03 - 0.0 = 2.90.$$

Substitution in Eq. 8.7 for the block estimation problem gives

$$e^2 = 2 \times 2.50 - 2.03 - 0.85 = 2.12.$$

Apart from illustrating the methodology of solving Eq. 8.7, this example also illustrates two important points that reappear:

(i) For a constant set of data, the estimation variance decreases as the volume to be estimated increases provided the block dimensions do not extend beyond the data array.

Table 8.1 Semivariogram values for estimations of Fig. 8.2 described in the text

Pair	d	$\gamma(h)$	$\gamma(h)^-$
$s_1 - x$	20.8	2.08	
$s_2 - x$	28.5	2.85	2.475
$s_1 - s_1$	0	0	
$s_1 - s_2$	40.5	4.05	2.025
$s_1 - +^1$	18	1.8	
$s_1 - +^2$	27	2.7	
$s_1 - +^3$	15	1.5	
$s_1 - +^4$	25	2.5	
$s_2 - +^1$	35.4	3.54	
$s_2 - +^2$	29	2.9	
$s_2 - +^3$	29	2.9	
$s_2 - +^4$	21.2	2.12	2.50
$+^1 - +^1$	0	0	
$+^1 - +^2$	10	1.0	
$+^1 - +^3$	10	1.0	
$+^1 - +^4$	14.1	1.41	0.85

(ii) It is common practice to determine the within-block average semivariogram value by discretizing the block by an array of points.

8.8: AUXILIARY FUNCTIONS

Auxiliary functions are a group of specialized functions that traditionally have been used for rapid manual determination of average semivariogram values for certain simple data configurations relative to the block or point to be estimated (their use requires that a mathematical equation be known for a semivariogram as discussed in Section 9.2.2). For example, the so-called chi function $\chi(L; l)$ is the mean value of the semivariogram when one end of the vector describes side L of an $L \times l$ rectangle and the other end of the vector describes the entire rectangle. Use of auxiliary functions arose relatively early in the evolution of geostatistical practice when access to efficient computing facilities was much more limited than today. Values of commonly used auxiliary functions can be read from published graphs that are based on the standardized spherical model with sill and range both equal to 1 (i.e., C_1 equals 1 and a equals 1) and no nugget effect. Thus, any spherical model can make use of graphs of the auxiliary function if distances

are considered as a proportion of the range (h/a) and the values obtained from the graphs are multiplied by the true C of the model. For the chi function mentioned previously, the sides of the $L \times l$ rectangle would be represented as L/a and l/a. Published graphs of many auxiliary functions relate to any spherical model in a manner analogous to that by which a standard normal distribution relates to any normal distribution. Of course, graphs of auxiliary functions can be developed for any semivariogram model.

Auxiliary functions were developed for highly regular arrays of data relative to a line, area, or volume being estimated. The functions provide the average gamma value for very specific geometric arrays. These average gamma values can then be incorporated manually into standard geostatistical formula to determine the estimation variance for a particular array of data relative to the entity to be estimated. The historical use of auxiliary functions is described in detail by Journel and Huijbregts (1978). The need for auxiliary functions has decreased in part because ideal data arrays are uncommon in practice, but largely because of ready access to extensive computing power. Today, it is more common to approximate parts of a mineral deposit by a grid of regular smaller blocks, and obtain both a grade estimate and an error estimate for each block using computer-based calculations. Consequently, most auxiliary functions are no longer used.

One function that remains important in present-day practice is the so-called F function, the average value of the semivariogram when the two ends of the vector (lag) take all possible positions within a block. This concept is illustrated schematically in two dimensions in Fig, 8.3 where a block is discretized by an array of closely spaced points. It is easy to see that an acceptable estimate of the average gamma value within a block can be obtained numerically by averaging all the discrete values that result from taking all possible pairs of point into account. The concept is illustrated in the worked example in Section 8.5. This discretized approach is incorporated in most modern software, when required. In addition, graphs of the F function are available in several widely used texts (e.g., David, 1977; Journel and Huijbregts, 1978). The F function is involved in the determination of the

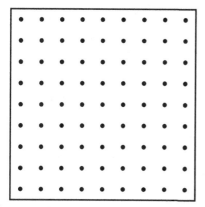

Figure 8.3: A realistic approximation of a block for purposes of numerically estimating the F function for the block. The array of dots approximates the infinity of points that define the block. An average gamma value can be determined by averaging the individual values obtained for each possible pair of points.

error resulting from the kriging estimation procedure (Chapter 10) and is an important component of the concept of *change of support* (e.g., procedures described later for estimating the distribution of block grades from a known distribution of sample grades). A chart giving standardized values of the F function for small volumes is reproduced in Fig. 8.4. In this case, the regular volumes are defined by lengths in three dimensions, each length being defined as a ratio of the true length (l_i) over the range (a_l) of the semivariogram (i.e., l_i/a_i).

8.9: DISPERSION VARIANCE

Dispersion variance is a measure of the spread of data. In general, the dispersion of a regionalized variable is less in small volumes than in large volumes. This fact stems from the characteristic shown by the semivariogram – that closely spaced samples are, on average, more similar in grade than are widely spaced samples. In other words, dispersion variance is a function of support. If we consider small samples to have support v, then their grades have a dispersion in a large block of volume V given by

$$D^2(v, V) = \bar{\gamma}(V, V) - \bar{\gamma}(v, v) \qquad (8.8)$$

where $D^2(v, V)$ is the dispersion variance of sample grades (volume v) in a larger volume V; $\gamma(V, V)$ is

the mean semivariogram value where the two ends of vector h take all possible positions; in V, and $\gamma(v, v)$ is the mean semivariogram value where the two ends of vector h take all possible positions in v. Note that the two average gamma terms in Eq. 8.8 are F functions. Dispersion variances are additive; therefore

$$D^2(s/M) = D^2(s/B) + D^2(B/M) \qquad (8.9)$$

where s represents small volumes (e.g., samples), B represents the intermediate volumes (e.g., blocks), and M represents a large volume (e.g., a mineral deposit). Equation 8.9 is known as *Krige's relation* after D. G. Krige, who first demonstrated it experimentally for data from South African gold deposits. The equation states that the variance of sample grades in a deposit can be considered as two components (i.e., the variance of sample grades in blocks plus the variance of block grades in the deposit). For practical purposes it is useful to represent Eq. 8.9 slightly differently, recognizing that $D^2(s/B)$ is the F function; hence

$$D^2(B/M) = D^2(s/M) - \bar{\gamma}(B, B). \qquad (8.10)$$

In this form (Eq. 8.10), the relation is known as the *volume–variance relation*. This is a particularly useful equation because it indicates that if both the semivariogram and the dispersion of sample grades are known, it is possible to determine the dispersion of block grades for any size of block. Hence, it is possible to construct a grade–tonnage curve for any appropriate size of selective mining unit, although an assumption regarding the form of the histogram of block grades is required (a topic considered further in Chapter 12).

8.10: A STRUCTURED APPROACH TO GEOSTATISTICAL MINERAL INVENTORY ESTIMATION

A geostatistical mineral inventory begins with extensive evaluation of geology, assays, and supporting chemical and physical data on which the inventory is based (cf. Table 1.5). The purely geostatistical component normally begins with variography, the determination of semivariogram (or other autocorrelation) models for the variables under study. The semivariogram model describes the autocorrelation

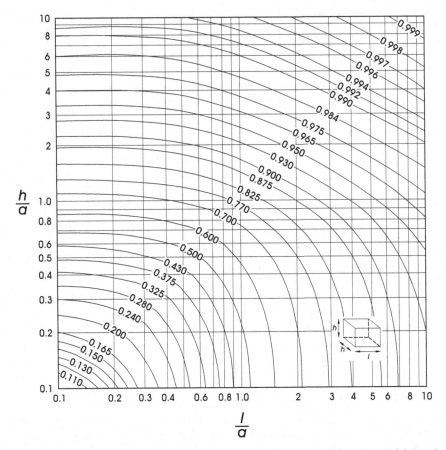

Figure 8.4: A graph of the F function for a three-dimensional block whose sides are $l/a, l/a$, and h/a, where l and h are absolute dimensions of the block, a is the range of a spherical semivariogram (in the same units as l and a) with zero nugget effect, and sill equals 1.0. Redrawn from Parker (1979). Also available in David (1977) and Journel and Huijbregts (1978).

character of a regionalized variable; that is, the semivariogram (or variogram) expressed as a mathematical model describes, on average, how similar in value two samples are as a function of their separation in space. This semivariogram model is then used as the basis for global or local resource estimation using kriging or the volume–variance relation. Kriging is a least squares procedure (with certain constraints) that provides minimum estimation variance when a particular set of data are used to make a specific point or block estimate. Additional studies can include an evaluation of the effects of additional data on error, the estimation of recoverable reserves, and considerations of the effects of internal dilution. Conditional simulation of the spatial distribution of grade can also be undertaken.

8.10.1: Applications of Geostatistics in Mineral Inventory Estimation

Applications of geostatistics in the mineral industry are wide ranging and include the following (cf. Blais and Carlier, 1968):

(i) Optimizing the precision of individual grade estimation
(ii) Kriging (optimal weighting) of a data array for estimation of points, blocks, or panels for global or local estimation
(iii) Calculation of the error of estimation of grade of a given block of ore, regardless of size
(iv) Design of optimum drilling patterns
(v) Calculation of the error of estimation of the volume or tonnage of an orebody

(vi) Conditional simulations for mill design, mine planning, and value continuity characterization

(vii) Optimal contouring with irregular data arrays.

8.10.2: Why Geostatistics?

As mentioned in the Introduction, geostatistics is not without its detractors (a matter for further discussion in Chapter 17). However, it also, suggests a number of reasons why geostatistical estimation procedures should be considered as one of the approaches to be used to develop resources/reserves for a mineral deposit (cf. Royle, 1979).

1. Geostatistics provides a theoretical base for estimation; no other method can make this claim.
2. Geostatistics quantifies and uses the random and structured components of variability that are recognized in grade dispersion patterns for estimation.
3. Geostatistics commonly improves on certain traditional estimators (especially polygonal) that have a history of serious overestimation.
4. Geostatistics can be applied successfully at an advanced stage of exploration when a substantial database exists. With forethought, exploration can be planned so that appropriate data are obtained earlier, rather than later!
5. Geostatistics is particularly amenable to grade control, especially in open-pit operations, where a substantial blasthole database exists.
6. Geostatistics provides the potential to optimize data collection once sufficient information has been obtained to develop appropriate semivariogram models.
7. Geostatistics is globally unbiased, and is an important contributor to minimizing conditional bias.
8. Geostatistics offers the potential of conditional simulation, a technique that permits a visualization of grade continuity during exploration, and can be used for mine and mill planning at more advanced stages of evaluation.

8.11: SELECTED READING

Clark, I., 1979, Practical geostatistics; Applied Science Pub., London, 129 pp.

Journel, A. G., 1975, Geological reconnaissance to exploitation – A decade of applied geostatistics; Can. Inst. Min. Metall. Bull., v. 68, pp. 1–10.

Matheron, G., 1971, The theory of regionalized variables and its applications; Les Cahiers du Centre de Morphologie Mathematique, Fontainebleau, France, 211 pp.

Matheron, G., 1963, Principles of geostatistics; Econ. Geol. v. 58, pp. 1246–1266.

Royle, A. G., 1979, Why geostatistics? Eng. Min. Jour., May, pp. 92–101.

8.12: EXERCISES

1. Calculate the F function for a $20 \times 20 \times 10 \, \text{m}^3$ block of ore for which the spherical semivariogram model is C_0 equals 0, C_1 equals 0.08, and a equals 40 m.

2. Calculate the estimation variance for a two-dimensional block $20 \, \text{m} \times 10 \, \text{m}$ using a centered datum and a datum located at one corner of the block. Use the semivariogram model from Question 1.

9

Spatial (Structural) Analysis: An Introduction to Semivariograms

It is well known that the values in veins tend to be concentrated in the form of ore shoots, which usually have lenticular outlines in the plane of the vein. Thus we are faced with the general probability that the values in veins have something of a zonal arrangement rather than a random distribution, and the question of how this will affect an estimate of the grade of an orebody in a vein. (Swanson, 1945, p. 327)

The *semivariogram* (or *variogram*), the fundamental tool of geostatistical analysis, is one of several commonly used measures of the spatial similarity of a regionalized variable. General procedures are explained for obtaining experimental semivariograms and fitting these experimental results by smooth, mathematical models (especially the spherical model). Considerable emphasis is placed on anisotropic models and their relation to directional geologic features. A variety of important topics related to the semivariogram are discussed, including the proportional effect, relative semivariograms, nested structures and the geometric meaning of many semivariogram models. Complexities in development of semivariogram models that are treated include: dealing with outliers, robustness of semivariograms, and semivariograms in curved coordinate space.

9.1: INTRODUCTION

Quantifying the similarity patterns (spatial characteristics) of grade or other regionalized variables pertaining to a mineral deposit normally is preceded by a critical examination of the geology of the deposit and a thorough data analysis. These studies provide the information base on which a thorough study of value continuity can be undertaken in an efficient manner. A detailed structural study of a regionalized variable is aimed at quantifying the spatial characteristics of the variable in a statistical sense. Such a quantification is an average model of value continuity for a domain or deposit and provides the basis for a wide range of geostatistical calculations concerned with mineral inventory estimation. The *variogram* (or *semivariogram*) is the principal measure of similarity (i.e., a structural tool) used for geostatistical purposes, although other autocorrelation functions (e.g., covariogram, correlogram) that are essentially equivalent (see Chapter 8) may equally well be used for calculation purposes. Autocorrelation studies in geostatistics are often referred to as *variography* because of this traditional emphasis on the variogram (or the semivariogram).

The variogram $2\gamma(h)$, defined in Chapter 8, is

$$2\gamma(h) = E\{[Z(x_i) - Z(x_{i+h})]^2\}$$

from which the semivariogram is given by

$$\gamma(h) = E\{[Z(x_i) - Z(x_{i+h})]^2\}/2$$

where $Z(x_i)$ is a value of a regionalized variable at

location x_i and $Z(x_{i+h})$ is a second value at a distance h from the first.

In general, $\gamma(h)$ is a vector and for constant h, $\gamma(h)$ can vary as a function of orientation. For example, $\gamma(h_1)$ determined in direction 1 (e.g., parallel to bedding in a stratiform deposit) can differ significantly from $\gamma(h_2)$ determined in a second direction (e.g., perpendicular to bedding); that is, $\gamma(h)$ can be anisotropic. For practical purposes, $\gamma(h)$ is estimated for a set of discrete values of h (lag intervals) and a plot of the resulting values are approximated by a smooth mathematical function that can be easily solved for gamma, given any corresponding h value.

9.2: EXPERIMENTAL SEMIVARIOGRAMS

Experimental semivariograms are those determined from a particular data set. Data available with which to estimate a semivariogram represent one of an infinite of possible samples of the population (one realization of the random function) under study (e.g., grade). Thus, a similarity measure (e.g., semivariogram) based on such data also can be seen as a sample of the true structural character of the underlying random function that defines the variable under study. Some variability in the detailed form of the semivariogram is to be expected if different subsets of data are used to estimate the autocorrelation character of a variable.

The concept of stationarity is introduced in Chapter 8. A domain is stationary if measurements from any of its parts are representative of the entire domain (i.e., the same statistic is being sampled throughout the domain). Viewed this way, stationarity is a function of scale because the statistic being considered must be measured over a discrete volume. Clearly, large-scale stationarity of the mean does not apply in the case of mineral deposits when large and systematic differences in grade are the rule. Instead, recourse is made to the assumption that the semivariogram model depends only on the distance h between sites and not on specific geographic locations of paired samples within the field of a regionalized variable. This means that many realizations of $\gamma(h)$ exist over the field for various values of h, thus

providing the possibility of estimating the semivariogram function for different sample separations (lags). The hypothesis (that $[Z(x) - Z(x + h)]$ is stationary) is less restrictive than assuming stationarity of $Z(x)$ itself (Agterberg, 1974; Journel and Huijbregts, 1978). The semivariogram is a function of differences in pairs of grades separated by distance h, and thus becomes a tool for the recognition of stationarity/nonstationarity (Fig. 9.3). The semivariogram can be used to define maximum distances over which stationarity can be assumed (i.e., to define the limits of local or quasi-stationarity).

In general, a field of data represents one realization (many spatially distributed samples) of a random function from which an experimental semivariogram can be calculated for various values of h, as follows:

$$\gamma^*(h) = \left\{ \sum [Z(x_i) - Z(x_{i+h})]^2 \right\} / 2n.$$

The estimated values of $\gamma^*(h)$ are plotted versus the corresponding values of h; the resulting plot defines the experimental semivariogram. The procedure is illustrated for an ideal, one-dimensional grid in Fig. 9.1. In brief, the semivariogram is half the mean squared difference of values separated by lag h.

Determination of the experimental semivariogram is an essential first step in defining a mathematical model for $\gamma^*(h)$. Consider a line of measurements at regular intervals as shown in Fig. 9.1; information is assumed to be known at regular grid intersections spaced a distance d apart. For $h = d$, there are $(n - 1)$ pairs with which to provide an estimate, $\gamma^*(d)$; for $h = 2d$, there are $(n - 2)$ pairs to provide an estimate, $\gamma^*(2d)$, and so on. Thus, the number of pairs on which a $\gamma^*(h)$ estimate is based decreases as the lag h increases, and there is a concomittent decrease in the quality of the $\gamma^*(h)$ estimates. A common rule of thumb is to accept $\gamma^*(h)$ estimates for an experimental semivariogram for values of h less than $L/2$, where L is the field width of the regionalized variable. In practice, L is not uniform for all directions throughout a field of data or even for a single direction, so discretion is needed when applying this rule.

It is important to realize that each value of $\gamma^*(h)$ calculated is only an estimate, and some error is involved in the estimation. Thus, experimental

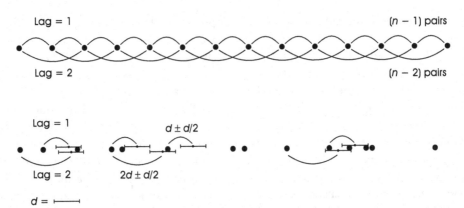

Figure 9.1: Regular (upper) and irregular (lower) one-dimensional data arrays illustrating how data are paired for the purpose of constructing an experimental semivariogram. In the irregular array, pairs of data must be collected in groups of approximately uniform h (separation distance) where some tolerance in h is allowed. In this example, the tolerance is $h' = h \pm d/2$.

semivariograms generally have a sawtooth pattern, as shown in Fig. 9.2. Such experimental patterns commonly are fitted by smooth, mathematical curves that more or less average the experimental fluctuations of the real data. In many cases, particularly when data are abundant, fluctuations are slight and the experimental semivariogram near the origin closely approximates a smooth, continuous curve.

Commonly, $\gamma^*(h)$ is an increasing function as lag increases, at least up to a particular value of h (the range) beyond which $\gamma^*(h)$ is a constant. This rela-

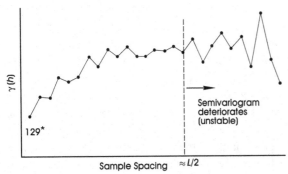

Figure 9.2: Hypothetical, experimental semivariogram illustrating general features that include: calculated, discrete, experimental values (dots), sawtooth chararacter (by joining dots), generally increasing $\gamma(h)$ values as sample spacing (distance) increases, and general increase in erratic character for distances greater than $L/2$, where L is the field width of data used to generate the experimental values. Number of data used to construct the experimental semivariogram is 129.

tion means that nearby samples are, on average, more similar than are sample pairs with greater separation. It is this range over which there is similarity that is the average structure of a regionalized variable. The lag at which $\gamma^*(h)$ becomes constant is the quantification of the concept of range of influence of a sample (cf. Matheron, 1963), although the influence of a sample on the surrounding volume is less for more distant parts of the volume. From a practical viewpoint, the range is a useful quantification of grade continuity.

Several common patterns of experimental semivariograms are shown in idealized form in Fig. 9.3; sawtooth patterns represent experimental semivariograms and the smooth curves represent mathematical models that approximate the experimental values. Figure 9.3a is an example for which all measurement separations give the same average variability (i.e., variability is more or less constant regardless of sample separation, a situation that indicates a random as opposed to a regionalized variable). If significant autocorrelation exists for such data, it is at a scale less than the minimum lag of the experimental semivariogram.

Figure 9.3b shows a discontinuity (C_0) at the origin, followed by a gradual increase in $\gamma^*(h)$ as the lag increases to $h = a$ (the range); for lags beyond the range ($h > a$), the experimental values of $\gamma^*(h)$ are roughly constant (C). Experience has shown that this is a relatively common semivariogram shape for

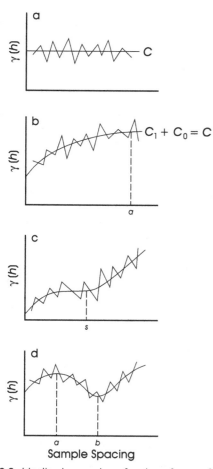

Figure 9.3: Idealized examples of various forms of experimental semivariograms (sawtooth curve) and their corresponding models (smooth curves). (a) Pure nugget effect, indicating a random variable, although autocorrelation is possible on a scale less than the minimum sample spacing. (b) Common pattern showing an intercept on the y axis and a gradual increase in $\gamma(h)$ up to a sample spacing a at which $\gamma(h)$ becomes uniform. This is a very common pattern in practice. (c) Similar to (b), except that beyond a sample spacing s, the semivariogram becomes parabolic. The parabolic increase in $\gamma(h)$ indicates the presence of a trend in the data that can be considered locally stationary over field width s. (d) A *hole effect*, centered on point b and indicating a cyclical character to the spatial distribution of the variable.

regionalized variables encountered in mineral inventory estimation.

Figure 9.3c shows a structure to $\gamma^*(h)$ for short sample spacings up to spacing s, beyond which a parabolic form is indicative of a trend. Such a pattern indicates that a trend (drift) is present in the data, but that local (quasi-) stationarity can be assumed over distances up to s. The identification of trends is essential because local stationarity (absence of trend locally) is required in many geostatistical calculations used in mineral inventory estimation. In this case, the semivariogram defines a distance over which the variable can be assumed to be locally stationary and represents a practical limit to a search diameter for use in selecting data to be used in making an estimate.

Figure 9.3d shows the semivariogram pattern that results when a periodic repetition (cyclical pattern) of a variable occurs (e.g., regular thickening and thinning of a vein, alternating high and low grades at a roughly regular interval). The resulting semivariogrm pattern is characteristic of a "hole effect."

9.2.1: Irregular Grid in One Dimension

Only rarely are data dispersed in the pattern of a perfectly regular grid. Thus, search of a data set for many pairs of samples separated by the exact distance h is doomed to failure. Instead, a search is made for all those pairs separated by $h \pm e$. In order to appreciate the methodology by which experimental semivariograms are obtained from real data and the limitations of these methods, it is useful to progress systematically from one- to three-dimensional data arrays. Consider a one-dimensional array of assay data (e.g., trench, drill hole, grid line) along which samples are dispersed at irregular intervals (Fig. 9.1). To accumulate sufficient data with which to estimate successive points of an experimental semivariogram, paired data are grouped as follows: for all data representing any lag h

$$\text{No. of data at lag } h = n.d \pm d/2 \qquad (9.1)$$

where h is the approximate average lag, d is a constant distance (equivalent to a regular grid spacing), and n has integer values 1, 2, 3. . . . Such a procedure is justified because semivariograms commonly are more or less linear over short intervals. However, when using this procedure, it is wise to calculate the real average distance between data pairs used for each point estimate of $\gamma(h)$ rather than assuming the distance to be exactly equal to $n.d$; in some cases the two distances h and $n.d$ can differ substantially. Moreover, it is useful to know the number of pairs used to determine each point of an experimental semivariogram; automatic grouping procedures commonly lead to some points (possibly spurious) that are based on a small number of pairs, and thus should be ignored or minimized in developing a semivariogram model. Both the grouping procedure and the number of pairs used to calculate each $\gamma(h)$ value are output in most commercially available software packages for determining experimental semivariograms.

Selection of an appropriate value for d in Eq. 9.1 is not always obvious; an inappropriate choice can lead to a highly erratic short range variations in $\gamma(h)$. A straightforward means of dealing with this problem is to produce a plot of the distance between paired values versus half the squared difference of the two values. An example of such a plot is given in Fig. 9.4 and clearly illustrates the ease of selecting an appropriate unit lag value (which maximizes the use of data) for the construction of an experimental semivariogram. In this example, an appropriate unit lag (d) is selected to maximize the number of pairs (n) used for a series of lags (nd).

9.2.2: Semivariogram Models

There has been some discussion in the literature as to why semivariogram models are necessary (e.g., Shurtz, 1991). The most fundamental reason is that an autocorrelation function (e.g., the semivariogram) quantifies the average three-dimensional continuity of grade and defines the isotropic/anisotropic nature of grade distribution, an essential concept regardless of which resource/reserve estimation method is to be used. As indicated in Chapter 8, various geostatistical calculations involve use of the semivariogram. In addition, resource/reserve estimation using kriging requires semivariogram (or comparable) values in the estimation procedure. Hence, semivariogram models are important for several reasons. The

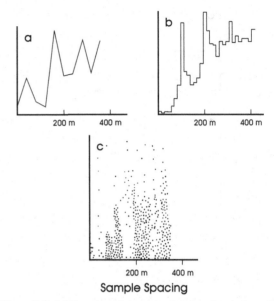

Figure 9.4: (a) Experimental semivariogram showing erratic character because of an injudicious choice of initial lag (20 m). (b) Bar graph of the relative frequency (ordinate) of paired data for various sample separation intervals (abscissa). (c) Plot of the squared grade differences (ordinate) versus sample separation (abscissa) for all data pairs. This latter graph demonstrates that an optimal initial lag is about 100 m. Redrawn from Armstrong (1984b).

semivariogram model must be "positive definite" in order to guarantee that positive variances result from its application in kriging equations. A smooth model can generally be justified also on the basis that local experimental values are subject to sampling error, whereas a model is more generally applicable because it smooths out these local sampling fluctuations. In general, a "sawtooth" experimental semivariogram is approximated by smooth mathematical functions (models).

Several mathematical models have been used historically in geostatistical applications, including linear, de Wisjian, exponential, Gaussian, and spherical (Matheron), as well as others (see Journel and Huijbregts, 1978). Without question, the most widely used model in mineral inventory applications is the spherical (or Matheron) model (cf. Figs. 9.3b and 9.6), and although that model is emphasized here, several other commonly used models also are described.

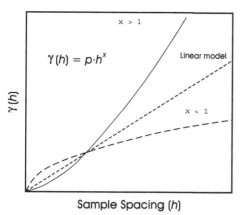

Figure 9.5: A general power model for a semivariogram; p is a constant, h is sample separation. The model becomes a straight line if the exponent, x, is 1.0. In this case there is no discontinuity at the origin, that is, no nugget component.

9.2.2.1: Linear Model

The *linear model* is a special case of a power model of the type (Fig. 9.5)

$$\gamma(h) = C_0 + p.h^x$$

where C_0 is the nugget effect; p is a constant (slope in the linear model); h is lag; and x is the exponent of h such that $0 \leq x \leq 2$ ($x = 1$, in linear case).

For the linear case, $x = 1$, p is the slope of the line, and C_0 is the y intercept. The linear model does not find much use in general application. However, because of its simplicity in terms of calculations, it can be used advantageously to work out simple calculations manually. It is realistic in some applications because the linear model closely approximates the more useful spherical model (see later) for distances up to about two-thirds the range of the spherical model (Section 9.2.2.4).

9.2.2.2: Exponential Model

The exponential semivariogram model is given by

$$\gamma(h) = C_0 + C[1 - \exp(-h/a)].$$

The exponential model has been used here and there in practical applications. It suffers the minor inconvenience that it attains a sill only at very great distances, although an effective sill is attained at $h = a' = 3a$.

9.2.2.3: Gaussian Model

The Gaussian model is given by

$$\gamma(h) = C_0 + C[1 - \exp(-h^2/a^2)].$$

The Gaussian model, like the exponential model, has found scattered use in resource estimation. Use of this model occasionally results in estimation problems because of the high degree of continuity inherent in the model. The sill is reached asymptotically, and a practical sill is defined by $h = a' = a.3^{1/2}$. The Gaussian model is relatively flat for low gamma values, a characteristic that leads to advantages for use in estimating variables such as topography and vein thickness. However, negative weights for some samples can result from using this model for estimation purposes.

9.2.2.4: Spherical (Matheron) Model

The spherical model (Fig. 9.6) is characterized by two components: (i) a purely random component referred to as the *nugget effect*, or C_0; and (ii) a structured component in which the structure is characterized by the range a of an autocorrelation function. The *range* is the distance over which average variability for the stuctured component rises from zero to C_1 (or from C_0 to C):

$$\gamma(h) = C_0 + C_1[3/2(h/a)$$
$$-1/2(h/a)^3] \quad \text{for } h < a \qquad (9.2)$$
$$\gamma(h) = C_0 + C_1 \qquad \text{for } h > a$$

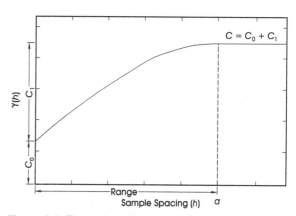

Figure 9.6: The spherical or Matheron semivariogram model. The model is defined by the nugget effect (C_0), the sill of the structured component ($C_1 = \gamma(\text{inf}) - C_0$), and the range ($a$).

where C_0 is the nugget effect, C_1 is the sill of the structured part of the model, a is the range (of influence), and h is sample spacing (lag). The range of a spherical model can be viewed as the average dimension of structures randomly located in the total field of data.

The nugget effect warrants comment. In general, C_0 represents a combination of factors including analytical variability; subsampling; sampling variability; and, in some cases, short-range (very local), real, geologic variability. Of these sources of variability, the combined analytical and sampling variance is easily determinable from a program of duplicate sampling (Chapter 5); differences between duplicate samples normally include a component of local geologic variability that is a function of the space between duplicate samples. Spatial variability that arises on a scale greater than duplicate sample spacing but less than the nominal sample spacing of an evaluation project must be estimated or assumed from other information; in many cases, this additional spatial contribution to the nugget effect is minor or negligible.

9.3: FITTING MODELS TO EXPERIMENTAL SEMIVARIOGRAMS

The art of fitting models to experimental semivariograms is fraught with subjective decisions. As Dubrule (1994) points out, there is uncertainty as to the general model to be used and in the estimation of the parameters of a particular model. Because the spherical model has found widespread application in mineral inventory studies, comments here are restricted to that model. One of the principal difficulties in many practical cases is that little reliable information exists for short sample spacings, which can lead to low confidence in estimating the range. An added complication is the sparsity of information on which to base an estimate of the nugget effect. Armstrong (1984b) notes that difficulties arise in modeling experimental semivariograms because of (i) a poor choice of distance classes, (ii) mixed populations inadvertently grouped together, (iii) outliers and skewed distributions, and (iv) artifacts. Operator error leading to artifacts is generally the result of carelessness and

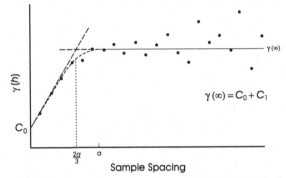

Figure 9.7: Fitting a model to an experimental semivariogram. The sill [γ(inf)] is generally easy to define even when there is much variability. The first few points of the experimental semivariogram commonly can be approximated by a straight line, which defines the intercept with the y axis (C_0), and an intersection with the sill, which occurs at two-thirds the range.

can be minimized by close attention to procedures and output. Outliers and mixed populations generally should be identified and considered separately, as suggested in Chapter 7. Choice of a distance class can be improved by examination of the data, as suggested by Armstrong (1984b) and Fig. 9.4.

The following procedures, although idealized, provide guidelines for maximizing confidence in a semivariogram model fitted to real data represented by an experimental semivariogram (Fig. 9.7):

1. The "sill" C (i.e., $C_0 + C_1$) of a spherical semivariogram is the variance of the data in the space under study. In practice, this rule has limitations with small amounts of data, with cyclical patterns in the data, where a trend exists in the data, or where the range of the semivariogram is large with respect to the field width. Barnes (1991) concludes, in part: (i) for large data sets for fields more than three times the semivariogram range, the sill of a spherical semivariogram closely agrees with the data variance; (ii) if data are from a field whose dimensions are the same order as the semivariogram ranges, the sill will be underestimated by the variance of the data; (iii) where "low-scale" trends exist in the data, the data variance will overestimate the sill of the semivariogram; and (iv) when the sill of a model differs from the data variance,

the model is suspect. In many practical cases, however, the data variance can aid in estimating the value of $(C_0 + C_1)$.

2. For an experimental semivariogram with "smooth" continuity through several points near the origin, a straight line through those points is an estimate of the tangent at $h = 0$ for the spherical model. The tangent intersects the y axis at the nugget effect and intersects the sill at a value of $2a/3$, to provide an estimate of the range (Fig. 9.7).

3. When the semivariogram has a sawtooth aspect near the origin, the tangent (at $h = 0$) can be estimated by a "visual best-fit" line that estimates the nugget effect and range.

4. Semivariograms should be derived separately for each support to test for effects of regularization (see Section 9.10). Different supports might give rise to different semivariogram models. Ordinarily, models for different supports are related in a systematic fashion (e.g., Huijbregts, 1971).

5. Semivariograms should be derived in a variety of directions (at least four directions in two-dimensional arrays of data) in order to test for anisotropy (see Section 9.4.1). In practice, exploration data are commonly arranged in preferential directions and it may be difficult to examine fully the possibility of anisotropy. In such a case, geologically important directions can be used as guides in the development of experimental semivariograms and their models. For example, obvious structural or stratigraphic directions generally are found to be important limiting cases for semivariogram modeling.

6. A plot of standard deviation versus mean grade for data subsets (e.g., individual drill holes) should be examined to test for a proportional effect (see Section 9.5). In the simplest case, a proportional effect can be accounted for by using a relative semivariogram (see Section 9.5). Separate experimental semivariograms for high- and low-grade subsets of data also are indicative of the presence of a proportional effect and the plausibility of using a relative semivariogram for modeling and estimation purposes.

7. When sufficient duplicate sampling data are available, it may be possible to use half the mean squared grades of paired duplicates to estimate a minimum nugget effect.

Models normally can be fitted to experimental semivariograms by eye without the added complexity of various best-fit procedures, such as those described by Cressie (1985). In many cases, least-squares fitting methods are inappropriate because they do not allow the models to integrate exact information easily that might be available independently (e.g., a limiting value of the nugget effect might be known from replicate sampling information, or the ranges of various attributes might be constrained geologically).

In general, confidence in a semivariogram model (i.e., a reasonable fit to experimental data) is essential because the model reflects the geologic character of the domain and is used in a variety of important calculations, especially kriging. Consequently, it is useful to have an appreciation of the magnitude of errors that can result if a semivariogram model is incorrect. Brooker (1986) documents kriging results using a spherical model and a regular square data array to illustrate the impact of errors in semivariogram model on resulting kriging errors. He adopts a known model, and then investigates the effect on kriging error of changing the parameters C_0 and a by various amounts. Brooker concludes that kriging is relatively robust to errors in semivariogram modeling. For errors of up to +25 percent of range, kriging variances change by less than 10 percent; for errors up to +25 percent of relative nugget effect $[C_0/C = C_0/(C_0 + C_1)]$, larger changes in kriging variance are possible, particularly when the relative nugget effect is low and the range is one to three times the data spacing. Underestimation of the nugget effect can lead to serious underestimation of kriging variance; thus, substantial attention should be paid to determining the true nugget effect.

9.4: TWO-DIMENSIONAL SEMIVARIOGRAM MODELS

In perfectly regular two-dimensional grids (e.g., Fig. 9.9a) it is possible to select parallel lines of data and

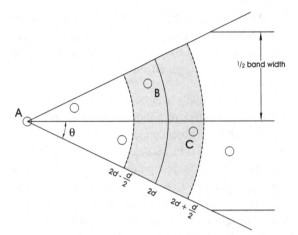

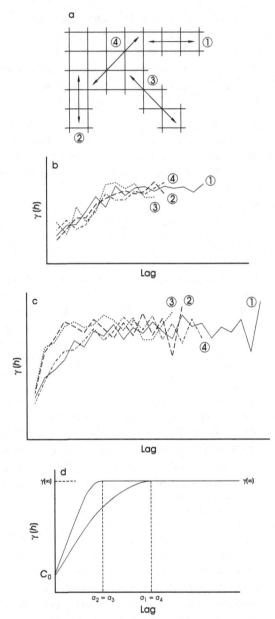

Figure 9.8: Illustration of how tolerance is applied in selecting pairs of data separated by distance $h \pm e$ in a two-dimensional field of data. For any one data point (e.g., A), there is both an angular tolerance ($\pm$) about the direction for which a semivariogram is being determined, a distance tolerance (e.g., $\pm d/2$), and a band width that limits the extent of the field to which the angular tolerance applies. For the example shown (distance $h = 2d \pm d/2$), the shaded area of tolerance indicates that two pairs of data would be selected, A–B and A–C.

determine the corresponding experimental semivariogram, as summarized in Eq. 9.1. Furthermore, sets of parallel lines with various azimuths can be selected, including the main grid directions and the principal diagonals, and experimental semivariograms can be determined separately for each direction. If the structure of the variable is isotropic, all of these directional, experimental semivariograms will be roughly the same within experimental error. Where there are anisotropies, the experimental semivariograms differ as a function of direction. Commonly, however, this simple ideal procedure for characterizing the experimental semivariograms is not possible because data are rarely located so regularly.

The general situation for determining experimental semivariogram values for an irregular data pattern in two dimensions is illustrated in Fig. 9.8. Here, the problem of grouping data in order to provide estimates for various lags is seen to be more complex than in the one-dimensional case. Commercial computer programs provide the ability to change the

Figure 9.9: (a) A regular grid showing the minimum of four directions in which the semivariogram should be determined in order to check for isotropy/anisotropy. (b) Idealized example of isotropy of four experimental semivariograms determined for our grid directions. (c) Geometric anisotropy of the semivariogram model demonstrated by different experimental semivariograms in different directions. Nugget effect and sill are identical in all directions, but the range changes as a function of direction. (d) Spherical models representing the experimental data of (c).

angular tolerance; large angular tolerances in general are not desirable, particularly with relatively small data sets, because all directions tend toward the same model, even when substantially different models exist. Consequently, the use of large angular tolerances can lead to the incorrect conclusion that the autocorrelation function is similar in all directions (i.e., that the semivariogram is isotropic). Of course, there are many cases in which essentially equivalent experimental semivariograms can be demonstrated (similar in range, C_0, and sill) along several different grid directions and isotropy can be accepted as illustrated in Fig. 9.9b. The term *omnidirectional*, which is now widely used in geostatistics, should not be confused with *proven isotropy*. *Omnidirectional* is a term recommended for data arrays in which spacing is very wide in one or two directions relative to the other, the common situation with exploration drill-hole sampling. In such cases, true isotropy might not be evident at the time of model fitting, although an isotropic model might well be assumed; the term *omnidirectional* should be used in such cases and implies that some uncertainty exists in the validity of an isotropic model (cf. Journel, 1985).

9.4.1: Anisotropy

There is no particular reason why a semivariogram in one direction will be the same as that in a second direction. For example, it is apparent that the variability along a sedimentary bed (e.g., iron formation) can be very different than the variability across strata. Two-dimensional data arrays require that the semivariogram model be characterized for several directions in order for a comprehensive semivariogram model to be established. When the experimental semivariograms for four or more directions (e.g., the principal grid directions and the two principal diagonal directions) are found to be essentially the same, an isotropic model (uniform in all directions) can be accepted. When the semivariogram varies with direction within a data array, the regionalization (structure) is said to be *anisotropic* (e.g., Fig. 9.9c). In general, it is not adequate to limit directional semivariograms to only

four directions in two-dimensional space; six or eight directions (e.g., each 22.5 degrees) provides sounder insight into the isotropic/anisotropic character of a variable (see Fig. 9.12).

Two types of anisotropy are recognized in mineral inventory applications – geometric and zonal. A further subdivision of zonal anisotropy into subcategories of sill, range, and nugget anisotropy is generally unnecessary (Zimmerman, 1993). In geometric anisotropy, the range varies as a function of direction, but the nugget effect and sill remain constant. A relatively simple example is illustrated in Fig. 9.9 (c and d) in which two directions show a common range, distinctly different from the range of the other two directions. In the more general case, of course, all directions might give different ranges. Geometric anisotropy is a relatively common feature in semivariograms of ore grades.

A two-dimensional, geometric, anisotropic structure can be visualized ideally as an ellipse whose variable diameters are the semivariogram ranges for various directions, as shown in Fig. 9.10. The ellipse is an estimate of the average structure that can be thought of as occurring randomly thoughout the two-dimensional field of data. Reality is approximated only by the elliptical model; locally real structures depart from the model in random fashion. The semivariogram model can be summarized in terms of the semivariograms representing the two principal axes of the ellipse. Because these models differ only in their ranges, it is sufficient to define a single isotropic model, say for the major axis of the ellipse, and then provide a multiplier (anisotropy ratio) that changes the range to that of the minor axis of the ellipse. For example, if ranges for the two principal axes of the structure are 50 m in a northeast direction and 25 m in a northwest direction, the model can be described by an isotropic component with a range of 50 m and an anisotropy ratio of 0.5 to take the northwest anisotropy into account.

In many cases, the directions of principal continuity of grade (long and short axes of the ellipse) can be (and should be) identified in advance from the geologic character of the mineralization. There are

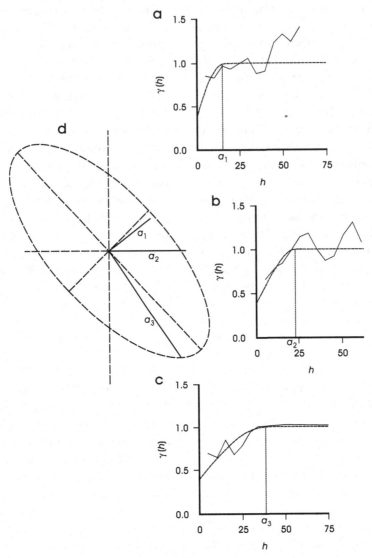

Figure 9.10: (a, b, and c) Experimental semivariograms and fitted models for three directions in a vertical section through the Warrego gold pod. Modified from Quinlan and Leahey (1979). (d) Ranges (a_1, a_2, and a_3) for the three directions, plotted to a common center, permit an ellipse to be constructed that represents the average structure in the two-dimensional field and clearly defines a geometric anisotropy.

many two-dimensional deposits for which the plane of the structure is well defined and the orientation of ore shoots within that plane is also known. In the more general case of three dimensions, the continuity model becomes an ellipsoid, the principal axes of which also correlate with geologic features having preferred orientation. In stockwork mineralization, for example, one direction of fracture control

commonly predominates to some extent, and this direction coincides with the long axis of the ellipse of anisotropy (e.g., Sinclair and Giroux, 1984; Sinclair and Postolski, 1999).

The second type of anisotropy, referred to as *zonal*, is characteristic of bedded sequences. In this type of anisotropy, all directions within the plane of the bedding are characterized either by an isotropic

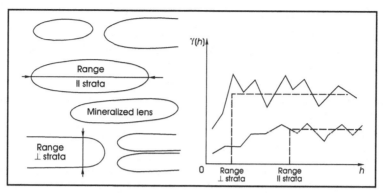

Figure 9.11: Zonal anisotropy in which the sill levels of semivariograms differ significantly with change in direction. Mineralized lenses shown on the left are parallel to stratigraphy. On the right, the experimental semivariograms show longer range and less variability (lower sill) parallel to stratigraphy, compared with a cross-stratigraphy orientation.

semivariogram or display geometric anisotropy. In contrast, the direction across bedding displays a much greater variability with shorter continuity than in directions parallel to bedding; thus, the model in this third direction has a shorter range and a higher sill compared with directions parallel to bedding, as illustrated in Fig. 9.11. For practical purposes, a zonal anisotropy can be approximated by two spherical components as follows:

$$\gamma^*(h) = \gamma^*(x, y, z) + \gamma^*(z)$$

that is, a component that is isotropic (or having geometric anisotropy) in three dimensions plus a component that depends only on direction z. Myers and Journel (1990) show that zonal semivariogram models can lead to problems in kriging; in some cases, the kriging equations contain noninvertible coefficient matrices because the semivariogram models are semidefinite rather than positive definite. This condition arises in a few restrictive geometries of data for which the kriging system is singular and probably occurs rarely in practice. Nevertheless, attention should be directed to this problem where zonal anisotropy models of semivariograms are used for estimation purposes with highly regular grids of data.

In examining data for anisotropy, a common practice is to determine experimental semivariograms along principal grid directions and the two principal diagonal directions. Ranges found in these directions

can be plotted on a common center and fitted by an ellipse to produce an idealized geometry of the "average" anisotropic structure of the data array. In many cases, the locations of data (drill holes, underground workings) dictate the data array and may limit the orientations for which acceptable experimental semivariograms can be determined (e.g., Fig. 9.10). As indicated previously, it is common that the principal directions of anisotropy coincide with geologically defined directions, which may well have controlled grids established during exploration. In other words, exploration grid axes may be parallel or subparallel to the axes of the continuity ellipse. When true northings and eastings are the principal coordinates, it would be coincidental for them to parallel the principal directions of anisotropy. Geology can commonly provide assistance in formulating an efficient approach to semivariogram modeling. In particular, when relatively few data are available, model development can be controlled by known directions of geologic control of mineralization (i.e., a model can be defined along the direction of principal geologic control and perpendicular to that direction). This possibility arises commonly in practice because of the manner in which data are collected for deposit exploration/evaluation (i.e., along the principal structural direction and perpendicular to that direction).

When principal directions of anisotropy are not coincident with principal grid directions, the angular relations between the two must be known. During estimation, a transformation to "anisotropy coordinates" is required so that estimation can be done as an isotropic exercise. As indicated previously, the procedure for checking for the presence of anisotropy involves determining experimental semivariograms in several different directions. If the results for individual directions are not significantly different, it is common practice to produce a weighted average semivariogram to which an isotropic model can be fitted. A particularly useful technique for gaining insight into the possible anisotropic nature of the semivariogram in two dimensions is to produce a semivariogram map. Commercial and academic software is available for this procedure (e.g., Deutsch and Journel, 1998). An example of a semivariogram map is shown in Fig. 9.12 for a two-dimensional array of data from a disseminated gold deposit (cf. Rendu, 1984). In this example, semivariogram values for various distance and direction, plotted outward from a common center,

are contoured. The elliptical form of the contours indicates a well-defined, anisotropic character to the deposit. Semivariogram maps also can be displayed as gray-tone maps on which low $\gamma^*(h)$ values are blank and increasing values become increasingly darker, finally becoming black for values near the sill. The use of gray-tone representation of semivariogram maps has the disadvantage that well-informed directions are not distinguished from directions for which data are limited or missing. In this respect, contoured maps such as Fig. 9.12 are a fairer representation of the state of knowledge. An additional limitation to some software for semivariogram maps is the fact that a common lag is used for all directions; in practice, of course, the same lag is not necessarily optimal for all directions.

9.5: PROPORTIONAL EFFECT AND RELATIVE SEMIVARIOGRAMS

A *proportional effect* is a systematic relation between variability of data and the mean value of the data. Presence of a proportional effect can be demonstrated by examining a plot of mean value versus corresponding standard deviation for clusters of data (e.g., individual drill holes). An example is shown in Fig. 9.13, in which the relation can be approximated by a straight line. In many early geostatistical publications, the proportional effect was approximated by simple equations describing C_1 and C_0, of the type $C = k.m^2$, where k is a constant to be determined by plotting various estimated C values versus their corresponding means m. More recently, however, there has been widespread recourse to the relative semivariogram to approximate variations in the semivariogram model as a function of the mean value of data to which the model is applied.

The need for taking the relation of mean and error into account can be demonstrated by an idealized example that illustrates a commonly encountered occurrence. Suppose that a particular geometry of data and an absolute semivariogram model provide a block estimate of 0.8 percent Cu with a kriging error (standard deviation) of 0.4 percent Cu. In a low-grade zone, the same geometry of samples might produce an estimate for a comparable block of 0.2 percent Cu; of

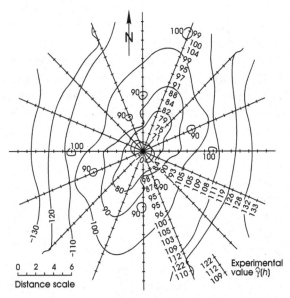

Figure 9.12: A semivariogram map characteristic of an anisotropic variable. Semivariogram values are determined for various sample spacings along various directions; results are plotted relative to a common center and contoured. Such diagrams are particularly useful in identifying the directions of principal grade continuity, as in this example for a disseminated gold deposit (redrawn from Rendu, 1984). Directions of principal grade continuity invariably have a geologic control.

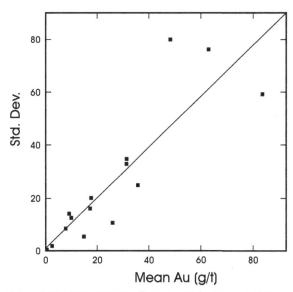

Figure 9.13: Proportional effect in which the variability of data is a function of the mean value of the data. These data are for the Snip mesothermal gold vein; values are in g/t. In such cases, the sill value of the semivariogram (equivalent to the variance of the data) also is a function of the mean value of the data.

course, the same absolute error of 0.4 percent would be produced and is a nonsensical result. However, if a relative semivariogram model had been used to provide a 0.8 percent copper estimate with a kriging error of 0.4 percent Cu (i.e., relative error of 50 percent), the low-grade block would have an estimated error of 0.1 percent Cu (i.e., 50 percent), a much more plausible result in view of the common dependency of error on concentration (e.g., Chapter 5).

The relative semivariogram is given by

$$\gamma^*(h)_r = \gamma^*(h)/m_h^2$$

where $\gamma^*(h)$ is the absolute semivariogram defined in Eq. 9.2 and m_h is the mean of data used to determine $\gamma^*(h)$. The estimation of relative semivariograms is a scaling procedure to produce uniform variance throughout the data array. Cressie (1985) demonstrated the general equivalence of using relative semivariograms and semivariograms of logtransformed data. Recognition of situations in which use of the relative semivariogram is appropriate can be made when linear patterns exist on plots of local values of s

versus m (i.e., standard deviation vs. mean value) as shown in Fig. 9.13.

The detailed procedure for estimating a relative semivariogram is important because different results can be obtained depending on how the relative gamma values are determined; in particular, whether a global mean or a local mean is integrated into the calculation of Eq. 9.2. In the past, it was common practice to simply divide $\gamma^*(h)$ values, determined as outlined previously, by their corresponding global mean value to obtain a global or general relative semivariogram. David (1988) states, "Experience shows that the general relative variogram ... overestimates the real relative variogram. Then dispersion and estimation variances derived from it are above, and sometimes far above, the real variances" (p. 43). Instead, the pairwise relative semivariogram is now used in which the squared difference of each pair of values is divided by the squared mean of the pair. This procedure has the effect of reducing the impact of very large differences. An example of a pairwise relative semivariogram for gold accumulation in a gold-bearing quartz vein is shown in Fig. 9.14. This example is for closely spaced stope samples and illustrates a small but definite regionalization for a case that had been described previously being pure nugget effect.

Relative semivariograms are a useful means of comparing and contrasting relative variability and continuity characteristics for different deposits, including those of various types. An example from Raymond (1982) is shown in Fig. 9.15, which compares the semivariogram of the Mt. Isa stratiform Pb–Zn deposit with those for two very different porphyry-type deposits. Of course, such comparisons should be made on samples of equivalent support.

9.6: NESTED STRUCTURES

In many cases it is not possible to make an adequate approximation of an experimental semivariogram by a single model. This situation can arise with the presence of more than one underlying structure in the data being considered. In other words, regionalization may be present at several scales. The concept of nested structures is illustrated in Fig. 9.16. Some nested

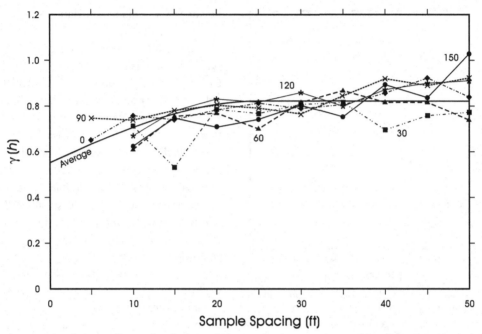

Figure 9.14: An example of six experimental, pairwise, relative semivariograms for a gold–quartz vein with a relatively large nugget effect and a fitted model (smooth curve – average). The variable is accumulation (opt × thickness). The experimental semivariograms are based on a rectangular array of 235 closely spaced stope samples for a vertical vein. Individual directions are labeled by azimuths on the assumption that on a longitudinal vertical projection, vertical is north and horizontal is east. Lag is in feet.

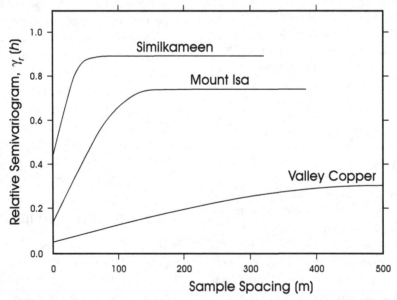

Figure 9.15: Relative semivariograms for three deposits illustrating the comparative difficulty in making grade estimates from one deposit to another. Even two porphyry-type deposits (Similkameen and Valley Copper) have dramatically different semivariogram models; grade data for Valley Copper are relatively continuous and the estimation problems are much less severe than for the Similkameen deposit. Redrawn from Raymond (1982).

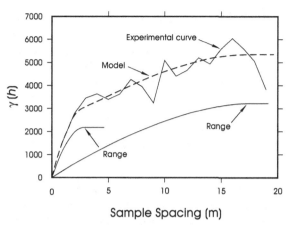

Figure 9.16: An example of a nested structure showing how two spherical structures (no nugget effect) combine to produce a model (dashed curve) that fits an experimental semivariogram for Au grades, Warrego gold pod. Redrawn from Quinlan and Leahey (1979).

structures relate to geometric characteristics that are easy to visualize. Figure 9.16 illustrates the presence of three scales of variability in metal accumulation values (grade × thickness), a very small nugget effect, and two separate scales of grade variability.

Nested structures commonly are indicated by breaks in slope of an experimental semivariogram. With limited data, the general fluctuations from point to point may mask these breaks in slope; consequently, a detailed semivariogram model may not be attainable with small amounts of data.

9.7: IMPROVING CONFIDENCE IN THE MODEL FOR SHORT LAGS OF A TWO- OR THREE-DIMENSIONAL SEMIVARIOGRAM

Often, an early goal in spatial sampling is to obtain data from which the semivariogram may be estimated. (Morris, 1991, p. 930)

A semivariogram model can be fitted to any set of points, no matter how many data are available or how valid the model is. Estimation of the semivariogram model is critical for many geostatistical purposes, and substantial effort must be directed to considering what constitutes an acceptable model (i.e., a model in which the user can have confidence). Empirical guidelines

(Journel and Huijbregts, 1978) indicate that at least 30 pairs are necessary for each lag of the experimental semivariogram and that no lag greater than $L/2$ should be accepted, where L is the average width of the data array in the direction for which the semivariogram is being estimated. These minimum requirements commonly are inadequate in practice, particularly in the case of deposits with extreme local variability (e.g., gold and uranium deposits).

Early in the data gathering process, one important goal is to design an array for data collection that includes enough closely spaced samples to provide confidence in a semivariogram for short lags. Such information can be provided from a number of sources, including drill holes oriented in various directions, surface exposures, trenches, and underground workings such as exploratory declines. Of course, in the general situation, closely spaced data are desirable in three dimensions because the semivariogram can be anisotropic. When exploration sampling is on a regular grid of spacing d, one part of the grid might be drilled at much closer spacing to provide the necessary information. The detail of information that is required might not be justifiable economically until the feasibility stage. When possible, such data should be selected optimally (i.e., as few data as possible to meet the need).

Idealized approaches to sampling for purposes of defining a semivariogram model confidently are seldom practical globally because, in a program of staged data collection, geology, personnel, cost, and time are overriding factors. However, idealized sampling concepts can be used to assist in localized sampling design to provide data for establishing semivariogram values for short lags, a factor too commonly overlooked during exploration to the detriment of subsequent resource/reserve estimation. Commonly, exploration programs involve drilling that produces closely spaced samples along the drill holes and relatively widely spaced samples in other directions. Local data arrays can be designed to offset this problem with respect to semivariogram modeling. For example, in a large domain characterized by one type of mineralization, a square surface (horizontal) sampling grid that is eight samples by eight samples with sample spacing equivalent to spacing in vertical drill holes might be ideal for characterizing short lags of the semivariogram in two dimensions at least. Such a grid would provide 56 pairs for lag 1 of a semivariogram parallel to a principal grid direction, 48 pairs for lag 2, 40 pairs for lag 3, and 32 pairs for lag 4. Somewhat fewer pairs would be obtained for diagonal lags. This type of sampling plan demands appropriate ore exposures (e.g., an outcrop or stripped area on the surface) with little in the way of surficial weathering. The same goal can be achieved in other ways – for example, a series of disjoint trenches that penetrate the weathering zone, sampling of underground workings, or preferred orientations to selected drill holes. Whatever approach is taken, it is most useful if samples are more or less of the same support as the great bulk of samples used for general semivariogram modeling.

When the opportunity exists to develop local sampling grids for the purpose of estimating short lags of the semivariogram, consideration might be given to triangular grids (unit cell is an equilateral triangle), which Yfantis et al. (1987) and Morris (1991) demonstrate to be the most efficient in covering a given area. Moreover, triangular grids provide the most efficient test of anisotropy in a two-dimensional plane, relative to square grids and hexagonal grids. It must be emphasized that design elements discussed here are not to provide samples to improve estimation, but to improve the quality of the semivariogram so that more confidence can be attached to all calculations that make use of the semivariogram.

9.8: COMPLEXITIES IN SEMIVARIOGRAM MODELING

9.8.1: Effect of Clustered Samples

It is not uncommon to have a 10:1 ratio of average grade in high-grade areas of orebodies compared to low-grade areas. With a proportional effect, variance from high-grade areas could differ by 100 times in comparison with values from low-grade areas. If the semivariogram is calculated by simply averaging

squared differences, the resulting model mainly reflects the semivariogram in high-grade areas because data generally are concentrated in high-grade zones. Generally, this problem can be minimized by determining the relative semivariogram, particularly the pairwise relative semivariogram.

9.8.2: Treatment of Outlier Values

Outliers should be considered and understood prior to semivariogram modeling through detailed evaluation and classification as described in Chapter 7. In particular, it is important to recognize the cause of outliers – Do they represent errors or fundamentally different geologic populations? When outliers represent a real geologic subpopulation, there is the added implication that each subpopulation has its own autocorrelation characteristics. When outliers are included accidentally in data used for semivariogram modeling, they can by recognized through the use of such diagrams as variogram clouds and h scattergrams (cf. Journel, 1987). Semivariogram clouds (e.g., Fig. 9.4c) can be presented in summary form as boxplots (e.g., Fig. 9.17) that indicate the concentrations (mean, range, and mid-50-percent range) of all individual gamma values for a given lag (h). For any h, an average gamma value that is far removed from the mid-50-percent range of the data on which it is based must be influenced by at least one outlier value; hence, the plots have been used by geostatisticians for the recognition of outliers. In general, however, outlier values should not appear in such plots, but should be recognized and dealt with independently prior to semivariogram modeling.

An h scattergram is an $x–y$ plot of the paired values used for the determination of a mean gamma value for a particular lag (h). The example shown in Fig. 9.18 illustrates the potential of such diagrams for identification of multiple subpopulations in the data, as well as recognition of a small percentage of outlier values as distinct departures from the main clustering of values. Once again, it is imperative to be aware that the presence of two geologic populations implies the likelihood of two "continuity" populations.

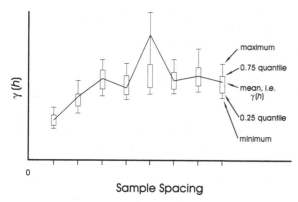

Figure 9.17: A "summary" of a semivariogram "cloud" diagram presented as a series of boxplots (box and tail), one for each lag. Each boxplot shows the mean, mid-50-percent range (box), and the total range of values (line) used to determine each point (i.e., value for each lag) for an experimental semivariogram. When a mean is substantially outside the corresponding box, the presence of at least one outlier value is indicated by data for the fifth lag. See also Fig. 9.4c for a complete semivariogram cloud diagram that illustrates how a few relatively large sample differences can control the form of the semivariogram.

Outliers generally have a detrimental affect on semivariogram modeling. In many practical cases, the sill of a spherical semivariogram is equivalent to the global variance of sample values. Hence, the impact of outliers on the semivariogram can be appreciated by considering two simple examples of mixed populations (cf., Zhu and Journel, 1991). Parameters of a mixed populations of two components are as follows (see Chapter 4):

$$\text{Mean value}, m = p \cdot m_1 + (1 - p) \cdot m_2$$
$$\text{Variance}, \sigma^2 = p \cdot \sigma_1^2 + (1 - p) \cdot \sigma_2^2$$
$$+ p(1 - p)(m_1 - m_2)^2$$

where m is mean value, $\sum$ is standard deviation, subscripts refer to populations 1 and 2, and p is the proportion of population 1. Substitute two hypothetical sets of parameters in the previous equations as follows:

$$\text{Case } 1 : m_1 = 0.5, \sigma_1 = 0.2; m_2 = 0.52,$$
$$\sigma_2 = 0.2; p = .7$$
$$\text{Case } 2 : m_1 = 0.5, \sigma_1 = 0.2; m_2 = 10,$$
$$\sigma_2 = 75; p = 0.99.$$

Case 1 exemplifies the situation in which two very similar populations are recognized, perhaps on geologic grounds, and provide parameters of the mixture ($m = 0.506$, $\sigma^2 = 0.0401$) that are roughly representative of the two component populations. Case 2 imitates an outlier situation and provides parameters of the mixture ($m = 0.545$, $\sigma^2 = 0.49$) that diverge markedly from those for both of the constituent populations. Clearly, a semivariogram with sill at 0.49 is far from representative of the sill for either of the two populations forming the mixture. Moreover, the added sampling variability of the mixed semivariogram leads to practical difficulty in defining ranges of the semivariogram model for the mixed populations, to the point that the range can be determined only with great uncertainty and perhaps with a large error that is not recognized.

9.8.3: Robustness of the Semivariogram

In geostatistical theory an assumption is made that the semivariogram is known or can be estimated reliably. For practical mining applications, however, this assumption is not generally safe, particularly when small data sets are concerned or highly skewed distributions are present. It seems evident that for a fixed area (or volume), the quality of an experimental semivariogram decreases as sample

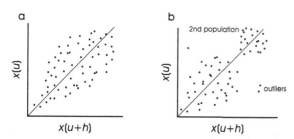

Figure 9.18: Pair of points [$x(u)$ and $x(u + h)$] used for the estimation of the experimental semivariogram value for one specific lag h is plotted as an h scattergram. Comparable diagrams can be constructed for all lags. (a) Common form of an h scattergram. (b) An h scattergram with complications viz. multiple subpopulations including outlier values.

spacing (lag) increases. This is because the geometric limitations of the data array are approached and there is a decrease in the number of pairs of data contributing to the estimate of $\gamma^*(h)$ for large values of h. A rule of thumb has emerged that an experimental semivariogram should not be considered for lags exceeding half the field width of the data array in any direction.

One practical problem in calculating an experimental semivariogram is the weight given to extreme values (so-called outliers) that generally represent a small proportion of the data. A common practice in geostatistical applications is to ignore outliers in developing a semivariogram model; however, these outliers are generally reinserted in the database for use during the actual estimation process (e.g., David, 1977). This procedure can be inappropriate in the unlikely situation that outliers have the same continuity model as do the more abundant low values. As discussed previously, outliers generally can be shown to represent a very different geologic feature than the great bulk of the data and, therefore, to have different continuity than the majority of data. Consequently, the outliers must be considered separately, both for semivariogram analysis and for estimation; to assign them the same continuity as the abundant lower values can result in serious grade and tonnage overestimation (cf. Sinclair et al., 1993).

The general procedure for estimating a point on the experimental semivariogram is to plot half the mean-squared difference versus lag. This use of the mean is optimal when the variable is distributed normally; strong positive skewness leads to lack of robustness. Omre (1984) lists distributional deviations (e.g., strong positive skewness), sampling deviations (e.g., biased sampling), and outlier deviations from normality as potential causes of lack of optimality in semivariogram estimation. Effects of these sources of variability can be minimized by improving data quality (including the minimizing of mechanical error), defining appropriate geologic domains within which to quantify value continuity (see Chapters 2 and 3), and using as much data as possible for estimating semivariograms, as opposed to using small subsets of data.

9.8.4: Semivariograms in Curved Coordinate Systems

There are a number of geologic environments in which it is advantageous to have a coordinate system that entails at least one curved coordinate. Consider the example of a series of curved veins at the Golden Sunlight deposit, Montana (Roper, 1986; Sinclair et al., 1984), where it was convenient to approximate the orientation of veins in plan by the arc of a circle whose center was found empirically. The three-coordinate system used in this case (Fig. 9.19) involved (i) distance along an arc, (ii) distance perpendicular to the arc (along a radius), and (iii) vertical distance. This procedure permitted the development of semivariogram models with geometric anisotropy, the greatest continuity of which paralleled the arcuate coordinate direction. Use of the curved coordinate system facilitated block kriging because it maintained the correct

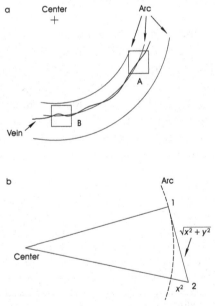

Figure 9.19: (a) The horizontal trend of an arcuate vein is approximated by the arc of a circle whose center was located by trial and error. The arc represents the direction of maximum continuity in the Golden Sunlight deposit; the lesser axis of grade continuity at any location is perpendicular to the arc. (b) Explanation of how distance components parallel to the arcuate trace are approximated. Redrawn from Sinclair et al. (1983).

spatial orientation of the semivariogram model to each block during kriging. A comparable procedure was attempted at the Climax molybdenum deposit (Noble and Ranta, 1984), where ore grades were concentrated in a cylindrical zone.

A variety of departures from a Euclidean coordinate system are possible. Dagbert et al. (1984) describe two such possibilities for dealing with folded strata. One suggestion is the "natural'" coordinate system, in which one coordinate is a digitized outline of folded bedding surfaces, a second is perpendicular to bedding, and the third is parallel to the fold axis.

9.8.5: The "Hole Effect"

Some directional experimental semivariograms rise from the origin (or nugget value) in the form of a spherical model and then decrease and increase alternately in the form of a wave, commonly of decreasing amplitude. These depressions, or "holes," in the sill give rise to the name *hole effect*. This form of semivariogram can be generated in one, two, or three dimensions by alternating high and low values on a perfectly regular grid of the appropriate dimension. For example, one could imagine a perfectly regular three-dimensional array of ellipsoids of high values separated in all dimensions by low values. Such an idealized repetition of high and low values is unlikely in practice; more probable is a hole-effect model in one dimension that crosses a regularly repeated geologic structure.

The hole-effect form of an experimental semivariogram is important to recognize because of the implied regular geometric repetition. This repetition is not to be confused with the occasional local semivariogram that can arise when a line of sampling intersects a single, well-defined structure in predominantly background values, or a long drill hole that intersects a short, high-grade zone in a field of low grade, for example. An example of a hole-effect experimental semivariogram and model for an indicator variable is given in Fig. 9.20. Note that the lows of the experimental semivariogram coincide with the distance of one complete repetitive structure (and simple multiples of this structure). Moreover, the range, as

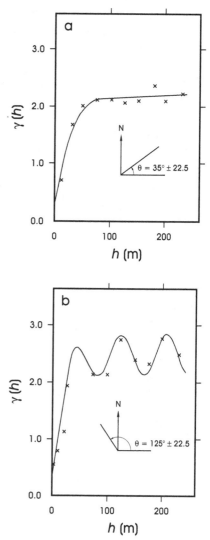

Figure 9.20: Example of a two-dimensional semivariogram model that is spherical in a 35-degree direction but has a pronounced hole effect in a 125-degree direction. An angular tolerance of ±22.5 degrees was used for both directions. Models are for tungsten data. Redrawn from Journel and Froidevaux (1982).

determined from the first rise, coincides with half the width of the structure. This idealized view provides some insight into the geometric information contained in a hole-effect semivariogram.

The hole-effect semivariogram model is introduced by Journel and Huijbregts (1978) and considered in greater detail by Journel and Froidevaux

(1982), who make the point that hole effects generally reflect complex anisotropies. A detailed three-dimensional structural analysis of an Australian tungsten deposit is described by Journel and Froidevaux (1978) using both indicator semivariograms and log (WO_3^+) semivariograms to deal with an extremely highly skewed distribution ($CV = 2.3$) and a complex anisotropy. Hand-contoured data (grade $\times$ thickness) produce a series of regularly distributed highs and lows strongly elongate along an approximate azimuth of 30 degrees. An indicator semivariogram across this direction is described by a hole-effect model (Fig. 9.20) of the form

$$\gamma(h_x, h_y, h_z) = C_0 + C_1\{1 - e^{-!h'!/a}.\cos(h_v/b)\}$$
$$\text{for } h > 0$$

where subscripted h is the distance along the principal coordinate directions such that true distance $h = (h_x^2 + h_y^2 + h_z^2)^{1/2}$; h is an effective distance that takes anisotropy into account; and h_v/b controls the frequency of the hole effect. This model is the nested sum of an isotropic nugget effect and a dampened, directional, hole-effect model. Dampening is controlled by the exponential component. The best-informed direction (e.g., downhole semivariogram) can be used to estimate the nugget effect and determine the parameters a and b. Complex models of this type can improve local estimation and may be warranted, particularly when the amplitude of the hole effect is a significant part of the sill. For global estimation, the hole effect can be approximated by a sill of a spherical model (cf. Journel and Froidevaux, 1978). Judicious selection of data (maximum intersample spacing less than the range) allows a spherical model to be used to approximate the first rise of the semivariogram, and this model might be adequate for local kriging and to avoid the complexity of modeling the hole effect.

9.9: OTHER AUTOCORRELATION FUNCTIONS

There are a variety of approaches to the study of autocorrelation, among which the covariance is fundamental. Consider a random function with expectation m. For each pair of values of this regionalized variable,

the covariance depends on separation h, as follows:

$$\text{Cov}(h) = E\{Z(x + h).Z(x)\} - m^2.$$

When there is stationarity, the following relations hold:

$$\text{Var}\{Z(x)\} = E\{[Z(x) - m]^2\} = \text{Cov}(0)$$

and

$$\gamma(h) = 1/2E\{[Z(x + h) - Z(x)]^2\}$$
$$= \text{Cov}(0) - \text{Cov}(h).$$

The semivariogram, thus, is tied directly to the covariance. Moreover, there is a direct relation with the correlation coefficient (perhaps more properly in this case, the autocorrelation coefficient), as follows:

$$r(h) = \text{Cov}(h)/\text{Cov}(0)$$
$$= 1 - [\gamma(h)/\text{Cov}(0)].$$

Plots of the (auto)correlation coefficient for various sample separations versus lag produce the correlogram. In practical applications of the correlogram $r(h)$, in geostatistics it is common to use $1 - r(h)$, which has the form of a semivariogram and to which a semivariogram model can be fitted.

9.10: REGULARIZATION

In practice, a true point support is not attainable with most regionalized variables. Reality necessitates that a variable be measured over a discrete volume (i.e., the sample volume). Consequently, the local variability of the regionalized variable is smoothed relative to that of the true point variable. This relation is simply a demonstration of the smoothing effect of increased support, as discussed previously (Section 6.3.1). The general relation for the semivariogram of a variable, based on a large support in terms of the semivariogram based on point data is

$$\gamma_v(h) = \gamma(v, v_h) - \gamma(v, v)$$

where $\gamma_v(h)$ is the semivariogram regularized over volume v; $\gamma(v, v_h)$ is the average semivariogram value for two points defining, respectively, volumes

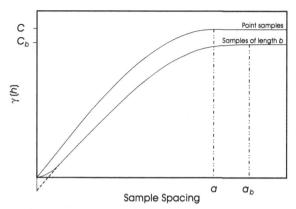

Figure 9.21: An example of regularization of a semivariogram. In general, if drill samples of length *l* are composited (regularized) over a greater length *b*, there is an increase in range of the semivariogram of (*b* − *l*). In addition, the regularization is accompanied by an increase in continuity near the origin. Redrawn from Clark (1979b).

v and v_h separated by distance h; and $\gamma(v, v)$ is the average semivariogram value for all pairs of points defining volume v. This is the so-called F function for which graphs exist (for a standard spherical model) in many texts (e.g., Clark, 1979; Journel and Huijbregts, 1978).

The general effect of regularization of the semivariogram is shown in Fig. 9.21; regularization decreases both the nugget effect and the sill of a spherical semivariogram. Theoretically, the range of a regularized semivariogram increases as the length of compositing increases, although this is not always clear in practice. The concept of regularization is important in comparing semivariogram models for data of different supports.

9.11: PRACTICAL CONSIDERATIONS

1. A three-dimensional study of autocorrelation is important because the development of a formal model quantifies the concept of grade continuity. This gives substance to the idea of range of influence as well as characterizing the isotropy/anisotropy of the mineralized system.

2. Practical estimation of a three-dimensional model for grade continuity commonly is plagued with deficiencies in the available data. This problem could be minimized if, during exploration, effort were directed toward obtaining close-spaced data along lines of various orientations (i.e., along trenches, underground workings, and drill holes with various orientations).

3. In tabular deposits, it is important to accumulate information with relatively high density along lines within the plane of the deposit. Ore shoots in veins, for example, commonly have a structural control that gives them an elongate form that could be indicative of an anisotropy to autocorrelation.

4. When dealing with large, three-dimensional ore bodies, it is useful to consider lines of data in various spatial orientations. A given data set generally constrains the directions in which experimental semivariograms can be obtained optimally. A general procedure that investigates experimental semivariograms along principal grid directions, oblivious to the orientations of linear sampling, is generally inadequate.

5. Developing semivariogram models appears easiest with modest amounts of data; this appearance is superficial. Limited data results in excessive variability that commonly masks important features of a regionalization. Consequently, the more data the better.

6. Ranges determined from experimental semivariograms for various orientations can be plotted about a common center to define a best-fit ellipse (two dimensions) or ellipsoid (three dimensions) that quantifies the average anisotropy of grade (if range is uniform in all directions, grade continuity is isotropic).

7. Geologic features invariably constrain the ellipsoid of continuity; the principal plane of an ellipsoid of continuity generally parallels a recognizable geologic feature, such as a shear direction, prominent direction of stockwork, or bedding. Such geologic features are generally recognized in advance and can be used to optimize the development of a semivariogram model.

8. Semivariogram modeling is time-consuming but warrants the effort regardless of the estimation method ultimately to be used. Because these models are so controlled by geologic features, domains

characterized by different styles of mineralization also have different semivariogram models.

9. In general, different semivariogram models are obtained for sample data of different support. Where support differences are small, the difference in semivariogram models might be negligible, as can be demonstrated by comparison of experimental semivariograms for the various supports.

10. Extreme complications to semivariogram modeling have not been dealt with at length here. Where such complexities as hole effects, widely varying supports, non-Euclidean coordinate space, and so on are encountered, reference should be made to the detailed literature on these topics.

9.12: SELECTED READING

Barnes, R. J., 1991, The variogram sill and the sample variance; Math. Geol. v. 23, no. 4, pp. 673–678.

Clark, I., 1979a, The semivariogram – part 1; Eng. Min. Jour., July, pp. 90–94.

Clark, I., 1979b, The semivariogram – part 2; Eng. Min. Jour., August, pp. 92–97.

9.13: EXERCISES

1. Determine a semivariogram model for the Similkameen blasthole data by determining experimental semivariograms in at least six directions (e.g., including principal grid directions and principal diagonal directions) using an available software package.

2. Calculate experimental semivariograms manually for lags up to 4 m using data from Fig. 6.3 as follows: (a) the 20 values of data of smallest support (1 m); and (b) the data of second smallest support (2 m). Note the effect of regularization.

3. An isotropic experimental semivariogram can be fitted by a linear model for lags up to 30 m, a spherical model for lags up to 45 m, and an exponential model for distances of at least 70 m. If blasthole data are available on a square grid with approximately 7-m spacing, $7 \times 7 \, m^2$ blocks are to be estimated and a maximum of 20 samples will be used for an estimate, which model(s) can be used to provide adequate block estimates? Which model is preferred?

4. Autocorrelation models, commonly semivariograms, are fundamental to all geostatistical studies. It is a useful exercise to critically evaluate how models are presented in geostatistical papers. Select a recent scientific/technical paper containing a semivariogram model fitted to experimental data and evaluate the model critically. Keep in mind such features as number of pairs for each plotted point on the experimental semivariogram, extent of erratic local variation of the experimental data, possible justification of the initial lag selected, adequacy of control points to constrain the model near the origin, justification of the nugget effect, justification of the likely stationarity of the domain represented by the model, and whether there is any attempt to relate the model to geologic features of the deposit.

10

Kriging

If geostatistics are to give improved reserve estimates, two conditions must be satisfied: geologists must be aware of the methods that are available to them to control the quality of the geostatistical study and geostatisticians must appreciate those areas in which geological input is required if credible results are to be obtained. (Rendu, 1984, p. 166)

Chapter 10 introduces the general concepts of kriging and introduces some of the more widely used procedures in the estimations of mineral inventory, particularly such methods as punctual kriging, ordinary kriging, and indicator kriging. A variety of concerns related to kriging are introduced, including negative weights, dealing with outlier values, and the presence of conditional bias in kriged results.

10.1: INTRODUCTION

Kriging is a generic term applied to a range of methods of estimation (punctual or block) that depend on minimizing the error of estimation, commonly by a least-squares procedure. The term was coined by G. Matheron and P. Carlier to honor D. Krige, whose empirical work on reserve estimation in South African gold mines was later expressed by geostatistical theory developed by Matheron. A brief history of the evolution of kriging is given by Cressie (1990). The methods have been referred to widely as best linear unbiased estimator (BLUE), but this simple public relations expression ignores the characteristics inherent in the wide range of techniques known as kriging, as well as the many complexities involved with practical applications. Nevertheless, kriging is an estimation procedure that is globally unbiased (i.e., unbiased, on average, over the entire data range). A conditional bias to kriging results can be significant and is discussed in a later section.

A number of specific methods are included in the general term kriging, including simple kriging (SK), ordinary kriging (OK), indicator kriging (IK), universal kriging (UK), probability kriging (PK), and multiple indicator kriging (MIK). All depend on the same general concepts – that the autocorrelation of a regionalized variable can be modeled by a mathematical function inferred from a realization (data) of the regionalized variable and used to assist in estimation. Here, attention is directed to a few specific kriging techniques that have been widely applied to resource/reserve estimation problems.

The general problem to be solved by kriging is to provide the best possible estimate of an unknown point or block from a discrete data set (samples), as shown schematically in Fig. 10.1. In this example, eight data are available with which to estimate block B, and there is an implicit assumption that the use of data outside as well as inside B will improve the

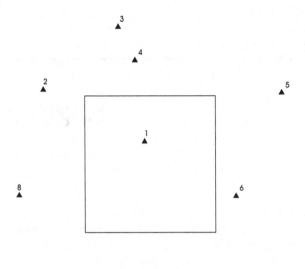

Figure 10.1: The general, block-estimation problem (in two dimensions); that is, to use a selection of included and nearby data with which to form an estimate of the mean grade of a block.

estimation. One can imagine that the eight data points could be weighted in some way (Eq. 10.1) to provide the best possible estimate of block B; the question is, how do we determine the weights in order to achieve this aim?

$$g_B^* = w_1 s_1 + w_2 s_2 + w_3 s_3 + w_4 s_4 + w_5 s_5$$
$$+ w_6 s_6 + w_7 s_7 + w_8 s_8 \qquad (10.1)$$

Simple averaging of the data is acceptable if the variable is random (i.e., all weights are equal), but is not optimal if the variable is regionalized (i.e., if significant autocorrelation exists over distances larger than the sample to block spacings). When autocorrelation is important, it is evident that a nearby datum should carry more weight than a more distant datum, although how much more is not clear.

10.2: BACKGROUND

In mineral inventory terms, the problem outlined previously is to determine sample weights in a manner that provides the best possible estimator of a variable

(e.g., grade, thickness) for a location or volume, the value of which is unknown and required. An obvious constraint on a solution to this problem is that the final estimate must be unbiased (i.e., $\sum w_i = 1$, where w_i is the weight of sample i). Furthermore, *best* must be defined. In kriging applications, "best" is taken to be the estimate that minimizes the estimation variance (i.e., the procedure of kriging is a least-squares procedure).

10.2.1: Ordinary Kriging

The estimation variance is given by

$$\sigma_e^2 = E\{[Z - Z_k^*]^2\}$$

where Z is the true value and Z_k^* is the kriging estimator. This expression can be represented in terms of the semivariogram, as follows:

$$\sigma_e^2 = 2\bar{\gamma}(B, s) - \bar{\gamma}(s, s) - \bar{\gamma}(B, B)$$

or, for a discrete data set with unknown weights on each datum, as

$$\sigma_e^2 = 2\left[\sum w_i \gamma(B, s_i)\right] - \sum\sum w_i w_j \gamma(s_i, s_j)$$
$$- \bar{\gamma}(B, B) \qquad (10.2)$$

where $\sum W_i \gamma(B, S_i)$ is the weighted average semivariogram value between all data points and the block to be estimated, $\sum\sum w_i w_j \gamma(s_i, s_j)$ is the weighted average semivariogram value between all possible pairs of data, and $\bar{\gamma}(B, B)$ is the average semivariogram value of all possible pairs of points within the block to be estimated.

Equation 10.2 can be minimized with the constraint that the weights must sum to 1 ($\sum w_i = 1$). This constraint is introduced into the minimizing procedure as an expression equivalent to zero, introducing a new unknown μ, the Lagrange parameter, into the system of equations (see Isaaks and Srivastava, 1989). The eventual equations that derive from this procedure, known as the *system of ordinary kriging*

(OK) *equations*, are as follows:

$$\begin{bmatrix} \gamma(s_1, s_1) & \gamma(s_1, s_2) & \cdots & \gamma(s_1, s_n) & 1 \\ \gamma(s_2, s_1) & \gamma(s_2, s_2) & \cdots & \gamma(s_2, s_n) & 1 \\ \gamma(s_3, s_1) & \gamma(s_3, s_2) & \cdots & \gamma(s_3, s_n) & 1 \\ \gamma(s_n, s_1) & \gamma(s_n, s_2) & \cdots & \gamma(s_n, s_n) & 1 \\ 1 & 1 & & 1 & 0 \end{bmatrix} \begin{bmatrix} w_1 \\ w_2 \\ w_3 \\ w_n \\ \mu \end{bmatrix}$$

$$= \begin{bmatrix} \gamma(s_1, B) \\ \gamma(s_2, B) \\ \gamma(s_3, B) \\ \gamma(s_n, B) \\ 1 \end{bmatrix}$$

where $\gamma(s_i, s_j)$ is the gamma value between any two data (Fig. 10.2), $\gamma(s_i, B)$ is the gamma value between a datum and the block to be estimated, and w_i is the

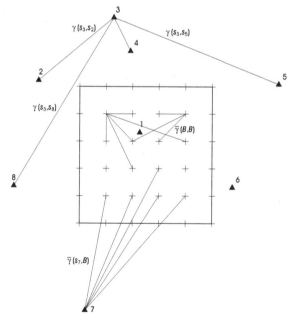

Figure 10.2: The block and data of Fig. 10.1, illustrating how various average gamma values are obtained in order to krige a single block. Lines extending from Sample 3 illustrate some of the pairs of data used to estimate the average gamma value between s_3 and all other data points; lines extending between discretetized block locations (+ signs) illustrate how pairs are selected to determine the average gamma value within the block; lines from sample s_7 to discretized block locations (+ signs) illustrate how paired values are obtained to determine the average gamma value between a sample and the block.

sample weights to be determined. The solution to this set of equations provides weights such that the block (or point) estimate is given by

$$Z_k^* = \sum w_i s_i.$$

The corresponding minimized estimation variance, known as the *kriging variance*, is given by

$$\sigma_k^2 = \sum w_i \gamma(s_i, B) + \mu - \bar{\gamma}(B, B).$$

Manual implementation of a set of kriging equations is generally impractical. Fortunately, various software packages are available at reasonable cost for a wide gamut of geostatistical calculations, including various types of kriging (e.g., Deutsch and Journel, 1998).

10.2.2: Simple Kriging

Ordinary kriging does not require that the mean value of the field of data be known. Consequently, OK is a common procedure of mineral inventory estimation. However, there are situations in which the mean value (of the field of data in which estimation in undertaken) is well known. In such cases, the kriging equations reduce to the situation of an unconstrained set of equations (i.e., the weights are not constrained to sum to 1), as follows:

$$\begin{bmatrix} \gamma(s_1, s_1) & \gamma(s_1, s_2) & \cdots & \gamma(s_1, s_n) \\ \gamma(s_2, s_1) & \gamma(s_2, s_2) & \cdots & \gamma(s_2, s_n) \\ \gamma(s_3, s_1) & \gamma(s_3, s_2) & \cdots & \gamma(s_3, s_n) \\ & \cdot & & \\ & \cdot & & \\ \gamma(s_n, s_1) & \gamma(s_n, s_2) & \cdots & \gamma(s_n, s_n) \end{bmatrix} \begin{bmatrix} w_1 \\ w_2 \\ w_3 \\ \\ \\ w_3 \end{bmatrix}$$

$$= \begin{bmatrix} \gamma(s_1, B) \\ \gamma(s_2, B) \\ \gamma(s_3, B) \\ \\ \gamma(s_n, B) \end{bmatrix}.$$

This system of equations is the simple kriging (SK) system. In general, the weights obtained will not sum to 1. They are made to do so (to assure nonbias) by assigning an appropriate additional weight to the mean value, as follows:

$$w_{n+1} = 1 - \sum w_I$$

where w_{n+1} is the weight to be applied to the mean value. Equations for both the optimized estimate and the kriging variance are the same as for ordinary kriging. Simple kriging has been applied extensively in making estimates for producing "Rand-type" gold deposits in South Africa, where very large amounts of data are available and the mean grade of a deposit or domain is well established (e.g., Krige, 1978).

10.3: GENERAL ATTRIBUTES OF KRIGING

The general procedure of kriging contains a number of important implications that are not particularly obvious to those with a limited mathematical background:

(i) Kriging is correct on average although any single comparison of a kriged estimate with a true value might show a large difference; on average, however, such differences generally are less for kriging than for other interpolation techniques.

(ii) Kriging of a location (point) for which information is included in the kriging equations results in a kriged estimate equivalent to the known data value (i.e., kriging reproduces existing data exactly).

(iii) Kriging takes into account data redundancy. In the extreme, a very tight cluster of several analyses carries almost the same weight as a single datum at the centroid of the cluster.

(iv) Kriging can be carried out as described, but on transformed data. An extensive literature exists for lognormal kriging, although this method has been found in practice to lead to serious biases in some cases (David, 1988). In general, if the transform function is not linear, the back transform will not produce an optimum estimator (see Journel, 1987). Other transformations are also common, including an indicator transform and a generalized normal transform.

(v) A mathematical criterion of the kriging system is that the semivariogram be represented by a "positive definite" mathematical function. This criterion normally ensures that the kriging variance is positive. Armstrong (1992) warns of situations in which positive definiteness is an insufficient criterion to ensure a positive variance; Journel (1992) suggests that such situations are extreme and not generally met by practitioners or are taken into account by practical methodology.

(vi) As described here, kriging depends on a stationary field of data. At the very least, local stationarity is required (i.e., stationarity must exist over a distance as great as the greatest sample separation to be expected). Where local stationarity is defined, the distance over which local stationarity occurs defines the maximum diameter of a search radius for selecting data with which to make an estimate.

(vii) Kriging is unbiased globally but generally contains some element of conditional bias (i.e., variable bias as a function of concentration), as do other estimation methods.

10.4: A PRACTICAL PROCEDURE FOR KRIGING

Kriging results are dependent on the autocorrelation model of the variable being estimated; hence, the semivariogram modeling process and the need for a high-quality model are of paramount importance. A sound semivariogram model that is integrated with the geologic model of a deposit is the foundation on which confident kriging is based. Once the semivariogram model is determined (see Chapter 9) the subsequent procedures have much in common with other estimation procedures. A practical approach to kriging involves a series of structured steps that include: (i) cross validation of the semivariogram model, (ii) criteria for selection of data for individual block estimates, (iii) specification of minimum and maximum numbers of data for kriging each block, (iv) constraining data in order to deal with specific problems (e.g., negative weights, strings of data, specific locales not estimated, block estimation at domain margins), followed by (v) systematic kriging of each block in the array to be estimated. Computerized systems normally work through the defined block array in a systematic manner, checking various criteria to determine if and how each block is treated.

Data must be selected for each point or block to be kriged. This normally entails a search of the data array for a group of data that meet several specified requirements. Efficient search procedures are discussed by Journel and Huijbregts (1978). Generally, all data within a particular search radius of the point (or block) being estimated are selected; the search volume may be spherical or ellipsoidal as the case demands. A maximum number of data is imposed on each block estimate so that the set of kriging equations for any single block estimate is relatively small and its solution is efficient. Decreasing the number of data points, perhaps by selecting the few nearest data in each quadrant or octant, has the added benefit that negative weights can be avoided and/or minimized. In addition, a minimum number of data is stipulated in order to avoid extremely large errors in which only local stationarity is guaranteed and ensure interpolation as opposed to extrapolation. It is normal to require that data be reasonably well distributed spatially (i.e., not all clustered). As an example, a two-dimensional search might select all samples within a search radius R from the point (block) being estimated. From these data, a maximum of two per quadrant are selected (i.e., the two samples closest to the point or block [center] to be estimated). In addition, a requirement might be that at least three quadrants contain at least one data point. This procedure necessarily sets a minimum number of data at three and a maximum of eight. Potential kriging situations that do not meet these requirements are not kriged. If many unkriged blocks arise, the kriging criteria might be inappropriate.

Clearly, the search radius and the spatial density of data control the number of data points available for estimating each block. For a particular data density, too small a search radius results in too few data being selected. Isaaks and Srivastava (1989) suggest that for irregularly located data, a minimum search radius can

be approximated by an average data spacing given by

$$\text{Avg. data spacing} = (\text{Area sampled})/n)^{1/2}.$$

Too large a search radius leads to large amounts of data being selected, with the result that computation time is increased, as are the likelihood of negative weights and the probability of exceeding the limits of local stationarity.

Block kriging is generally appropriate when the block size is comparable to or larger than the spacing of the data array (Journel and Huijbregts, 1978). When blocks are small relative to data spacing, estimates have large errors and conditional bias can dominate the block estimates. As a general rule, it is not wise to estimate blocks whose dimensions are less than half the sample spacing.

Cressie and Zimmerman (1992, p. 57) conclude that, "the geostatistical method is surprisingly stable.... However, we do not advocate the blind use of geostatistics, but in the hands of a practitioner, well-informed about its limitations, it can be a powerful too."

10.5: AN EXAMPLE OF KRIGING

An experimental semivariogram model has been determined for a bulk, low-grade, epithermal gold deposit in the southwestern United States. Data are average oz/t for 30-ft composites from several hundred vertical drill holes. The deposit occurs in a nearly flat-lying volcanic sequence that has broad, lateral, geologic continuity but systematic changes with depth. Bench height is 30 ft, and locations of exploration drill holes closely approximate a 50×50 ft grid. The two-dimensional relative spherical semivariogram model developed for horizontal directions in the uppermost variable level is isotropic with parameters $C_0 = 1.00$, $C_1 = 2.1$, and $a = 325$ ft.

This model can be used to examine a variety of estimation situations. Consider the problem of estimating a $50 \times 50 \times 30$ ft^3 block with a central drill hole and the eight data items in the first aureole of nearby data (Fig. 10.3). This pattern of data reduces to a set of kriging equations with three unknown weights

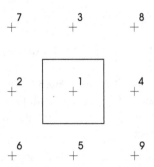

Figure 10.3: A square grid of data to be used to estimate a block array defined so that each block contains a centrally positioned datum. This array is that of the $50 \times 50 \times 30$ ft^3 block estimation described in the text using ordinary kriging.

Table 10.1 Gamma values for kriging example

Symbol	Gamma
$\gamma(s_1, s_1)$	0
$\gamma(s_1, s_2)$	1.48
$\gamma(s_1, s_3)$	1.67
$\gamma(s_2, s_2)$	1.32
$\gamma(s_2, s_3)$	1.76
$\gamma(s_3, s_3)$	1.54
$\gamma(s_1, B)$	1.21
$\gamma(s_2, B)$	1.48
$\gamma(s_3, B)$	1.70
$\gamma(B, B)$	1.273

because of the symmetry of the situation (i.e., w_1 applied to sample 1; w_2 to be distributed evenly to samples 2, 3, 4, and 5; and w_3 to be distributed equally to samples 6, 7, 8, and 9). The general kriging equations that must be solved follow:

$$\begin{bmatrix} \gamma(s_1, s_1) & \gamma(s_1, s_2) & \gamma(s_1, s_3) & 1 \\ \gamma(s_2, s_1) & \gamma(s_2, s_2) & \gamma(s_2, s_3) & 1 \\ \gamma(s_3, s_1) & \gamma(s_3, s_2) & \gamma(s_3, s_3) & 1 \\ 1 & 1 & 1 & 0 \end{bmatrix} \begin{bmatrix} w_1 \\ w_2 \\ w_3 \\ \mu \end{bmatrix}$$

$$= \begin{bmatrix} \gamma(s_1, B) \\ \gamma(s_2, B) \\ \gamma(s_3, B) \\ 1 \end{bmatrix}.$$

All the mean gamma values as determined from the semivariogam model and the known geometric configuration of the data are listed in Table 10.1. Note that these values can be estimated directly from the semivariogram model; in some cases, the use of auxiliary functions is a possible alternative to numeric approximation. Solving these equations with appropriate substitutions of values gives the following weights:

$$w_1 = 0.224, \quad w_2 = 0.527, \quad w_3 = 0.249,$$
$$\mu = 0.0142.$$

Remember that w_2 applies to four data items; thus, each item has a weight of $w_2/4 = 0.132$. Similarly, $w_3/4 = 0.062$ is the weight to be assigned to each of samples 6, 7, 8, and 9. With these weights and the Lagrange multiplier, the equation for kriging variance

(Eq. 10.3) can be solved as follows:

$$\sigma_k^2 = 0.224 \times 1.21 + 0.527 \times 1.48 + 0.249 \\ \times 1.70 + 0.0142 - 1.273 = 0.216$$

or

$$\sigma_k = 0.46.$$

Because the semivariogram model is relative, the calculated kriging variance is actually a proportional error. In other words, the value 0.46 must be multiplied by the corresponding block grade in order to determine the absolute error. Even without knowing the block grade, it is of considerable interest to realize that the regular, 50-ft grid of data produces block estimates with an error (one standard deviation) of 46 percent. There can be little wonder that large errors are made in ore/waste classification and that great disparities exist in many metal reconciliation studies.

10.6: SOLVING KRIGING EQUATIONS

The solution of a system of kriging equations commonly is routine and a variety of subroutines exist commercially. Some aspects of this topic are discussed by Journel and Huijbregts (1978). However, problems can arise as a result of ill-conditioned covariance matrices or numeric instability of the algorithm used to solve the matrix equations (O'Dowd, 1991). Davis et al. (1978) provide an algorithm that has been used widely in the mining industry for local estimation from blasthole data.

Table 10.2 Cross-validation results, Au accumulation (oz/st × ft), Silver Queen deposit

Data set	Method	Estimates mean	s^2	Residuals mean	s^2
DDH ($n = 112$)	OK-200	0.61	0.12	0.01	0.37
	OK-150	0.62	0.16	−0.01	0.43
	POLY	0.62	0.37	−0.02	0.51
	ISD	0.61	0.14	−0.01	0.38
	TRUE	0.59	0.40		
DRIFT ($n = 365$)	OK	0.54	0.15	0.007	0.114
	POLY	0.54	0.28	−0.011	0.178
	ISD	0.53	0.15	0.00	0.11
	TRUE	0.53	0.25		

Source: After Nowak (1991).

A matrix is ill conditioned if small errors (rounding or in the original data) lead to large errors in results that are produced. Small and large are defined in relation to a specific problem. Numeric stability is the quality of a particular algorithm in obtaining a result. The conditioning number of a symmetric matrix, $k(A) = l_{max} / l_{min}$, where l_{max} and l_{min} are the largest and smallest eigenvalues, is an indication of the possibility of a matrix being sensitive to error. O'Dowd (1991) demonstrates that an ill-conditioned covariance matrix leads to an ill-conditioned kriging matrix. In addition, he shows the following:

(i) For a pure nugget effect, as the sill decreases in value, the conditioning number increases.

(ii) For the spherical model, the conditioning number is principally a function of the sill value. Large conditioning numbers can be reduced simply by scaling. As O'Dowd (1991, p. 731) says, "for most purposes, scaling the model so that it has a sill value of 1 to $1,000/n^{1/2}$ will reduce the condition number to tractable levels." This effect is the same as scaling the data themselves. It is useful to note that scaling the semivariogram or covariance model to a sill of one is equivalent to using the correlogram in the kriging system.

10.7: CROSS VALIDATION

Cross validation is a widely used procedure in which point data are successively extracted individually from a data array and each is then estimated by a group of the neighboring data (the leaving one out method of Davis, 1987). Thus, any particular estimation method can be used, and the results of point (sample) estimates can be compared with true (known) values. When many such estimates can be made and compared with reality, an estimation method can be evaluated for global bias and an average error can be determined (as can the histogram of errors). Estimated values can also be plotted versus known values and the resulting x–y plot can be evaluated for conditional bias (i.e., bias dependent on the value of the estimate).

The procedure of cross validation is particularly useful in comparing results by several estimation methods. Consider as an example, the case of Silver Queen polymetallic epithermal deposit studied by Nowak (1991). The deposit is a steeply dipping vein system for which 112 diamond-drill-hole pierce points exist at a spacing of roughly 100 ft or more. Also available are 365 channel samples, each taken across the vein at 5- to 10-ft intervals along an exploratory drift. Autocorrelation models have a range about the same as the diamond-drill-hole spacing (Nowak, 1991). Known values have been estimated independently for each data set by three methods: ordinary point kriging (OK), polygonal (nearest-neighbor) estimation (POLY), and inverse squared distance estimation (ISD). Each known point also was estimated by OK, first for a search radius of 150 ft and then for a search radius of 200 ft. Results are summarized in Table 10.2. The 200-ft search radius produced marginally better results (i.e., an error dispersion of

0.37 versus 0.43 for a 150-ft radius), so a search radius of 200 ft was selected for estimation purposes.

The POLY diamond-drill-hole cross-validation results most closely reproduce the global characteristics of the raw diamond-drill-hole data (Table 10.2). However, the residuals are high in comparison with ISD and OK results, indicating that POLY errors are relatively large; nonetheless, the errors must be compensating because the mean error (residual) is essentially zero. ISD and OK cross validation of diamond-drill-hole data show the two methods to be essentially equivalent in quality for these point data. Both methods are globally unbiased, but are substantially smoothed relative to original data (dispersions of estimates are much less than the dispersion of real data). For block estimation purposes, kriging is

preferred relative to ISD because kriging takes block size into account, whereas ISD does not.

The drift, cross-validation results illustrate the advantage of an abundance of closely spaced data at a relatively early stage of deposit evaluation. The comparative results of Table 10.2 and Figs. 10.4 to 10.6 show that all three estimation methods appear to slightly overestimate the true mean value, that the POLY method approximates the dispersion of real data but incorporates relatively large random errors, and that ISD and OK are essentially equivalent in quality as point estimators. In this case, the similarity of ISD and OK estimates results from the fact that both methods place most weight on nearby samples that are highly correlated with the point being estimated. In such circumstances, there is little opportunity for

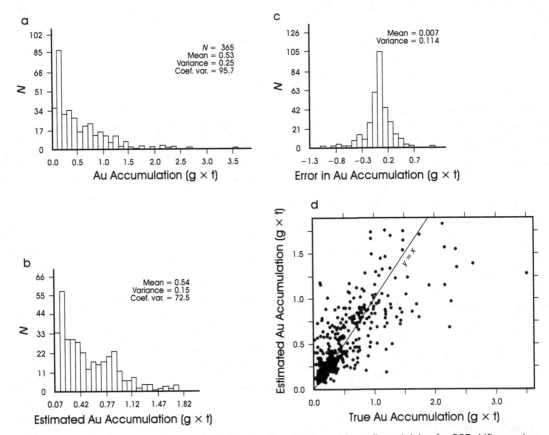

Figure 10.4: Cross-validation of Au accumulation (grade × thickness) by ordinary kriging for 365 drift samples, Silver Queen epithermal vein. (a) Histogram of original values. (b) Histogram of kriging estimates. (c) Histogram of kriging errors (estimate − true value). (d) Scatterplot of kriging estimates (y) versus true values (x).

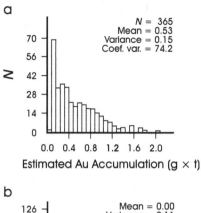

a

N = 365
Mean = 0.53
Variance = 0.15
Coef. var. = 74.2

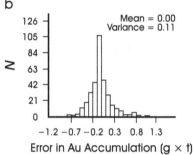

b

Mean = 0.00
Variance = 0.11

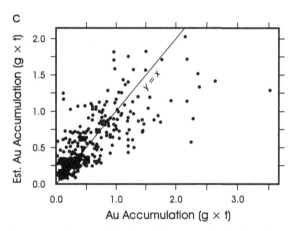

c

Figure 10.5: Cross-validation of Au accumulation (grade × thickness) by inverse distance squared for 365 drift samples, Silver Queen epithermal vein. (a) Histogram of estimated values. (b) Histogram of estimation errors (estimate − true value). (c) Scatterplot of IDW estimates (y) versus true values (x).

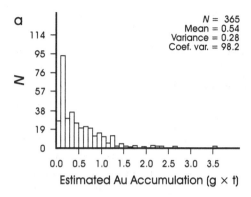

a

N = 365
Mean = 0.54
Variance = 0.28
Coef. var. = 98.2

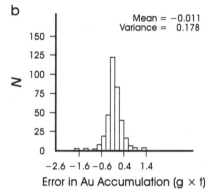

b

Mean = −0.011
Variance = 0.178

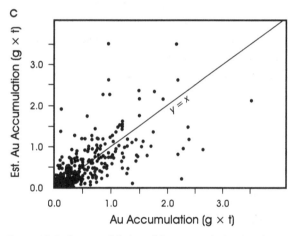

c

Figure 10.6: Cross validation of Au accumulation (grade × thickness) by polygonal estimation (nearest-neighbor) for 365 drift samples, Silver Queen epithermal vein. (a) Histogram of estimates. (b) Histogram of errors (estimate − true value). (c) Scatterplot of polygonal estimates (y) versus true values (x).

kriging to operate to its best advantage. Nevertheless, as concluded from the diamond-drill-hole cross-validation results, OK is preferred for block estimation because change of support is automatically taken into account, whereas this is not the case with ISD.

It is common practice to conduct cross validation only at a scale (spacing) for which most estimates are to be made. For example, if only exploration drill data are available and a sample in a drill hole is to be cross validated, no other sample in the same drill hole is used in the cross validation. This procedure ignores (i) the fact that in some localities, data for estimation might be closely spaced, even where most data are widely spaced; (ii) in block estimation, some blocks are very close to or contain some of the data; and (iii) a comparison of estimation methods is incomplete unless such a comparison is made both for widely and closely spaced data. In general, it is advantageous to conduct cross validation at more than one scale, as illustrated by the Silver Queen example, if for no other reason than the difficulty of distinguishing relative quality of estimation methods where data are dispersed in space with variable density.

Cross-validation results can be examined usefully on scatter diagrams of estimated values versus true values (Figs. 10.4 to 10.6) because outliers and patterns indicating conditional bias are commonly evident. When large data sets are cross validated by several techniques, it is common to describe the results by a least-squares line fitted to the data on an x–y plot. In such a case, a traditional, least-squares procedure (all error in the estimate) should be used. The statistics of such lines are useful to compare cross-validation results by various methods. For example, the more the slope departs from 1, the greater is the conditional bias. The relative scatter about the line is also a useful comparative tool, as is the correlation coefficient.

10.8: NEGATIVE KRIGING WEIGHTS

10.8.1: The Problem

Negative weights are a peculiarity of certain data geometries of kriging systems combined with a high degree of continuity (including a low to negligible nugget effect) in the semivariogram model. They are acceptable in estimations involving some data types. With topographic data, for example, negative weights permit values that are outside the limits of the data used to make an estimate. However, with assay data,

in some cases they can lead to enormous estimation errors, particularly when relatively few data are involved in an estimate. Negative weights create problems that can be illustrated by several simple examples, as follow:

- Example 1: Consider a block to be estimated by five data, one of which is one-third the average grade of the others and has a weight of -0.1. Consequently, the sum of all the other weights is 1.1. Assuming grades of 1 and 3 g/t, the average grade estimated is $(-0.1 \times 1) + (1.1 \times 3) = 3.2$ g/t, a value that is higher than any of the data used in making the estimate. A negative weight on a low grade leads to an overestimate.

- Example 2: Consider a block to be estimated by five data, one of which is three times the other four data and has a weight of -0.1. Consequently, the sum of all the other weights is 1.1. Assuming grades of 1 and 3 g/t, the average grade estimated is $(-0.1 \times 3) + (1.1 \times 1) = 0.8$ g/t, a value that is less than any of the data used in making the estimate. A negative weight on a high grade leads to an underestimate.

- Example 3: Assume the situation of Example 2, except that the negative weight applies to an outlier grade of 75 g/t. Hence, the average grade estimated is $(-0.1 \times 75) + (1.1 \times 1) = -6.4$ g/t, an impossible negative grade!

- Example 4: Assume the situation of Example 3 except that the negative weight that applies to the outlier is very small, for example, -0.01. Hence, the average grade estimated is $(-0.01 \times 75) + (1.01 \times 1) = 0.26$ g/t. This low positive result could send a block of ore to waste!

It is evident from the foregoing examples that negative weights can be a serious problem. Of course, the problems illustrated are alleviated if (i) outliers are dealt with separately in the estimation procedure, and (ii) negative weights are much smaller in absolute value than are those in the examples cited. However, even small negative or positive weights present a serious estimation problem if applied to outlier values.

Table 10.3 Kriging weights for "rings" of data when a block to be kriged is situated in a regular, square data array (nugget effect present)

Ring	w_1	w_2	w_3	w_4	σ_k^2	Remark
1	1.0				1.2	Data at four corners of block
2	0.6	0.4			0.84	Two rings of data
3	0.59	0.35	0.06		0.83	Three rings of data
4	0.59	0.34	0.06	0.01	0.83	Four rings of data

10.8.2: The Screen Effect

The common problem of negative weights in kriging is illustrated in Fig. 10.7, where an outer fringe of samples (shown as filled triangles) with small negative weights are screened from the block to be estimated by samples (+ signs) with positive weights. In such cases, the negative weights tend to be very small values relative to other weights (commonly 1/100 or less), provided there are many samples to be weighted and the distribution of samples is not strongly clustered. In cases of a substantial amount of data and an absence of outliers, negative weights normally can be ignored, because in total they represent only up to a few percent of the total weights and they are applied to grades of the same magnitude as are the positive weights. The situation in which negative weights become detrimental to the estimation process is when relatively few data are used to make an estimate or a relatively high value (perhaps but not necessarily an outlier) receives a negative weight. In the extreme situation a negative grade estimate could result.

An idealized example of negative weights is provided by Brooker (1975) and is summarized in Fig. 10.8 and Tables 10.3 and 10.4. The block in

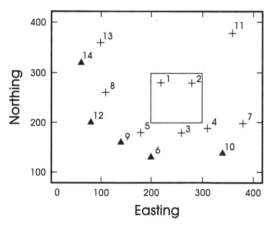

Figure 10.7: An example of the common problem of negative weights in block kriging (redrawn from Baafi et al., 1986). The block is kriged using all the data locations shown and a semivariogram model that, unfortunately, is not summarized in the text. Data locations shown as + have positive weights; data locations shown as filled triangles have negative weights. Many such trials indicate that negative weights are common for (i) data locations that are screened by other data relative to the block to be estimated, and (ii) a semivariogram model that has a high degree of continuity.

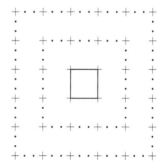

Figure 10.8: A regular grid of data (+) illustrating a block to be estimated. Ring 1 of data is at the block corners ring 2 is the next square "halo" of 12 data, and ring 3 is the next "square" array of 20 data. Block dimensions are the same as data spacing. This pattern leads to negative kriging weights in the outer fringes of data if the semivariogram model has a high degree of continuity (low nugget effect and long range) as shown in Tables 10.3 and 10.4.

Table 10.4 Kriging weights for "rings" of data when a block to be kriged is situated in a regular, square data array (no nugget effect)

Ring	w_1	w_2	w_3	w_4	σ_k^2	Remark
1	1.0				0.25	Data at four corners of block
2	1.07	−0.07			0.25	Two rings of data
3	1.07	−0.03	−0.04		0.25	Three rings of data
4	1.06	−0.01	−0.03	−0.02	0.25	Four rings of data

Fig. 10.8 is to be estimated by various numbers of rings of data; ring 1 is the four data at the corners of the block, ring 2 is the next halo of data, and so on. A spherical semivariogram model ($C_0 = 3.8$, $C_1 = 12.2$, and a range of $11.4a$, where a is the dimension of the square block to be estimated) is used for one series of estimates (Table 10.3); then a similar model (except with $C_0 = 0$) is used for a second series of estimates (Table 10.4). Results in these two tables indicate that negative weights can be avoided or minimized if a nugget effect exists or if relatively few rings of data are used for block estimation. The widely used procedure of limiting data to two per octant is one practical means of minimizing the likelihood of negative weights. As a sidelight, it is interesting to note from Table 10.4 that the absence of a nugget effect leads to little practical advantage to adding many haloes of data.

Baafi et al. (1986) conducted a variety of kriging experiments to investigate negative weights and understand better the situations in which they occur; the principal results of this study are as follows:

(i) An increase in the number of negative weights occurs as the spatial continuity increases as demonstrated by kriging experiments in which the ratio of $C_0/(C_0 + C_1)$ is changed.

(ii) As the range of influence increases, other parameters remaining constant, the number of negative weights increases.

(iii) In all experiments, several haloes of data about the block being estimated were used. In all cases, negative weights were assigned only to sample points that are screened by other samples located closer to the block being estimated.

(iv) "A practical way of deciding the number of samples to use during kriging to avoid negative kriging weights is to use the 'first layer' of samples only." (Baafi et al., 1986, p. 438)

There are other ways of dealing with the problem of negative weights. Barnes and Johnson (1984) propose *positive kriging*, in which an additional constraint is imposed beyond the normal kriging constraints (i.e., all weights w_i are positive). This solution to the problem requires rewriting kriging routines (see also an example by Schaap and St. George, 1981). Other less onerous solutions include

(i) Requiring appropriate amounts and spatial distribution of data (e.g., use an octant data search and accept a maximum of two data items per octant); this decreases the risk of screened data occurring

(ii) Adopting a simple routine to test weights obtained for a particular point or block kriging estimation; if negative weights are found, the corresponding data are omitted and the remaining data are used to krige again

(iii) Removing outliers so that they are not available for weighting should negative weights arise.

In general, negative weights are small in relation to other weights, perhaps representing up to a few percent of the total weight. Thus, if only a small proportion of total samples in any one kriging array carry negative weights and outliers are absent, the effect of negative weights is negligible. In general, however, outliers are a matter of great concern and must not be included in routine kriging arrays.

10.9: DEALING WITH OUTLIERS

The problem of outliers is of utmost importance. An entire chapter has been devoted to the topic (Chapter 7) and a discussion presented of the problems caused by outliers during semivariogram modeling. A further indication of their importance to kriging estimation procedures has been considered in connection with negative kriging weights; a negative weight attached to an outlier can, in some cases, lead to a negative grade estimate! A common practice among geostatisticians has been to ignore outliers during semivariogram modeling but retain them for purposes of mineral inventory estimation (e.g., David, 1977). This procedure fails to recognize the common geologic situation that outliers represent a separate grade population characterized by its own continuity; generally, the physical continuity of high grade is much less than that of the much more prevalent low grades. Thus, serious overestimation of both tonnage and average grade above a cutoff grade can result if a general model, normally dominated by the lower, more continuous grades, is applied to very high-grade values. The problem is acute when high grades are isolated in a field of lower values. When many high-grade values cluster spatially, it may be possible to establish a separate domain with its own continuity model. Practical examples dealing with the problem of multiple continuity populations are considered by Champigny and Sinclair (1984, 1998b), Parker (1991), and others.

10.9.1: Restricted Kriging

Several approaches to kriging have been proposed to deal with the problem of outliers. An important method known as *indicator kriging* has found significant application in the mining industry and is discussed at length in a later section of this chapter. An alternative kriging approach to dealing with outliers is *restricted kriging* (Journel and Arik, 1988; Pan, 1994; Pan and Arik, 1993), which makes use of an indicator variable to recognize an outlier sample and then introduces the restriction that the weight of a high grade (outlier) is equal to the probability of encountering a high-grade (outlier) value within the search

radius. For example, if n values including p outliers are within a search radius, the proportion is estimated as $f = p/n$. The resulting form of the kriging equations is as follows:

$$
\begin{bmatrix}
\gamma(s_1,s_1) & \gamma(s_1,s_2) & \cdots & \gamma(s_1,s_n) & 1 & j_1 \\
\gamma(s_2,s_1) & \gamma(s_2,s_2) & \cdots & \gamma(s_2,s_n) & 1 & j_2 \\
\gamma(s_3,s_1) & \gamma(s_3,s_2) & \cdots & \gamma(s_3,s_n) & 1 & j_3 \\
 & & \cdot & & & \cdot \\
\gamma(s_n,s_1) & \gamma(s_n,s_2) & \cdots & \gamma(s_n,s_n) & 1 & j_n \\
1 & 1 & & 1 & 0 & 0 \\
j_1 & j_2 & & \cdots j_n & 0 & 0
\end{bmatrix}
$$

$$
\times
\begin{bmatrix}
w_1 \\
w_2 \\
w_3 \\
\cdot \\
w_n \\
\mu_1 \\
\mu_2
\end{bmatrix}
=
\begin{bmatrix}
\gamma(s_1,B) \\
\gamma(s_2,B) \\
\gamma(s_3,B) \\
\cdot \\
\gamma(s_n,B) \\
1 \\
\phi
\end{bmatrix}
$$

where symbols are as described previously, except j_i is an indicator with a value of 1 if s_i is an outlier and zero otherwise, and Φ is the probability of outliers in the particular search radius (Φ can be determined on a block-by-block basis in advance, or can be estimated as the ratio of outliers/total samples within a particular search radius). The value of Φ can be estimated globally with reasonable confidence, but it is very difficult to estimate locally with a high degree of confidence. The net effect of accepting Φ as the ratio of outliers to total samples in the search volume is that the outlier grades affect many blocks (however many blocks occur in a search radius centered on the outlier). This is not consistent with the limited spatial extent of outlier grades.

If a single outlier is present in a search volume for the random case, the effect of restricted kriging is to weight it the same, regardless of location relative to the block or point being estimated. Consequently, an outlier is spread with the same weight to all blocks whose centers are separated from the outlier by a distance less than the search radius. This is contrary to the view expressed in Chapter 7 – that true outliers generally have restricted physical continuity and do not extend beyond the block within which they are contained. Hence, restricted kriging in the treatment of high-grade outliers can be seen as an empirical technique that arbitrarily reduces the impact of high-grade values and thus is analogous to cutting or capping of values. Pan (1994) also describes the incorporation of low-grade outliers into the kriging system using a similar approach.

Because outliers represent such a small proportion of the total data (and volume) of a deposit, their characteristics are relatively poorly known. Of course, an unbiased histogram provides an estimate of their true relative abundance, but with a large error. Estimation methods can be tailored to provide block estimates that globally sum to the correct amount of metal, based on an unbiased histogram. However, individual block estimates can be seriously in error because outlier occurrences not intersected by sampling cannot be taken into account during estimation, and existing outlier values will certainly be extended to some inappropriate blocks in order to produce sufficient metal.

10.10: LOGNORMAL KRIGING

The mathematical background to lognormal kriging is given by Journel and Huijbregts (1978) and a summary, with cautionary remarks regarding practical applications, by David (1988). In practice, data are logtransformed and ordinary kriging is then carried out on the logtransformed values. This procedure is applied when the data are lognormally distributed; consequently, the value estimated is the mean logtransformed value, the backtransform of which is the geometric mean. In lognormal distributions, the geometric mean is substantially less than the arithmetic mean; consequently, the arithmetic mean and associated error dispersion must be calculated from the estimates of log parameters as illustrated in Chapter 4. Of course, this calculation assumes a perfect lognormal distribution. Consider a variable $Y(x) = \text{Ln}[Z(x)]$ (i.e., Z represents the raw data and Y the logtransformed data).

David (1988, p. 119) discusses an example of lognormal kriging in which variations in the semivariogram parameters lead to two widely different estimates of orebody grade of 2.38 percent metal and 1.54 percent metal, using the same database. Such differences arise when the semivariogram is not well defined and there is substantial latitude in choice of a sill level. Differences in sill levels can translate into differences in kriging variances, which affect the transformation from the estimated geometric mean to the corresponding average value, a transformation that is extremely sensitive to the kriging variance. David (1988) concludes, "a dishonest person can get the grade he wants, simply by changing the variogram parameters. This does not make lognormal kriging a recommendable tool" (p. 119). In general, lognormal kriging should be avoided and an alternative, equivalent procedure such as ordinary kriging with a relative semivariogram should be used.

Of course, in special cases, particularly where a thorough cross validation of lognormal kriging has been demonstrated, the method may find acceptable use as indicated by Clark (1999), who discusses some of the practical difficulties and misconceptions that have arisen in the application of lognormal kriging, including backtransformation of estimates and conservation of lognormality. Kriging of logtransformed values produces a block estimate that, if backtransformed to its original arithmetic base, produces the geometric mean rather than the desired arithmetic mean. The correct value to be backtransformed is

$$L_k + \sigma_1^2/2 + \gamma(B, B) - \mu$$

where

L_k	is the kriging estimator from logtransformed data
$\sum_1^2$	is the kriging variance from logtransformed data (semivariogram sill)
$\gamma(B, B)$	is the F function for the block estimated
μ	is the Lagrange multiplier determined in solving the kriging equations.

The involvement of the kriging variance in this backtransformation function clearly requires confidence in the sill of the semivariogram as emphasized by David (1988). Clark (1999) also recognizes that although no theoretical reason exits for "conservation of lognormality" in moving from data distribution to block distribution, such conservation "appears to be the case in many practical applications" (p. 409).

An easy way to avoid the concerns mentioned regarding lognormal kriging is to krige without a transform, but using a relative semivariogram. Cressie (1985) demonstrated the general equivalence of the two procedures.

10.11: INDICATOR KRIGING

Indicator kriging (Journel, 1983) involves the transformation of data to zeros or ones depending on (i) the position of a value relative to a designated threshold or (ii) the presence or absence of a geologic feature. For example, samples of a barren dyke can be attributed a value of zero and mineralized samples assigned a value of 1; similarly, samples above an arbitrary threshold grade can be assigned 1, whereas samples below the threshold are assigned 0. Of course, a semivariogram model must be determined for the newly transformed variable, after which kriging can

be carried out exactly as described previously. The result of kriging with indicator values is to estimate a proportion p_k that can be interpreted in two ways:

(i) The probability that condition "sample = 1" prevails at the point kriged
(ii) The proportion of condition "1" that exists in the vicinity of the point kriged (the proportion of condition "0" that exists is $1 - p_k$).

Indicator kriging is simply the use of kriging (simple or ordinary) to estimate a variable (e.g., gold in g/t) that has been transformed into an indicator variable. In this case, each value of g/t gold is transformed into 0 or 1, depending on whether the value is above or below an arbitrary threshold called the *indicator threshold* or *indicator cutoff*. For example, assume that in a data set all values greater than 0.75 g/t will be transformed to 1 and all those values less than or equal to 0.75 g/t will be transformed to 0, as shown in a histogram of gold grades (Fig. 10.9) and on a plot of a small data array (Fig. 10.10). The transformed data (Fig. 10.10b) now represent the probabilities of being above or below the threshold. If the grade of the sample is above the threshold, it is assigned a 100 percent probability

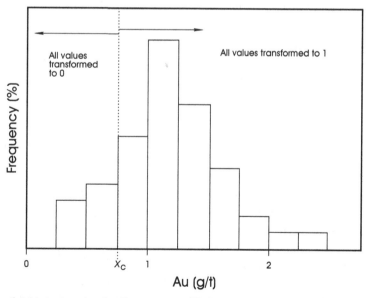

Figure 10.9: Histogram of gold grades showing the concept of indicator transformation. For a threshold of 0.75 g/t (i.e., $x_c = 0.75$), all higher values are assigned a value of 1; all equal or lower values are assigned a value of 0. Hence, a transform of 1 indicates certainty that the value is greater than the threshold.

Figure 10.10: Location of a small subset of the data of Fig. 10.9. (a) Posting of original values. (b) Indicator transforms of data in (a) relative to a threshold of 0.75 g/t. Point A can be estimated by kriging using the transformed data and an appropriate indicator semivariogram model to determine the probability that the point has a value greater than 0.75.

(1.0) of being above the threshold; if the grade of the sample is below the threshold, it is assigned a 0 percent probability of being above the threshold. It can be seen that what has been done with the indicator transformation is change the variable that is estimated from a grade variable to a probability variable. A kriged estimate is now the estimate of the probability of being above the threshold, not an estimate of the average grade. Of course, assignment of an indicator assumes that the values are known perfectly. This is not generally true with ore grades, where the error of an individual sample can be large; herein lies one of the principal sources of unknown error in the use of indicator-transformed data for estimation purposes.

10.11.1: Kriging Indicator Values

Consider point A on Fig. 10.10b. Kriging, using the surrounding data and an appropriate semivariogram model, produces an estimated value that lies in the range from 0 to 1. Assume that kriging produces an estimate of 0.27 for point A. This estimate can be interpreted in two ways:

1. 0.27 is the probability that the true value of the indicator at point A is equal to 1. In this case, there is a 27 percent chance that point A has a grade greater than 0.75 g/t gold; hence, there is a 73 percent chance that A has a grade less than 0.75 g/t gold.
2. Assuming there is a cluster of samples centered on point A, 0.27 is the proportion that have grades above the indicator threshold of 0.75 g/t gold.

It is this latter interpretation that is important in applications of indicator kriging to resource/reserve estimation. This shift in thinking, in going from considering the estimate to be an estimate of probability to considering it an estimate of proportion, is subtle but valid. If the probability is applied to a large enough cluster of samples (over a large enough volume – generally, one block of the block model), it is reasonable to consider the estimate as being a proportion of the block.

Such simple applications of indicator kriging have found widespread use in mineral inventory estimation because of their simplicity. Variography must be carried out for the indicator-transformed data and point estimation is then by ordinary kriging of the transformed values. This indicator kriging procedure has been used to estimate relative proportions of mineralized versus unmineralized ground (David, 1988), the proportion of barren dykes within a mineralized zone (Sinclair et al., 1993), the presence or absence of a particular geologic feature such as a barren dyke within a deposit as opposed to outside a deposit (Sinclair et al., 1993), and so on. However, the method deteriorates as one of the proportions becomes small, in part because the corresponding semivariogram model is poorly defined.

10.11.2: Multiple Indicator Kriging (MIK)

To this point it has been shown that indicator kriging can be used to estimate the proportion of samples with values greater than a specific threshold. This is equivalent to defining a point on the local cumulative

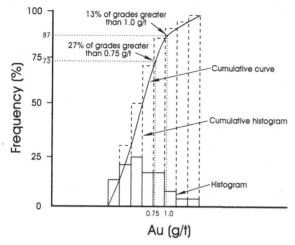

Figure 10.11: The indicator kriging value for point A, based on a threshold of 0.75 g/t, is equivalent to determining a point on the local cumulative curve for small sample grades. In this case, a kriging estimator of 0.27 indicates that 27 percent of the samples in the vicinity of A are higher than 0.75 g/t. A second indicator kriging, relative to a threshold of 1.0 g/t, provides an estimate of another point on the local cumulative curve at 0.13 (or 13 percent). Repetition of this procedure leads to an estimate of many points on the cumulative curve; some assumptions are required for the tails of the distribution.

curve for grades of point samples (Fig. 10.11). Exactly the same data could be transformed to a new set of indicators relative to a different indicator threshold (e.g., $x_c = 1.0$ g/t). The data in Fig. 10.10a would then be transformed into a slightly different set of indicators than those shown in Fig. 10.10b. A semivariogram model could be determined for this new set of indicators and point A could be kriged a second time using this new model and the corresponding indicator values to determine the proportion of samples around point A, with grades greater than 1.0 g/t. Suppose this were done and a value of 0.13 was obtained by kriging. Now a second point can be added to the cumulative frequency curve for locality A, as shown in Fig. 10.11. Thus far, estimates indicate that 27 percent of the data are above 0.75 g/t and 13 percent of the data are above 1.0 g/t. Hence, the difference (14 percent) occurs in the interval between 0.75 to 1.0 g/t.

It is apparent that the forgoing procedure can be repeated for many different indicator values so that

points along the entire cumulative distribution can be estimated. This procedure is referred to as *multiple indicator kriging* (MIK), and the result is an estimate of the entire cumulative curve (except for the tails); that is, the frequency distribution shown on Fig. 10.11. Experience suggests that the cumulative curve can be estimated satisfactorily from about 8 to 12 points (i.e., from 8 to 12 thresholds). Estimation of the tails of the distribution is arbitrary, but in many cases a practical procedure is to assume that the lower tail extends to zero frequency at zero metal content in linear fashion. Definition of the high tail is more critical because it has a significant impact on estimates of metal content. Barnes (1999) suggests two practical ways of delimiting the upper end of the cumulative curve (z_{max}) as follows:

$$z_{max} = z_{cn} + 3(z_n - z_{cn})$$
$$z_{max} = z_{cn} + [2^{1/2}/(2^{1/2} - 1)](med_n - z_{cn})$$

where

z_{cn}	is the uppermost indicator threshold
z_n	is the mean of values greater than z_{cn}
med_n	is the median of values greater than z_{cn}.

Having estimated the distribution of sample grades in a block, it is desirable to be able to calculate the average grade of the block. Also, having the distribution of grades in the block allows the application of a cutoff grade to the distribution such that the proportion of the block above the cutoff and the average grade above the cutoff can be calculated.

The average grade of the distribution (the estimated grade at point A) can be determined if the average grade of each indicator interval (class mean) is known. In the previous example, the average grade of 14 percent of the samples in the interval between 0.75 g/t and 1.0 g/t might be estimated to be 0.88 g/t from the available data within the class, or the mid-value of 0.875 g/t might be accepted arbitrarily. With the class means, the average grade of the distribution can be determined as a weighted average of the class means, with the weights being the estimated proportion of samples in each class.

It must be remembered that the estimated cumulative distribution is for sample grades. If the data are to be used for block-estimation purposes, the data distribution must be corrected in some fashion to produce a distribution for block grades of a specified size. This procedure is known as *change of support*, and several methods have been widely used in practice. One particularly straightforward method, the *affine correction*, is considered here. In this method, each of the estimated points on a point distribution can be squeezed toward the mean according to the relation

$$z_b = (z_s - m)\sigma_b/\sigma_s$$

where

z_b	is a value on the block distribution
z_s	is a value on the sample grade distribution
σ_b	is the dispersion of block grades
σ_s	is the dispersion of sample grades
m	is the mean grade of the distribution.

The dispersion of block grades (σ_b) is determined from the well-established, volume–variance relation

$$\sigma_b = \sigma_s - \bar{\gamma}(B, B)$$

where $\gamma(B, B)$ is the average value of the semivariogram when two points describe all possible positions in a block of the size being considered. At this stage, a cutoff grade can be applied to the block-grade distribution. If the block is a selective mining unit (SMU), the distribution shows the probability that the block is above the cutoff grade. If the block is much larger than a selective mining unit, the distribution and the cutoff grade provide an estimate of the proportion of blocks above cutoff grade and their average grade.

This procedure also has found widespread use in mineral inventory estimation because it is nonparametric and avoids some of the problems of dealing with multiple populations. Moreover, the methodology is identical to ordinary kriging once a simple in-

dicator transformation has been done. However, the method does have limitations (e.g., on the upper tail of the estimated distribution when the indicator semivariogram generally is poorly defined). In addition, a variance correction involving assumptions must be applied to transform point estimates of the cumulative distribution function to block estimates. An excellent application to the Page–Williams mine (C-zone) is described by Froidevaux et al. (1986).

10.11.3: Problems in Practical Applications of Indicator Kriging

A practical concern with multiple indicator kriging is that a substantial amount of variography is required, one model for each indicator threshold, although Journel (1983) suggests that for a "quick solution," it is possible to use the same indicator semivariogram model (i.e., the model for a cutoff equivalent to the median value) for all indicator transforms. Such a solution would rarely be possible because high thresholds generally give indicator semivariograms with short ranges and high nugget effects relative to low thresholds.

Some assumptions must be made about the tail of the cumulative distribution function being estimated in order that the grade distribution above the highest indicator threshold is defined. This is a critical decision because tails of grade distributions are poorly known relative to ranges closer to the mean, and assumptions about the tails can have very serious impact on the calculated, average grade above a particular cutoff, particularly if the cutoff grade is high relative to the mean value of the distribution. As a consequence of this problem, multiple indicator kriging does not completely obviate concerns regarding outliers.

A second concern is the "order relations" problem, which arises in a small proportion of estimates. This problem is the situation in which the proportion estimated above one cutoff is not compatible with the proportion estimated from an adjoining cutoff (e.g., a high cutoff may lead to a cumulative proportion higher than that estimated by the next lower cutoff), an impossibility for a cumulative curve. For example, the indicator kriging procedure for one site might lead to an estimate of 10 percent of material above a cutoff of 3.0 g Au/t in contrast with an estimate of 12 percent above a cutoff of 3.25 g Au/t; obviously, this is an impossible situation. A common solution to this order relations problem is to check for such inconsistancies, and when they exist, average the two conflicting estimates and apply the same average estimate to each of the two cutoffs in question.

10.12: CONDITIONAL BIAS IN KRIGING

Conditional bias is a term used to indicate bias that varies as a function of absolute value of a variable (i.e., a bias that is conditional on value). Block estimates in resource/reserve estimation can be unbiased on average (globally unbiased), but might suffer from different amounts of bias at different absolute grades. The general problem has been long recognized in South African gold mines and was first quantified by Krige (1951), whose early considerations are well summarized by Krige (1978). Krige developed an empirical quantitative model (linear) to explain the established fact that panel estimates determined from assays, on average, overestimated high-grade panels and underestimated low-grade panels. Krige assumed this relation to be linear (for logtransformed gold grades) and developed a linear regression system to quantify conditional bias, as illustrated in Fig. 10.12.

Figure 10.12 is an hypothetical plot of true grades of blocks (y) versus estimates based on assay data (x). The linear model of conditional bias passes through the mean values of x and y and has a slope less than 1 on the diagram and represents the regression of y on x. The $y = x$ line is called the *first bisectrix* and represents the regression of x on y. The equation of the conditional bias line is

$$y = B_1(x - m) + m$$

where x and y are as above, m is the mean value of x, and B_1 is the slope of the line.

The general equation for the slope of a linear model is

$$B_1 = s_{xy}/s_x^2$$

where s_{xy} is the covariance of x and y, and s_x and s_y are the standard deviations of x and y, respectively.

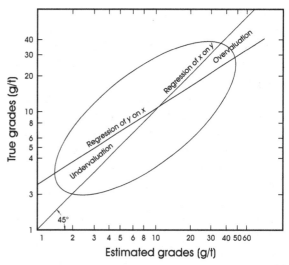

Figure 10.12: A traditional view of conditional bias in which high-grade values are, on average, overestimated and low-grade values are, on average, underestimated. True block values are along the ordinate; estimated block values are along the abscissa. Scales are logarithmic. An ellipse encompasses a field of plotted points (not shown). The 45-degree line is the desired relation; the regression line for y on x is the actual relation. The two lines intersect at the mean value of the distribution, where conditional bias is zero. If a cutoff grade is applied to the estimates (abscissa) above the mean value of the distribution, gross overestimation can result, as commonly happens with polygonal estimation. Modified from Krige (1978).

The correlation coefficient r of two variables x and y is given by

$$r = s_{xy}/s_x s_y.$$

Consequently, the slope of the first bisectrix (cf. Fig. 10.12) is given by

$$1.0 = r(s_x/s_y) \tag{10.4}$$

and, the slope of the linear model for conditional bias (cf. Fig. 10.12) is

$$B = r(s_y/s_x). \tag{10.5}$$

Combining Eqs. 10.4 and 10.5 leads to the important relation

$$B = s_y^2/s_x^2 < 1.0. \tag{10.6}$$

Equation 10.6 embodies the fundamental concept of

conditional bias as described by Krige and expressed as a linear model (i.e., the bias is a function of the variance of true, block values [relatively small] and the variance of block estimates based on data [relatively large], specifically, the ratio of variances is an estimate of the slope of the linear conditional bias model). Slopes near 1 have small conditional bias, whereas slopes much different from 1 have large conditional bias. For very low cutoff grades (lower than a high proportion of the data), conditional bias might not be of great concern. However, where the bias is large and/or the cutoff grade is relatively high (e.g., greater than the mean value) conditional bias can result in serious estimation problems; in particular, metal produced from a given volume is less than metal estimated to be present in that volume. The practical problem is expressed elegantly by Journel and Huijbregts (1978, p. 457) as follows: "The recoverable tonnage must be estimated from the distribution of the estimator. . . . But once this selection is made, the mill receives the true values, not the estimates; more precisely, it receives the true values conditioned by the fact that their estimators are greater than the cutoff."

Matheron (1971) provided a partial physical explanation of Krige's account of conditional bias as follows: it is common practice to estimate locations of unknown, irregular ore/waste margins by smooth surfaces (e.g., in Fig. 10.13, line AC estimates the irregular margin $W_1O_1, \ldots W_4O_4$). If "high-grade" assays along drives AB and CD are used to estimate the "ore" block $ACDB$, the result is an overestimate of grade because dilution caused by protrusions of waste into the block at W_1, W_2, W_3, and W_4 is not taken into consideration. In contrast, if the block $ACDB$ is viewed as a low-grade block surrounded by higher-grade values, then assays along AB and CD underestimate the block because high-grade protrusions into the block are not considered.

In the literature on conditional bias, there is substantial implicit reference to a linear model, first proposed by Krige. Philosophically, there is no reason why the bias must follow a linear model, and many publications recognize this either implicitly or explicitly (e.g., Guertin, 1987; Journel and Huijbregts, 1978).

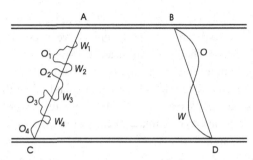

Figure 10.13: Matheron's (1971) explanation of the conditional bias described by Krige in South African gold mines. The trapezoid $ACBD$ of ore is defined based on data on the two drifts shown. Ore/waste margins are interpreted to have simple (straight line) geometry that departs from reality. In fact, the ore trapezoid is not all ore but contains protrusions of waste (W_i); hence, on average, the trapezoid is overestimated when complex margins occur. Similarly, protrusions of ore (O_i) extend into waste, so ore is lost. Modified from Matheron (1971).

Matheron (e.g., 1971) first proposed that linear kriging was an optimal approach to minimizing conditional bias. More recently, Journel and Huijbregts (1978) emphasized kriging as an optimal estimator in this regard. In particular, they point to the problem of minimizing conditional bias as being one of minimizing errors of incorrect estimation (e.g., on Fig. 10.12, minimizing the undervaluation and overvaluation areas by reducing the two sources of scatter, parallel to both the x and y axes).

More specifically, the problem of minimizing the conditional variances is to obtain estimators that minimize both the accuracy criterion and the dispersion criterion, as accomplished by kriging and illustrated in Eq. 10.7:

$$E\{[Z_v - Z_v^*]^2\} = E\{[Z_v - f(Z_v^*)]^2\} + E\{[f(Z_v^*) - Z_v^*]^2\}.$$
(10.7)

"By minimizing the sum of these two criteria, kriging represents a compromise between satisfying one or the other, and because of this, it is preferred to any other linear estimator for selection problems" (Journel and Huijbregts, 1978, p. 459).

David et al. (1984) describe estimates by several methods using a simulated database with a known mean and spatial structure and a guaranteed stationarity arising from the simulation procedures. They are able to compare various estimates with known (simulated) values and conclude the following:

1. For a normally distributed variable, simple kriging is somewhat better than ordinary kriging for block estimation.
2. A procedure that uses ordinary kriging to estimate a global mean, which is then used in simple kriging, is almost as good as is simple kriging with a known mean.
3. For a simulated, postively skewed distribution (Coef. of Variation = 2), ordinary linear kriging produces block estimates with a more severe conditional bias than does simple linear kriging.

10.12.1: Discussion

Conditional bias as described by Krige (1978) leads to average overestimation of high-grade blocks and average underestimation of low-grade blocks:

1. The relation is nonlinear in general, but commonly is approximated by a linear relation between true and estimated grades, with the linear model passing through the mean grade (i.e., zero conditional bias at the mean grade). The slope of a line (p), which incorporates both the classic idea of conditional bias and the smoothing that results from kriging small blocks with widely dispersed data, can be determined for each block estimate as follows (Pan, 1998):

$$\rho = \left[\sigma_0^2 - \sigma_k^2 + \mu\right)\right]/\left[\sigma_0^2 - \sigma_k^2 + 2\mu)\right]$$

where

σ_0^2 is the declustered variance of the data
σ_k^2 is the kriging variance of the block estimated
μ is the Lagrange multiplier obtained from kriging.

2. The classic definition of conditional bias attributes it to differences in variances of block estimators and true block grades (i.e., it is a function of the volume–variance relation, so well established in geostatistics).

3. Correct documentation of conditional bias requires a comparison of estimates with true values, an unlikely situation at the feasibility stage but one that can occur for deposits that have produced for many years, as in the case of South African gold mine experience, or for the case in which grades are known for many subvolumes of a bulk sample.

4. Because the general nature of conditional bias is understood, there are a number of actions that can be taken to reduce possible conditional bias, generally to an unknown extent, including
   (i) Use of separate continuity domains with characteristic grade distributions and semivariogram models
   (ii) Use of composites rather than individual assays of small support
   (iii) Special treatmemt of outlier values.

5. Note that all these procedures have the effect of reducing the spread (variance) of the information used to make block estimates.

6. Conditional bias generally is not resolved "by a mere change of variogram model. . . . Rather the prior decision of stationarity (averaging) over A and/or the estimation algorithm itself (to determine the xs) may have to be reconsidered" (Journel, 1987, p. 101).

7. There is an implicit (rarely explicit) assumption that perfect stationarity exists in the deposit being considered, or at least, within certain defined zones of a deposit. This may be true for some deposits, but it is unlikely to be so in many deposits, including porphyry copper and large, epithermal gold deposits. Smoothed grade profiles along individual drill holes may give some idea of the scale on which local high to low variations in grade occur. Note that if blocks are small relative to distances over which grades vary systematically, then use of the global mean introduces a conditional bias. If block size is about the same as the wavelength of grade variations, then use of the global mean in simple kriging produces block estimates that are conditionally unbiased.

8. Simulations show that simple kriging is better than ordinary kriging when simulated data are the basis of comparison, but simulations are highly idealized relative to real grade distributions, and the comment cannot be generalized. Simulated data sets are strictly stationary, whereas real data generally are not.

9. One step toward minimizing conditional bias is to add more data by increasing the search radius. Unfortunately, this solution leads to other problems such as oversmoothing and the introduction of negative kriging weights. Some compromise among these factors must be considered.

10.13: KRIGING WITH STRINGS OF CONTIGUOUS SAMPLES

Certain linear data configurations lead to ordinary kriging weights that are peculiar because the samples furthest from the point/block being estimated carry the largest weights (cf. Rivoirard, 1984). Also, samples that are roughly centrally located in the data string can take on large negative weights if they are positioned slightly off the line of data (see Chapter 17, Section 17.4; Shurtz, 1994, 1998). Both of these situations arise commonly in mineral inventory estimates – contiguous drill-hole samples are the fundamental data of many studies and slightly offline samples can arise in the "linear" array at breaks in the linear trend that arise from intermittent down-hole survey locations. The common geometry of the problem is illustrated in Fig. 10.14. Deutsch (1993) studied the problem at length using simple examples involving a linear array of 11 samples used to estimate a point along a line through the central datum and perpendicular to the linear array. For example, the point is distance a from the center of the string of samples and kriging weights are determined for three different ratios of nugget effect to total sill, that is, $C_0/(C_0 + C_1)$, of 0, 0.2, and 0.8. In the highly continuous case ($C_0 = 0$), the paradoxical result of extremely high weights being assigned to the two end samples is evident; specifically, the two extreme samples account for about 67 percent of the total weight, despite the

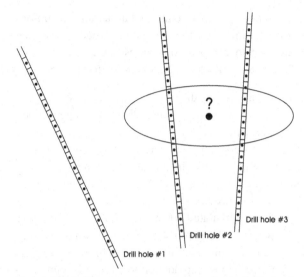

Figure 10.14: The common occurrence of using strings of data for mineral inventory estimation. When the database is contiguous samples along drill holes, a search ellipsoid necessarily locates strings of data for block or point estimation. Redrawn from Deutsch (1993).

fact that each of the other nine samples is closer to the point being estimated. The weights clearly appear more sensible as the nugget effect increases (Fig. 10.15).

In an additional test, Deutsch kriged a point at different distances from the same string using the semivariogram model with $C_0/(C_0 + C_1) = 0.2$. Results show that as the point gets closer to the string of data, the concern about excessive weights on distant samples lessens. According to Deutsch (1993), "Kriging yields this type of weighting because of the implicit assumption that the data are within an infinite domain – the outermost data inform the infinite half-space beyond the data string and hence receive greater weights" (p. 42). However true this explanation might be, such weights are not satisfactory in local estimation. Solutions that emerge from the study by Deutsch (1993) include the following:

(i) Using two of the data in a string and rejecting the other data. This solution does not take full advantage of the available data. One practical means of adapting this rule is to use relatively

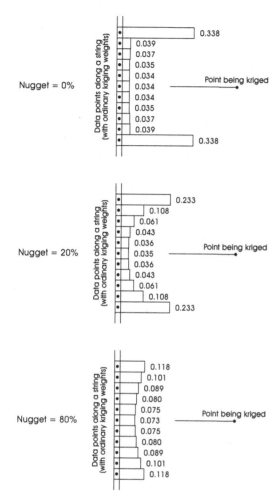

Figure 10.15: An example of kriging weights for a single string of data used to estimate a single point. As the nugget effect of the semivariogram model increases, the high distant weights become of less concern. Redrawn from Deutsch (1993).

long composites of the string of samples as the basis for kriging, and then selecting the two nearest composites for estimation purposes.

(ii) Extending the data string to include a dummy data point at each end, then excluding the two end weights and recalculating the remaining weights to sum to 1. Practical implementation is awkward.

(iii) Using simple kriging rather than ordinary kriging (see also Rivoirard, 1984). In general, simple

kriging requires a knowledge of the mean value. Because only local stationarity exists in most mineral deposits and the local mean is not well known, ordinary kriging is the appropriate kriging procedure in many cases.

(iv) Unacceptable weight distributions arise when only one or two data strings are involved in an estimation. Kriging experimentation shows that the problem of distant high weights quickly disappears in the presence of additional strings. Even for three strings, the problem is minimal. Consequently, the simple requirement that a minimum of three quadrants or five octants contain data for block estimation is a practical solution to the problem.

Rivoirard (1984) has a useful explanation of the phenomenon (i.e., he explains ordinary kriging as a combination of simple kriging and kriging of the mean). The relatively high, more distant weights are related to estimating the mean value, whereas the simple kriging weights represent the more local effect. Such an explanation is consistent with a more theoretical discussion by Deutsch (1993).

10.14: OPTIMIZING LOCATIONS FOR ADDITIONAL DATA

Because geostatistics was first widely promoted as an improved basis for mineral inventory estimation, it has been touted (e.g., David, 1977) as providing a quantitative approach to optimizing the locations of additional data with which to improve the quality of estimation. It is important to understand exactly what *optimizing* means in this context. Generally, in such studies, *optimization* means the location of additional samples in such a pattern that a specified minimum number of new samples decrease the estimation variance to a predefined, acceptable maximum. Numerous examples exist in the public domain; one of the clearest is that of Dowd and Milton (1987), who consider the affect of various sampling arrays of information on errors of several types, including error in surface area, error in tonnage, error in metal quantity, and error in grade. Such an approach permits a critical assessment of individual sources of error.

It is a relatively simple matter to (i) add specific sampling sites to a data array (without actually going to the cost and effort of taking the samples), (ii) perform kriging with the new array on the assumption that the global mean will not change, and (iii) estimate the effect of the new data on the estimation (kriging) variance. In the great majority of cases, such a procedure is adequate. In this way, a variety of new data patterns can be kriged and their relative effects on both local and global errors compared, leading to an informed decision as to which data array is best for the purpose at hand. An example is given in Fig. 10.16 for a stope at the Giant Mine, Yellowknife, Northwestern Territories, Canada. In this case, a stope (Fig. 11.3) was defined using 31 drill intersection through a near-vertical, tabular, mineralized zone. Gaps were filled with various amounts of well-placed additional drill holes to examine their impact on errors; similarly, data were extracted in a few cases. For each data array, the global errors for thickness and gold accumulation were determined and the results are shown on Fig. 10.16. These results show a pattern common for such studies (i.e., for small numbers of data, the addition of a few additional data decreases the global estimation error significantly; the same is not true when large numbers of data already exist). In this example, there is little improvement in estimation error beyond a total of about 40 data.

The more general problem of optimizing the individual locations of m additional samples in a data array in order to minimize the kriging variance is a constrained, nonlinear, programming problem considered by Chou and Scheck (1984). The problem has been dealt with at length by Barnes (1988) and Aspie and Barnes (1990). A simple example provides considerable insight into the problem (Aspie and Barnes, 1990, p. 523): "[T]here are over 75 million different ways to select 5 new samples from 100 candidate locations." Obviously, the potential computer time required to check and compare benefit measures of all possible locations is exhorbitant and alternative strategies are necessary. Two possibilities are presented by Aspie and Barnes (1990): the "Greedy algorithm" and a

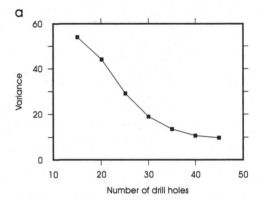

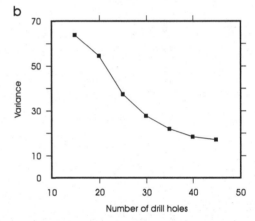

Figure 10.16: Example of the systematic decrease in global kriging error versus number of data for a stope in the Giant Mine, Yellowknife, Northwest Territories, Canada (after Shahkar, 1995). (a) Error variance versus number of data for global mean-thickness estimator. (b) Error variance versus number of data for global accumulation (gold × thickness) estimator. Both error variances decrease little, even for large increases in number of data beyond $n = 40$.

"sequential exchange algorithm." With the Greedy algorithm, all possible new data locations L are checked to locate the one that meets the guiding criterion/criteria best (e.g., minimizes global estimation variance). This location is accepted and the next best location that makes the best pair with the first is located. The next location that makes the best trio with the two previous locations is then identified, and so on until a predetermined number of new data locations M has been identified. This procedure does not optimize the positions of the M new sample locations, but provides a good first estimate.

Output from the Greedy algorithm may be adequate, or it can be used as input for the more sophisticated sequential exchange algorithm (cf. Aspie and Barnes, 1990). Although there may be situations in which such optimization is desirable, the practical demands exerted by geology on sample locations generally overrides such sophisticated optimizing procedures that implicitly assume uniformity of geology. The hypothetical example used by Aspie and Barnes (1990) is a case in point: the five new "optimal" sample sites ignore the concept of verifying the high-grade core. Only one of the five samples is on the edge of the high-grade center, the others being dispersed in the low-grade fringe. In fact, for relatively few additional samples, judicious evaluation of contoured maps of grade and local estimation variance in conjunction with a detailed geologic map can provide the essential controls for defining new sample locations. At all times, however, it is important to keep in mind the stage of data collection and the likelihood that a substantially denser data array might be necessary to improve the certainty of geologic continuity and to provide adequate quality for local estimation. For many operations it is the location of the boundary between ore and waste (i.e., the cutoff grade contour) that is critical, and extra drilling and additional drilling to better define this contact is essential.

10.15: PRACTICAL CONSIDERATIONS

1. *Kriging* is an optimal block- or point-estimation procedure, optimal in the sense that weights for samples are unbiased and are determined using a least-squares procedure such that the estimation variance is minimized. Confidence in the semivariogram (value continuity) model is an essential prerequisite to successful kriging.

2. Ordinary kriging of raw data or composites (local mean is unknown) is the most widely used and perhaps most fundamental kriging procedure.

3. Simple kriging requires a knowledge of the mean value; consequently, it is less widely used than ordinary kriging. In certain cases, a useful operational procedure is to krige a large block by ordinary kriging to estimate the local mean, and to use this mean with simple kriging to obtain more detailed estimates.

4. Three-dimensional arrays of blocks approximating the geometry of a particular geologic domain are kriged block by block in systematic fashion. For each block, a specified search volume (sphere or ellipsoid) centered on the block contains the data used for kriging. This search volume must be consistent with the semivariogram model, the data array, and the geologic character of the domain. It is common to define minimum and maximum amounts of data required for kriging to be carried out. Similarly, it is common to ensure that sufficient quadrants (at least three) or octants (at least five) contain data, so that the estimate is by interpolation rather than extrapolation.

5. Both ordinary and simple kriging can be done on transformed data. Thus, lognormal kriging is simply kriging of logtransformed data, a process that is equivalent to kriging of raw data using a relative semivariogram. Similarly, indicator kriging is simply ordinary or simple kriging of data that have been transformed to indicator values (0 or 1) that represent the presence or absence of a condition (i.e., mineralized ground = 1; unmineralized ground = 0).

6. The indicator concept can be applied to a continuous distribution of grades in which 1 is above any arbitrary threshold and 0 is equal to or below the threshold. In this latter case, indicator kriging produces a point on the local cumulative curve of samples. Repeated indicator kriging for different thresholds, a process known as *multiple indicator kriging*, allows the local cumulative curve to be estimated, from which the local mean can be determined, a block distribution can be estimated, and the proportion of blocks above cutoff grade (and their average grade) can be estimated. Multiple indicator kriging is widely applied to seemingly erratic values, such as those common to many gold and uranium deposits.

7. Multiple indicator kriging has been promoted as a means of dealing with outlier values, but it is not particularly effective for such a purpose because the procedure interpolates substantial gold

that does not exist in and around true outliers. Outliers, as a rule, have very different continuity than the much more abundant lower grades; hence, outliers should not be incorporated in kriging equations that use the continuity model of those lower grades. The mining industry combines cutting of grades with ordinary kriging as a practical, if arbitrary, solution to this problem.

8. Restricted kriging is another approach suggested for dealing with outliers. Application of the method is somewhat arbitrary and likely to spread outlier values far beyond their true physical extent.

9. Negative weights resulting from kriging occur in specific situations that can generally be avoided or the effects minimized by certain precautionary moves: (i) ensure that data in the search volume are not screened by other data that are nearer the block/point being estimated, (ii) use the method of positive kriging, (iii) check for negative weights following an initial kriging, reject those data with negative weights, and re-krige the remaining data, and (iv) as a safeguard, deal with outliers separately – even small negative weights applied to outlier values can produce extreme estimates and in some cases even negative grade values!

10. Strings of data can lead to peculiar kriging results in cases in which a semivariogram model is highly continuous. To avoid or minimize such problems, (i) use composites of as great a length as is reasonable, (ii) limit to two the number of composites selected from each string, (iii) ensure that at least three strings are represented in the search volume, and (iv) use simple kriging wherever possible.

11. Conditional bias (i.e., bias that is a function of grade) is inevitable to some extent in resource estimation. Such bias is particularly extreme in polygonal estimation, and generally is somewhat less extreme if estimates are based on increasing amounts of data. What is traditionally termed *conditional bias* involves overestimates of high grades and underestimates of low grades. This feature is counteracted by a process known as *smoothing*, illustrated by widely spaced data that produce similar estimates for all small blocks in a local cluster of blocks. Both situations are improved by increasing the amount of data. However, in many deposits, a practical maximum search volume exists because only local stationarity exists.

12. In using kriging to optimize the location of additional data for evaluating mineral deposits, it is essential to understand the basis of the optimization. Within a homogeneous domain for which a semivariogram model is known, it is possible to add potential new sampling sites and estimate the impact such sites will have on the kriging variance. Such calculations can be done for many hypothetical data arrays, and the corresponding kriging errors can be evaluated to determine which array is best in a specific case. The controlling parameters are a desirable (relatively low) error and a corresponding reasonable cost to obtain the data that provides that error. "Best" is an acceptable compromise between data density and cost, always ensuring that geologic integrity is maintained during the locating of additional samples.

10.16: SELECTED READING

Barnes, T., 1999, Practical post-processing of indicator distributions; *in* Dagdelen, K. (ed.), Proc. 28th APCOM symposium, Golden, Colo., pp. 227–238.

Brooker, P. J., 1979, Kriging; Eng. Min. Jour., September, pp. 148–153.

Clark, I., 1979, Practical geostatistics; Applied Sci. Pub. Ltd., London, 129 pp.

Cressie, N., 1990, The origins of kriging; Math. Geol. v. 22, no. 3, pp. 239–252.

David, M., 1976, The practice of kriging; *in* Guarascio, M., M. David, and C. Huijbregts (eds.), Advanced geostatistics in the mining industry, Nato Advanced Study Inst., Series C, D. Reidel Pub. Co., Dordrecht, the Netherlands, pp. 31–48.

Dowd, P. A., and D. W. Milton, 1987, Geostatistical estimation of a section of the Perseverance nickel deposit; *in* Matheron, G., and M. Armstrong (eds.), Geostatistical case studies, D. Reidel Pub. Co., Dordrecht, the Netherlands, pp. 39–67.

Matheron, G., 1971, The theory of regionalized variables and its applications; Les Cahiers du Centre de Morphologie Mathematique, Fontainebleau, France, No. 5, 211 pp.

10.17: EXERCISES

1. Krige the central point in each of the triangular and square grid patterns of Exercise 4 at the end of Chapter 5 (Fig. 5.23). In both cases, assume the distance from the block or triangle center to sample is $a = 5$ m, and assume a spherical semivariogram model with $C_0 = 0.2$, $C_1 = 0.8$, and range of $a = 20$ m. (Note: If the problem is being done manually, it becomes easier if the spherical model is approximated by a linear model with $\gamma(h) = 0.2 + 0.058h$. Note also that if one aureole of data is used for the estimate, all data take equal weights. The point of interest here is a comparison of the kriging variances.)

2. Evaluate the kriging errors for point estimation using the two data arrays of Question 5.4 (Fig. 5.23) with a spherical model: $C_0 = 0.2$, $C_1 = 0.8$, and $a_1 = 35$ m. Use one aureole of data. Repeat with two aureoles of data. In all cases, estimate a central point within a block/triangle. Assume that distance a in Fig. 5.23 is 21.2 m.

3. Four data are located at the corner of a square cell. The central point in the cell is to be estimated by $1/d_2$ (ISD) and ordinary kriging (OK).

(a) Assuming that grade continuity is isotropic, which of the two estimates will be the best?

(b) Suppose that grade continuity is anisotropic, with a range along one diagonal that is twice the range of the other diagonal. Which estimation method is best?

4. Construct an arbitrary data array for estimation purposes that generates negative kriging weights. Recall that screening of data and a highly continuous semivariogram model contribute to the ease with which negative weights occur. Use any available kriging routine to demonstrate the negative weights.

5. Construct an arbitrary linear array of 11 sample locations that simulate sample sites along a drill hole or a trench. The centroids of each sample are separated by 1 unit. Estimate a point that is 10 units away from the middle datum along a line that is perpendicular to the line of samples. Use any available routine for ordinary kriging with all the available data to arrive at your estimate. Assume a spherical semivariogram model with $C_0 = 0$, $C_1 = 1.0$, and $a = 10$ units.

6. An absolute semivariogram model produces grade estimates of 0.8 g/t and 3.1 g/t for two blocks. In each case, an identical data array was used for the estimation and the identical kriging error (standard deviation) of 0.6 g/t was estimated. Comment on the relevance of the error estimates.

11

Global Resource Estimation

While it is well known that a smoothing estimation method such as kriging is not suitable to produce realistic estimates in blocks which are far away from the sampling information, it [kriging] is unfortunately too often used for this purpose. (Ravenscroft and Armstrong, 1990, p. 579)

Many estimation procedures, traditional and geostatistical, have been described in Chapters 1 and 10, respectively. Chapter 11 introduces a variety of procedures concerned with the estimation of very large volumes commonly referred to as *global resources*. The concept of global resources is introduced and such estimates are considered in terms of regular and irregular data arrays. A common case, concerning tabular deposits that can be treated as two-dimensional, is considered at length. Specialized procedures for the estimation of co- and by-products are also considered.

11.1: INTRODUCTION

Global estimation is directed toward mineral inventory estimates for very large volumes, perhaps an entire deposit or all or most of a mineralized zone or domain. Such estimates are generally possible with substantially less data than are required for confident local estimation; thus, global estimates generally can be done earlier in the exploration and evaluation of a deposit than can local estimates. The principal purposes of global estimation are

(i) To quantify the likely target as a guide to further data-gathering decisions relatively early in the evaluation of a deposit
(ii) To summarize the resources/reserves of a deposit as a basis for medium- and long-term planning in a feasibility or production situation.

Global estimates are included in such terms as in situ estimates, geologic estimates, and others, although they form only a part (commonly the least well-known part) of those categories. Because global estimates generally are based on widely spaced exploration data and, thus, are known with relatively low confidence, they can be classed at best as "inferred resources" and, in most jurisdictions, cannot be used as a basis for mine or financial planning.

Estimation of global resources can be by any of the traditional or geostatistical methods described in Chapters 1 and 10, depending on the geologic situation and the nature of available data. In addition to those well-known estimation approaches, there are a variety of more specialized procedures that, although not necessarily restricted to global estimation, can be applied usefully in cases of global estimation, including (i) simple statistics applied to simple spatial arrays of data, (ii) composition of terms, (iii) volume variance, and (iv) estimates involving irregular arrays of data. Special procedures are required in cases in which data are widely dispersed relative to the size of blocks that form the basis of the estimate.

11.2: ESTIMATION WITH SIMPLE DATA ARRAYS

11.2.1: Random and Stratified Random Data Arrays

Purely random arrays of data (Fig. 11.1a) are rarely encountered in the estimation of mineral inventory. Treatment of such an ideal situation is useful, however, as part of the conceptual development of methods by which to treat more realistic situations. In general, the optimal estimate of grade of a volume V (the domain), estimated by n randomly distributed samples (of uniform support) is the arithmetic mean of the samples, and the estimation error is the standard error of the mean. This error, expressed as a variance, is equivalent to the variance of a sample within the volume V divided by n (number of data). When the variable is characterized by a spherical semivariogram whose range is small relative to the dimensions of V, the error is equivalent to the sill of the semivariogram divided by n:

$$\sigma_n^2 = [\sigma^2(s/V)/n]$$

where $\sigma^2(s/V)$ is an unbiased estimate of the variance of samples in the deposit.

Random stratified data arrays are approximated in some practical cases. Consider a large volume V subdivided into n smaller volumes, each of size v and each containing a single sample that is randomly positioned in v. In this case, the estimated grade is the arithmetic mean of the n samples and the estimation variance is the average error for a single volume v divided by n:

$$\sigma_n^2 = [\sigma^2(s/v)/n].$$

For a random stratified grid (Fig. 11.1b), the term $\sigma^2(s/v)$ refers to the extension variance of randomly located samples in a block, which is given by

$$\sigma_n^2 = 2\bar{\gamma}(s, v) - \bar{\gamma}(s, s) - \bar{\gamma}(v, v)$$
$$= \bar{\gamma}(v, v) \tag{11.1}$$

where

$\bar{\gamma}(s, v)$	is the average semivariogram for two points that take all possible positions in s and v. In this particular case, the term is equivalent to the F function
$\bar{\gamma}(s, s)$	is the average semivariogram for a single point and is equal to zero
$\bar{\gamma}(v, v)$	is the F function (see Chapter 8).

11.2.2: Regular Data Arrays

In the case of regular grids of data with few or no gaps (Fig. 11.1c) the global mean is the arithmetic mean of

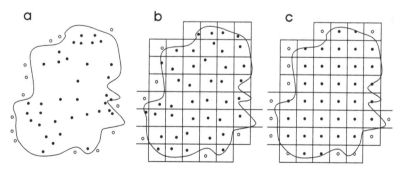

Figure 11.1: A mineralized zone explored by various data arrays. Black dots indicate mineralized samples; open circles are samples that are unmineralized. (a) Data are randomly distributed throughout the zone. (b) Each datum is randomly distributed within a cell of a regular array of cells. (c) Each cell within a regular array of cells contains a centered datum.

the sample values. If samples are located at the centers of grid cells the global variance is given by

$$\sigma_g^2 = \sigma_c^2/n$$

where n is the number of sampled cells and σ_c^2 is the extension variance of a central point extended to a block and can be determined from the semivariogram model of the variable for a block of any desired dimension in the same manner illustrated by Eq. 11.1. For a regular grid, however, the term $2\gamma(s, v)$ is much less than for the random stratified grid described in Eq. 11.1.

11.3: COMPOSITION OF TERMS

In estimating global averages of mineral domains or deposits, it is common practice to subset the data into groups, each of which has a systematic arrangement relative to a subvolume of the domain for which a global estimate is being determined. The relative sizes of the subvolumes then become the weights applied to the corresponding data in order to obtain an unbiased global estimator of a variable.

11.3.1: An Example: Eagle Vein

When data are preferentially and unevenly distributed, individual values must be weighted in proportion to their lengths or areas of influence in order to obtain unbiased global estimators. In the case of the Eagle vein, northern British Columbia, Sinclair and Deraisme (1974) estimate the mean grade of a Cu vein and use a composition of terms procedure to determine the error to attach to the mean value. A longitudinal section of the near-vertical vein (Fig. 11.2) shows the four weighting areas (blocks) relative to the exploration drifts for which 693 samples of vein width were obtained. The gap between blocks 1 and 2 is a highly fractured and oxidized part of the vein, to be omitted from the global estimation. Because samples are approximately equally spaced (about 8–10 ft apart) along the drift in each of these large blocks, average thickness and accumulation values were determined separately for each block. These averages were weighted by the corresponding block areas in order to determine global average grade for the vein. Weighted mean values of 17.0 ft and 4.02 ft were obtained for accumulation and thickness, respectively,

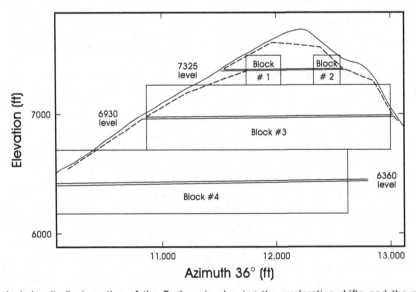

Figure 11.2: A vertical, longitudinal section of the Eagle vein showing the exploration drifts and the rectangular areas constructed about the drifts to define weights for global estimation. In this case, it has been possible to approximate the area of the vein by four areas with centered strings of data. The gap in the upper drift arises because it represents a zone of fractured oxidized, low-grade vein material that is to be omitted from the global estimate.

and from these the mean Cu grade was determined as follows: $17.0/4.02 = 4.23$ percent Cu.

The error on the mean-grade estimate can be determined by the method of composition of terms for which components of error are determined separately for a random component, extension error in one dimension (extending sample grades to corresponding drift), and extension error in the second dimension (extending drift grades to corresponding blocks). These three components of error are assumed to be independent, hence, they are additive and can be expressed as follows:

$$\sigma_e^2 = C_0/n + \Sigma\left[l_i^2\sigma_i^2(l_i)\right]/(\Sigma l_i)^2$$
$$+ \Sigma[(D_i h_i)^2]\sigma_2^2/(\Sigma D_i h_i)^2 \qquad (11.2)$$

where

C_0	is the nugget effect
n	is the number of samples
l_i	is the length of vein represented by an individual sample
$D_i h_i$	is the area of a block with dimensions D_i and h_i
$\sigma_i^2(l_i)$	is the extension variance of a sample in line l_i
σ_2^2	is the extension variance of a central line of data in a block $D_i h_i$.

The relative weighting factors are lengths squared $[(l_i^2)/(\Sigma l_i)^2]$ for extension to one dimension, and area squared $[(D_i h_i)^2/(\Sigma D_i h_i)^2]$ for the second dimension (slice term). For Eagle vein thickness, $C_0 = 0$, so the nugget term does not exist; the line term is estimated as the extension variance of a point centered in a line 10 ft long divided by the number of samples, and is so small as to be negligible; the slice term is the weighted sum of four slice terms, one for each block. Each of these four slice-term components is the extension variance of a central line to a rectangle (Fig. 11.2), each weighted by the squared area of the respective rectangle; for Eagle vein, the slice term is 0.038. The error variance for accumulation, determined in a comparable manner to that for thickness, has a significant nugget term (from the semivariogram) of 0.09, a line term of 0.014, and

Table 11.1 Summary of global estimators for the Eagle vein

Variable	Mean grade	Error variance	Relative error variance (%)
Thickness (ft)	4.02	0.038	4.8
Accumulation ($g \times t$)	17.0	1.33	6.8
Grade (% Cu)	4.23	0.072	6.3

Source: After Sinclair and Deraisme (1974).

a slice term of 1.23, to give an estimation variance of 1.33.

Errors determined for thickness and accumulation are used to determine error for grade as follows:

$$\sigma_g^2/g^2 = \sigma_a^2/a^2 + \sigma_t^2/t^2$$
$$- 2r_{at}(\sigma_a/a)(\sigma_t/t) + \cdots \qquad (11.3)$$

where

σ_g^2/g^2	is the relative variance for grade
σ_a^2/a^2	is the relative variance for accumulation
σ_t^2/t^2	is the relative variance for thickness
g	is the mean grade
a	is the mean accumulation
t	is the mean thickness
r_{at}	is the correlation coefficient for thickness and accumulation.

Using the mean values of grade, thickness, and accumulation as given and $r = 0.45$ for Eagle vein data, the absolute error for the global grade estimate can be calculated from Eq. 11.3. Results are summarized in Table 11.1. The Eagle example is for a two-dimensional case because of the tabular form of the vein, but the composition of terms procedure can be extended to three dimensions by the addition of another term to Eq. 11.2. This additional term takes into account the extension error from the slice into the third dimension.

11.4: VOLUME–VARIANCE RELATION

The *volume–variance relation* is a fundamental geostatistical relation that permits calculation of the dispersion variance of average grades for various volumes using raw data and a knowledge of the semivariogram model (e.g., Parker, 1979). The dispersion variance of small volumes v in a larger volume V is given by (Journel and Huijbregts, 1978):

$$D^2(v/V) = \bar{\gamma}(V, V) - \bar{\gamma}(v, v) \qquad (11.4)$$

where

$D^2(v/V)$	is the dispersion variance of grades of small volumes v in much larger volumes V
$\bar{\gamma}(V, V)$	is the mean semivariogram value in large volume V (F function)
$\bar{\gamma}(v, v)$	is the mean semivariogram value in small volume v (F function).

If s represents samples, SMU is the selective mining unit, and D is the deposit, Eq. 11.4 can be written in two ways, as follows:

$$D^2(s/\text{SMU}) = \bar{\gamma}(\text{SMU}, \text{SMU}) - \bar{\gamma}(s, s)$$
$$(11.5)$$
$$D^2(\text{SMU}/D) = \bar{\gamma}(D, D) - \bar{\gamma}(\text{SMU}, \text{SMU}).$$
$$(11.6)$$

Adding Eqs. 11.5 and 11.6 gives Krige's relation

$$D^2(s, D) = D^2(s/\text{SMU}) + D^2(\text{SMU}/D)$$
$$(11.7)$$

or, rearranging Eq. 11.7

$$D^2(\text{SMU}, D) = D^2(s, D) - D^2(s, \text{SMU}).$$
$$(11.8)$$

The development of Eq. 11.8 demonstrates that the dispersion of block grades is a function of the semivariogram and the dispersion of sample grades. Thus, if the semivariogram model is well established for samples, the dispersion of SMU grades can be calculated from the previously mentioned equations. Both blocks and samples have the same mean value; consequently, the only information missing to establish the relative histogram of SMU grades is the shape of the distribution. In practice, an unbiased histogram of the data is obtained and an assumption is made that the same general form is retained by the distribution of SMU grades. Thus, if the form can be assumed and

the dispersion of block grades can be determined (as mentioned), the estimated distribution of block grades can be determined. This general procedure is known as *change of support*. Several methods for calculating change of support are in common use, including the affine correction and the indirect lognormal approach (see Journel and Huijbregts, 1978). The affine correction is discussed in Chapter 12. Attaching errors to estimates based on the volume–variance relation is difficult at best (e.g., Froidevaux, 1984b), and is not done routinely.

A cutoff grade can be applied to the distribution of block grades to determine the proportion of blocks that are ore, and, of course, the average grade of these ore blocks can be determined from the histogram. In order to proceed to an estimate of tons of ore, it is necessary to know fairly precisely the volume to which the estimated block distribution applies. When this is possible, the proportions of ore that arise from the volume–variance approach can be transformed into volumes of ore, and, in turn, volumes can be transformed to tons using an appropriate bulk density. The general procedure and a simple example are described in Chapter 12, Section 4.

11.5: GLOBAL ESTIMATION WITH IRREGULAR DATA ARRAYS

Global estimation is possible by any of the traditional or geostatistical methods discussed elsewhere. Statistically based methods have an advantage over empirical techniques in that an estimate of error can be generated to accompany the global mean estimator. Here, the procedure is illustrated using ordinary kriging. Shahkar (1995) conducted a retrospective evaluation of production and various mineral inventory estimation techniques for a near-vertical, shear-controlled gold deposit, the C7-04 stope at the Giant Mine, Yellowknife, Northwest Territories. As part of that study he approximated the stoped outline on vertical section by 24 blocks, each 25×25 ft^2 (Fig. 11.3). Gold grades and true thicknesses for 31 diamond-drill-hole intersections were then used to krige each of the 24 blocks for thickness and gold accumulation. Mean thicknesses and associated kriging errors (standard deviation) are shown on Fig. 11.3. Grades

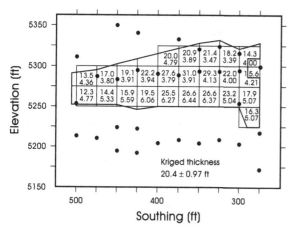

Figure 11.3: Longitudinal vertical projection of the C7-04 stope, Giant Mine, Yellowknife, Northwest Territories. The stope outline is approximated by 24 cells, each 25×25 ft^2. Numeric values in each cell are kriging estimates for zone thickness (upper) and error as a standard deviation (lower). Dots are drill-hole pierce points. Average thickness and the associated global error (standard deviation) are listed on the lower part of the diagram. Redrawn from Shahkar (1995).

and associated errors were then determined using Eq. 11.3, with a correlation coefficient of 0.845 for thickness versus accumulation. The resulting global estimate of grade is 0.72 opt $\pm$ 0.023 with error as a standard deviation. This simple example is useful because it uses all the drill-hole data available at the time the stope was planned and demonstrates the large local errors involved, yet gives a clear idea of a central thick "ore" zone that thins to the north and south. This example illustrates quantitatively the generalization that for a common data array, errors for estimates of large volumes are less than errors for estimates of small volumes. Other estimation methods could be used to estimate global means in a comparable manner, but empirical techniques could not be used to quantify the corresponding error. The main practical points to be made here are as follows:

1. Estimation of an array of blocks of uniform size is a routine procedure common to virtually all commercially available estimation software.
2. When many blocks are involved, the block size should approximate the sample spacing, although the exact block size selected is not critical.
3. Using a relatively small block size (approximately

equivalent to sample spacing and no less than half the sample spacing) clearly indicates areas where data are scarce (i.e., have high kriging variance) and where added data could most effectively improve the quality of the global estimate.

11.5.1: Estimation with Multiple Domains

Mineral deposits for which estimates are required commonly are divisible into two or more domains. The need for these domains generally relates to different geologic control from place to place, but in some cases might be the result of very different quality and/or arrays of data from one part of a deposit to another. Champigny and Sinclair (1984) found two grade populations for the Cinola gold deposit, Queen Charlotte Islands, British Columbia. A rare (< 1 percent of assays) high-grade subpopulation is superimposed on dominant widespread lower grade mineralization. Champigny and Sinclair (1984) determined a semivariogram model and kriged the abundant low-grade data to estimate large panels ($100 \times 100 \times 8$ m^3); the rare high-grade population was assumed to be a random variable, and the mean value was combined with the ordinary kriging result for each block estimate as a weighted average (weights were the global proportions of the two subpopulations). Later work showed that block estimates could have been improved by incorporating the high-grade subpopulation only in the geologically defined zone of quartz veinlets where outlier values were concentrated. In such cases, a popular alternative solution to block estimation (e.g., Clark, 1993) is to define a threshold between the two populations (cf. Chapter 7), krige each grade population separately, and weight the two estimates by using the results of an indicator kriging for the proportions of the two grade populations. Deutsch (1989) used such an approach in determining reserves for the Eastmain gold-bearing vein deposit in northern Quebec.

Characterization of lithologic domains requires thorough geologic mapping and data analysis to establish significant differences with respect to continuity. Sinclair et al. (1983) demonstrate widely different semivariogram models for five mineralized rock types at the Golden Sunlight deposit, Montana. Use of the corresponding semivariogram models produced block estimates based on exploration data that were used both for global estimates and for block estimates used in mine planning. Subsequent work showed that these early estimates were within 5 percent of production figures (Roper, 1986). Lithologies commonly have a marked control over continuity for several reasons, as follow:

(i) Various lithologies respond differently to a structural regime.

(ii) Ore fluids react variably with wallrocks depending on their textures and compositions.

(iii) Some genetic processes are characterized by fundamentally different styles of mineralization from place to place during the formation of a deposit (e.g., a layered massive sulphide deposit with an underlying stockwork feeder zone).

These simple examples illustrate the importance of incorporating the concept of geologic (continuity) domains into the procedure for global estimation.

11.6: ERRORS IN TONNAGE ESTIMATION

11.6.1: Introduction

Errors in tonnage estimates are paid lip service but are not commonly quantified. In too many cases, tonnage estimates are quoted to five or six significant figures and give an impression of a high level of accuracy even when errors are large; in other cases errors in tonnage can be negligible. Whatever the situation, it is useful to know the level of error. Here, such errors are considered in relation to in situ or geologic resources rather than for local reserves.

11.6.2: Sources of Errors in Tonnage Estimates

Tonnages of mineral inventory are determined from a volume component and a conversion factor that takes rock density into account, according to the relation

$$T = V \times f$$

where T is tons, V is volume (m^3), and f is average rock density. For tabular deposits, V can be considered

$$V = t \times S$$

where t is average thickness (m) and S is surface area in the plane of the vein (m^2). Errors can arise in the estimation of each of the terms t, S, and f. These errors can be expressed as relative errors and can be assumed to be independent; thus, they are additive as variances as follows:

$$(\sigma_T/T)^2 = (\sigma_S/S)^2 + (\sigma_t/t)^2 + (\sigma_f/f)^2$$

where

S	is surface area
t	is thickness
f	is bulk density
σ_x^2	is the error variance of the parameters indicated by various subscripts.

This equation is in the form for application to a two-dimensional deposit; however, the product of the t and S terms could be replaced by a single volume term. In some practical cases, the volume term has negligible error, for example, a month's production from the center of a porphyry copper deposit.

11.6.3: Errors in Bulk Density

Even when average rock density is known, systematic variations can exist that lead to errors in tonnage estimates of large blocks of ore. Such variations might be related to systematic zonations of ore and gangue minerals or abrupt changes in mineralogy from one place to another in a deposit. Several examples of the impact of such errors are documented in Chapter 15.

Highly significant errors can occur even in seemingly uniform ore such as that encountered in porphyry-type deposits. Consider the difference in specific gravity of two granitic ores that contain 1 percent pyrite and 10 percent pyrite, respectively. If the bulk densities of granite and pyrite are 2.7 and 5.1 g/t, respectively, the two mixtures have bulk densities of

2.72 and 2.94 (assuming zero porosity), a difference of about 8 percent.

This hypothetical case indicates the importance of obtaining representative estimates of bulk density so that appropriate values can be used for purposes of global estimation of tonnage. However, it is apparent that the potential for errors in local estimates of tonnage is high because common practice is to use a single global average value for bulk density. Clearly, a simple means of monitoring or estimating local bulk density is desirable. Of course, estimates of bulk density can be determined in the same way as any other variable providing sufficient data are available; this is generally not the case. One practical approach, useful when direct measurements of bulk density are limited in quantity, is to develop a simple or multiple regression relation between bulk density and metal assay values, as described in Chapter 15. An example for the Craig nickel mine in the Sudbury area (Bevan, 1993) illustrates the procedure:

Bulk density $= 0.297 \times \%Ni + 2.839$.

This relation is a best fit with errors of about ± 1 percent for Ni grades and about ± 0.2 g/cc for bulk density values. When appropriate data are available, the potential error in tonnage that emerges from errors in bulk density estimates can be determined.

11.6.4: Errors in Surface (Area) Estimates

The general concept of error in the estimation of an area based on widely spaced sampling points is illustrated in Fig. 11.4. Clearly, different locations of the sampling grid relative to the surface being estimated lead to somewhat different area estimates. For a fairly regular grid of data the surface area estimate (two-dimensional limits) is approximated by the number of small cells that approximate the true outline

$$S = n \cdot a = n \cdot l \cdot h \qquad (11.9)$$

where a is the area of regular cell of dimensions $l \times h$ and n is the number of cells that defines the surface. Error in a surface estimate can be determined with the following equation (e.g., Matheron, 1971):

$$(\sigma/S^*)^2 = [(l/n^2)[N_2/6 + 0.06(N_1)^2/N_2 \ldots]$$

where $(\sigma/S^*)^2$ is the relative surface estimation variance, n is the number of positive (ore) intercepts on a more or less regular sampling grid (i.e., the number of cells that approximate the surface), N_1 is half the total unit-cell edges defining the perimeter parallel to

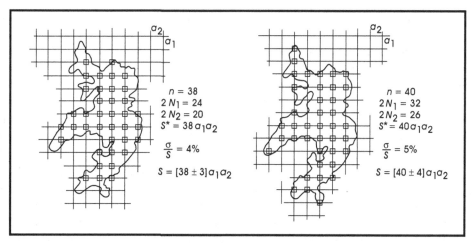

Figure 11.4: Illustrations of the concept of error in surface area estimation. Two different grid positions are illustrated, relative to the same irregular target. In one case, the target is approximated by 38 cells; in the other case, the target was approximated by 40 cells. The associated errors in each case, determined using Eq. 11.9, are shown. Redrawn from Journel (1975).

direction one, and N_2 is half the number of unit cell edges defining the perimeter in direction two; directions must be labeled such that $N_2 \leq N_1$. The equation applies in the specific case in which a regular sampling grid has been superimposed on an ore zone, such that limits of the ore are known to occur between "hits" and "misses" in the regular sampling pattern. The precise locations of the boundary between adjacent hit–miss pairs are assumed to have an even distribution (i.e., all positions of the boundary between a hit–miss pair are equally probable). Hence, the formula applies in cases in which a deposit has relatively sharp margins.

11.6.5: Surface Error – A Practical Example

A practical example of the application of the formula concerns a statistical study of the thickness of a clay layer by Wolfe and Niederjohn (1962). An exploration concession was covered by a regular grid of drill holes on 300-ft centers to evaluate the average thickness and extent of a clay layer that was to be used as a source of argillaceous material in making cement. The outline of the concession and the drilling grid are shown in Fig. 11.5. In this case, the total area of the concession is known, as is the area of the northern part where the layer has eroded. Furthermore, the clay layer is known to extend beyond the limits of the concession to the east, west, and south, so no surface error is attributed to these boundaries. A maximum surface area of the clay layer can be determined precisely (S_T). Within the concession there are local areas shown by drilling, in which the clay layer has eroded (incised sand deposited by a former river that cut through and removed part of the clay layer); this area of eroded clay (S_0) can be estimated, as can the error associated with the estimate. Hence, the surface area underlain by clay (S_c) can be determined as

$$S_c = S_T - S_0.$$

Pertinent information relative to an error estimate of S_0 can be determined from Fig. 11.5: $n = 27$ cells, $N_1 = 22$, $N_2 = 16$, and $a_1 = a_2 = 300$ ft. These data substituted in Eq. 11.6 give a relative error of 0.00694 (i.e., 0.7 percent). Because $S_0 = n \cdot a_1 \cdot a_2 = 2,430,000$ ft^2, the absolute error is 17,000 ft^2.

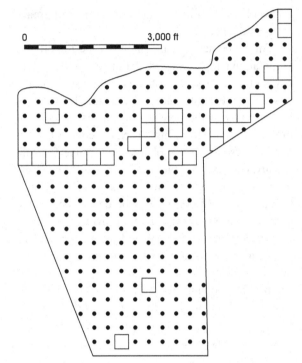

Figure 11.5: Distribution of vertical bore holes (dots) that intersect a horizontal clay layer in a concession with east, west, and south boundaries shown as solid and dashed straight lines. The clay layer is absent (eroded) north of the sinuous northern boundary. Blank squares indicate cells in which the clay layer is absent. Data from Wolfe and Niederjohn (1962).

The geologic nature of the sand "strings" that indicate erosion of clay, combined with abundant drill information close to the concession boundaries, suggest that the "internal" sand is the only significant contributor to error in the lateral extent of clay. If this is an acceptable assumption, the absolute error in sand is also the absolute error in clay. The total area of the concession (556 acres) is 24,219,360 ft^2; thus, $S_c = S_T - S_0 = 24,219,360 - 2,430,000 = 21,789,360$ ft^2. Hence, the relative error is $17,000/21,800,000 = 0.0008$, or 0.0000064 as a variance. Wolfe and Niederjohn (1962) found an average thickness of 9.51 ft with a standard error of the mean of 0.275 (i.e., a relative error of $0.275/9.51 = 0.0289$, or 0.000835 as a variance). The relative error (as a variance) for total volume of clay is $0.000853 + 0.0000064 = 0.00086$ as a variance, or 0.03 as a relative standard deviation. With a negligible error in bulk density, the error for both volume and tonnage of clay is about 3 percent.

11.6.6: Errors in Thickness

Thickness is a regionalized variable, and related error estimates rely on an appropriate semivariogram model and subsequent mean and error estimates by a variety of procedures, including kriging and composition of terms. Where data are widely spaced, as in the example in Section 11.6.5, the error on global thickness is given by the standard error of the mean.

11.7: ESTIMATION OF CO-PRODUCTS AND BY-PRODUCTS

Jen (1992) defines the *principal (metal) product of a mine* as "the metal with the highest value of output, in refined form, from a particular mine, in a specified period" (p. 87). A *co-product* is "a metal with a value at least half . . . that of the principal product" (p. 87); a *by-product* is "a metal with a value of less than half . . . that of the principal product" (p. 87). By-products are subdivided into *significant by-products*, which are "metals with a value of between 25% and 50% . . . that of the principal product" (p. 88) and *normal by-products*, which are "metals with a value of less than 25% . . . that of the principal product" (p. 88).

Estimation of co-products and by-products commonly is done by techniques similar to that of the principal commodity (e.g., inverse distance weighting, kriging). In these cases, each estimate of the two or more products is made totally independently of the other, with the implicit assumption that no significant correlation exists among them. This procedure is time-consuming and costly, and information might not be used to its best advantage if correlations exist but are not taken into account in the estimation process.

11.7.1: Linear Relations and Constant Ratios

In the extreme case of a very strong correlation among co-products it may be reasonable to estimate only one variable independently and determine the others as a

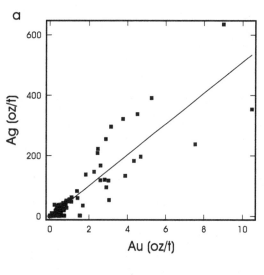

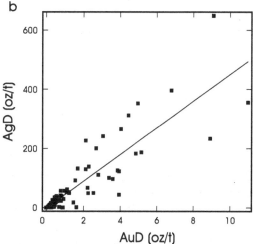

Figure 11.6: Scatterplots of Au versus Ag analyses of run of mine samples, Shasta gold deposit. (a) Original data. (b) Reject analyses. Statisical parameters for the two linear models (all error in y) are listed in Table 11.2. The two models differ by a small but significant proportion.

function of the first. Consider the example of Au versus Ag grades for the Shasta epithermal gold deposit (Fig. 11.6). Statistical and regression (Ag regressed on Au) parameters for Au and Ag for an original set of analyses ($n = 81$) and for duplicate reject analyses are given in Table 11.2. The results are of interest because they illustrate the method by which one variable can be estimated by another. However, they are also informative because the two estimates (pulps and

Table 11.2 Statistical and regression parameters for Au and Ag, Shasta epithermal gold deposit, northern British Columbia

Data	Au		Ag					
	x	s	x	s	r	Slope[a]	Intercept[a]	Dispersion
Original	1.477	1.953	74.0	113.1	0.883	51.1(0.42)	−1.53(1.02)	53.5
Duplicate	1.637	2.168	73.1	112.4	0.870	45.1(0.385)	−0.73(1.04)	55.8

[a] Numbers in parentheses are one standard deviation error.

rejects) actually provide significantly different results. In particular, the slopes are seen to be significantly different and give Ag estimates that differ by about 10 percent. A sampling problem is indicated that must be resolved in order to adopt a correct linear model for co-product estimation using this modeling procedure. Apart from that, 81 samples is a relatively small sample base with which to develop a model to be used for estimation purposes.

As a futher illustration, assume that the model for the original data is correct, as follows:

$$Ag(g/t) = -1.5 + 51.1 Au(g/t) \pm 53.5.$$

Gold is most important from an economic point of view and is the variable that should be measured directly; hence, Au is the independent variable of the regression equations. Because of the remarkably strong correlation ($r = 0.88$) and the relatively lesser economic importance of Ag, it is possible to consider assaying all samples for Au and, perhaps, to assay only a proportion for Ag as a monitor on the correlation. The linear relation then can be used to estimate Ag for all samples not assayed for Ag. In practice, if a correlation exists on a sample scale, it is even stronger on a block scale, so the same linear relation can be used to provide Ag block estimates from their corresponding Au estimates. In addition, a linear model that passes through the origin indicates a constant ratio of the two metals, equivalent to the slope of the line.

The linear model provides an easy route to monitoring the relative values of the two commodities as metal prices change. Suppose that metal prices are Au = US\$280/oz and Ag = US\$5/oz. For a gold grade of 10 g/t, the linear model gives a corresponding Ag

grade of about 510 g/t. The relative gross value of the metals is about US\$122/t for Au and US\$82/t for Ag, approximately 60:40.

11.7.2: A General Model for Lognormally Distributed Metals

Models for the average grade and tonnage above cut-off for both normally and lognormally distributed variables were introduced in Chapter 4, Sections 4.4.3 and 4.4.4. The case of two correlated metals derived by Matheron (1957) is summarized by David and Dagbert (1986) as follows. Assume that block grades have the same correlation as sample grades (probably a conservative assumption) and define the following parameters:

$$K = m(x0) \cdot \exp\left(u_x^2\right)$$
$$L = m(x0) \cdot \exp(r \cdot u_x \cdot u_y)$$
$$R(x, y) = \Phi\left[(Ln(x))/y + y/2\right]$$

where

r	is the correlation coefficient of logtransformed variables x and y
$m(x0)$	is the mean value of logarithms of x (metal 1)
$m(y0)$	is the mean value of logarithms of y (metal 2)
u_x^2	is the variance of logtransformed x values
u_y^2	is the variance of logtransformed y values
$\Phi[\cdot]$	is the integral of the normal distribution from $[\cdot]$ to infinity (see Eq. 4.6).

If the volume and bulk density are well constrained, the total tonnage is known, in which case $T(x_c)$, the tonnage above cutoff grade (x_c) (applied to the principal commodity) is given in Eq. 11.10:

$$T(x_c) = T_0 R(x_c/m_x, u_x) \qquad (11.10)$$

where T_0 is the tonnage for zero cutoff.

The quantity of the principal commodity x above cutoff grade x_c is

$$Q_x(x_c) = m(x0) \cdot T(x_0) \cdot R(x_c/K, u_x). \qquad (11.11)$$

The quantity of the co- or by-product y above cutoff grade x_c is

$$Q_y(x_c) = m(y0) \cdot T(x_0) \cdot R(x_c/L, u_x). \qquad (11.12)$$

Corresponding average grades of each metal can be determined by dividing total tonnage by the appropriate quantity of metal. A detailed example is shown by David and Dagbert (1986), who illustrate the use of a chart to assist with calculations such as those of Equations 11.10 to 11.12.

11.7.3: Equivalent Grades

An *equivalent grade* is one that is a combination of two or more grade variables in an arbitrary manner to produce a single variable for estimation purposes. For example, in an Au deposit containing some Ag, an equivalent grade might be obtained as follows:

$$Au_{eq} \ (g/t) = Au \ (g/t) + k \cdot Ag(g/t)$$

where k is a parameter that, in this case, generally is taken as the ratio of Ag price to Au price (e.g., $k = 1/56$ if Au and Ag values are \$280/oz and \$5/oz, respectively). Equivalent grades are used commonly to simplify the problem of mineral inventory estimation by estimating a single variable, rather than the two or more variables from which the single variable (equivalent grade) is derived. In general, the use of equivalent grades should be discouraged. Fluctuations in metal prices can change k significantly, and each new value requires a completely new mineral

inventory estimation. Perhaps even more serious is the common problem that recoveries of individual metals generally are not taken into account (e.g., Goldie, 1996; Goldie and Tredger, 1991) and can vary significantly as a function of grade. Recoveries of individual metals generally cannot be interpreted from estimates of equivalent grades.

11.7.4: Commentary

When correlations are used as a basis to assist in the estimation of a co- or by-product, it is important to ensure that the correlation exists throughout the volume being estimated. This can be ascertained by conducting a monitoring program in which the correlation is tested in about every tenth sample. Moreover, different correlation (geologic) domains may exist as a function of variations in the mineralization style. A linear model for one domain might not be applicable in another domain. This can be true even in adjacent and superficially similar deposits. For example, Cu/Au ratios for various mineralized zones of the Similkameen porphyry copper camp (cf. Nowak and Sinclair, 1993; Raymond, 1979) differ by up to 50 percent.

With adequate approximations to lognormal distributions for all the variables involved, co- and by-products can be estimated from the corresponding value of the principal commodity using appropriate equations. In some cases, equivalent metal grades can provide increased efficiency. In general, equivalent metal grades can pose problems arising from their dependency on metal prices and should be avoided.

11.8: PRACTICAL CONSIDERATIONS

1. Global estimates of mineral deposits are used for two main purposes – (i) to justify further expenditures and investigations and (ii) summarize the results of more detailed estimates.
2. Several techniques of global resource/reserve estimation do not provide a specific estimate for a specific block (e.g., volume variance); such methods

can be used to define the general magnitude of a target that must be proved to exist through additional exploration.

3. Where a grade cannot be assigned to a selective mining unit with reasonable confidence, an estimation method must be considered global rather than detailed.

4. The general procedure of approximating a large ore volume by an array of relatively small and regular blocks is more useful than a widely used procedure involving an array of blocks of various sizes, each with a roughly centrally located datum.

5. Attention should be directed toward calculation of errors for tonnage estimates. At the very least, such an endeavor leads to an appreciation of the principal sources of error (e.g., thickness, bulk density) and how the quality of the estimate could be improved.

6. Estimation of co- and by-products indirectly by their relation to the principal product is possible when a strong correlation exists between the co-product/by-product and the principal product. Constant ratios or linear relations of by-product/principal product are common in practice, although they can vary from one geologic zone to another in the same deposit. When such simple relations are demonstrated, they can save considerable time in the estimation process.

7. When simple mathematical relations such as ratios and linear trends are used to estimate co- and by-products, it may be acceptable practice to save on assay cost by not assaying all samples for the co-product/by-product. However, the relation between co-product and principal products should be monitored, perhaps by assaying every tenth sample for the variable(s) that are to be estimated by the relation.

8. Equivalent metal grades are commonly estimated as a cost-saving procedure, but the procedure is generally false economy and should be avoided.

11.9: SELECTED READING

David, M., and M. Dagbert, 1986, Computation of metal recovery in polymetallic deposits with selective mining and using equivalent metals; Can. Geol. Jour., Can. Inst. Min. Metall., v. 1, no. 1, pp. 34–37.

Dowd, P. A., and D. W. Milton, 1987, Geostatistical estimation of a section of the Perseverance nickel deposit; *in* Matheron, G., and M. Armstrong (eds.), Geostatistical case histories, D. Reidel Pub. Co., Dordrecht, Holland, pp. 39–67.

Jen, L. S., Co-products and by-products of base metal mining in Canada: facts and implications; Can. Inst. Min. Metall. Bull. v. 85, no. 965, pp. 87–92.

11.10: EXERCISES

1. A stope on a vein dipping 65 degrees is planned to extend from the 855-m level to the 910-m level and has a strike length of 45 m. The true vein thickness ($\pm$ standard deviation) of 2.3 ± 1.1 m is based on 38 more or less evenly distributed drill intersections. Determine the tonnage of the global (in situ) resources if 10 local, large grab samples provide a density estimate ($\pm$ standard deviation) of 2.98 ± 0.15. Also, determine the error attached to the tonnage estimate.

2. The volume–variance method can be applied to any unbiased histogram of data representing a known volume for which the bulk density is known. Of course, an implicit assumption is that ore/waste selection can be done on the same support as the data on which the histogram is based. Select an appropriate histogram from the technical literature and develop a corresponding grade–tonnage curve.

Grade–Tonnage Curves

A frequently used decision tool is the so-called grade–tonnage relationship, which relates the tonnage of mineable ore to its average grade, for a given set of economical conditions. It is in the estimation of grade–tonnage relationships that most errors and biases occur for lack of a clear definition of the necessary concepts. (Huijbregts, 1976, p. 114)

Chapter 12 deals with various derivations of grade–tonnage curves including unbiased histograms of data (and block estimates) and continuous probability density functions of unbiased data (and block estimates). A case history illustrates some of the various grade–tonnage curves that can result from a mineral inventory estimation project. The effect of conditional bias on kriged block estimates can be taken into account globally, but local estimates are more difficult to correct. Thus, grade–tonnage curves based on widely spaced data can approach, but rarely attain, reality.

12.1: INTRODUCTION

Grade–tonnage curves are one of the more useful means of summarizing mineral inventory information. Generally, two curves are involved: one is a graph of tonnage above cutoff grade versus cutoff grade, and the other is the average grade of tonnage above cutoff versus cutoff grade. A third graph, quantity of metal versus cutoff grade, can be derived from the other two. These graphs are constructed either from a tabulation of grade estimates or from smooth curves fitted to histograms of grade estimates. An example is given in Fig. 12.1 for the South Tail zone, Equity Silver deposit, northwestern British Columbia (Giroux and Sinclair, 1986); data for the curves are listed in Table 12.1. In this example, two separate tonnage curves are indicated: an optimistic estimate and a conservative estimate of the tonnages to which the curves apply, reflecting somewhat different criteria for interpolating and extrapolating data. In the conservative case, there are several gaps within the rectangular field of data that are not considered as potential ore. For the optimistic case, ore continuity is assumed through those internal gaps.

Grade–tonnage curves are useful at several stages of deposit evaluation. During the exploration stage, they can be an important component of target characterization (i.e., they provide a quantification of resources based on exploration data and are an initial rough estimate of the general size [tons of ore or quantity of metal] of a resource for various cutoff grades). At this early stage, estimates are neither resources nor reserves in the strictest sense, because a detailed economic analysis has not been performed; therefore, an appropriate cutoff grade might not have been defined. Furthermore, because selectivity has not been considered in detail, the total quantity of metal is likely to be optimistic. At the feasibility stage

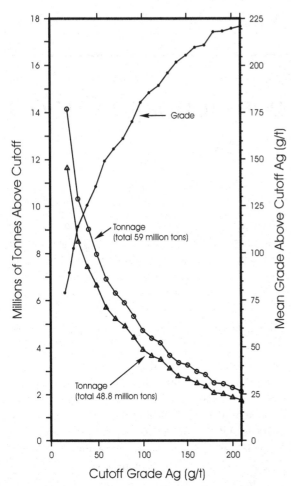

Figure 12.1: Grade–tonnage curves for South Tail zone, Equity Silver Mine. Two tonnage curves are shown: an upper curve, based on relatively optimistic extrapolation and interpolation; and a lower more conservative curve. The x axis is cutoff grade; the y axis is millions of tons (left) and average grade of ore (right). After Giroux and Sinclair (1986).

Table 12.1 Summary of grade–tonnage information for silver, South Tail zone, Equity Silver Mine

Cutoff (g/t)	Grade (g/t)	Tonnage using $T_0 = 59$ million[a]	Tonnage using $T_0 = 48.8$ million[a]
20	89.9	14,056,000	11,626,000
30	114.1	10,293,000	8,513,000
40	125.1	9,025,000	7,465,000
50	135.8	7,983,000	6,603,000
60	149.0	6,877,000	5,688,000
70	156.1	6,327,000	5,233,000
80	161.7	5,917,000	4,894,000
90	170.3	5,331,000	4,409,000
100	180.2	4,711,000	3,897,000
110	185.7	4,389,000	3,630,000
120	188.6	4,221,000	3,491,000
130	196.5	3,754,000	3,105,000
140	203.3	3,368,000	2,786,000
150	206.1	3,212,000	2,657,000
160	210.0	2,988,000	2,471,000
170	212.7	2,823,000	2,335,000
180	218.3	2,469,000	2,042,000
190	218.7	2,440,000	2,018,000
200	220.1	2,291,000	1,895,000
210	220.9	2,162,000	1,789,000
220	221.4	2,028,000	1,678,000

[a] $T_0 =$ tonnage for zero cutoff grade.
Source: After Giroux and Sinclair (1986).

economic controls are much more stringent, and grade–tonnage curves are useful in contrasting grades and tonnages for various operating scenarios (e.g., large-scale production at low cutoff grade versus smaller-scale production at higher cutoff grade). During planning and production, grade–tonnage curves are useful for summarizing quantities of tonnes and metal to be mined during a specific time interval or from a particular part of a deposit (e.g., one bench) and provide rapid insight into changes in ore recov-

ery that could result from significant changes in cutoff grade (e.g., in response to changes in metal prices or smelter contractual arrangements).

The use of grade–tonnage curves to summarize "recoverable reserves" contrasts with their use during the exploration stage for quantifying a somewhat idealized target that remains to be proved. For their various applications, grade–tonnage curves can be approximated in several ways and it is essential that the means by which such curves have been generated is well understood. Histograms or probability density functions (e.g., normal and lognormal density functions) of sample grades provide an overly optimistic approach to determining grade–tonnage curves because they are based on samples of small volume, far smaller than any practical ore/waste selective mining unit (SMU). Nevertheless, such information can be used to estimated grade–tonnage characteristics of a deposit and provide important information on which to base decisions regarding continued evaluation. It is important to recognize that spatial location is not known for ore implied by grade–tonnage curves, as established from histograms and normal and lognormal density functions for sample data. More realistic uses of sample histograms or probability density functions, for derivations of grade–tonnage curves involving recoverable reserves, require a knowledge of the dispersion characteristics of average grades for blocks approximating a realistic selective mining unit. *Recoverable reserves* is a widely used term for the tonnage (and its average grade) that is above cutoff grade; geostatistical estimates of recoverable reserves may be somewhat optimistic because they do not take into account factors other than block grades (e.g., blocks that might not be milled for reasons of technology and mine planning).

12.2: GRADE–TONNAGE CURVES DERIVED FROM A HISTOGRAM OF SAMPLE GRADES

An unbiased histogram of grades contains much of the necessary information with which to construct grade–tonnage curves provided the samples are representative of the deposit in question. As David (1972, p. 90) pointed out, "mathematical treatment cannot increase the quantity of information." For purposes here, suppose that a large block of ore has been delimited by adequate sampling and the volume V is well defined and known. An unbiased histogram of sample grades for this volume contains information about

(i) The proportion of grades above any specified cutoff grade
(ii) The average of all grades above any cutoff grade.

That is to say, an unbiased histogram of assay data (or composites) contains grade–tonnage curve information, albeit for very small samples each of volume v; hence, such curves represent an idealized view of the proportions of ore and waste based on selection units the size of samples. Proportion of grades above

cutoff ($p_{>c}$) is the discrete estimator

$$p_{>c} = N_{>c}/N_t$$

where $N_{>c}$ is the number of grades greater than cutoff and N_t is the total number of grades. If values have been weighted to produce an unbiased histogram, then $N_{>c}$ is replaced by the sum of weights above cutoff $[\sum w_j(\geq c)]$, and N_t is replaced by the sum of all weights $[\sum w_i(t)]$ to give

$$p_{>c} = \sum w_j(>c) / \sum w_i(t)$$

Where volume V and appropriate bulk density factors (d) are known, the proportions of tonnage (i.e., $p_{>c}$ values) can be transformed into amounts of absolute tonnage greater than cutoff grade ($T_{>c}$):

$$T_{>c} = T_0 \times p_{>c}$$
$$= V \times d \times p_{>c}.$$

If a histogram is based on very large N_t, it is more convenient to use frequencies of class intervals as weights. For the case of equal weighting of values, the average grade of the proportion above cutoff is easily determined as the arithmetic mean of the individual items above cutoff. With weighted information for an unbiased histogram, a weighted average must be calculated. This procedure can be repeated for a variety of cutoff grades (commonly chosen as the values that bound class intervals for the histogram) to produce a tabulation of data that can be plotted as grade–tonnage curves. A simple example is illustrated in Fig. 12.2, in which data are represented by a histogram with frequency as percent; the necessary data for constructing the corresponding grade–tonnage curve (as calculated from the histogram information) are summarized in Table 12.2. Note that grade–tonnage calculations such as these can be done conveniently using readily available spreadsheet software.

This procedure for producing a grade–tonnage curve is global (rather than local) in character because there are no implications as to where, in the relevant volume V (the deposit), the higher grade vs (samples) occur. Moreover, the histogram and the grade–tonnage curve derived from it can be attributed only to the rock volume represented by the samples; only

Table 12.2 Tabulated grade–tonnage information for the data in Fig. 12.2

$g(c)^a$	$T_{>g(c)}{}^b$	$g(m)_{>c}{}^c$
0.5	1	2.72
1.0	0.99	2.74
1.5	0.92	2.85
2.0	0.81	3.01
2.5	0.6	3.28
3.0	0.35	3.66
3.5	0.17	4.1
4.0	0.08	4.5
4.5	0.03	4.92
5.0	0.01	5.25

[a] Cutoff grade.
[b] Proportion of tonnage above cutoff grade.
[c] Average grade (% metal) of tonnes above cutoff grade.

limited extrapolation may be possible. It is useful to be aware that an unbiased histogram provides grade–tonnage information, because such knowledge opens the possibility relatively early in the exploration of a deposit of estimating grade–tonnage curves, and thus estimating potential target size. However, it must be remembered that data distribution is a function of sample size (volume) and that grade–tonnage curves generated from sample data, however large, must be viewed as optimistic because of the volume–variance relation (cf. Chapter 8).

12.3: GRADE–TONNAGE CURVES DERIVED FROM A CONTINUOUS DISTRIBUTION REPRESENTING SAMPLE GRADES

Just as unbiased histograms can be used to create grade–tonnage curves, continuous distributions can be used in the same manner. This idealized approach is particularly useful when examination demonstrates that a normal or lognormal distribution characterizes an unbiased distribution of the grades in question. For example, consider the parameters of a lognormal distribution of blasthole grades for the Bougainville porphyry copper deposit (David and Toh, 1988); the arithmetic mean and standard deviation of the distribution are 0.45 percent Cu and 0.218, respectively, and the mean and standard deviation of natural logarithms of the data are −0.9035 and 0.458, respectively. The application of Eq. 4.14 for a series of different cutoff grades provides the proportion of tonnage above each cutoff grade. This information, along with Eqs. 4.15 and 4.16, can be used to determine the average grade of material above each cutoff grade. Results of a number of calculations are listed in Table 12.3 and are shown as grade–tonnage curves in Fig. 12.3. If data fit a lognormal distribution, the ease with which grade–tonnage curves can be established is apparent.

It is clear that similar calculations for a normal distribution can be done using the appropriate

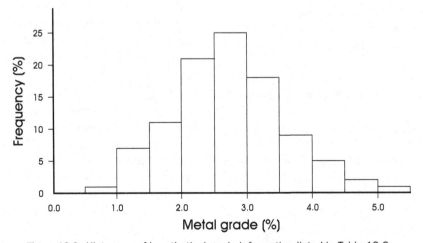

Figure 12.2: Histogram of hypothetical grade information listed in Table 12.2.

Table 12.3 Tabulated grade–tonnage information for the lognormal distribution in Fig. 12.3

Cutoff	$P_{>c}$	$R_{>c}$	$g(m)_{>c}$
0	1.0	1	0.45
0.1	0.9993	0.999	0.45
0.2	0.9415	0.97999	0.468
0.3	0.74486	0.86958	0.525
0.4	0.51124	0.68688	0.605
0.5	0.3228	0.50041	0.698
0.6	0.1944	0.34472	0.798
0.7	0.1139	0.23006	0.909
0.8	0.0657	0.15028	1.03
0.9	0.0376	0.09684	1.16
1.0	0.0215	0.06188	1.30
1.2	0.0070	0.02498	1.61
1.4	0.0024	0.01009	1.89
1.6	0.0008	0.00412	2.32

equations in Chapter 4. The assumption that a normal (or lognormal) distribution describes a real data set breaks down on the tails of the distribution. Thus, estimates of both proportion (tonnes) above cutoff grade and average grade of those tonnes (and, therefore, quantity of metal) can be seriously in error, particularly with a cutoff grade on the upper tail of the distribution.

12.4: GRADE–TONNAGE CURVES BASED ON DISPERSION OF ESTIMATED BLOCK GRADES

12.4.1: Introduction

For feasibility and production purposes, grade–tonnage curves must be based on dispersion of grades of blocks of a realistic size for purposes of classification as ore or waste (i.e., an SMU must be assumed or defined and that information integrated into the determination of grade–tonnage curves). Two approaches are possible:

(i) The histogram (or continuous density function) of samples can be modified to take into account the change in dispersion of grades of a different support (i.e., the SMUs).

(ii) Grades can be estimated for individual SMUs, and these data can be accumulated into grade–tonnage curves.

First, consider the correction to be applied to an unbiased histogram of sample grades in order to derive a histogram of grades averaged over a larger support, the SMU. The volume–variance relation (cf. Huijbregts, 1976; Parker, 1979) serves this purpose in part (cf. Fig. 12.1):

$$D^2(b/V) = D^2(s/V) - D^2(s/b) \qquad (12.1)$$

where

$D^2(b/V)$	is the dispersion variance for SMU (b) grades in the deposit (V)
$D^2(s/V)$	is the dispersion variance of sample grades (s) in the deposit (V)
$D^2(s/b)$	is the dispersion of sample grades (s) in the SMU (b).

The term $D^2(s/V)$ represents the unbiased dispersion variance of sample grades and can be determined from the data themselves or from the sill of the semivariogram. The term $D^2(s/b)$ represents the dispersion variance of sample grades in volume b and can be determined knowing the block size and the semivariogram model; the term $D^2(s/b)$ is also equivalent to the F auxiliary function of many authors and can be determined from graphs available in a variety of texts (e.g., Clark, 1979; David, 1977; Journel and Huijbregts, 1978) or with published computer programs (e.g., Clark, 1976). With the forgoing information, Eq. 12.1 can be solved for the dispersion variance of block grades throughout the deposit.

A dispersion variance of block grades also can be expressed in terms of average semivariogram values, as in the following example:

$$D^2(b/V) = \overline{\gamma}(V, V) - \overline{\gamma}(b, b) \qquad (12.2)$$

where $\gamma(x, x)$, for $x = b$ or $x = V$ is the average of all gamma values obtained for all possible pair separations in volume x (see Fig. 10.2). A good-quality semivariogram is essential to the determination of dispersion variances and the F function. It is clear from Eq. 12.2 that the dispersion variance of average grades

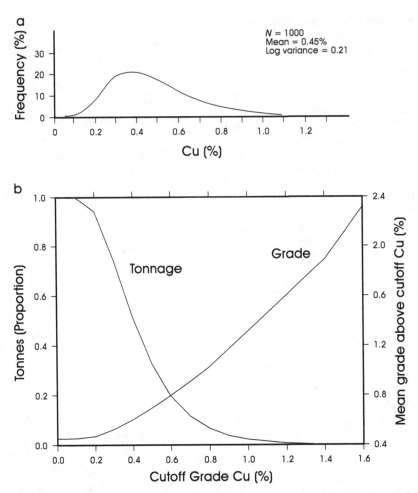

Figure 12.3: (a) Lognormal distribution of grades for the Bougainville copper mine (after David and Toh, 1988), and (b) grade–tonnage curve derived from the lognormal distribution of (a). Note that tonnage is given as a proportion because the tonnage at zero cutoff grade is not available.

of SMUs as a function of the size of the SMU. Thus, decisions relating to mining method must be made so that an appropriate SMU can be used in developing a grade–tonnage curve. At this stage, if an assumption also can be made about what happens to the shape of the probability density function in the change from a data distribution to a block–grade distribution (e.g., conservation of normality or lognormality), grade–tonnage curves can be estimated as described in Section 12.3.

When a histogram approach is to be used to construct grade–tonnage curves, the histogram of sample grades must be transformed to an estimate of the histogram of block grades based on a knowledge of

$D^2(b/V)$ (determined from Eq. 12.1) and assumptions concerning the form of the histogram of grades of blocks. A simple assumption that is commonly reasonable is the *affine correction*, which produces the histogram of grades of blocks from the histogram of sample grades by the following transform:

$$g_b = (g_s - m)[D^2(b/V)/D^2(s/V)]^{1/2} + m. \tag{12.3}$$

By this transform, each sample grade (g_s) is moved toward the mean (m) as a function of the ratio of the two dispersion standard deviations to produce a corresponding block grade (g_b). Other transforms can be used (see Journel and Huijbregts, 1978); for example, an assumption of the conservation of lognormality, which, although not true theoretically, has been shown to be useful in practice (e.g., David, 1972). Once the histogram of block grades is estimated, a grade–tonnage curve can be determined, as described in Section 12.2.

12.4.2: Grade–Tonnage Curves from Local Block Estimates

Conceptually, the simplest means of producing grade–tonnage curves is from grade estimates for an array of selective mining units of equal size. The great advantage of this procedure is that block locations are known and can be integrated into a particular mining procedure. Tonnage above any cutoff grade is easily determined by arranging the mineable SMUs in order of decreasing grade; tonnage above cutoff is $T \times n$, where T is tonnes per SMU (corrected for variable specific gravity if necessary) and n is the number of SMUs above cutoff. Similarly, average grade of

Table 12.4 Tabulated grade–tonnage information for the text example (Section 12.4.2) shown in Fig. 12.4

$g(c)$	z	$P_{>c}$	$Z[z]$	$g(m)_{>c}$
Data				
0.0		1.0		0.76
0.2	−2	0.98	0.054	0.775
0.4	−1.286	0.903	0.1745	0.814
0.6	−0.571	0.717	0.3389	0.892
0.8	0.143	0.443	0.3949	1.01
0.96	0.714	0.237	0.309	1.126
1.0	0.857	0.194	0.276	1.158
1.2	1.571	0.055	0.116	1.351
1.4	2.286	0.009	0.029	1.662
Blocks				
0.0		1.0		0.76
0.2	−2.333	0.992	0.0262	0.765
0.4	−1.5	0.936	0.1295	0.793
0.6	−0.6666	0.748	0.3195	0.862
0.8	0.1666	0.434	0.3934	0.978
0.96	0.8333	0.201	0.2819	1.096
1.0	1.0	0.157	0.242	1.13
1.2	1.8333	0.030	0.0743	1.349
1.4	2.666	0.003	0.0114	1.773

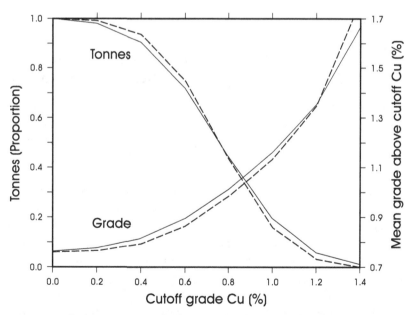

Figure 12.4: Solid curves are grade–tonnage curves for the normal grade distribution of hypothetical Cu sample data (mean and standard deviation are 0.76 percent Cu and 0.28 percent Cu, respectively) used as an example in the text. Dashed curves are grade–tonnage curves derived for the normal distribution of hypothetical block average grades (mean and standard deviation are 0.76 percent Cu and 0.24 percent Cu, respectively) used as an example in the text.

tonnes above cutoff is simply the average grade of all SMUs above cutoff. These two parameters, determined for a range of cutoff values, can then be plotted as grade–tonnage curves.

This seemingly simple procedure is not without its problems, particularly if applied at the feasibility stage. In general, early planning is based on relatively widely spaced sampling information, such as diamond-drill-hole samples; in contrast, production is based on much closer control that in many cases gives rise to selectivity of ore and waste on a much finer scale than exploration data spacing. To attempt to estimate relatively small SMUs from widely spaced exploration data results in extensive smoothing according to the relation (e.g., Huijbregts, 1976)

$$D^2(\text{SMU}^*) = D^2(\text{SMU}) - \sigma_k^2 + 2\mu \quad (12.4)$$

where $D^2(\text{SMU}^*)$ is the dispersion of estimated (kriged) grades of SMUs, $D^2(\text{SMU})$ is the dispersion variance of true grades of SMUs, σ_k^2 is the average kriging variance of the SMUs, and u is the average Lagrange multiplier (a product of kriging). When the mean grade is well known, μ is negligble and the expression reduces to

$$D^2(\text{SMU}^*) = D^2(\text{SMU}) - \sigma_k^2. \quad (12.5)$$

The foregoing relation quantifies the smoothing that results from block kriging. With limited data and small blocks, the kriging variance is large; with abundant and high-quality data, the kriging variance is smaller and dispersion of kriged block grades more closely approximates the truth. The variances considered in Eqs. 12.4 and 12.5 are global or averages and thus can be applied as corrections only to global estimates. Consequently, they can be taken into account in grade–tonnage curves, but, of course, individual blocks are not corrected.

The importance of establishing grade–tonnage curves for block estimates rather than sample estimates can be illustrated by an artificial example. Such calculations illustrate why mineral inventory estimation procedures that depend on sample-grade dispersion (e.g., polygonal estimation) rather than block-grade dispersion can result in large errors. As-

sume a normal distribution of sample data with $g_m = 0.76$ percent Cu, $D(s/V) = 0.28$, $D(b/V) = 0.24$, and $g_c = 0.96$ (Fig. 12.4). Substitute $z = (0.96 - 0.76)/0.28 = 0.714$ in Eq. 4.6 and solve to estimate $P_{>c}$ (samples) $= 0.237$ (i.e., 24 percent of the total tonnage is above cutoff grade). Substitute in Eq. 4.9 to determine the average grade of material above cutoff, $g_{m(>c)} = 1.126$ percent Cu.

Similar calculations can be done using the block-grade dispersion of 0.24 to provide $P_{>c}$ (blocks) $= 0.201$ and an average grade of 1.096 percent Cu. Suppose that these calculations apply to a block of 200,000 tonnes. The overestimation of tonnes above cutoff by using the data distribution rather than the distribution of block grades is $(0.237 - 0.201)200,000 = 7,200$ tonnes of ore. Much larger errors can arise with skewed distributions. The error in estimated metal contained in material above cutoff grade using data rather than blocks is $(0.01126 \times 47,400) - (0.01096 \times 40,200) = 533.7 - 440.6 = 93.1$ tonnes of Cu metal, a 21 percent overestimation relative to ideal metal recovery by blocks. This example is not at all unrealistic; much larger errors are possible, particularly when positively skewed distributions are present and the cutoff grade is further toward the upper tail of the distribution.

12.5: GRADE–TONNAGE CURVES BY MULTIPLE INDICATOR KRIGING

The product of multiple indicator kriging is a series of estimates on a cumulative frequency distribution for which mean and variance can be determined. For example, the resulting distribution might represent the distribution of sample grades in a $30 \times 30 \times 10$ m^3 panel. When the semivariogram is known, the volume–variance relation can be used to alter this sample distribution to a distribution representing grades of blocks of a specified size (e.g., $5 \times 5 \times 10$ m^3). Note that there are 36 such blocks in the larger panel. A cutoff grade can be applied to the small block distribution in order to estimate the proportion of such blocks that are above cutoff grade and the average grade of those "ore" blocks. Moreover, the information from similar block distributions for many large panels can be combined to produce a grade–tonnage curve for a very large volume such as the deposit itself. The general procedure of multiple indicator kriging is considered in greater depth by Journel (1985).

12.6: EXAMPLE: DAGO DEPOSIT, NORTHERN BRITISH COLUMBIA

The Dago deposit is a stratabound, massive to disseminated sulphide concentration with important precious metal content that occurs within cherty tuff and bleached andesite of the Lower Jurassic Hazelton Group, northern British Columbia (Dykes et al., 1983). Pyrite, sphalerite, galena, and lesser chalcopyrite occur as narrow, discontinuous sulphide lenses that are locally massive, but more commonly are 5 to 15 percent sulphides (by volume). The deposit has been evaluated by diamond drilling: 120 drill holes scattered through an area of about 135×270 m, which is of interest here for the gold content of core samples (5-m composites). Declustering with a variety of cell sizes (Fig. 12.5) indicates that a 45-m square cell is optimal with which to produce an unbiased histogram of gold grades. The cumulative histogram of 5-m composites (in the form of grade–tonnage information) is given in Table 12.5. Comparable information is given for $5 \times 5 \times 5$ m^3 and $10 \times 10 \times 5$ m^3 blocks in Table 12.6 and for

Table 12.5 Tabulation of grade–tonnage curve data, 16.4-ft composites, Dago deposit (see Fig. 12.6)

Cutoff (oz/t)	Average grade above cutoff	Short tons
0.010	0.037	2,158,000
0.015	0.047	1,552,000
0.020	0.054	1,266,000
0.025	0.064	969,000
0.030	0.077	724,000
0.035	0.084	624,000
0.040	0.091	541,000
0.045	0.093	522,000
0.050	0.096	488,000
0.055	0.098	461,000
0.060	0.107	394,000
0.065	0.115	333,000
0.070	0.138	225,000
0.075	0.142	212,000
0.080	0.149	191,000

kriged $10 \times 10 \times 5$ m^3 blocks in Table 12.7. Note that a high level of smoothing is to be expected in the kriging results because the data density (two dimensional) is one "sample" per 300 m^2. The corresponding grade–tonnage curves are given in Figs. 12.6 and 12.7.

It is informative to consider the magnitude of differences among the various grade–tonnage curves for

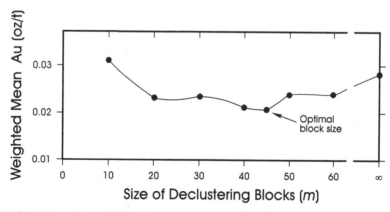

Figure 12.5: Declustered means versus cell (block) size, exploration drill data, Dago deposit. Size of square declustering blocks refers to length of a side. The weighted mean is based on the total samples in each block having a constant weight, regardless of the number of samples in the block. Optimal block size is the size corresponding to the lowest weighted mean, assuming high grades are clustered. After Giroux and Sinclair (1983).

Table 12.6 Tabulation of grade–tonnage curve data by volume variance, Dago deposit (see Figs. 12.6 and 12.7)

Cutoff grade (oz/t)	$16.4 \times 16.4 \times 16.4\ ft^3$ blocks		$32.8 \times 32.8 \times 16.4\ ft^3$ blocks	
	Tons above cutoff	Average grade above cutoff	Tons above cutoff	Average grade above cutoff
0.010	2,548,000	0.031	2,624,000	0.030
0.015	1,663,000	0.042	1,685,000	0.041
0.020	1,250,000	0.050	1,251,000	0.049
0.025	932,000	0.059	925,000	0.059
0.03	693,000	0.071	686,000	0.070
0.035	584,000	0.078	576,000	0.077
0.04	529,000	0.082	526,000	0.080
0.045	494,000	0.085	489,000	0.083
0.05	461,000	0.087	447,000	0.087
0.055	373,000	0.096	359,000	0.095
0.06	299,000	0.106	271,000	0.107
0.065	219,000	0.123	215,000	0.121
0.07	198,000	0.128	192,000	0.127
0.075	176,000	0.135	171,000	0.134
0.080	160,000	0.141	158,000	0.139

the Dago example. Consider a cutoff grade of 0.05 oz Au/st in the following table:

Volume unit	Tons	Average grade oz Au/st	Contained Au (oz)
Volume variance			
5-m composite	487,700	0.096	46,800
$5 \times 5 \times 5\ m^3$ blocks	461,200	0.087	40,100
$10 \times 10 \times 5\ m^3$ blocks	446,900	0.087	38,900
Kriged estimates			
$10 \times 10 \times 5\ m^3$ blocks	490,000	0.0711[a]	34,800

[a] Au-equivalent grade (includes Ag).

Of the estimates by the volume–variance method, the 5-m composites produce a result that is high (as expected) by nearly 7,000 oz relative to the maximum expectation for $5 \times 5 \times 5\ m^3$ blocks. Recall that the block estimate is optimistic vis-à-vis production; that is, it is based on the assumption that all the material above cutoff grade will be located (e.g., by blasthole data). A somewhat larger selective mining unit ($10 \times 10 \times 5\ m^3$ blocks) leads to a minor reduction in tonnage (−4 percent) with no significant change in average grade. The grade–tonnage curves obtained by kriging (Fig. 12.7) are not strictly comparable because they are based on an equivalent gold grade that includes the associated silver information. Both Au and Ag were kriged (ordinary kriging) independently to obtain an average Au equivalent for each block; these data are incorporated in Fig. 12.7. In general, kriging is based on an average data spacing (horizontal plane) of one sample per 300 m^2 and a spherical semivariogram model with a large nugget effect and a range of about 25 m. This produces a substantial amount of local averaging when data are clustered, with the result that tonnage above high cutoff grades is underestimated, as is the average grade above cutoff. Local estimates, regardless of the method used, generally have a large error in this case.

In summary, assuming a $10 \times 10 \times 5\ m^3$ SMU, the true reserves are somewhat less than the "ideal" volume–variance results for such blocks (446,900 tons at 0.087 oz Au/st). Ore/waste selection should not be based on the kriging results determined from exploration data, but should depend on the much more abundant data that is obtained from blastholes during

Table 12.7 Tabulation of grade–tonnage curve data by block kriging $32.8 \times 32.8 \times 16.4$ ft³ blocks, Dago deposit (see Fig. 12.7)

Cutoff (oz/t)	Average Au grade (oz/t)	Average Ag grade (oz/t)	Au equivalent (oz/t)	Tonnage
0.010	0.0306	0.58	0.044	1,651,000
0.015	0.0386	0.73	0.056	1,236,000
0.020	0.0452	0.87	0.066	985,000
0.025	0.0512	0.99	0.075	820,000
0.030	0.0563	1.10	0.083	706,000
0.035	0.0603	1.20	0.089	629,000
0.040	0.0639	1.28	0.094	569,000
0.045	0.0672	1.34	0.099	521,000
0.050	0.0709	1.41	0.104	489,000
0.055	0.0741	1.48	0.109	413,000
0.060	0.0774	1.53	0.114	413,000
0.065	0.0805	1.58	0.118	381,000
0.070	0.0844	1.65	0.124	343,000
0.075	0.0902	1.73	0.131	300,000
0.080	0.0945	1.87	0.139	262,000

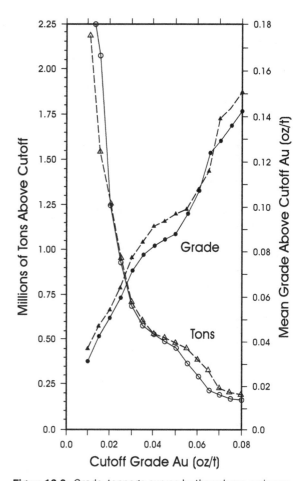

Figure 12.6: Grade–tonnage curves by the volume variance method for the Dago deposit. Triangles are for 5-m (16.4-ft) composites; circles are for $5 \times 5 \times 5$ m³ ($16.4 \times 16.4 \times 16.4$ ft³) blocks. Data are tabulated in Tables 12.5 and 12.6. After Giroux and Sinclair (1983).

production. For purposes of obtaining a more realistic "global" grade–tonnage curve for kriging estimates, the curves of Fig. 12.7 could be corrected for smoothing (Eq. 12.5) by a method comparable to a change of support (e.g., the affine correction of Eq. 12.3).

12.7: REALITY VERSUS ESTIMATES

All grade–tonnage curves contain some error, even those based on an abundance of closely spaced information. The better the quality of data, the better are the estimates and the grade–tonnage curves that result. All estimation techniques produce conditional bias to some extent. In the case of the application of a cutoff grade, these conditional biases can become important. Block estimates are based on data that generally have been smoothed to some extent. This smoothing compensates to an unknown extent for conditional bias and might overcompensate.

Grade–tonnage curves should be accompanied by explanatory information summarizing how they have been obtained and indicating likely sources of error. One error that deserves specific mention is analyti-

cal and sampling error. Selection is not based on true grades but on estimated grades. A significant sampling error can be introduced that can lead to substantial differences between what is estimated and what is produced, especially as regards local estimates. With relatively few data at the exploration stage, a large sampling and analytical error can have a significant impact on a grade–tonnage curve, generally leading to an overestimation of high-grade tonnage.

A further problem can be encountered when block selection is based on the presence of two or more value measures (e.g., Cu and Mo, in the case of many

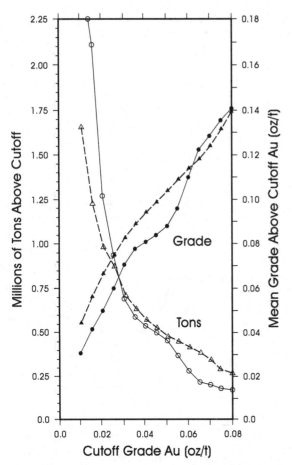

Figure 12.7: Grade–tonnage curves for $10 \times 10 \times 5$ m³ ($32.8 \times 32.8 \times 16.4$ ft³) blocks at the Dago deposit by ordinary kriging (triangles) and by volume–variance method (circles). Redrawn from Giroux and Sinclair (1983). Data are tabulated in Tables 12.6 and 12.7.

porphyry copper deposits). If the estimation procedure smooths the data, both variables are smoothed; that is, low values of both variables are overestimated. This situation leads to the problem that both variables can contribute to the "grade" of some waste-block values being inadvertently raised above the cutoff grade (*grade* here means the combined value of the two variables in question). This source of error may be significant when much of the ore and waste occur in blocks close to the cutoff grade; but is difficult to correct, other than arbitrarily. The problem has less impact when the two metals are randomly related and greater impact when the two metals are highly correlated.

12.8: PRACTICAL CONSIDERATIONS

1. Unbiased histograms of grade contain the information necessary for construction of a "relative" grade–tonnage curve (i.e., a grade–tonnage curve showing proportion of tonnes rather than absolute tonnes). When the volume represented by data and appropriate bulk density information are available, the relative curve can be made absolute.
2. The type of information (e.g., sample data, block estimates) used in the construction of a grade–tonnage curve should be documented clearly.
3. When histograms (or probability density functions) are involved, clear evidence of how they have been "unbiased" should be provided. If probability density functions are used, an evaluation of why these functions are representative of the deposit must be given. If the volume–variance relation has been used, then the details of determining block dispersion from data dispersion should be given.
4. When block estimates are used to generate a grade–tonnage curve, the block size and data density should be specified and the method of block estimation should be outlined.
5. Virtually all methods of developing grade–tonnage curves contain error, the "direction" of which might be known. A discussion of the nature of likely error in a grade–tonnage curve is a useful guide to appreciating the quality of the curve.

12.9: SELECTED READING

David, M., 1972, Grade–tonnage curves: use and misuse in ore-reserve estimation; Trans. Inst. Min. Metall., Sect. A, pp. A129–A132.

Huijbregts, C., 1976, Selection and grade–tonnage relations; *in* Guarascio, M., et al. (eds.), Advanced geostatistics in the mining industry, pp. 113–135, D. Reidel Pub. Co., Dordrecht, the Netherlands.

12.10: EXERCISES

1. Cu assays for the 1,181 blasthole samples from one bench (bench height $= 40$ ft) of the Similkameen Mine (Raymond, 1979) are given

in data files SIMILK.dbf. (a) Determine grade-relative tonnage curves for these data based on the histogram (i.e., using the method outlined in section 12.2). (b) The SIMILK.dbf file can be read by P-RES, software supplied through the publisher's website and that can be used to view the distribution, establish its form (e.g., normal or lognormal), and obtain simple statistical parameters. Knowing the normal or lognormal form, the parameters can be used in the appropriate equations from Chapter 4 to establish an idealized grade–tonnage curve; determine such a curve and compare results with the results from (a). Note that the grade–tonnage curve for specified normal or lognormal distributions can be generated directly by P-RES.

2. An adequate spherical, two-dimensional semivariogram model for blasthole Cu assays for the Similkameen Mine is isotropic with $C_0 = 0.068$, $C_1 = 0.069$, and $a_1 = 200$ ft. Coordinates in the SIMILK.dbf file are in feet. Produce grade–tonnage curves for $45 \times 45 \times 40$ ft^3 blocks using statistics of the block-grade distribution. The dispersion variance of samples in the deposit is known from Exercise 1 and should be in close agreement with the sill of the semivariogram. The semivariogram model can be used to estimate the F function [i.e., $D^2(s/b)$ of Eq. 12.4] using graphs (see Fig. 8.4).

3. Any unbiased histogram of grades can be converted into a corresponding grade–tonnage curve. Assume that the histogram of Ni grades for the Cerro Matoso deposit (Stone and Dunn, 1994,

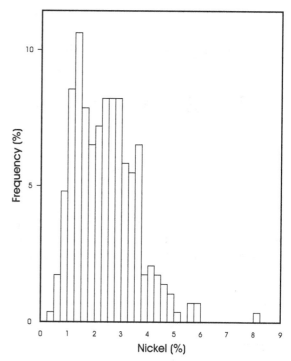

Figure 12.8: Histogram of Ni grades, Cerro Matoso nickel laterite deposit, Columbia. Redrawn from Stone and Dunn (1994).

p. 53), reproduced in Fig. 12.8, is representative of the deposit (unbiased) and pertains to a volume containing about 24 million tons. Construct a grade–tonnage curve from the histogram (recognizing that a histogram of sample grades gives an optimistic view). The same reference contains a listing of the polygonal "preliminary" reserves with from which a grade–tonnage curve could be constructed for comparison with the histogram results.

13

Local Estimation of Resources/Reserves

The comparatively low grades of many . . . mineral deposits, the escalation of mine operating costs, and the volatility of metal prices in recent years, has required improvements in short- and long-term mine planning. One of the prerequisites to improved planning systems is the need for accurate grade prediction. (Norrish and Blackwell, 1987, p. 103)

Chapter 13 is concerned with the application of geostatistical procedures to the specific problem of local resource/reserve estimation. First, *local* is defined in terms of both the autocorrelation characteristics of the variable being estimated and the size of block estimated. Some of the estimation problems/situations that are particularly pertinent to local estimation are discussed, including data location, the use of outlier values, adapting estimation to individual domain characteristics, and domain boundaries. Local estimation, particularly at the preproduction stage, involves smoothing. Some consideration is given to the problem of correcting local estimates to provide a realistic estimate of recoverable reserves.

13.1: INTRODUCTION

The term *local estimation* is not defined rigidly, but is used in the general context of point estimation or the estimation of small blocks or units on the scale of a selective mining unit (SMU). A volume within which local estimates of a mineral inventory are to be determined is generally divided into a two- or three-dimensional array of blocks of uniform size. Each block within this array is evaluated within the constraints of a specific problem. For example, a two- or three-dimensional semivariogram model is essential for geostatistical estimates; cross validation normally has been done to verify that the model is unbiased and assist in defining the search field to be used to isolate data for each local block estimate. A two-dimensional example illustrates the general problem of approximating an irregular geometric form (an ore deposit) by a regular block array (Fig. 1.1). Marginal blocks consisting of both ore and waste lead to dilution and ore loss, matters discussed further in Chapter 16.

Local estimation is normally directed toward specific goals such as detailed mine planning and grade control during production. Mine planning might be conducted without the ultimate database that will be available during exploitation, whereas *grade control* (ore identification and selection during mining) is an integral part of production operations and uses the final database available for distinguishing ore from waste. For the purposes of local estimation, these are two fundamentally different situations.

13.2: SAMPLE COORDINATES

The general introduction to kriging in Chapter 10 clearly indicates the advantage that kriging has in providing not only an optimized estimate of a block's mean grade, but also an estimate of the error that can be attached to the mean value. In global estimation, the assumption commonly is made that no errors exist in the location of samples and that all variability arises because of errors in the variable in question. Although this may be a reasonable assumption for global estimation, it is clearly not the case in local estimation when grades might be assigned to specific blocks for a final ore/waste decision. The excessive errors in sample locations that can arise because of "wandering" diamond-drill holes are legion; both dip and lateral deviations occur, and surveying of long drill holes is essential. As a rule, vertical diamond-drill holes deviate less than do inclined holes. Cummings (1980) discusses factors that contribute to drill-hole deviation and lists controls that can be implemented to minimize such deviation. Large discrepancies between planned and true drill-hole locations can occur; a 200-m drill hole whose plunge is off by only 5 degrees will have the end-of-hole displaced horizontally by about 17.4 m from its expected position. If locations are not corrected, the use of the end of the hole sample value for estimation purposes could lead to serious errors of local estimation. For example, a $20 \times 20 \times 10 \text{ m}^3$ block, thought to be centered on the end-of-hole sample of the previous example, would actually be estimated by a sample that lay well outside the block, and the estimated value might bear little relation to the true average grade of the block. It is clear that weights assigned to such seriously misplaced samples are meaningless; to counteract this problem, surveying of diamond-drill holes along their lengths is necessary if the location and assay data are to be used for local estimation.

Even blastholes can have significant location errors (cf. Konya, 1996). In particular, the burden might not be constant, blasthole spacing can vary substantially from planned distances along lines, the drilling angle might not be vertical, and a hole might be either too deep or not deep enough. Factors that contribute to location errors are not necessarily compensating in the case of local estimation. A 17-m blasthole drilled at 15 degrees from the vertical has its centroid misplaced by about 1.1 m vertically and 4.4 m laterally from its planned position. Such a sample might be re-

moved significantly from its anticipated position for local block estimation, depending on the autocorrelation model and other data available. Any weighting method, including kriging, could be carried out with weights that are seriously in error if samples are not located well in space. Local estimates based on both blasthole and exploration data are particularly susceptible to errors that arise from incorrect locations of data, if the block to be estimated is very small and the incorrect locations are closest to the block.

13.3: BLOCK SIZE FOR LOCAL ESTIMATION

> While it is well known that a smoothing estimation method such as kriging is not suitable to produce realistic estimates in blocks which are far away from the sampling information, it is unfortunately too often used for this purpose. (Ravenscroft and Armstrong, 1990, p. 579)

Linear estimation, particularly that based on widely spaced data relative to the size of block to be estimated, is subject to two competing effects, conditional bias and smoothing (Pan, 1998), topics discussed in detail in Chapter 10. Of these, smoothing imposes limits on the size of block that can be estimated effectively because sparse, widely spaced data produce comparable estimates for small adjacent blocks that do not reflect the true local variability of small block grades.

Local reserve estimation at the feasibility stage commonly uses widely spaced exploration data as a basis for estimating average grades of relatively small blocks. This undertaking is almost too easy, with the ready access to kriging software for personal computers, because it can lead to substantial errors if estimates are made for blocks whose dimensions are small relative to sample spacing. Armstrong and Champigny (1989) address the following question: How small a block size is too small? They examine variations in the kriging variance, dispersion of estimated grades, slope of the regression of the true but unknown grades on estimated values, and the coefficient of correlation between actual grades and

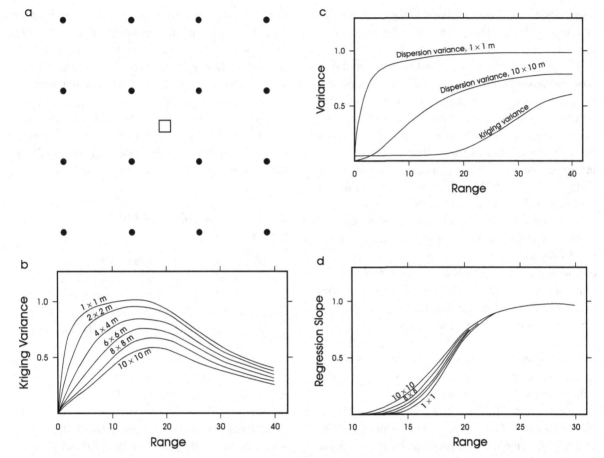

Figure 13.1: (a) A square data array (dots with a 10 × 10 m² unit cell) with a central small square, the size of which can vary. This fixed data array is the basis for calculations discussed in the text. (b) The block kriging variance has been estimated for many block sizes, centered at the same point using a semivariogram model (described in the text) for which the range was varied. (c) Dispersion variance of the kriged estimators and kriging variance as a function of range of the semivariogram. (d) Slope of the linear regression of the actual grades on the estimated grades for various block sizes. Redrawn from Armstrong and Champigny (1989).

estimates. These four quantities were calculated for several simple, two-dimensional geometries of data versus block (e.g., Fig. 13.1) using a spherical semivariogram model with a sill of 1.0, no nugget effect, and a range varying from 0.1 to 40 m (twice the sample spacing). Important generalities from their study are as follows:

(i) The kriging variance for block estimates changes significantly as a function of block size and range of the semivariogram.

(ii) The kriging variance differs fundamentally from

the dispersion variance and the two should not be confused.

(iii) The dispersion variance of block grades varies widely, depending on block size and the range of the semivariogram.

(iv) The slope of the regression of actual grades Z on the estimated grades Z_k^* and the correlation coefficient should both approach 1.0. Results show this to be the case only if the semivariogram range is at least as great as the average sample spacing. When the range is small relative to sample spacing, small-block estimates are "virtually useless for distinguishing ore from waste" (p. 129) and "should not be used for predicting recoverable reserves" (p. 133).

(v) Improvements in quality of block estimates can be achieved by using increased amounts of surrounding data for block estimations if the semivariogram range is close to the sample spacing (the difference in kriging error from adding a few fringing data is minor).

(vi) When the range is greater than 1.5 times the sample spacing, block estimates (for block sizes at least half the sample spacing) are adequate to distinguish ore from waste.

These results by Armstrong and Champigny (1989) lead to the following guidelines for successful small-block kriging:

(i) Data spacing should be substantially less than the range of the semivariogram, perhaps no more than two-thirds the range.

(ii) Block dimensions should be as large as possible and not much smaller than the average data spacing. A rule of thumb is that each block to be estimated should contain at least one sample (cf. Ravenscroft and Armstrong, 1990).

Ravenscroft and Armstrong (1990) document an example in which 10-m spaced data, extracted from a set of real data spaced at 2 m, are used to estimate 2-m blocks. Kriged estimates of 2-m blocks using all available data are assumed to represent the "real" block values to which estimated values are compared. In this case, they show that for cutoff values below the mean, kriging overestimates tonnage; for cutoff values above the mean, kriging underestimates tonnage. At relatively high cutoffs, kriging estimates 5 percent of blocks as ore instead of the 12 percent known to be ore! Moreover, kriging consistently underestimates average grade above cutoff, regardless of cutoff value. Ravenscroft and Armstrong (1990) conclude the following:

[K]riging of small blocks from sparse data will always over-estimate the recoverable tonnage for a cut-off value below the mean, and underestimate this tonnage for a cut-off higher than

the mean. The method will also always underestimate recovered mean grade, whatever the cut-off value. (p. 586)

An important consequence of the forgoing results is that kriging generally is not a sound basis for the direct estimation of individual selective mining units (SMUs) from widely spaced exploration data. Instead, a variety of techniques have been developed to determine "recoverable" reserves from relatively limited exploration data. One of the general approaches to this problem involves obtaining a kriging estimate of a large panel and a separate estimate of the dispersion of small-block SMU grades within this large panel. This dispersion of block grades is then centered on the kriged mean to provide an estimate of the distribution of small-block SMU grades. Of course, the locations of these individual ore-grade SMUs within the large panel are not known, nor is the form of the probability density function of the SMU grades; assumptions are necessary.

13.4: ROBUSTNESS OF THE KRIGING VARIANCE

Kriging is an estimation procedure that minimizes the estimation variance. In many estimation projects, estimates are required before data gathering (exploration/evaluation) is complete, and the semivariogram model may be determinable only within limits. In such cases, it is important to have an understanding of the impact that errors in semivariogram parameters have on the kriging variance. Brooker (1986) conducted a study of the robustness of kriging variance to errors in these parameters. For his work, he assumed the following:

(i) An isotropic spherical model with a sill level of 10 units (i.e., $C_0 + C = 10$)

(ii) A regular square data array (5×5 units) with which to estimate a central block containing a central datum.

Kriging variances were then determined for semivariogram models corresponding to a broad spectrum of orebody models by varying the relative nugget effect (C_0/C) from 0 to 2 and by varying the range a from 1 to 10 units where these units are equivalent to the data-array units. With the resulting values as a base, specific percentage changes can be imposed on any one model parameter and a percentage change in kriging variance determined. An example of Brooker's results is shown in Fig. 13.2 that illustrates some of his principal conclusions:

> [K]riging variance is robust to most errors likely to be made in semivariogram models selection for the widely used spherical model and this [described previously] block-sample geometry. However, if a nugget effect near zero is selected instead of a proper nonzero value, kriging variances can be understated significantly. Thus the extra effort needed to define this parameter well, perhaps by a series of closely spaced holes, or by deflection holes, can be warranted. (Brooker, 1986, p. 487)

The magnitude of possible errors in range and relative nugget effect are dependent on the nature of available data. When data are collected with a view to aiding the semivariogram modeling process, large errors in the parameters of a model are unlikely. Unfortunately, this is not always the case, and there are far too many examples of highly approximate semivariogram models being used as the basis for geostatistical studies. Clearly, it is essential to take full advantage of all available information. Exploration commonly results in a variety of data types, all of which can contribute to the modeling process. In particular, trench assays and close-spaced data along exploration drives can provide insight into the nugget effect and the range in specific directions, even when most data for estimation purposes are of another type (e.g., diamond-drill core). Similarly, the common practice of defining a semivariogram model for a few fixed-grid directions (e.g., principal grid directions only) is generally inappropriate because it commonly ignores two important contributors to model quality: geology and preferential directions of linear samples (e.g., along drill holes that cross the direction of principal geologic continuity).

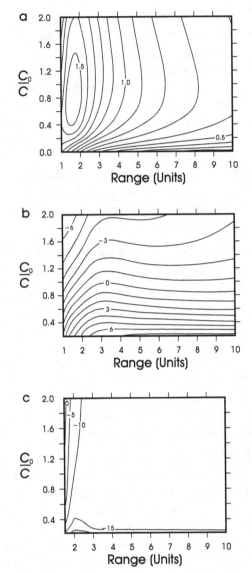

Figure 13.2: Examples of the robustness of the kriging variance. (a) Kriging variance as a function of range and relative nugget effect. (b) Percentage change in kriging variance if relative nugget effect of the semivariogram model is changed by +25 percent. (c) Precentage change in kriging variance if range is changed by +25 percent. See text for details. Redrawn from Brooker (1986).

13.5: BLOCK ARRAYS AND ORE/WASTE BOUNDARIES

Where a block array of SMUs is superimposed on an ore/waste boundary, individual blocks, in general,

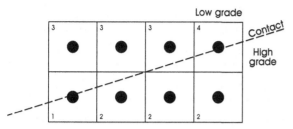

Figure 13.3: A block array in two dimensions superimposed on a sharp ore/waste boundary. Blocks straddling the boundary are composed of both ore and waste. Their classification as ore leads to dilution; their classification as waste leads to ore loss.

are a combination of both ore and waste material (Fig. 13.3). This configuration leads to dilution and loss of ore, both of which result in less metal recovered per tonne mined. The problem is minimized by using the smallest blocks possible, as illustrated by the following example (Fig. 13.4). Consider a single block, $20 \times 20 \times 10$ m^3 (ca. 12,000 tonnes), a principal diagonal of which is the ore/waste contact. If the half that is ore has a grade more than twice the

cutoff grade, the block is classed as ore when diluted by half a block of zero grade. In this case, dilution results in a doubling of tonnes and a 50 percent loss of grade. If the half that is ore is between one and two times the cutoff grade, the overall block grade is below cutoff and the block is classed as waste, in which case 6,000 tonnes of ore is lost to waste.

Now, assume that the selection size is $5 \times 5 \times 10$ m^3 (ca. 750 tonnes each). Four such blocks aligned along the ore/waste contact lead to much less dilution or ore loss than does the larger block. For one such SMU, the dilution amounts to 375 tonnes (i.e., $4 \times 375 = 1500$ tonnes for the four blocks that cover the ore/waste boundary). Similarly, if all four blocks were classed as waste, the ore loss would total 1,500 tonnes, much less than the 6,000 tonnes for the larger block. This example applies to sharp and simple ore/waste boundaries; the situation is somewhat more complicated when boundaries are gradational.

A variety of methods have been introduced to characterize gradational ore/waste or domain boundaries (cf. Chapter 2), including individual profiles of linear sample grades that cross a boundary, average profiles, geochemical contrast, and x–y plots of grade pairs located at comparable distances from the contact but one in ore and the other in waste. An idealized grade profile across an ore/waste boundary is shown in Fig. 13.5, where the width of the ore/waste boundary zone is indicated. Information of this sort

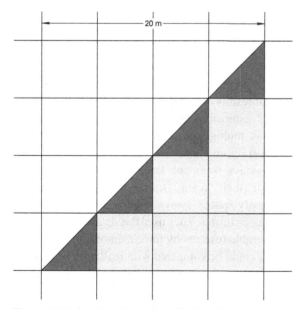

Figure 13.4: As the block size of the selective mining unit decreases, dilution and loss of ore decrease at an ore/waste boundary. This example, described in the text, compares dilution and ore loss for 20×20 m^2 blocks (pale plus dark patterns) and 5×5 m^2 blocks (dark pattern).

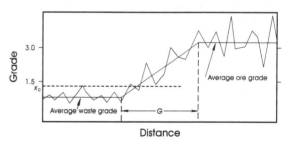

Figure 13.5: An idealized grade profile across a gradational ore/waste boundary. The sawtooth curve represents individual grades that have been smoothed to the average linear models shown in each domain. G is the width of the gradational boundary zone. How data are treated during block estimation depends on the relative dimensions of G and the block.

is commonly available from diamond-drill holes that extend beyond the limits of a domain. For practical purposes, the average grade profile across a gradational boundary zone is more informative because the model for a domain boundary zone must be integrated into an estimation procedure that is acceptable for all locations of the boundary. Where the width of the gradational boundary is small relative to the block size being estimated, the boundary can be treated as a sharp boundary (i.e., data on one side of the boundary cannot be used to estimate blocks on the other side). Where the gradational contact is very wide relative to the dimension of the blocks to be estimated and is characterized by substantial local fluctuations of grade, a single estimation procedure is acceptable across the zone. Where a well-defined gradational zone is wider than the blocks to be estimated, it might be appropriate to attempt to define the gradational boundary as a transitional zone in which blocks are estimated using only data within the zone.

The foregoing discussion is idealized. Practicing geologists recognize that it might be easy to define an ore/waste (domain) boundary zone in individual drill holes; a common problem is to predict confidently the position of such boundary zones between control points that are widely spaced both horizontally and vertically (e.g., between drill sections). This interpolation problem results in geometric errors of interpretation, considered in Chapter 2.

13.6: ESTIMATION AT THE FEASIBILITY STAGE

13.6.1: Recoverable "Reserves"

Recoverable reserves are defined as that mineable volume for which each block (normally the SMU) is above a specified cutoff grade. Many methods are described in the literature for estimating recoverable reserves; the problem can be treated locally or globally. The smoothing of block grades (i.e., generating groups of blocks with very similar grades) that is endemic in many estimation procedures using widely spaced exploration data is well documented in the literature and led to the development of a variety of methods that try to reproduce local block-grade

distributions that are realistic. These procedures are known collectively as *estimation of recoverable reserves* (e.g., Journel, 1985d). The terminology is somewhat unfortunate because the methods have been widely applied to what are now referred to as *resources*. The smoothing problem decreases as the data spacing decreases (i.e., adding infill data) and the block size increases (i.e., increasing the scale of selection).

Concern about smoothed estimates is particularly relevant at the feasibility stage of deposit evaluation because it is then that estimates form the basis of major investment decisions. For kriged results, the net impact of smoothing (and conditional bias) can be quantified (e.g., Pan, 1998). Many studies, including those of Armstrong and Champigny (1989) and Ravenscroft and Armstrong (1990), as discussed previously, emphasize the impracticality of estimating grades of very small blocks using relatively sparse data. Alternative approaches to offsetting the smoothing effect include the following:

  (i) Volume variance
 (ii) Conditional probability (Raymond, 1979)
(iii) Multiple indicator kriging (Journel, 1983)
(iv) Disjunctive kriging (Matheron, 1976).

Guibal and Remacre (1984) use blasthole data from a porphyry copper deposit to demonstrate the similar results obtained from several nonlinear methods of estimating recoverable reserves (disjunctive kriging, multigaussian kriging, and uniform conditioning). From a database consisting of 2,095 blasthole assays from one level of a porphyry copper deposit (their "reality"), they selected 173 roughly uniformly spaced assays to serve as a database for making estimates. They used this database to estimate recoverable reserves by three methods, the results of which could be compared with reality. In brief, their principal conclusions are as follows:

  (i) Tonnage is slightly overestimated for very low cutoff grades and overestimated for high cutoff grades.
 (ii) Quantity of metal is underestimated for all cutoff grades, but disparity with reality is greatest for high cutoff grades in rich areas of the deposit.

(iii) Overall results are generally acceptable except for very high-grade blocks. The disparity between estimates and reality for high-grade blocks is attributed to a lack of strict stationarity in the deposit.

This study is important for three reasons: (i) it shows that theoretical approaches can overcome the smoothing problem and provide realistic estimates of recoverable reserves, (ii) it indicates the consistency of several independent methods of estimating recoverable reserves, and (iii) it emphasizes the fact that mineral deposits generally are not "strictly" stationary, and departures of estimates (of recoverable reserves) from ideal models are to be expected.

Of the many methods of estimating recoverable reserves, multiple indicator kriging is described in Chapter 10, and other nonlinear methods such as disjunctive kriging are beyond the aims of this text. Comment here is restricted to two topics, the volume-variance approach used to provide conceptual insight into procedure, and conditional probability as practiced by Raymond (e.g., 1979).

13.6.2: Volume–Variance Approach

A simple procedure for estimating recoverable reserves involves the estimation of the mean value of a large panel, a corresponding estimate of the dispersion of sample grades in the panel, and a change of support procedure to produce a block-grade distribution within the panel. The procedure is similar to that described in Chapter 12 as a global resource-estimation method. Following is a brief summary of the procedure directed toward the estimation of local recoverable resources.

1. Consider a single large panel, perhaps $80 \times 80 \times 5$ m^3, for which a mean value is estimated by ordinary kriging using contained and immediately surrounding data.
2. Choose an appropriate SMU (e.g., $10 \times 10 \times 5$ m^3). Thus, the panel contains 64 blocks in this example.
3. Assume an appropriate form for the unbiased histogram of data (e.g., lognormal). Assume that log-

normality is preserved in deriving the distribution of block grades, an assumption that is not strictly true but that has been found adequate in many practical cases.
4. Determine the F function for a $10 \times 10 \times 5$ m^3 block (F_b). This requires that a high-quality semivariogram model be established for the deposit.
5. Calculate the dispersion of block grades as in equation 12.5.
6. In general, a change of support model (e.g., affine correction) is required to transform the sample distribution to a block-grade distribution. In the simplest case, such as assumed here in Step 4, the change of support is incorporated in the assumption that the blocks have a lognormal distribution.
7. Now that the block distribution is fully defined, a cutoff grade can be applied and the proportion of blocks above cutoff identified. Moreover, the average grade of these ore blocks can be estimated (appropriate equations are in Chapter 4).
8. Repeat the procedure for all panels.

Suppose that when a cutoff grade is applied to the distribution of block grades in a panel (Step 8), 23 percent of blocks are indicated as being above cutoff grade. This is equivalent to $0.23 \times 64 = 19.3$ blocks with an average grade that can be estimated from equations in Chapter 4. Note that a similar calculation can be done on the same panel for many cutoff grades so that a grade–tonnage curve can be estimated for the panel. Combining comparable grade–tonnage curves for other panels allows a grade–tonnage curve for the entire deposit to be constructed.

Of course, it must be realized that this approach is probabilistic, and that the positions of individual $10 \times 10 \times 5$ m^3 ore blocks within a panel are not known. From a practical point of view, the exact locations of ore blocks and their grades might not be critical. Where precise block values are required for some pit-optimization programs, one solution is to simply assign block grades arbitrarily such that the known distribution of block grades is reproduced.

Parker et al. (1979) present a detailed description of a somewhat more complex approach that can be compared to the foregoing procedure: an application to the estimation of recoverable reserves for the Imouraren uranium deposit, central Niger.

13.6.3: "Conditional Probability"

Raymond (1979, 1982, 1984) developed a practical approach to estimation using widely dispersed exploration data. His method, referred to as *conditional probability*, is an empirical attempt to estimate local, sample-grade distributions that approximate expected recoveries. The method itself is not of concern here; the interested reader is referred to a detailed account concerning the Valley porphyry copper deposit (Raymond and Armstrong, 1988).

However, an important concept in Raymond's work is worth illustrating – the idea of a *conditional distribution*. Consider an array of $10 \times 10 \times 10$ m^3 cells superimposed on a zone of trial surface mining for which widely spaced exploration data exist. Each node of the array of cells can be estimated from two different data sets, the exploration data, and the ultimate database (blastholes for trial mining). Figure 13.6a is such a plot for many hundreds of paired estimates. From the figure, it is evident that blasthole estimates have a much larger spread of values than do the exploration estimates (an example of smoothing). Consider a slice from Fig. 13.6b for a narrow range of exploration estimates, 0.425 ± 0.025 percent Cu (Fig. 13.6a). Remember that the blastholes represent the ultimate data on which ore/waste selection is made. Hence, the slice represents a close approximation of the true grade distribution, conditional on the kriging estimate being 0.425 ± 0.025 percent Cu. This "conditional" distribution is shown as a histogram in Fig. 13.6d. Suppose a cutoff grade of 0.4 percent Cu is applied to this distribution. About 11 percent of the blocks classed as ore using the exploration estimates are seen to be waste based on the blasthole estimates.

13.7: LOCAL ESTIMATION AT THE PRODUCTION STAGE

The procedure followed for routine grade estimation at the production stage (part of grade control) is that

outlined in Section 10.4. Estimation, although generally based on relatively abundant data (e.g., blasthole assays), is not routine everywhere and constant monitoring is necessary in order to recognize and deal with special circumstances. A few of these special situations are discussed in the following subsections.

13.7.1: Effect of Incorrect Semivariogram Models

The effect of incorrect semivariogram models on the quality of block estimates is strongly dependent on the local situation and the nature and degree of the "incorrectness." The robustness of the semivariogram is discussed in a previous section, and the general conclusion is that substantial errors can occur in semivariogram modeling without producing drastic changes to the quality of block estimates. Nevertheless, there are practical situations in which estimates can be greatly improved during production with improved quality of the semivariogram model. This is particularly the case when semivariogram models are anisotropic.

Sinclair and Giroux (1984) document an interesting case for the South Tail zone of the Equity Silver deposit, central British Columbia. Blasthole data for the 1310 bench were used to construct a general, two-dimensional, semivariogram model for the bench; this strongly anisotropic model was used to demonstrate a production-kriging procedure for the mine. In the course of this study, it became apparent that the northerly direction of greatest overall continuity contrasted sharply with geologic evidence and grade contours for the northern part of the zone. Independent semivariogram modeling was possible in this small zone because of abundant blasthole data; the direction of greatest continuity was found to be roughly at right angles (i.e., easterly) compared with the general model (see Fig. 3.3).

A rough idea of the impact of using the incorrect, semivariogram model as opposed to the correct model to make block estimates can be appreciated with a specific example. A subset of the real data for the small northern domain is shown in Fig. 13.7, with a superimposed array of 25 blocks. Each block is

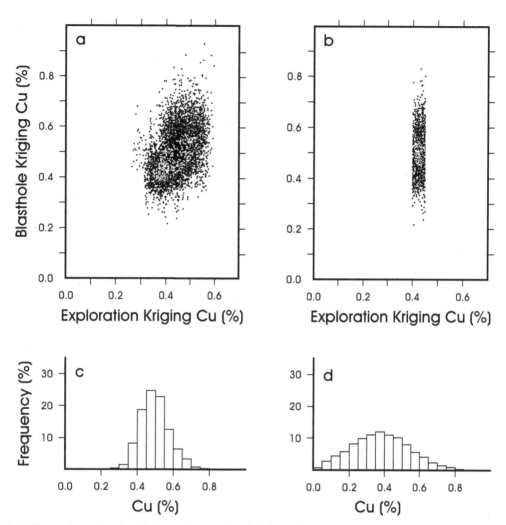

Figure 13.6: (a) Comparison of exploration-based versus blasthole-based estimates by ordinary kriging of a three-dimensional array of nodes spaced at 10 m and coinciding with a trial mining zone, Valley copper deposit. Redrawn from Raymond and Armstrong (1988). Note the difference in dispersion of the two types of estimates. (b) A slice from (a) for a narrow range of grades (0.425 ± 0.025 percent Cu) estimated by ordinary kriging, using exploration data. Note the large dispersion of true grades determined by kriging using blasthole data. The true distribution is conditional on the exploration estimate being 0.425 ± 0.025 percent Cu. (c) A histogram of the data in (a) projected on the x axis. (d) A histogram of the data in (b) projected on the y axis (i.e., the conditional distribution of true grades, given that the exploration kriged estimate is 0.425 ± 0.025 percent Cu).

$5 \times 5 \times 5$ m³ and is equivalent to about 360 metric tonnes. The blocks can be estimated using the general (incorrect) and the correct semivariogram models; results are shown in Fig. 13.7b and 13.7c, respectively. Of course, 25 blocks hardly represents a large sample of this northern domain of the South Tail zone; nevertheless, a rough idea of the impact on estimation

of an incorrect semivariogram model can be obtained by a simple deterministic calculation summarized in Tables 19.3 and 19.4. In brief, with a cutoff grade of 50 g Ag/t, the incorrect model selects nine blocks as ore, with an estimated mean grade of 65.2 g Ag/t. A better estimate of these nine blocks is obtained with the correct semivariogram model, which gives an

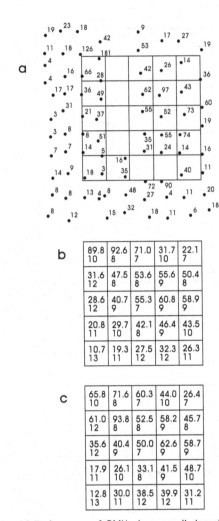

b

89.8 10	92.6 8	71.0 7	31.7 10	22.1 7
31.6 12	47.5 8	53.6 8	55.6 9	50.4 8
28.6 12	40.7 9	55.3 7	60.8 9	58.9 9
20.8 11	29.7 10	42.1 8	46.4 9	43.5 10
10.7 13	19.3 11	27.5 12	32.3 12	26.3 11

c

65.8 10	71.6 8	60.3 7	44.0 10	26.4 7
61.0 12	93.8 8	52.5 8	58.2 9	45.7 8
35.6 12	40.4 9	50.0 7	62.6 9	58.7 9
17.9 11	26.1 10	33.1 8	41.5 9	48.7 10
12.8 13	30.0 11	38.5 12	39.9 12	31.2 11

Figure 13.7: An array of SMUs in a small domain at the north end of the South Tail zone of the Equity Silver Mine, Houston, British Columbia (a) Locations of 5-m blasthole data (g Ag/t) and block outlines. (b) Block kriging estimates (top) and number of blasthole samples used (bottom) based on incorrect semivariogram model. (c) As in (b), except that the correct semivariogram model was used for kriging.

are in common with classification using the incorrect model. These 10 blocks have an estimated mean value of 63.5 g Ag/tonne. It is possible to evaluate the metal profit that results from using the correct semivariogram model rather than the incorrect model. Calculations, summarized as a metal balance in Table 19.4, show that for the 25 blocks ($25 \times 360 = 9,000$ tonnes) considered here, the correct model provides a net gain (increase in true operating profit relative to the nine blocks identified as ore by the incorrect model) of more than 21,000 g Ag for the 10 blocks classed as ore. Note that operating costs have been taken into account by subtracting the cutoff grade from the block-average grades.

A second example of the effect of different semivariogram models is shown in Fig. 13.8, in which two comparisons are made for different approaches to block estimation at the Cinola deposit, Queen Charlotte Islands, British Columbia (cf. Champigny and Sinclair, 1984). The upper diagram compares two two-dimensional kriging estimation procedures; one assumes isotropy, the other takes account of anisotropy. The two two-dimensional approaches can be seen to give equivalent results. However, if anisotropy in the third dimension is taken into account, the two-dimensional estimates are seen to contain a proportional bias. Postolski and Sinclair (1998a) describe a similar case history for the Huckleberry porphyry copper deposit, central British Columbia.

13.7.2: Spatial Location of Two-Dimensional Estimates

In many two-dimensional cases (e.g., veins, beds) involving estimation of thickness and accumulation, there is the added complexity that the locations in space of the estimated quantities are not known and are not a product of the estimation of thicknesses, accumulations, or grades. The problem is easily appreciated if the horizontal trace of a tabular deposit is sinuous on two levels; because of a comparable sinuosity between the two levels, the exact location of small blocks comprising the deposit is not known. For example, sudden or systematic changes in orientation of the ore sheet are not evident purely from thickness

estimated mean value of 58.4 g Ag/t; the wrong model appears to overestimate blocks classed as ore by more than 10 percent and can lead to problems in reconciling production with estimates. In this case, the two kriging estimates of Ag content differ in profit by about 22,000 g of silver.

However, the correct semivariogram model leads to selection, as ore, of a slightly different subset of the blocks; 10 blocks are classed as ore, 8 of which

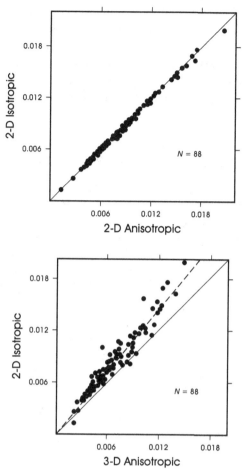

Figure 13.8: A comparison of different approaches to ordinary block kriging for the Cinola epithermal gold deposit, Queen Charlotte Islands, British Columbia. (a) Two-dimensional block estimates assuming horizontal isotropy versus estimates based on an anisotropic semivariogram model. (b) Two-dimensional block estimates versus estimates based on a three-dimensional semivariogram model. Note the tendency for a proportional overestimation by the two-dimensional estimation approaches.

and accumulation estimates. This contrasts with the estimation of blocks representing a two-dimensional bench in which the position in space of each block estimated is fixed and known.

This problem can be met in a variety of ways. For example, a deterministic interpolation (e.g., linear, quadratic) can be used that experience has indicated to be adequate. Such an approach generally is acceptable where the sinuosity of the tabular deposit is not pronounced. Various interpolation methods (e.g.,

such as those used for grade interpolation) are also possible; kriging of the distance from block centers to a reference plane is one possible solution. In flat-lying bodies, the reference plane may be horizontal; in steeply dipping bodies, the reference plane might be vertical (e.g., a particular longitudinal section). In the general case of a moderately dipping, tabular body, the reference plane might be one with the mean orientation of the tabular sheet, passing near or partly through the tabular sheet. A complete additional estimation undertaking is required that involves variography and kriging so that the process is not a trivial one. The previously mentioned approach could be applied separately to the hangingwall and the footwall, doubling the computational effort and leading to the possibility that the two estimated margins might cross in space. More complicated approaches to this problem have been attempted (e.g., Dowd, 1990), but it is not clear that they are warranted, unless there are special circumstances.

13.7.3: Planning Stopes and Pillars

Local estimates are important for mine planning and an infinite number of situations can be imagined. Here, a simple case history is described to illustrate concepts, methods, and problems. The example is for the Whitehorse copper deposit, Yukon, a roughly vertically dipping tabular zone of copper skarn that was nearing the end of productive life. Planning was underway to develop the lowest known extension of the deposit (Sinclair and Deraisme, 1976). The general scenario is illustrated in the vertical projection of Fig. 13.9. Alternating stope and pillar positions were to be estimated from the information base of limited diamond-drill data. For example, estimates could be made for the stope–pillar array shown in Fig. 13.9. Estimates can be made by shifting the stope–pillar pattern horizontally to various locations and for each position, the estimated copper content of all the stopes can summed. Hence, the particular stope–pillar array that maximizes the copper content of stopes can be determined. The important concept here is that production design can be optimized by comparing the estimated metal contents for various design scenarios. The practical problems in this example are twofold

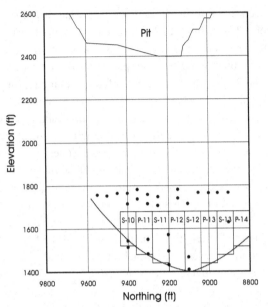

Figure 13.9: A planned, alternating sequence of stopes and pillars for the deepest part of the Whitehorse copper deposit. Dots are nearly horizontal drill holes, the assay data for which are the assay base for estimating metal content of the stopes and pillars for planning purposes. The dotted line is the assumed extent of the copper skarn. Modified from Sinclair and Deraisme (1976).

(see Fig. 13.9): First, the drill database is so sparse that errors for individual stopes are large; second, the limit of mineralization is highly speculative. Clearly, more data are required.

13.8: POSSIBLE SIMPLIFICATIONS

13.8.1: Block Kriging with Bench Composites

A common procedure in determining reserves for an open-pit mine is to produce bench composites of drill-hole samples to be used as the information base for estimates and use two-dimensional kriging to obtain block–grade estimates. The procedure entails the formation of composites that extend from the top to the floor of the bench; hence, for an inclined hole subdivided into many contiguous samples, all samples totally contained within the bench plus parts of the two extreme or encompassing samples are weighted to form the composite value. This procedure is compli-

cated, introduces a small component of oversmoothing of the composites, and is unnecessary (see also Chapter 5). A sounder and more efficient approach is to decide on a standard length of composite (preferably a multiple of the sample length and less than or equal to the bench height) and use three-dimensional kriging for block-estimation purposes.

13.8.2: Easy Kriging with Regular Grids

When a highly regular data grid (e.g., blasthole array) exists, a block can be defined by data at the block corners and an acceptable kriging estimate for the block is the average of the four corner samples. David (1988, p. 115) states that such a procedure is acceptable even with locally missing data; such missing data can be simulated by averaging the three nearest neighbors.

13.8.3: Traditional Methods Equivalent to Kriging

It is commonly possible to determine, fairly efficiently, an inverse-distance weighting (IDW) estimation procedure that is closely equivalent to kriging. For example:

(i) Select a trial block array (e.g., $5 \times 5 \times 4$) and estimate each block by kriging and by a variety of IDW approaches (i.e., various exponents).
(ii) Produce x–y plots of each set of IDW estimates versus the corresponding kriging estimates.
(iii) Fit a reduced major-axis regression to the results.

An exponent that leads to a slope of 1 with little scatter about the line is a reasonable IDW estimator that produces results comparable to kriging.

Raymond (1979) found that for the Similkameen porphyry copper deposit, an inverse cubed distance estimator based on weights that were a function of $1/(d^3 + k)$, where d is distance (sample to block center) and k is an empirically determined constant, provided block estimates are essentially equivalent to block kriging estimates. Clearly, trial and error is necessary to arrive at such a procedure.

13.9: TREATMENT OF OUTLIERS IN RESOURCE/RESERVE ESTIMATION

Outliers have been treated in an infinity of ways during resource/reserve estimation (e.g., Bird, 1991; David, 1979; Journel and Arik, 1988). Among the procedures are a group known as *cutting* or *capping*, in which all values above a particular threshold are reduced (cut) to that threshold value and are then incorporated into the estimation process. In gold deposits, for example, it is common to reduce extreme assays to an arbitrary value, perhaps 30 g/t. An arbitrary procedure of this nature requires some production data so that an estimation versus production reconciliation can be done. If the reconciliation is poor (e.g., metal produced is very different from metal estimated), a different cutting factor might be necessary. The philosophy of cutting is described in detail in Chapter 7. In practice, a variety of totally arbitrary threshold selection procedures have been recommended (e.g., Bird, 1991), and there is no practical way to choose the most useful of these until production data are available for comparison with estimates.

Geostatisticians have considered the treatment of outlier values in estimation (as outlined in Chapter 10). Traditionally, their methods have not relied heavily on geology. The widely used procedure of extracting outliers from a data set to develop a semivariogram model and reinserting the outlier values for estimation by kriging is flawed geologically. Restricted kriging is useful as an automated procedure, but assigns arbitrary weights that can lead to large errors in local block estimates. Even multiple indicator kriging does not apply to true outliers. The geologic characteristics of outliers suggest the following procedure for their incorporation into mineral inventory estimation:

1. Verify that outlier values are not errors. Examine the drill core and rock faces from which outlier values were obtained. Attempt to characterize outlier values in terms of geologic features. Also characterize the three-dimensional shape of outliers in both geologic and sample terms geometry. Remember that a 1-cm-thick vein of native gold is spread through the sample of 1 m. Hence, the physical dimension of the outlier is 1 m. The aim of this study is to determine an average geometry of outlier values and the average grade of the outlier population.

2. Experience suggests that average outlier geometries are much smaller than the dimensions of blocks to be estimated. Hence, outlier grades should be extended only to the blocks in which they occur, unless field examination suggests otherwise.

3. A block model should be estimated by whatever interpolation method is to be used but omitting the outlier values.

4. All blocks containing outlier assays must now be corrected. A weighted average for such a block is obtained as follows:

$$g_w = (B - v_o) g_b + v_o g_o$$

where

g_w	is weighted block grade
B	is block volume
g_b	is block grade estimated by selected interpolation method
v_o	is average volume of an outlier
g_o	is average grade of outlier population.

The procedure assumes that both outlier and lower-grade material have the same bulk density. In addition, there is an assumption that the specific outlier grade is not a good measure of the contribution the outlier makes to a block grade; instead, the mean value of the population is attributed to the outlier. Finally, a single intersection of an outlier provides little information as to volume (although exceptions exist), so the average size of an outlier is used for estimation purposes.

It is evident that the use of average outlier character (size and grade) lead to substantial errors in some higher-grade block estimates. However, the chances are that such blocks will be above cutoff grade, so ore is not likely to be lost because of these errors. In addition, some blocks contain outliers that have not been intersected (e.g., between drill sections), and thus are not represented in the sample database; on average, such blocks are underestimated, and some of these blocks might be ore that is inadvertently sent to waste because of the underestimation. The principal advantage of the procedure lies in the fact that outlier values are not spread widely to produce large quantities of false metal not found during production.

13.10: PRACTICAL CONSIDERATIONS

1. Data location is of fundamental importance in local estimation. Surveying positions of sample sites can be essential to success in estimation.
2. Selection of block size can be a conundrum. Selectivity of the mining method commonly places a lower limit on block size. However, the disposition of data themselves is commonly such that only relatively large blocks can be estimated with an acceptably low estimation error.
3. Where blocks impinge on ore/waste margins, the smaller the block size, the less the contact dilution.
4. Where data are especially widely spaced, it is likely that local estimation of individual blocks will be highly smoothed and unrepresentative of local grade variations. In such cases, the estimation method must be tailored to the situation if a reasonable estimate of potentially recoverable resources is to be obtained. One widely used approach is a variation of the volume–variance method described in Chapter 11.
5. It is important that semivariogram models be monitored during production. A model developed early in the production history of a deposit might be inappropriate for a part of the deposit developed later. Use of incorrect semivariogram models can lead to significant differences in the blocks selected as ore during production.
6. There are cases in which relatively simple, empirical estimation techniques have been shown to be generally equivalent to more labor-intensive kriging results. Kriging can be replaced by more traditional methods when there are data to demonstrate that the traditional methods are acceptable.
7. Outliers have an extraordinary impact on individual block estimates. There are two common problems: (i) the high-grade value is spread to many surrounding blocks and leads to overestimates

of both tonnage and average grade above cutoff; and (ii) some existing outliers are not intersected by sampling and are not taken into account during estimation, leading to underestimation. Consequently, a simple reconciliation of production with estimates is not necessarily a fair indication of the quality of the estimation procedure. This arises because the two sources of error are, in part, compensating.

13.11: SELECTED READING

Huijbregts, C., 1976, Selection and grade–tonnage relations; *in* Guarascio, M., M. David, and C. Huijbregts (eds.), Advanced geostatistics in the mining indusry, D. Reidel Pub. Co., Dordrecht, the Netherlands, pp. 113–135.

Parker, H. M., A. G. Journel, and W. C. Dixon, 1979, The use of conditional lognormal probability distribution for the estimation of open-pit ore reserves in stratabound uranium deposits – a case study; Proc. 16th APCOM Symp., Amer. Inst. Min. Eng., New York, pp. 133–148.

13.12: EXERCISES

1. The two patterns of Exercise 4, Chapter 5 (Fig. 5.23) can be examined for quality of estimation by kriging for any specified autocorrelation (semivariogram) model. Compare the quality of ordinary kriging estimates of the central point in each array for the following two linear semivariogram models: $\gamma(h) = k \cdot h$ and $\gamma(h) = C_0 + k \cdot h$, with $k = 0.5$ and $C_0 = 0.5$.

2. The following 26 Au values (g/t) are for contiguous 3-m composites from a drill hole that crosses an ore/waste boundary. Assume the data are representative of the boundary. Develop an estimation procedure for an array of $5 \times 5 \times 5$ m^3 blocks, assuming that holes are drilled approximately on a square grid with spacing of 20 m.

   Au data values (g/t): 0.61, 1.02, 0.48, 0.67, 0.39, 0.95, 0.53, 1.01, 1.48, 0.77, 1.45, 1.09, 1.73, 1.58,

1.33, 1.55, 1.13, 2.37, 1.74, 2.21, 1.31, 2.52, 1.01, 2.75, 1.79, and 2.61

3. Seven outlier values have been identified in a data base of 3,422 samples for a Ag–Pb–Zn deposit. The outliers have been examined in outcrop and drill core and found to be late-stage vertical veins, rich in sulphides, that average about 0.3 m in width and 6 m in horizontal extent. Their vertical extent is uncertain. Grades of 1-m samples that include these late veins are about one order of magnitude greater than most of the mineralization, which averages 0.8 percent Pb, 2.3 percent Zn, and 148 g Ag/t. Consider a two-dimensional estimation problem (a bench) for which the SMU is 10 × 10 × 5 m^3 and diamond-drill data are available for a square grid with spacing of 40 m. Discuss the practical problems of developing a grade-estimation procedure for an array of SMUs with emphasis on the presence or absence of outlier values.

14

An Introduction to Conditional Simulation

[W]hile the real surface $z_0(x)$ is known only at a limited number of locations $x_i, \ldots$ the simulated surface can be known at almost every point x of the deposit. It is then possible to apply to the simulation the various processes of extraction, hauling, stockpiling, etc., to study their technical and economic consequences and by feedback to correct these processes. (Journel, 1974, p. 673).

Two- and three-dimensional arrays of values, having the same statistical and spatial characteristics as grades of a mineral deposit or domain, are becoming increasingly useful in the design of advanced exploration/evaluation programs, as well as in mine and mill planning at the feasibility and operating stages. Chapter 14 provides insight into the development of arrays of conditionally simulated values and their use specifically for purposes related to improving the quality of resource/reserve estimation.

14.1: INTRODUCTION

Simulation in a mining context means imitation of conditions. Simulation, as it relates specifically to estimation, generally involves an attempt to create an array of values that has the same statistical and spatial characteristics as the true grades; however, values are generated on a much more local scale than that for which true grade information is available. Although reference here is restricted to grade, other variables can be simulated. A simulation is not an estimate; it is rather a two- or three-dimensional set of values having the same general statistical character as the original data.

It is important to realize why simulations are necessary when both data and estimates exist. The reason is best illustrated in a simple diagram showing grade profiles across a mineralized zone (Fig. 14.1). True, local fluctuations in the spatial distribution of grade are smoothed by most estimation methods, such that estimates do not reflect the local grade variations. In two dimensions, one can imagine a complex surface that represents the true grade distribution and a much smoother surface that represents the distribution of estimates. Simulated arrays of values are constructed to vary on the same scale as the true variations of sample grades. In estimation, we are concerned with minimizing the error variance; in simulations, we are concerned with reproducing the dispersion variance of the real data.

Simulation procedures can be designed that reproduce histograms of any shape (e.g., Agterberg, 1974), including normal, lognormal, and three-parameter lognormal distributions. In the simplest case (a single process), the procedures simply involve random draws from a defined population. In more complicated cases, several random processes can contribute to a final distribution. Such simulations have limited

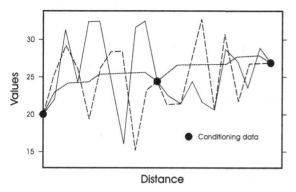

Figure 14.1: A comparison of idealized profiles of true grade (solid line), simulated grade (dashed line), and kriged estimated grade (dotted line) on a common profile. Note the similar variability of the true and simulated grades in contrast to the smoothed pattern of estimated grades. All profiles pass through the known data points. Redrawn from Journel (1975).

application in a mining context because they do not take into account the spatial characteristics of the variable under study. Simulated data arrays that retain the same density distribution (histogram) and autocorrelation character and that are linked spatially to reproduce the existing data are called *conditional simulations*. Conditional simulations are the type normally used for applications involving ore/waste grades. A conditional simulation and reality can be considered realizations of the same random function. Journel (1975) states, "Reality and simulation can be considered as two variants of the same mineralized phenomenon" (p. 8).

14.2: AIMS OF SIMULATION

Simulations serve a variety of purposes in the mineral industry, including (see Journel, 1979) the following:

(i) Study of grade continuity (e.g., Nowak et al., 1993)
(ii) Optimizing sampling plans for advanced exploration
(iii) Evaluation of resource/reserve estimation methods (e.g., Dowd and David, 1976)
(iv) Mine planning (e.g., Blackwell et al., 1999)

(v) Mill optimization (e.g., Journel and Isaaks, 1984)
(vi) Financial risk analysis (e.g., Ravenscroft, 1992; Rossi, 1999)
(vii) Any combination of the purposes listed here.

Here, emphasis is placed on the use of conditional simulations for characterizing grade continuity and as a means of evaluating estimation methods.

Simulations produce values at the nodes of an extremely fine grid (i.e., very closely spaced nodes relative to the distance separating the conditioning data) such that the character of a simulated deposit or domain is almost perfectly known by a large set of punctual values. These are the values used for the various purposes listed previously. Because so many punctual values can be simulated within blocks of a practical size (e.g., selective mining unit [SMU]), the average of all simulated values within a block can be taken as a close estimate of the true average value of the block. This opens the way for a number of practical applications of simulations (e.g., a mining sequence can be imposed on the blocks based on estimated grades to determine the true impact of production on such procedures as ore/waste selection, haulage scheduling, stockpiling, blending, and mill efficiency. Here, emphasis is directed to simulations as they relate directly to the problem of obtaining high-quality block estimates in a production scenario.

Various estimation methods can be applied to blocks whose simulated true grades are known. Estimates by each method can be compared with true block grades and the best method selected for use in practice. When complex data distributions occur, it has become common to use multiple indicator kriging to obtain grade estimates. One of the necessary steps in this procedure involves a change of support operation (see Chapter 12) that derives a block-grade distribution from a sample-grade distribution. Simulations can provide examples of the types of block-grade distributions that are encountered in practice that can be compared with distributions obtained by change of support procedures; simulations can also be used to verify change of support procedures (e.g., Rossi and

Parker, 1994). An important but little-used application of simulation involves an understanding of the spatial distribution of grade subpopulations (i.e., simulations of the pattern of physical continuity in space of various grade populations).

14.3: CONDITIONAL SIMULATION AS AN ESTIMATION PROCEDURE

In certain respects, a simulated value is an estimate of the true value at the site of the simulation. However, the simulated value is not a best estimate, and the related estimation error is twice that of the corresponding kriging error. Consequently, conditional simulation has not been viewed as a reliable or accepted estimation procedure. Of course, simulation and estimation have different purposes. Simulation allows local variations in values of a variable to be examined, particularly with regard to what impact these local variations have on sampling plans, estimation procedures, mine and mill planning, and financial matters. Estimates, however, are made ultimately for the purpose of identifying ore and waste on a scale that physical separation can be achieved.

There has been a trend toward the use of conditional simulations as a realistic approach to estimation, particularly for grade control during production (e.g., Blackwell et al., 1999). The general procedure involves the production of n simulations (for example, 10) of the same locations in space. The 10 values simulated at each point define the probability distribution for the grade at that point. These distributions can be used in a variety of ways (e.g., to estimate the probability that grade is above a cutoff at a point). Alternatively, all such distributions in a specified block can be combined to estimate the grade of the block or to estimate the probability that the block grade is above a cutoff grade.

14.4: A GEOSTATISTICAL PERSPECTIVE

Consider a single point, x, among many in space. This point has a real value of $z_0(x)$ and a kriged value of $z_k^*(x)$ such that

$$z_0(x) = z_k^*(x) + [z_0(x) - z_k^*(x)].$$
(kriging error)

The expected value of this kriging error is zero. Now consider a random function $Z_0(x)$ independent of $z(x)$ but isomorphic (having similar structure) to it. That is, $Z_s(x)$ has the same covariance $C(h)$, as $Z(x)$. Then the same kriging procedure leads to a comparable result (i.e., or any sampling of the random function):

$$Z_s(x) = Z_{sk}^*(x) + [Z_s(x) - Z_{sk}^*(x)].$$

Because Eqs. 14.1 and 14.2 are *isomorphic equivalents* (realizations of the same random function), it is possible to mix components on their right sides to produce a new realization of the same random function, as follows:

$$Z_c(x) = z_k^*(x) + [Z_s(x) - Z_{sk}^*(x)]$$

where $Z_c(x)$ is constructed by combining the residuals of the second random function $Z_s(x)$ with the kriging $Z_k^*(x)$ of the initial random function.

Thus, simulation can be seen to be the addition of a simulated-error factor to a kriged value. Hence, kriging is a product of the overall simulation procedure. Equation 14.3 also demonstrates that the simulation is conditional on the original data (i.e., reproduces the original data) because, for the conditioning sample sites, kriging reproduces the sample value, and the error term disappears.

14.5: SEQUENTIAL GAUSSIAN SIMULATION

Of the several simulation procedures in use (e.g., turning bands, LU decomposition, sequential Gaussian), only the sequential Gaussian procedure is considered here. The method is based on the principal that an appropriate simulation of a point is a value drawn from its conditional distribution given the values at some nearest points. The sequential nature rests on the fact that subsequent points that are simulated make use not only of the nearby original conditioning data, but also the nearby previously simulated values. Software for this purpose is readily available in Deutsch and

Journel (1998). The general procedure is as follows:

(i) Select a set of conditioning data.

(ii) Transform conditioning data to equivalent normal scores.

(iii) Develop a semivariogram model for the transformed data.

(iv) Check transformed data for bivariate normality by comparing sample indicator semivariograms for different thresholds (lower quartile, median, upper quartile) to a theoretical bivariate model.

(v) Proceed with sequential Gaussian simulation routine (from Deutsch and Journel, 1998).

(vi) Conduct several checks to demonstrate that simulation has honored the desired constraints.

It is important to compare a histogram of the simulated values with a histogram of the conditioning data and compare experimental semivariograms of the simulated values with the semivariogram model determined for the normal scores of the conditioning data. A contour plot of the simulated values is also useful for comparison with a contoured plot of the conditioning data.

14.6: SIMULATING GRADE CONTINUITY

There are many practical mineral exploration programs in which the physical continuity of ore-bearing rock is less than the spacing between drill sections that form the basis of data collection. In this situation, a particular mineralized structure intersected on one drill section is not represented on adjoining drill sections (e.g., Fig. 3.10). The Shasta epithermal precious-metal deposit (see Section 3.5.2) is such an example. In this case, the lengths of many of the mineralized quartz veins within an altered zone are much less than the spacing between drill sections. Hence, information from drill sections is not a sound basis for interpolation of grade between sections. Two horizontal and vertical simulations (Nowak et al., 1993) clearly indicate the scale on which additional data are required in order to provide data that can be used to interpolate with confidence (Fig. 3.11).

This general situation (widely spaced data sites relative to the average physical continuity of local ore-bearing structures) is common to many deposits; in particular, shear-controlled deposits such as many gold-bearing shear veins. When the exploration of such deposits has advanced to a stage that includes underground workings (raises and drifts within the structure) in addition to widely spaced piercement by diamond-drill holes, information can be collected to construct a semivariogram model for gold grade or accumulation and thickness. Underground access provides the database with which to construct a semivariogram model; diamond-drill intersections provide the conditioning data with which to develop simulations.

14.7: SIMULATION TO TEST VARIOUS ESTIMATION METHODS

14.7.1: Introduction

A set of 302 Cu grades of blasthole cuttings for a part of one bench (bench height $= 40$ ft) of the Similkameen porphyry copper deposit (cf. Raymond, 1979) is used as an example of the use of conditional simulation to compare the effectiveness of various estimation methods. These data are largely at a spacing of 20 to 25 ft. The aim of the study is twofold:

(i) To produce a closely spaced two-dimensional array of values with the same mean, standard deviation, and spatial structure as the original data

(ii) To test various block estimation techniques (nearest neighbor, inverse distance weighting, ordinary kriging) that use the original data to estimate $45 \times 45 \times 30$ ft^3 blocks whose mean grades are known (average of all contained simulated values).

14.7.2: Procedure

A sequential Gaussian method of conditional simulation was selected for illustrative purposes (see Section 14.4) using software from the widely available GSLIB software package (Deutsch and Journel, 1998). In

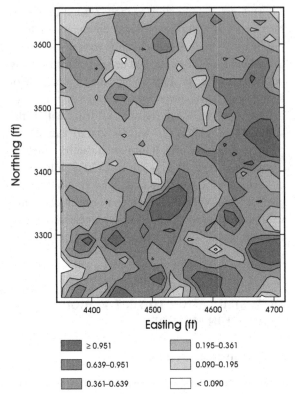

Figure 14.2: Contours for 302 blasthole Cu values (Similkameen deposit). Data locations are omitted for clarity but values are more or less uniformly distributed over the field, with a general spacing of about 20 to 25 ft. Contours demonstrate a preferential trend in a roughly northeasterly direction.

brief, the simulation procedure involves the following steps:

(i) Select a specific area for which the simulation will be performed (Fig. 14.2).

(ii) Transform the original 302 data to a normal distribution using a generalized normal transform (e.g., Journel and Huijbregts, 1979).

(iii) Determine a semivariogram model for the transformed data (Figs. 14.3 and 14.4 and Table 14.1).

(iv) Check for bivariate normality (Fig. 14.4).

(v) Conduct sequential Gaussian simulation of 18,000 points using the SGSIM program (Deutsch and Journel, 1998).

(vi) Check the validity of the resulting simulation.

(vii) Repeat steps (v) and (vi) for as many additional simulations as are required.

The test program reported here, based on Similkameen blasthole Cu grades as conditioning data, involved 10 independent simulations.

14.7.3: Verifying Results of the Simulation Process

Histograms of the conditioning data and the simulated values are compared in Fig. 14.5. It is evident both from the forms of the two histograms and from the accompanying means and standard deviations that the simulation process has faithfully reproduced the basic statistical character of the conditioning data. The contoured plot of the simulated values (Fig. 14.6) shows

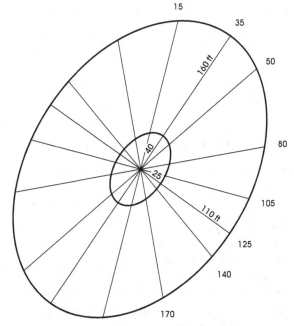

Figure 14.3: Pictorial summary of the semivariogram model derived for normal score transforms of the 302 blasthole Cu grades (conditioning data) representing a part of one bench (bench height = 30 ft) of the Similkameen porphyry copper deposit. The two ellipses represent anisotropic ranges of two structures. The radii shown are those used to construct experimental semivariograms to develop the model. The model differs slightly from that obtained using the entire 1,181 blastholes for the bench.

Table 14.1 Two-dimensional semivariogram model for 302 blasthole copper values, Similkameen porphyry-copper deposit, Princeton, British Columbia

Structure	Range (ft)[a]	C	Anisotropy ratio
Nugget	0	0.3	1
Spherical 1	40	0.4	0.625
Spherical 2	160	0.3	0.688

Note: [a]The indicated longest range is along a 035 azimuth.

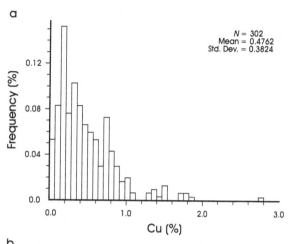

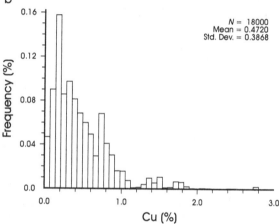

Figure 14.5: Histograms for (a) 302 blasthole Cu values (conditioning data) and (b) 18,000 simulated Cu values, Similkameen porphyry copper deposit. Note the similarity of the two histograms.

the same general character as the contoured conditioning data (Fig. 14.2), except, as anticipated, the relatively few and more widely spaced conditioning data are incapable of showing the local detail contained in the simulated values. A more formal test of the similarity of spatial character is illustrated by a comparison of the experimental semivariogram of the simulated values with the semivariogram model that characterizes the conditioning data. Figure 14.7 clearly demonstrates that the model for the conditioning data could just as well serve as a model for the experimental semivariogram of the simulated values.

Clearly, this simulation has been successful in reproducing the statistical and spatial character of the conditioning data.

14.7.4: Application of Simulated Values

As a practical application of the simulated values generated for part of one level of the Similkameen porphyry copper deposit, consider their use as a standard against which to measure the effectiveness of various estimation procedures. A 45×45 ft² grid is superimposed on the 360×450 ft² field (Fig. 14.2) of simulated values (recall that the values represent blasthole assays for a bench height of 40 ft). Each

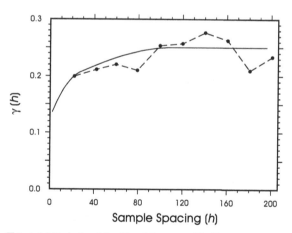

Figure 14.4: A check for bivariate normality. Dots are an experimental indicator semivariogram for the 302 conditioning data using a cutoff grade of 0.385 percent Cu (median grade). The smooth curve is a fit by a Gaussian theoretical model calculated using normal scores for the 302 conditioning data. This example for an azimuth of 125 degrees is typical of the fits obtained.

Table 14.2 Parameters for traditional least-squares linear models describing estimates versus true block values, Similkameen test

Estimation method	Correlation coefficient	y Intercept[a]	Slope[a]	Scatter
Ordinary kriging	0.894	0.0999 (0.0791)	0.818 (0.151)	0.0951
$1/d^2$	0.819	0.0304 (0.0320)	0.941 (0.0604)	0.1093
$1/d^3$	0.754	−0.0994 (0.0480)	1.229 (0.0902)	0.1677
$n = 80$				

[a] Bracketed number is one standard deviation error.

cell (block) of the resulting array of 80 cells (8 × 10) is then estimated by several methods; in this case, nearest neighbor, various inverse-distance weighting procedures, and ordinary kriging using only the original data. The true value of each block is taken to be the average of all simulated points within the block. Individual estimation techniques can then be compared against these true values on an x–y plot. Because the expectation is for unbiased results with each estimation technique, on average, producing the correct result, the data on x–y plots can be approximated by a linear model. Three examples are shown in Fig. 14.8 for ordinary kriging, inverse-distance weighting – exponent 2; and inverse-distance weighting – exponent 3. Statistical and linear model parameters are summarized in Table 14.2.

This test demonstrates the relative quality of various estimation methods as applied to an idealized array of block values generated from a parent population with statistical and spatial characteristics

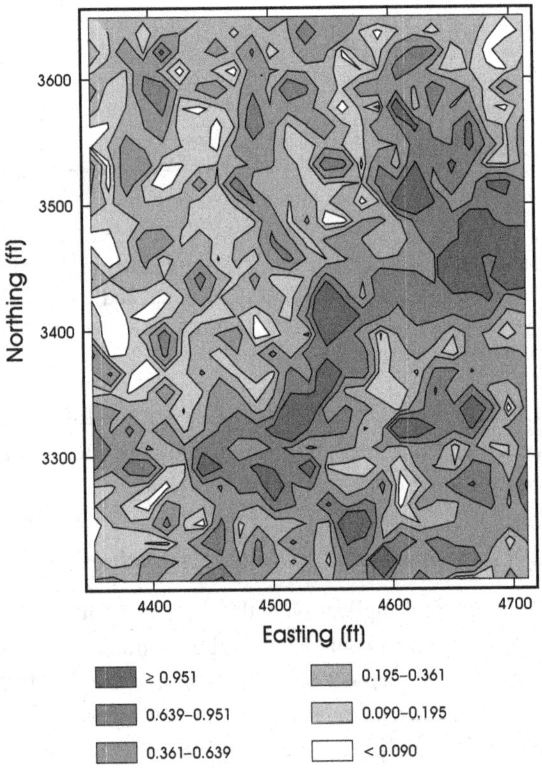

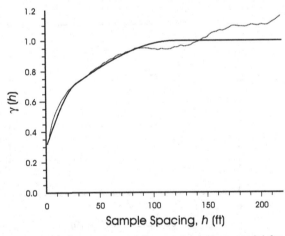

Figure 14.7: Comparison of the semivariogram model for 302 blasthole Cu values (smooth solid line) and the experimental semivariogram for the 18,000 simulated values (dotted, irregular line). Note that the model based on the conditioning data could serve equally well for the simulated data.

Legend:
- ≥ 0.951
- 0.639–0.951
- 0.361–0.639
- 0.195–0.361
- 0.090–0.195
- < 0.090

Figure 14.6: Contoured plot for 18,000 simulated Cu values for the same area shown in Fig. 14.2.

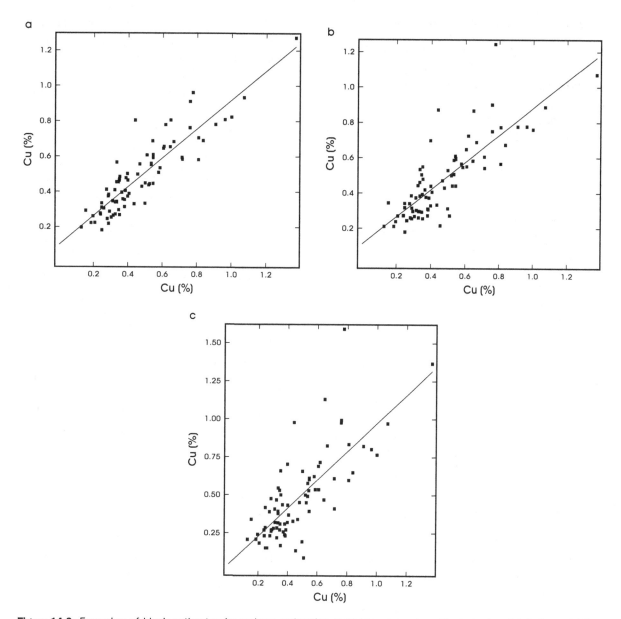

Figure 14.8: Examples of block estimates by various estimation methods versus true block grades. (a) Ordinary kriging estimates versus true block grades (x axis) (b) Inverse squared distance estimates versus true block grades. (c) Inverse cubed distance estimates versus true block grades. Filled squares are individual $45 \times 45 \times 30$ ft^3 blocks. Linear models are traditional least-squares fits, with error entirely in the ordinate (i.e., the y axis, which represents the estimation method). Parameters for the linear models are listed in Table 14.2.

comparable to those of a part of the Similkameen porphyry copper deposit. In particular, kriging and $1/d^2$ are clearly better than $1/d^3$, as indicated by the statistics in Table 14.2. Specifically, the conditional bias (slope) and scatter of values (error) are greatest for $1/d^3$. The choice between kriging and $1/d^2$ is not so clear, although, statistics aside, the kriging results show much less scatter than do the $1/d^2$ results.

14.7.5: Sequential Indicator Simulation

Another method also available in Deutsch and Journel (1998) is sequential (multiple) indicator simulation (SISIM). This does not require a Gaussian distribution of the data and uses multiple indicator kriging, as described in Section 10.11.2, to model complex grade distribution patterns. The general procedure is as follows:

(i)   Select a set of conditioning data.
(ii) Select a set of cutoff (threshold) grades and transform to a set of indicator (0 or 1) data.
(iii) Develop semivariogram models for the various sets of indicator-transformed data.
(iv) Proceed with the simulation routine (from Deutsch and Journel, 1998).
(v)   Conduct several checks to demonstrate that the simulation has honored the desired constraints.

In this case, the simulation routine randomly selects blocks for multiple indicator kriging (MIK) using the original data and any nearby blocks already simulated. A random number (0 to 1) is selected and applied to the MIK probability of being between two of the specified thresholds. The grade is interpreted linearly between the two threshold grades defined by the random number selected.

14.8: PRACTICAL CONSIDERATIONS

1. To be useful in the mineral industry, most simulations of grade distributions must be conditional (i.e., in addition to reproducing the mean and dispersion of the original data, the simulated values must also reproduce the spatial characteristics of the original data).
2. Generating a conditional simulation is a nontrivial undertaking that requires a reasonable level of geostatistical sophistication. Software to conduct conditional simulations is readily available (e.g., Deustch and Journel, 1998), but its implementation requires a fundamental understanding of a range of geostatistical procedures.
3. Conditional simulation is far more warranted than the limited practical applications in the technical literature would suggest. Most reported practical applications relate to production scenarios, but substantial scope for the use of conditional simulations exists prior to exploitation at the advanced exploration and feasibility stages of deposit development.
4. When detailed semivariogram models are obtained confidently by the judicious collecting of data early in an exploration program, conditional simulation can provide remarkable insight into grade continuity and how continuity can influence the development of an adequate sampling program, especially as regards sample spacing.

14.9: SELECTED READING

Deutsch, C., and A. Journel, 1998, GSLIB, Geostatistical software library and user's guide; Oxford University Press, New York, 369 pp.

Journel, A. G., 1979, Geostatistical simulation: methods for exploration and mine planning; Eng. Min. Jour., December, pp. 86–91.

Journel, A. G., and E. H. Izaaks, 1984, Conditional indicator simulation: application to a Saskatchewan uranium deposit; Math. Geol., v. 16, no. 7, pp. 685–718.

14.10: EXERCISES

1. Simulate a grade profile of 20 5-m composites. The mean grade is 2.3 g/t and the standard deviation is 3.1 g/t. Assume a lognormal distribution.

   Step 1: Calculate the mean logarithmic value m_l and the logarithmic variance s_l^2 from the arithmetic data provided, using equations in Chapter 4.

   Step 2: Draw 20 values from a table of random z scores for a normal distribution (e.g., from a mathematics handbook).

   Step 3: Transform each randomly selected z score into a corresponding logarithmic value using the relation $z = (x - m)/s$ (i.e., $x = zs_l + m_l$).

   Step 4: Plot the profile of 20 successive x values

at sites 0 to 19, inclusive. Note that this procedure produces a profile of a variable with a pure nugget effect.

2. Assume that at sites 0, 9, and 19 in Question 1, the values are known (and are to be used to condition the remaining values). The deposit is known to have alternating high- and low-grade zones on a scale of 40 to 50 m. In this example, *conditioning* means that the position of intermediate samples relative to a line joining two random end points that coincide with conditioning sites is maintained relative to a line joining the two conditioning points.

Suppose the random values for Sites 0 and 9 are 1.8 and 3.1, respectively, and the known values at these sites are 1.2 and 4.1, respectively. Suppose further that a random value at Site 2 is 1.1. The conditioned value at Site 2 is determined as follows. A line based on the two random points has the equation $y = 1.8 + (3.1 - 1.8) \times /9$. For site $x = 2$, the corresponding line value of y is 2.1 (1.0 higher than the random value of 1.1). This difference must be maintained relative to a line through the two known conditioning points. The equation for the conditioning line is $y' = 1.2 + 2.9 \times /9$ which, for site $x = 2$, gives $y' = 1.84$. The conditioned value for Site 2 must be 1.0 lower than this point on the conditioning line; hence, the conditioned value at Site 2 is $1.84 - 1.0 = 0.84$. Produce a profile of conditioned values from the randomly selected values of Question 1.

15

Bulk Density

A more detailed investigation of the specific gravity of the two types of ore was made in an effort to determine more accurate tonnage factors. The results of this work showed that there was considerable variation in the specific gravities throughout the orebody and that there was no acceptable average tonnage factor. (Brooks and Bray, 1968, p. 178)

Bulk density determinations of ores and waste require close attention because they directly affect the conversion of volumes to tonnages. The factors that control bulk density are variations in mineralogy and porosity, and the scale of variations in these factors controls the scale at which bulk densities, must be recorded. Practical methods of dealing with the complex interplay of mineralogy and bulk density are discussed.

15.1: INTRODUCTION

Density, defined as mass per unit volume (e.g., pounds per cubic foot or grams per cubic centimeter), is an important rock and ore characteristic used to transform measured volumes into tonnages. Practical consideration of ore and waste in the mineral industries commonly deals with volumes of in situ or broken material with a significant porosity (void space within the volume) that reduces the effective density of the volume below that of the solid and perhaps nonporous material. A density that takes voids into account is termed a *bulk density*. Of course, when porosity is negligible,

density of the solid material and bulk density are essentially equivalent.

Specific gravity, a term that is widely used interchangeably with density, is relative density, and is therefore unitless. A specific gravity of 2.0 indicates that a substance is twice the weight of an equal volume of water (at standard temperature and pressure). In its mineralogic use, *specific gravity* applies to solid, nonporous materials; consequently, use of the term should be avoided in general reference to ore and waste, with deference to the term *bulk density*. Of course, when porosity is negligible, the numeric value of specific gravity is identical with the numeric value of density.

Potential ore and related waste material generally are defined initially by volume and then transformed to tonnage by calculations that use an appropriate conversion factor that reflects the average bulk density of the volume under consideration. It is common practice in the mineral industry to determine an average volume (tonnage) factor for very large volumes; in the extreme, a single conversion factor for an entire deposit. Such an approach is rarely appropriate because of the variations in mineralogy and physical character that are so prevalent throughout a single mineral deposit.

There are a variety of methods for estimating bulk density that can be categorized into the following general procedures:

(i) Archimedes-type measurements on whole core (in the field) or core or rock fragments (in the lab)

(ii) Lab measurements of pulverized material (generally unsatisfactory in resource/reserve estimation)

(iii) Physical measurements for long lengths of whole core on-site (dimensions and weight) (This method provides small-scale bulk densities, but can be impossible when appropriate core cannot be obtained or cumbersome when appropriate core is available.)

(iv) Stoichiometry based on whole-rock chemical analyses

(v) Gamma–gamma logging of drill holes (spectra from this geophysical procedure must be calibrated against physical measurements of corresponding core segments)

15.2: IMPACT OF MINERALOGY ON DENSITY

Mineral zoning is a feature of many mineral deposits and is the principal reason why there are systematic differences in bulk density within a single deposit. Consider the following example, summarized from Sinclair (1978, p. 130):

Suppose that the central part of a layered, massive sulfide deposit consists of 10(vol)% sphalerite (s.g. = 4.0), 45(vol)% pyrrhotite (s.g. = 4.6) and 45(vol)% pyrite (s.g. = 5.1). The (bulk density) of the ore is 4.77 (g/cc) assuming no porosity. The tonnage conversion factor in the British system is 6.71 (ft³/ton). If the fringes of the same deposit consist of 60(vol)% argillite (s.g. = 2.7), 30(vol)% pyrite, 4(vol)% sphalerite and 6(vol)% galena (s.g. = 7.6) the ore has a (bulk density) of 3.77 g/cc (tonnage conversion factor = 8.49 ft³/ton). Note that the two bulk densities differ by more than 20%. Even if an average bulk density of 4.27 g/cc were used to

make tonnage estimates, estimates for both zones would be in error by $100(0.5/4.27) = 11.7\%$, one zone being overestimated, the other underestimated. In the second case containing argillite, an increase in the argillite content to 70 vol.% and a corresponding decrease in pyrite to 20 vol.% results in a 7% reduction in bulk density, which translates into a 7% error in tonnage estimates.

A practical example is provided by Brooks and Bray (1968) in their discussion of tonnage factors for massive sulfides forming a core zone at the Geco deposit, northern Ontario, surrounded by "an envelope of disseminated sulfides" (see the quote at beginning of this chapter).

Bulk density varies among rock types at the Mactung tungsten skarn deposit, Northwest Territories, ranging from an average of 2.9 g/cc for hornfels to 3.4 g/cc for pyrrhotite skarn (Fig. 15.1). Mustard and Noble (1986, p. 75) describe how geologic mapping of skarn types improves the local estimation of an

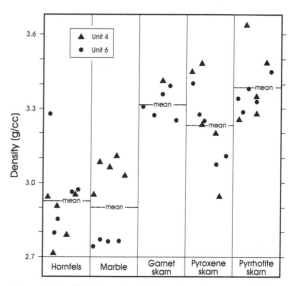

Figure 15.1: Bulk density of tungsten skarns as a function of skarn types, Mactung scheelite deposit, northern Canada. Note the substantial density differences for waste (hornfels, marble) and tungsten mineralization (garnet, pyroxene, and pyrrhotite skarns). Redrawn from Mustard and Noble (1986).

appropriate bulk density, leading to high-quality local tonnage estimates: "Definition of the spatial distribution of skarn and rock type allows more accurate local estimation of (bulk density) than can be achieved with an overall average value."

Improved local tonnage estimates based on geologic mapping and an appreciation of the spatial distribution of bulk density values lead to closer reconciliation of local estimates with production.

15.3: IMPACT OF POROSITY ON BULK DENSITY

Voids in a rock mass have zero specific gravity and must be weighted appropriately in the determination of bulk density. Consider the following example, in which samples have been "image analyzed" to produce the mode (vol %) in column one of Table 15.1. Column two contains the specific gravities of the minerals obtained from a mineralogic text. Note that the specific gravity of sphalerite can vary substantially, and an average figure has been used. The carbonate is assumed to be calcite. In practice, measurements of the specific gravities of sphalerite and carbonate can be obtained so that figures appropriate for a specific deposit are used. With the information in columns one and two, a number of calculations are possible, as follows:

$$\text{Bulk density of dry ore} = 378.4/100$$
$$= 3.78 \text{ g/cc}.$$

Note that the specific gravity of ore material, assuming zero voids, can also be determined using information derived from columns one and three, as follows:

$$\text{Specific gravity of ore (zero porosity)}$$
$$= 378.4/(100 - 8.3) = 4.13 \text{ g/cc}.$$

The specific gravity (voids not taken into account) is numerically 8 percent higher than the true bulk density. If these figures were representative for a deposit and specific gravity rather than bulk density were used to convert volume into tonnes, this 8 percent error

Table 15.1 Modal analysis of a Pb–Zn–Ag ore

	Vol.%	SG	Vol. × SG	Wt.%
Galena	24.0	7.5	180.0	47.6
Sphalerite	18.2	3.5	63.7	16.8
Argentite	0.2	7.3	1.5	0.4
Carbonate	49.3	2.7	133.2	35.2
Voids	8.3	0.0	0.0	0.0
TOTAL	**100.0**		**378.4**	**100.0**

would translate directly into an 8 percent overestimation of "ore" tonnage.

As a matter of interest, metal assays can be determined from the estimates of weight percent metal using the following relation:

$$\text{metal wt.\%} = \text{mineral wt.\% (molar wt. metal/}$$
$$\text{molar wt. mineral)}$$
$$\text{Pb\%} = 47.6(207.21/(207.21 + 32.06)) = 41.2\%$$
$$\text{Zn\%} = 16.8(65.38/(65.38 + 32.06)) = 11.3\%$$
$$\text{Ag\%} = 0.4(215.8/(215.8 + 32.06)) = 0.35\%$$
$$= 3.5 \text{ kg/t}$$
$$= 3,500 \text{ g/31.103(2,000/2,204.62)}$$
$$= 102.1 \text{ troy oz/t}.$$

If similar calculations are done for modal data obtained from grain or point counts, it is possible to assign an error to the volume estimates, the mineral weight estimates, and the metal weight estimates.

15.4: IMPACT OF ERRORS IN BULK DENSITY

Consider Purdie's (1968, p. 187) comment regarding the A zone at Lake Dufault, Quebec:

The volume factor for the A zone was calculated from an estimate of the average mineral content of the massive sulfides and the specific gravities relative to each mineral constituent. This was checked during surface exploration of the A zone by weighting several core samples of massive

Table 15.2 Sample lengths, densities, and assays

Sample length (L)	Bulk density (D)	L.D	Assay % Cu (T)	L.T	L.D.T
3.0	3.0	9.0	1.0	3	9.0
4.0	4.2	16.8	30.0	120	504.0
4.0	3.5	14.0	3.0	12	42.0
2.0	4.0	8.0	1.0	2	8.0
2.0	3.0	6.0	1.0	2	6.0
15.0			53.8	36.0	139
	569.0				

sulfides and calculating a volume factor from these weights. The results were

(i) Estimated by mineral content: 7.6 ft^3/t @ 5% Cu, 12% Zn
(ii) Average of 490 feet of massive sulfide core: 6.9 ft^3/t @ 5.7% Cu, 11.8% Zn.

These two estimation methods produce tonnage (volume) factors that differ by about 10 percent, a difference that is translated directly to tonnage estimates. Estimate (i) is based on the average mineralogy for the entire A zone; estimate (ii) is based on a number of direct local core measurements. The results clearly indicate substantial local variability in bulk density. Clearly, an understanding of the spatial variability of bulk density of ore (and waste) is necessary in order to make confident local estimates of tonnage.

Dadson (1968, p. 4) provides several examples comparing the differences in use and nonuse of bulk density in determining average grade, one of which is reproduced in the following. Data are given in Table 15.2. "Example: 'Assume copper ore, a mixture of massive and disseminated sulfides in siliceous gangue; drill core . . . samples are taken.' " The average grade calculation using data of Table 15.2 is

(i) Weighted by bulk densities: $m = 569/53.8 = 10.58$ percent Cu
(ii) Weighted by length: $m = 139/15 = 9.27$ percent Cu.

The difference in estimated grades by the two methods is about 10 percent; in this case, the incor-

rect method that does not take density into account produces an underestimation of the true grade.

15.5: MATHEMATICAL MODELS OF BULK DENSITY

Consider an ore that is a mixture of quartz and sphalerite and has zero porosity. The bulk density (ρ_b) of the ore can be expressed in terms of the specific gravities of the individual minerals as follows:

$$\rho_b = 2.7 f_q + 7.8 f_g$$

where $f_q + f_g = 1$ are the volume proportions of the minerals quartz and galena, respectively. If a small variable component of porosity is present, the model becomes

$$\rho_b = 0 f_p + 2.7 f_q + 7.8 f_g \pm e \qquad (15.1)$$

where $f_p + f_q + f_g = 1$ and e is a random measurement error. Note that the first term of Eq. 15.1 is zero; therefore

$$\rho_b = 2.7 f_q + 7.8 f_g \pm e \text{ but} (f_q + f_g) \leq 1. \qquad (15.2)$$

When f_p is very small, the relation $f_q = 1 - f_g$ holds approximately, and Eq. 15.2 reduces to

$$\rho_b = K + 5.1 f_g \pm e.$$

Hence, in this simple, mineralogic system, bulk density has a simple linear relation with the volume abundance of galena. Because the weight/percent of galena is directly related to the volume/percent, a linear relation also exists between bulk density and weight/percent galena (Eq. 15.3):

$$Gn_w = 100\{7.8 f_g / [7.8 f_g + 2.7(1 - f_g)]\}$$
$$= 100\{7.8 f_g / [5.1 f_g + 2.7]\} \qquad (15.3)$$

where Gn_w is the weight/percent galena and other symbols are as shown. Because a direct relation exists between weight percent galena and weight/percent Pb, a linear relation also holds between metal assay and bulk density. The relation between weight/percent

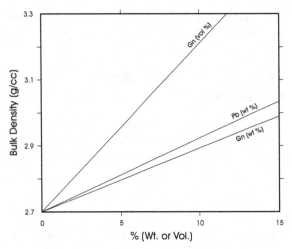

Figure 15.2: Linear models of bulk density versus volume/percent galena, weight/percent galena, and weight/percent lead for mixtures of galena and quartz with zero porosity.

galena and metal/percent, based on atomic weights of constituent atoms, is as follows:

$$\%\text{Pb} = Gn_w(207.21/239.27) = 0.866Gn_w.$$

All three of the foregoing relations are illustrated in Fig. 15.2 and data are listed in Table 15.3. This same type of simple linear pattern can exist when small amounts of other minerals are present (see Fig. 4.17); the presence of additional minerals contributes to the constant K and the magnitude of the random error e. Of course, as the number and amounts of additional minerals increase, the relation gradually breaks down.

It is surprising how commonly a simple linear model provides an adequate definition of bulk density. However, ore mineral assemblages are generally much more complex than the simple example described. Bevan (1993) describes such a case for Cu-Ni ores of the Sudbury camp, where he demonstrates bulk density to be a quadratic function of sulphur content.

A more fundamental approach to developing a mathematical model for bulk density is the use of multivariate models, such as multiple regression. This arises because bulk density is an exact function of mineralogy and porosity. For example, the bulk density of the mineralogic data of Table 15.1 can be

Table 15.3 Bulk density values and corresponding weight and volume percentages of galena and Pb assays (see Fig. 15.2)

Vol.% Qtz	Vol.% Gn	Wt.% Gn	Wt.% Pb	Bulk density
100	0	0	0	2.70
99	1.0	2.84	2.46	2.751
97	3.0	8.20	7.50	2.853
95	5.0	13.2	11.43	2.955
93	7.0	17.9	15.5	3.057
91	9.0	22.2	19.2	3.159
89	11.0	26.3	22.8	3.261

expressed as a multivariate equation as follows:

$$\begin{aligned}
\rho_b &= f_g x \rho_g + f_s x \rho_s + f_a x \rho_a \\
&\quad + f_c x \rho_c + f_v x \rho_v \\
&= 7.5 x f_g + 3.5 x f_s + 7.3 x f_a \\
&\quad + 2.7 x f_c + 0 x f_v
\end{aligned} \qquad (15.4)$$

where ρ is specific gravity and f is volume proportions (fractions) of minerals; subscripts are g, galena; s, sphalerite; a, argentite; c, carbonate; and v, voids.

Substituting the volume abundances of Table 15.1 in Eq. 15.2 gives

$$\begin{aligned}
\rho_b &= 0.24 \times 7.5 + 0.182 \times 3.5 \\
&\quad + 0.002 \times 7.3 + 0.493 \times 2.7 = 3.78.
\end{aligned}$$

More generally, if only Pb and Zn metal abundance (assays) are to be used to estimate bulk density, Eq. 15.4 would take the form

$$\rho_b = b_0 + b_1 \text{Pb}\% + b_2 \text{Zn}\% \pm e \qquad (15.5)$$

where b is constant. Obviously, more complex models (involving more variables) are possible. In practice, an equation such as Eq. 15.5 would be determined using a least-squares fitting procedure applied to samples for which both grades and bulk densities were known. In cases where the error term is small, the model is an acceptable means of estimating bulk density without going to the cost and effort of physical measurements. If the error term is too large, the model is inadequate. Commonly, a linear, least-squares model is sufficient.

15.6: PRACTICAL CONSIDERATIONS

1. Bulk density differs from specific gravity in taking account of porosity in an otherwise solid mass. Idealized estimates of bulk density can be calculated when (a) the mineral content and the porosity are known and (b) specific gravity of each mineral component is available.

2. Bulk density is a variable much like grade. Hence, bulk density can vary significantly with spatial location.

3. Variations in bulk density arise from variations in mineralogy, porosity, and the chemical composition of minerals that permit a wide range of solid solution. In many deposits, mineralogic variation is the principal control on bulk density; thus, mineralogic zonation is commonly a practical guide to systematic variations in bulk density.

4. It is important that bulk density measurements be conducted early in the exploration of a deposit. Commonly, it is sufficient to make direct measurements on large pieces of core before the core is split or deteriorates. For many ore types, this is a simple undertaking; for others, difficulties can be encountered, mostly arising from lack of cohesiveness of core or porosity on a scale larger than can be represented by core. In some difficult cases, the use of bulk samples may be required. In such cases precise measurements of both the volume and weight of a bulk sample (e.g., a drift round) must be obtained.

5. In many cases, a systematic relation exists between bulk density and one or more metal abundances. When such a model exists, it provides a simple way of incorporating density into the estimation procedure at the level of individual blocks. Average grades estimated for the blocks can be inserted into the model to provide a bulk-density estimate that translates into local tonnage estimates. This procedure improves on the widespread use of a single average density factor everywhere in a large domain, a procedure that can lead to substantial disparity in estimated tonnage versus produced tonnage over short periods.

6. With improvements in miniaturization of downhole geophysical instruments, continuous estimates of bulk density are becoming more readily available in exploration drill holes for mineral-deposit evaluation (e.g., Mutton, 1997). Such information is desirable wherever possible because of their abundance, a spatial distribution that coincides with much of the assay information, and a uniformity of quality of bulk density measurements.

7. In reconciling production tonnages with estimates, it can be important to consider wet tonnes that are being moved through the production phase; these may differ by as much as 5 percent relative to tonnes in place (e.g., Parrish, 1993).

15.7: SELECTED READING

Bevan, P. A., 1993, The weighting of assays and the importance of both grade and specific gravity; Can. Inst. Min. Metall. Bull., v. 86, no. 967, pp. 88–90.

Hazen, S. W., Jr., 1968, Ore reserve calculations; Can. Inst. Min. Metall. Spec. v. 9, pp. 11–32 (esp. pp. 29–30)

15.8: EXERCISES

1. Table 15.4 is a grain count of a concentrate from a beach sand that is under evaluation (grains approximately 2 mm in diameter). (a) Using the available data, calculate the specific gravity of the solid matter. (b) Assuming a reasonable porosity for sand grains, determine the bulk density of the concentrate. (c) Assuming an ideal mineral composition for zircon, calculate the Zr assay for the concentrate.

2. Vein mineralogy changes abruptly along strike from a quartz-sulfides assemblage (bulk density $= 3.15$ g/t) to a quartz-carbonate-sulfides assemblage (bulk density $= 3.48$ g/t). Suppose the incorrect density were used to estimate tonnage of a 147×55 m stope in quartz-carbonate-sulfide

Table 15.4 Grain count of a beach sand concentrate

Mineral	Grains (no.)	Specific gravity
Ilmenite	762	4.8
Altered ilmenite	330	4.72
Manaccanite	14	5.0
Rutile	160	4.5
Rutile/gangue	8	3.5
Zircon	184	4.68
Monazite	14	5.1
Silicates	140	3.0
Unidentified	14	4.0
Plucked grains	12	4.5

ore with an average thickness of 1.83 ± 0.13 m, what would the error in tonnage estimate be? Is the error an overestimate or an underestimate? What are the tonnage and grade if an additional thickness of overbreak dilution (zero grade, bulk density $= 3.05$ g/t) were added to the vein material?

3. If ore is crushed, there is a very significant increase in volume because of the pore space produced. This pore space commonly amounts to about 35 percent of the total volume of crushed material. What is the bulk density of a crushed ore whose in situ bulk density is 3.73 g/t?

4. Transported crushed material commonly becomes wet with handling and in transit. Consequently, a dry bulk density of crushed material is an inappropriate basis for determining shipping weight. The amount of water absorbed by crushed material prior to shipping is variable and should be estimated in advance by comparing weights of wet and dry material. Suppose that one-seventh of the pore space were replaced by water retained in the crushed material, calculate the increase in bulk density (and thus increased shipping costs) for the crushed material of Question 2. This question does not consider the loss of fine material that commonly occurs during transport.

Toward Quantifying Dilution

Ore losses and dilution are present at all stages of mining and while several models can investigate the influence of dilution it is its quantification that poses the most serious challenge. (Pakalnis et al., 1995, p. 1136)

Chapter 16 considers dilution, either internal or external to ore. Both categories of dilution can be further subdivided on the basis of geometric considerations about the deposit itself or of the diluting material. External dilution involves that related to minimum mining width, contact dilution, and overbreak of wallrock relative to planned mining margins. Internal dilution can be considered from the perspective of volumes of barren ground within an ore zone or the inherent diluting effect resulting from either increasing the size of SMUs or the effect of misclassification of blocks arising from sampling and analytical errors occurring in grade control.

16.1: INTRODUCTION

Dilution, the "contamination" of ore with waste material, is an important consideration in all planned or operating mines because some amount of dilution is virtually unavoidable in most practical mining operations (e.g., Elbrond, 1994; Scoble and Moss, 1994). Dilution arises from a variety of causes (Figs. 1.7 and 16.1) that include deposit form (especially highly irregular deposit contacts), geomechanical properties of ore and country rock, determination of cutoff grade, sampling error, and the mining method itself. For

purposes here, it is convenient to consider two broad categories of dilution: *external dilution* and *internal dilution*. *External dilution* refers to all those types of dilution originating external to defined ore zones. *Internal dilution* refers to those types of dilution that occur within the ore zones themselves. These definitions imply that it is possible to define the geometry of ore. An important concept not always appreciated is that for some kinds of dilution (e.g., dilution from block estimation error and dilution due to highly sinuous ore/waste contacts), there is a corresponding loss of ore material as well as the addition of waste to ore.

16.2: EXTERNAL DILUTION

External dilution (e.g., wall slough, contact dilution, minimum mining width dilution, overbreak) is not easily quantifiable in totality, and commonly is approximated on the basis of experience. As Stone (1986) suggests, the estimation of dilution is too commonly done in retrospect and is "designed to force a reserve calculation to match the observed mill feed" (p. 155). Highly formalized empirical methods of estimating dilution in operating underground mines are summarized by Pakalnis et al. (1995) and involve an extensive study of the physical characteristics of a deposit and the development of a deterministic model that relates these characteristics to dilution.

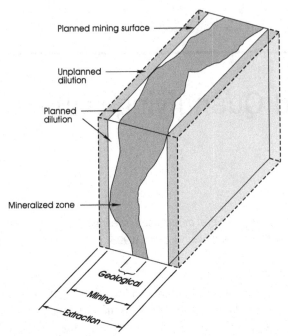

Figure 16.1: Conceptual view of dilution associated with "tabular" deposits. Note that the planned mining surface cannot be duplicated exactly during production; there is always a tendency toward overbreak. Redrawn from Scoble and Moss (1994).

Such detailed modeling is difficult enough in operating mines, but is intractable for deposits that are undergoing development or for which a feasibility study is planned or in progress. However, some components of external dilution can be estimated geometrically (Fig. 16.2); thus, it is useful to consider various types of external dilution as follows:

$$V_m = [V_{ore} - V_{rem}] + V_{mmw} + V_c + V_{ob}$$

where

V_m	is the volume mined
V_{ore}	is the volume of ore
V_{rem}	is the volume of ore (remaining) not mined (e.g., underbreak)
V_{mmw}	is the volume of dilution due to thicknesses less than minimum mining width
V_c	is the contact dilution arising from sinuous contacts relative to more regular mining margins

V_{ob} is the volume resulting from overbreak of country rock, relative to planned mining margins.

16.2.1: Vein Widths Partly Less Than Minimum Mining Width

In certain cases, a significant component of external dilution can be estimated with reasonable confidence in advance of production, particularly at the feasibility stage, when there is some underground access. Consider the case of a continuous vein that can be mined to sharp boundaries, except in those parts of the vein where vein width is less than the minimum mining width w_m (Fig. 16.2). An unbiased histogram of vein widths for a large part of the vein (Fig. 16.3) provides an estimate of the proportion of the vein that has widths less than the minimum mining width p_c. The average vein thickness w_c of this proportion p_c can be determined from the unbiased histogram as a weighted average of the appropriate class intervals (i.e., those class intervals less than w_m) and can be compared with the minimum mining width. The difference between the two widths ($w_m - w_c$) is the average ideal (point) dilution perpendicular to the plane of the vein over that portion of the vein that is less than minimum mining width (i.e., $S_o \cdot p_c$, where S_o is the area in the plane of the vein represented by the data). In this way, the point estimate of dilution is modified to a volume of diluting material (Eq. 16.1):

$$v_d = S_o \cdot p_c(w_m - w_c) \qquad (16.1)$$

where v_d is the volume of diluting material and S_o is the known area extension of the vein in the plane of

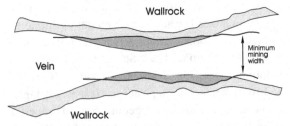

Figure 16.2: Idealized plan of a vein, part of which (proportion p_c) is less than the minimum mining width (w_M) leading to unavoidable dilution (darkest pattern).

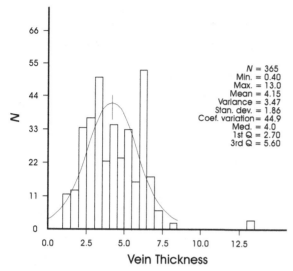

Figure 16.3: Histogram of vein thickness measurements in mine drift, Silver Queen vein deposit, central British Columbia. A normal curve with the same parameters as the data ($n = 364$; see Table 16.1) has been fitted to the histogram. Frequency is in absolute numbers of measurements; thickness is in feet. Data from Nowak (1991).

the vein. Consequently, the new volume V_p of mined (produced) material is

$$V_p = V_o + v_d \qquad (16.2)$$

where V_o is the original vein volume. This new volume V_p has a weighted average grade

$$g_p = (V_o \cdot g_o + v \cdot g_w)/(V_o + v_d) \qquad (16.3)$$

assuming that the vein material and the diluting country rock have the same bulk density. When densities differ, Eq. 16.4 becomes

$$g_p = (d_o V_o \cdot g_o + d_d v_d \cdot g_w)/(d_d V_o + d_d v_d).$$

In many practical situations, g_w is approximately zero.

A comparable approach can be used where an unbiased histogram is approximated by a normal or lognormal distribution. When the histogram has the form of a normal distribution, the proportion of vein that is thinner than the minimum mining width (p_c in Eq. 16.1) can be estimated with Eq. 4.6 and the average thickness of that portion below minimum min-

ing width (w_c in Eq. 16.2) can be determined with Eq. 4.9. Application of Eqs. 4.6 and 4.9 requires that the minimum mining width be transformed to a standard normal score (z value). If z is positive, the equations apply directly; if z is negative, they require some attention. Comparable equations and procedures exist for the case of a lognormal distribution (see Chapter 4).

16.2.2: Silver Queen Example

As an example of the application of the forgoing procedures, consider 365 thickness determinations taken at short equal intervals along the 2600 level of the Silver Queen epithermal vein deposit (Nowak, 1991). For the sake of example, these data are assumed to be representative of a large portion of the vein. A histogram of the data is shown in Fig. 16.3. A normal curve with the same mean (4.13 ft) and standard deviation (1.74 ft) as 364 of the values has been fitted to the histogram (a single outlier value of 13 has been omitted).

Assume that the minimum mining width is 4 ft. Using the absolute frequencies of the class intervals below 4 ft (Fig. 16.3), it is straightforward to determine that $p_c = 0.46$ and $w_c = 2.75$ (calculations are summarized in Table 16.1). The values of p_c and w_c can be used in Eqs. 16.1 to 16.3 to quantify dilution resulting from vein widths less than the minimum mining width for a particular area S_o within the plane of the vein (or other tabular deposit). The estimate for w_c assumes that all measurements in each class interval could be approximated by the mean of the class interval. The potential error from this assumption is probably much less than the error inherent in the assumption that the histogram represents the vein thicknesses perfectly over area S_o.

A comparable estimate for external dilution at Silver Queen, resulting from some vein widths being less than minimum mining width, can be made on the assumption that the vein widths have a normal distribution. The standard score for the minimum mining width is $z = (4.0 - 4.13)/1.74 = -0.08$. Substitution of this value in Eqs. 4.6 and 4.9 gives $p_c = 0.468$ and $w_c = 2.42$ ft (Fig. 16.3). These estimates are close to those obtained from the histogram analysis

Table 16.1 Summary, minimum mining width dilution, Silver Queen vein deposit, central British Columbia

Thickness interval (ft)	Mean thickness (ft)	Frequency (no.)	Cumulative ($t \times f$)
1.0–1.5	1.25	12	15.0
1.5–2.0	1.75	13	37.75
2.0–2.5	2.25	34	114.25
2.5–3.0	2.75	36	213.25
3.0–3.5	3.25	50	375.75
3.5–4.0	3.75	23	462.0

$P_{<4.0} = 168/364 = 0.462$
$t_{<4.0} = 462.0/168 = 2.75$ ft
Vein thickness parameters, Silver Queen drift

n		x	s
365		4.15	1.86
364[a]		4.13	1.74

[a] Ignores one large value of 13 ft.

discussed previously. The value for w_c based on the assumption of normality may be a slight underestimate because a small portion of the normal tail extends to low values not reflected in the histogram and the lowest extremity of the tail extends to negative values that are physically impossible. Nevertheless, the similarity of the two approaches, both of which are approximate, demonstrates the viability of each approach.

16.2.3: Dilution from Overbreaking

In general, it is virtually impossible to mine exactly to a planned mining limit or a contact, however sharp; instead, there is a tendency for overbreak with the obvious result of attendant dilution by some amount of wallrock. This effect is clearly a function of wallrock properties and mining procedures (cf. Pakalnis et al., 1995); when rock is highly competent and the mining process highly selective, the effect can be relatively minor. In some cases, limited experience gained from exploratory and development work provides information that permits reasonable estimation of the average overbreak to be anticipated. For example, in the case of the Silver Queen deposit, the wallrock is mostly highly competent and overbreak will be minimal, perhaps on the order of 1 ft of added thickness throughout the vein. This additional dilution amounts to

$$V_{ob} = f \cdot S_o \text{ ft}^3$$

where V_{ob} is a volume of overbreak material, easily transformed to tonnage with an appropriate bulk density factor. The total volume of mined material determined from Eq. 16.2 is thus increased to $(V_p + f \cdot S_o)$, and a corresponding weighted, average grade can be determined using a relation comparable to Eq. 16.4. Such a procedure is used at the Lupin gold mine (Bullis et al., 1994), where "reserve dilution is estimated by adding 1 m, at assay grade, on both sides of the ore in the Centre zone. In the West zone, reserve dilution is estimated by the addition of 0.5 m, at assay grade, on both sides of the ore outline" (p. 75). Bullis et al. (1994) document the practical difficulty of reconciling production with estimates, because of the common uncertainty of quantifying dilution accurately.

16.2.4: Contact Dilution

Stone (1986) describes an empirical deterministic approach to the estimation of a type of external dilution that he refers to as *contact dilution*. This particular component of external dilution results from intermittent protrusions of wallrock that penetrate beyond smooth, interpreted mining margins into ore. Stone's estimation procedure for this form of dilution is complex and difficult to implement because appropriate information is rarely available with which to make an estimate to a reasonable level of confidence. A simpler

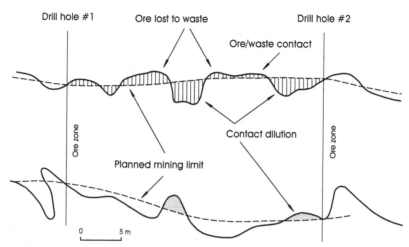

Figure 16.4: An approach to measuring "contact dilution" as defined by Stone (1986). Vertical lines extending from the irregular deposit margin to the smooth interpreted margin along the upper margin contact of the hypothetical deposit have been superimposed on the original diagram and their lengths measured to characterize the ore loss and dilution populations relative to the planned mining unit. An arbitrary scale (32 m between drill holes) has been assumed for the purposes of illustration. See text for details.

approach that meets his aim can be applied when exploration exposes a representative length of deposit–wallrock contact such that it can be examined and characterized with confidence. Given sufficient information that the ore/waste contact is known in detail in one or more parts of a deposit, it is a simple matter to estimate the extent of waste protrusions into ore on detailed sections and extend this estimate to the third dimension. When such detail is known only at a few, local parts of a deposit, an alternative procedure is useful.

Consider Fig. 16.4 (modified from Stone, 1986), for which a scale has been assumed for purposes of illustration. Lines have been superimposed on Stone's original diagram parallel to the bordering section traces and are sites used to measure distances from the interpreted average contact or mining margin (dashed line) to the true ore/waste contact. Negative values are ore that is lost to wallrock; positive values represent dilution by lobes of wallrock extending across the mining margin into ore. A histogram of these distances is prepared and serves as a basis for estimating the proportion of length between control points, where waste penetrates across the mining margin into ore. The average thickness of such waste extending into ore can be determined as a weighted average, using class frequencies as weights for the mid-values of corresponding class intervals. If the distribution of measurements is found to be approximately normally distributed, then Eqs. 4.6 to 4.9 can be used to determine the proportion of positive values and the average thickness of the positive values. Of course, the length over which positive values occur can be measured directly from the figure. The thickness data measured along the lines shown on Fig. 16.4 are presented as a histogram in Fig. 16.5. The positive values represent 44 percent of the measurements (this compares with 43 percent of the actual measured length of the mining limit assumed in Fig. 16.4) and have a weighted mean value of 1.4 m. For the assumed scale, the proportion of positive values is the proportion of the 32-m length between control points that is represented by waste protruding into ore, a length of 14.1 m. The product of this length and the average thickness of protrusions is a two-dimensional estimate of the quantity of contact dilution along one mining margin, in this case about 19.7 m^2, or about 4.7 percent of the area mined.

A comparable estimate made along the opposite margin shows that the total contact dilution in this hypothetical example is about 5.1 percent of the area outlined for mining. Several such estimates,

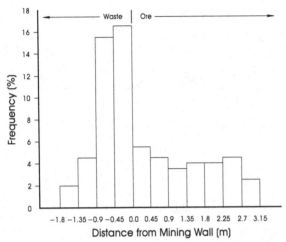

Figure 16.5: Histogram of measurements of thickness of contact diluting ground obtained from the hypothetical example in Fig. 16.4. There are 67 equally spaced measurements in total. Negative measurements are ore beyond the mining limit (i.e., ore lost to waste); positive values represent wallrock dilution.

averaged, provide a global estimate. The method has the advantage that it uses point estimates and thus can combine information from another dimension (e.g., when both a drift and a raise provide information on the nature of the true ore/waste contact, each can be used for independent estimates of contact dilution or they can be combined to a single estimate).

16.3: INTERNAL DILUTION

16.3.1: A Geostatistical Perspective

David (1988) and David and Toh (1989) discussed internal dilution from a geostatistical perspective. The principal causes of inherent internal dilution relate to

(i) Size of the selective mining unit (SMU)
(ii) Density of sampling
(iii) Small-scale ore continuity
(iv) Position of the cutoff grade relative to the mean of the probability density function.

A component of internal dilution is considered inherently in the concept of the changing dispersion of average grades with changing support. This is the

so-called block size effect on dilution (David and Toh, 1988). As the SMU size increases, selectivity decreases, and there is a corresponding increase in tonnage mined with a decrease in average grade. Such dilution can be considered automatically with the application of the volume–variance relation of geostatistics (cf. Journel and Huijbregts, 1978):

$$D^2(\text{SUM}, D) = D^2(v, D) - D^2(v, \text{SMU})$$

$$(16.4)$$

where $D^2(a, b)$ is the dispersion variance of small volume a in large volume b. The parameters in Eq. 16.4 can be determined with a knowledge of the semivariogram. For example, $D^2(v, D)$, the dispersion of sample grades in the deposit, is the sill of the semivariogram, and $D^2(v, \text{SMU})$, the dispersion of sample grades in an SMU, is the F function of geostatistics and can be determined from graphs (e.g., David, 1977; Journel and Huijbregts, 1978) or by an appropriate computer program (e.g., Clark, 1976). This effect can also be viewed in terms of regularization (i.e., the smoothing of grades as support increases). For example, the nugget effect varies as a function of volume over grade, which is determined, as follows:

$$C_0(v) = A/v$$

where $C_0(v)$ is the nugget effect for grades of sample volumes v and A is a constant. As v becomes very large, A/v becomes negligibly small, and the nugget effect disappears. There is a corresponding decrease in the sill of a spherical model for a variable regularized over increasing volumes, and thus a concomitant decrease in the dispersion variance. Normally, the SMU volume is controlled by the mining method. Sample size v can be variable and theoretically may have a minor impact on Eq. 16.4. For this reason, it is useful to use a database with as uniform a support as is practical. For unbiased data of common support, the dispersion variance is equivalent to the sill of the semivariogram

$$\gamma_v(\infty) = D^2(v, D).$$

The information effect also produces an apparent dilution (David, 1988). Selection is made on estimated values (e.g., by kriging blocks), which imposes smoothing on results according to an established relation (Journel and Huijbregts, 1978):

$$D^2(SMU^*, D) = D^2(SMU, D) - \sigma_k^2 \quad (16.5)$$

where

$D^2(SMU^*, D)$ is the dispersion of estimated grades of SMUs

$D^2(SMU, D)$ is the dispersion of true grades of SMUs

σ_k^2 is the average kriging variance of SMUs.

Clearly, the better the estimate, the less the smoothing effect and the fewer errors made during selection. Note that the dispersion of estimated-block (SMU) grades is greater during mining than for estimates based on exploration diamond drilling because more information available at the mining stage (e.g., blastholes) means the average kriging variance is reduced. Equation 16.5 can be used during the exploration stage to estimate the average kriging variance from an expected blasthole array; hence, it is possible to calculate the estimated dispersion of grades at the production stage.

The continuity model has an important effect on dilution in the case of geostatistical block estimates because with better continuity, fewer classification errors are made and less internal dilution results. Thus, highly continuous variables such as sedimentary iron values suffer little in the way of internal dilution, relative to less continuous grade distributions such as epithermal gold deposits.

The relation of cutoff grade to the mean of the density distribution function also affects dilution. In general, as the cutoff grade increases, the number of classification errors (ore or waste) increases, and more dilution results. For a situation of constant error, the absolute number of classification errors increases as the cutoff grade approaches the peak of the distribution because of the increased frequency of values.

16.3.2: Effect of Block Estimation Error on Tonnage and Grade of Production

Block-estimation error can have a dramatic impact on production tonnage (e.g., Postolski and Sinclair, 1998b). Clearly, if some blocks of ore are inadvertently classed as waste, tonnes are lost, metal is lost, and profit is decreased relative to expectations. Similarly, if blocks of waste are included in ore, total tonnage is increased but average grade is decreased (i.e., dilution occurs), and operating profit is less than expected. The problem is illustrated clearly in Fig. 16.6, where error curves are superimposed on very narrow grade ranges both above and below cutoff grade and the proportions of misclassified blocks are indicated. Of course, if the error curve had greater dispersion (e.g., 20 percent error instead of 10 percent error), the proportion of misclassified blocks would be larger for any specific narrow-grade range.

When block-grade distributions can be approximated by a normal or lognormal distribution, a cutoff grade can be defined and the error can be taken to be normally distributed. The equations in Chapter 4 allow rapid calculation of (i) the proportion of a distribution that is above (or below) the cutoff grade (this is equivalent to the proportion of tonnes that are misclassified) and (ii) the average grade of that proportion above the cutoff grade. This procedure can be repeated for many narrow contiguous-grade ranges in the manner illustrated (Fig. 16.6) and the total number of misclassified blocks can be calculated. Because manual application of this procedure to many class intervals is time-consuming, tedious, and thus subject to error, Postolski and Sinclair (1998) produced a computer program, GAINLOSS, to carry out and summarize such calculations (see publisher's website for GAINLOSS software).

As an example, consider the case of the Bougainville porphyry-copper deposit (David and Toh, 1988). The distribution of blasthole grades (lognormal with a mean of 0.45 percent Cu and natural logvariance of 0.21) is assumed to represent the true distribution of grades (parameters are summarized in Table 16.2). Small departures from this assumed distribution have a negligible impact on the principal

Table 16.2 Parameters of blasthole grade distribution, Bougainville copper mine[a]

	x	s²	s
Raw data	0.45	0.0473	0.218
Natural logs	−0.9035	0.21	0.458

[a] Cu metal in wt.%.
Source: After David and Toh (1988).

conclusions that follow. The cutoff grade is quoted as 0.215 percent Cu; therefore, Eqs. 4.14 to 4.17 can be used to demonstrate that 91.9 percent of true grades are above cutoff grade, 97 percent of the contained metal is in the material above cutoff grade, and the average grade of material above cutoff grade is 0.475 percent Cu.

For this example, an ideal situation is assumed wherein each blasthole is considered to be centrally located within a block of ore and the average grade assigned to the blasthole is used to classify the block as ore or waste (i.e., a classic polygonal estimate). For 1,000 such blocks, only 919 are truly ore, although classification of blocks as ore or waste is in error to some extent if the true grades are near the cutoff grade. The likelihood of misclassification is thus seen to be a function of the true estimated grade and the error distribution curve. Consider a single short-grade interval of 0.195 to 0.205 percent Cu, assumed to be centered on 0.20 percent Cu and an estimation error of 10 percent (i.e., error as one standard deviation is 0.020 percent Cu). Equation 4.14 can be used to estimate the proportion of blocks with a true grade of 0.20 percent Cu that will be reported with a grade above cutoff

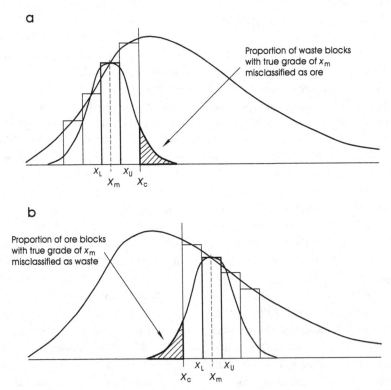

Figure 16.6: A normal curve representing 10 percent random error is superimposed on a narrow class interval (X_L to X_U). This interval is one of many such intervals that define a block-grade distribution represented by the positively skewed (lognormal) distribution. (a) The normal distribution is below cutoff grade X_C and a proportion of waste blocks (hachured area) of average grade X_m are included in ore. (b) The normal distribution is above cutoff grade X_C and a proportion of ore blocks (hatched area) of average grade X_m are included in waste.

Table 16.3 Number of waste blocks mistakenly included in ore due to various levels of error, Bougainville porphyry deposit, for a cutoff grade of 0.215 percent Cu

Grade interval center	Freq. in 1,000 blocks	10% error $P > c$	$N > c^*$ 10% error	20% error $P > c$	$N > c^*$ 20% error	30% error $P > c$	$N > c^*$ 30% error
0.11	1.1					0.000	0.000
0.12	1.8					0.003	0.005
0.13	2.7			0.000	0.001	0.012	0.033
0.14	3.9			0.003	0.010	0.034	0.131
0.15	5.2			0.013	0.067	0.071	0.373
0.16	6.7	0.000	0.001	0.040	0.268	0.124	0.833
0.17	8.4	0.003	0.024	0.090	0.754	0.187	1.568
0.18	10.1	0.023	0.232	0.164	1.647	0.258	2.592
0.19	11.8	0.091	1.075	0.255	2.994	0.330	3.883
0.20	13.4	0.226	3.032	0.354	4.751	0.401	5.390
0.21	15.0	0.406	6.104	0.453	6.807	0.468	7.044
Sum of misclassified waste blocks			10.468		17.298		21.854
Average true grade of misclassified waste blocks			0.204		0.198		0.194

* $N > c$ is the number of waste blocks per 1,000 blocks mined, mistakenly included in ore due to the indicated percentage error and a cutoff grade of 0.215 percent Cu.

grade; that proportion is 0.226. Also, Eq. 4.14 can be used to estimate the cumulative proportion of grades from infinity to each side of the grade interval; the difference in these two cumulative percentages is the frequency within the interval and is estimated to be 0.013. Thus, for the 1,000-block example, 13 blocks will have true values between 0.195 and 0.205 percent Cu, and $13 \times 0.226 = 2.9$ (round to 3) of these waste blocks will be misclassified as ore if the error is 10 percent. A similar procedure can be followed for many contiguous short-grade intervals below cutoff grade, and the number of misclassified blocks can be determined in each case. Results of such a calculation using the GAINLOSS software are summarized in Table 16.3 for assumed errors of 10 percent, 20 percent, and 30 percent. The table includes the average total number of diluting blocks (per 1,000 blocks) and their true average grade for each of the three error secenarios (e.g., 10.5 blocks of waste, averaging 0.204 percent Cu incorrectly classed as ore in the case of 10 percent error). The true average grades were determined as a weighted average of the central grade of each grade interval, weighted by the number of diluting blocks.

Of course, errors of misclassification also apply to grades above but near cutoff grade; some ore blocks are inadvertently classed as waste. In this case, for any short grade interval it is possible to determine the proportion of ore blocks that will be classed incorrectly as waste, using a procedure comparable to that described previously. For each short-grade interval above cutoff grade, it is possible to estimate the proportion of blocks that will be classed incorrectly as waste due to any specified error. Calculations of this nature for the Bougainville Cu distribution are summarized in Table 16.4 for errors (as standard deviations) of 10 percent, 20 percent, and 30 percent. In the case of the 10 percent error scenario, an average of 17.3 blocks of low-grade ore are classed as waste. These blocks average 0.233 percent Cu, substantially above cutoff grade, and the loss of profit is evident.

The results are of considerable significance for several reasons. Return to the 1,000 block and 10 percent error scenario; 919 blocks are truly above cutoff grade. Of these 919 blocks, 17 are inadvertently classed as waste, leaving 902 blocks with average

Table 16.4 Number of blocks of ore mistakenly classed as waste due to verious levels of error, Bougainville porphyry deposit, for a cutoff grade of 0.215 percent Cu

Grade interval center	Freq. in 1,000 blocks	10% error P<c	N<c*	20% error P<c	N>c*	30% error P<c	N<c*
0.22	16.6	0.410	6.787	0.455	7.527	0.470	7.776
0.23	17.9	0.256	4.602	0.372	6.676	0.414	7.427
0.24	19.2	0.147	2.819	0.301	5.776	0.364	6.989
0.25	20.3	0.078	1.580	0.241	4.896	0.320	6.499
0.26	21.3	0.039	0.821	0.192	4.084	0.282	5.984
0.27	22.1	0.018	0.400	0.152	3.361	0.248	5.467
0.28	22.7	0.008	0.185	0.121	2.738	0.219	4.962
0.29	23.2	0.004	0.082	0.095	2.212	0.193	4.481
0.30	23.6	0.002	0.036	0.075	1.777	0.171	4.029
0.31	23.8	0.001	0.015	0.060	1.420	0.152	3.612
0.32	23.9	0.000	0.006	0.047	1.132	0.135	3.229
0.33	23.9	0.000	0.003	0.038	0.900	0.120	2.881
0.34	23.8	0.000	0.001	0.030	0.715	0.108	2.567
0.35	23.7	0.000	0.000	0.024	0.568	0.097	2.285
0.36	23.4			0.019	0.451	0.087	2.032
0.37	23.1			0.016	0.358	0.078	1.807
0.38	22.7			0.013	0.285	0.071	1.606
0.39	22.2			0.010	0.228	0.064	1.428
0.40	21.7			0.008	0.182	0.058	1.270
0.41	21.2					0.053	1.129
0.42	20.6					0.049	1.005
0.43	20.0					0.045	0.895
0.44	19.4					0.041	0.797
0.45	18.8					0.038	0.711
0.46	18.2					0.035	0.634
0.47	17.6					0.032	0.566
0.48	17.0					0.030	0.506
0.49	16.3					0.028	0.452
0.50	15.7					0.026	0.405
0.51	15.1					0.024	0.363
0.52	14.5					0.022	0.325
0.53	13.9					0.021	0.292
Sum of misclassified ore blocks			17.339		45.284		84.410
Average true grade of misclassified ore blocks			0.233		0.261		0.298

* $N<c$ is the number of ore blocks per 1,000 blocks mined that are mistakenly classed as waste for the indicated percentage error and a cutoff grade of 0.215 percent Cu.

grade (g_{902}), as follows:

$$902 \times g_{902} = 919 \times 0.475 - 17 \times 0.233$$
$$g_{902} = 0.480\% \text{ Cu.}$$

Adding the dilution resulting from the 10 blocks of waste (cf. Table 16.3) that are incorrectly classed as

ore, the resulting average grade is

$$912 \times g_{912} = 902 \times 0.480 + 10 \times 0.204$$

to give $g_{912} = 0.477\%$ Cu.

The important point to be made is that although dilution has occurred, the average grade of mined ore is slightly higher than the overall average grade above cutoff (0.477 percent Cu vs. 0.475 percent Cu), because the effect of losing 17 blocks of low-grade ore slightly overshadows the diluting effects of including 10 blocks of relatively high-grade waste. Thus, a loss in tonnage has resulted in somewhat higher grade than expected from all true ore blocks. This effect occurs in reverse if the cutoff grade is on the upper tail of the distribution of real grades (i.e., if a high absolute number of waste blocks are included with ore relative to the number of ore blocks lost as waste). In the latter case, the effect of dilution predominates over the effect of losing ore to the waste dump, and the mean grade of material classed as ore is less than the mean grade of all ore blocks. Moreover, the tonnage of material classed as ore is greater than the tonnage of true ore.

The effects of error are likely to be overlooked if they are on the scale indicated by the 10 percent error case discussed previously because there is a minimal improvement in grade and a relatively small loss in tonnes, although the actual metal loss is significant. However, as the level of error increases, the impact becomes more significant. The results for 20 percent and 30 percent errors applied to the Bougainville example are summarized as metal accounting in Table 19.5 and clearly demonstrate the serious loss of metal as the level of sampling plus analytical error increases. For the 30 percent relative error scenario, the loss of metal is 150 tonnes of Cu metal (block size is 2,000 tonnes).

The effect on grade of waste material classed as ore can be calculated, as was done above for the 10 percent error scenario. For example, the effect of 84 blocks of lost ore is

$$919 \times 0.475 - 84 \times 0.299 = 835 \times g_{835}$$

from which g_{835} is found to be 0.493 percent Cu. Now, add the 22 blocks of dilution

$$0.493 \times 835 + 0.194 \times 22 = 857 \times g_{857}$$

from which g_{857} is found to be 0.485 percent Cu. Note that the average grade of material classed as ore is significantly higher than the expected average (0.475 percent Cu); however, the loss of tonnes is $(919 - 857)2000 = 124,000$ tonnes per 1,000 blocks

mined. For distributions for which the cutoff grade were on the high-grade tail (rather than the low-grade tail, as is the case here), the average grade would be below the expected grade because the effects of diluting with waste would be greater than the relatively small loss of ore. Although idealized, these calculations provide useful insight into the need for high quality in both sampling and assaying so that block-estimation errors are minimized.

16.4: DILUTION FROM BARREN DYKES

In many producing mineral deposits, the presence of barren dykes represents a significant source of actual or potential dilution. Although the geologic control of the dykes may be understood in a general way, the estimation of where dykes occur and their quantitative impact on grade is commonly not well understood in detail, and is estimated in highly subjective ways based on local geologic knowledge involving time-consuming manual methods. Practical situations involving dilution of ore by barren dyke material in two different geologic and production environments are discussed for the Snip mesothermal gold vein (underground) and the Virginia Cu–Au porphyry-type deposit (open pit). In each of these cases, indicator kriging has been an aid to local estimation of dilution by barren dykes (e.g., Sinclair et al., 1993).

16.4.1: Snip Mesothermal Deposit

The Snip Mine is about 110 km northwest of Stewart, northwestern British Columbia, Canada. Published reserves are about 870,000 tonnes grading approximately 28 g Au/mt with minor amounts of silver and copper. The ore deposit is a small, complex, mesothermal gold vein system known as the *Twin Zone* in a sequence of stratified rocks, mainly feldspathic graywackes. A longitudinal section is shown in Fig. 16.7. The metasedimentary unit is intruded by lamprophyre dykes that dip about 85 west and crosscut the gold-bearing structure. Rhys and Godwin (1992) indicate that these dykes were emplaced late in the mineralization and structural history of the Twin Zone. Some of these dykes, locally referred to as the

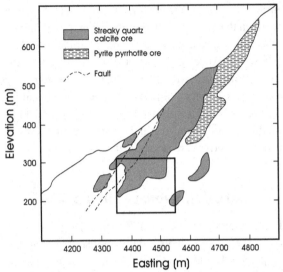

Figure 16.7: Longitudinal section of the Snip mesothermal gold–quartz vein, northern British Columbia, showing the test panel for which a study was done of the impact of barren dyke dilution on ore grade. Redrawn from Sinclair et al. (1993).

biotite spotted unit (BSU), are spatially associated with the Twin Zone; one prominent dyke roughly parallels the Twin Zone partly in the adjoining hanging-wall, but in places internal to and subparallel with the Twin Zone (Fig. 16.8), dividing the vein into two tabular sheets (hence the name *Twin Zone*). Mining

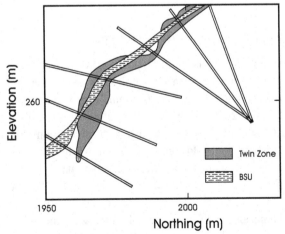

Figure 16.8: Illustration of the relation of barren dykes (BSU) to mineralized vein, Snip mesothermal gold–quartz deposit. Redrawn from Sinclair et al. (1993).

procedures require that, where present within the Twin Zone, the barren BSU must be mined with vein material. Thus, for any position on the longitudinal section of the vein, it is important to determine where BSU exists and, if present, to estimate the amount of BSU.

In a recent geostatistical study of the Snip deposit, Sinclair et al. (1993) developed a procedure for the sequential estimation of reserves that includes

(i) Estimating presence or absence of BSU dyke material using indicator kriging
(ii) Where BSU is present internally within the Twin Zone, estimating the thickness of the BSU by ordinary kriging to quantify dilution.

In this way, the BSU is handled as an estimation problem independent of the estimation of vein material. Thus, where appropriate, internally diluted reserves can be determined in blocks (or stopes) in longitudinal section as a weighted average of a vein estimate and barren dyke (BSU) estimate. Relatively few thin horses of weakly mineralized graywacke are included in Twin Zone intersections; hence, this small source of dilution is included in what are referred to as *undiluted reserves*.

Indicator kriging results were used to produce a contour map of the probability that condition (i) prevails (i.e., the probability that the BSU occurs within the Twin Zone; Fig. 16.9). Overlaying a block model on such a contour map allows those blocks with internal BSU to be identified. The amount of internal dilution by BSU depends on its average thickness where it occurs within the Twin Zone; BSU thickness was estimated by ordinary kriging.

Diluted block grades are weighted grades over the combined (vein + BSU) thickness. Results for 15 m × 15 m blocks (in longitudinal section) in the test panel under study have been estimated. The total undiluted reserves (cutoff = 13 g Au/t) for the test panel are 115,000 tonnes at an average gold grade of 34.4 g/t and an average thickness of 2.47 m. Corresponding internally diluted reserves are 134,000 tonnes averaging 30.3 g/t gold and 3.08 m thickness (Sinclair et al., 1993). The internal dilution is thus about 16 percent of in situ tonnes.

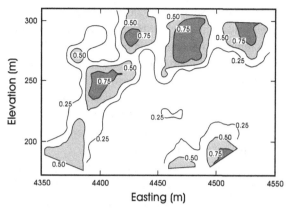

Figure 16.9: Contour plot (for test panel) of estimated probabilities that BSU (barren dyke) occurs within the Snip vein. Redrawn from Sinclair et al. (1993).

16.4.2: Virginia Porphyry Cu–Au Deposit

The Virginia copper–gold porphyry-type deposit is in the Copper Mountain Camp near Princeton, British Columbia. Mineralization occurs at the contact of the Lost Horse Complex and the Nicola Group. The Lost Horse Complex is a composite body of dykes, sills, and irregular stocks that range in composition from syenite to diorite; Nicola Group rocks consist of coarse-grained agglomerates, tuff breccias, tuffs, and flow rocks. Mineralization occurs mainly as a variety of magnetite–pyrite veins, some of which contain Cu (mainly chalcopyrite) and significant amounts of Au. Veins strike roughly easterly and dip steeply to vertically. Numerous felsite dykes occur within the proposed pit; these are unmineralized and represent a serious dilution problem during mining. Dykes strike roughly northerly and have near vertical dips; they are a few feet to a few tens of feet thick.

The effect of dilution by felsite dykes on ore grade and tonnage for production purposes was estimated at nodes ($10 \times 25 \times 10$ ft^3 or $3.05 \times 7.62 \times 3.05$ m^3) of a horizontal grid by indicator kriging (Nowak and Sinclair, 1993). The estimated value at each node represents the probability of occurrence of a felsite dyke; if the probability is high (e.g., >0.5), the presence of a dyke is indicated. These data were used to estimate the total volume of dyke material in a much larger volume of ore/waste (one bench) to provide an

estimate of the proportion of dyke material present in a large panel (e.g., 3540 bench). A manual estimate by mine staff is in close agreement with the results by indicator kriging. The proportion of the volume to be mined represented by dykes is estimated manually to be 13 percent, which compares favorably with the 14 percent estimate from the more automated procedure of indicator kriging.

16.4.3: Summary: Dilution by Barren Dykes

For the Snip mesothermal gold–quartz vein, dilution by BSU dyke material within the vein structure (so-called internal dilution) was estimated by a two-step procedure. Average BSU thickness for each block (defined on a long section through the deposit) was estimated by ordinary kriging. To determine which ore blocks were to be diluted, indicator block kriging was performed, using an indicator for presence or absence of the BSU within the mineralized zone. Blocks for which the probability of BSU occurrence within the mineralized zone was equal to or greater than 0.5 were diluted by the estimated amount of BSU taken at zero grade.

The Virginia study shows that indicator kriging can be used as a practical estimation method of the position and the proportion of dykes in a large volume (e.g., one bench). Thus, the technique could be implemented as a practical procedure for estimating dilution at the mine site or could be used as a supplement to the manual method (i.e., as a checking procedure). One of the strengths of the kriging approach is that it can be implemented as an automated method based on exploration data, and thus can serve as a basis for early mine planning. It is also less time consuming and in turn cheaper than the manual approach, although there is certainly room for potentially dangerous errors if care is not taken in establishing an appropriate autocorrelation model. Such a model must represent anisotropies and ranges of continuity that truly reflect reality.

Alternative approaches can be used to deal with internal dilution of large, easily mappable geologic units. Sinclair et al. (1994) describe a simple method that they used for the south-pit mineral inventory estimates at the Nickel Plate skarn gold deposit, southern British Columbia. In this case, large diabase dykes cut the ore, but can be mapped easily in drill-core logging and can be interpolated onto plans and sections with confidence because of their regularity. A three-dimensional block array is superimposed on the geologic interpolations and for each block an estimate is made of the proportion that is dyke and the proportion that is nondyke (see Fig. 3.13). The method used for these estimates was to digitize the dyke outlines and use Techbase software to merge this digitized outline with the block array in order to estimate the required proportions. Subsequently, interpolated grade estimates for each block were used to produce a final weighted block estimate as follows:

$$g_{reported} = f_m \times g_m + f_b \times g_b$$

where $g_{reported}$ is the grade estimate reported for a block, g_m is the interpolated grade of the block, g_b is the grade of diluting material (in this case, barren dyke; hence, $g_b = 0$), f_m is the fraction of the block that is mineralized, and f_b is the fraction of the block that is barren dyke.

16.5: PRACTICAL CONSIDERATIONS

1. A component of internal waste is incorporated into block estimates that are formed by averaging. This process is automatic for kriging estimates and increases as block size increases because of the decrease in selectivity with increasing block size. The effect cannot be quantified easily for estimates based on empirical weighting procedures.

2. In some cases, it is possible to estimate dilution at the margins of deposits resulting from (i) deposit thickness less than minimum mining thickness, (ii) sinuous ore/waste contacts, and (iii) overbreak.

3. Measurements at short regular intervals of (i) the thickness of waste included in ore and (ii) the thickness of ore included in waste can be used to produce a histogram that quantifies both contact dilution and lost ore, respectively, for local accessible parts of a deposit. Geology indicates if such estimates are representative of large parts of

a deposit. Use of this procedure achieves the same purpose aimed for by Stone (1986), but avoids his complicated geometric approach.

4. An unbiased histogram of vein thicknesses for a well-sampled part of a deposit provides information on the proportion of the deposit below minimum mining width and the average thickness of diluting material over the surface represented by the measurements.

5. Overbreaking is not easily estimated with confidence from exploration data without a basis of actual measurements. In some cases, such information can be obtained from exploratory underground workings and development workings.

6. The common example of large barren dykes cutting ore zones can be dealt with in a variety of ways, including (i) mining such dyke material as waste and (ii) independent estimates of both the grade of the mineralized part of a block and the proportion of dyke material present in a block. Using these two pieces of independent information prior to an ore/waste decision, appropriate blocks are diluted.

7. Dilution as a function of random error of block-grade estimates can be quantified in a general way if the level of error is known. Although estimation errors can never be reduced to zero, it is possible to determine the effect of various errors on a specific grade distribution and estimate how worthwhile it is to improve (decrease) the level of sampling and analytical error.

16.6: SELECTED READING

Elbrond, J., 1994, Economic effects of ore losses and rock dilution; Can. Inst. Min. Metall. Bull., v. 87, no, 978, pp. 131–134.

Pakalnis, R., R. Poulin, and S. Vongpaisal, 1995, Quantifying dilution for underground mine operations; presented at the Ann. Gen. Mtg., Can. Inst. Min. Metall., Halifax, N. S., May 14–18, 8 pp. plus diagrams.

Scoble, M. J., and A. Moss, 1994, Dilution in underground bulk mining: implications for production management; *in* Whateley, M. K. G., and P. K.

Harvey (eds.), Mineral Resource Evaluation II: methods and case histories, Geological Society Spec. Pub. No. 79, pp. 95–108.

16.7: EXERCISES

1. The program GAINLOSS is available on the internet at the publisher's website. This program provides a rapid means of estimating both dilution and loss of ore that accompanies a random block estimation error (combined sampling and analytical error). Recall that dilution arises because some waste blocks are incorrectly identified as ore; ore loss occurs because some ore blocks are incorrectly classed as waste. The program requests user input, including normal or lognormal distribution, mean and standard deviation of true grades (e.g., estimated from blasthole distribution), units, percentage error to be assumed, and cutoff grade, and then outputs a detailed table and summary of dilution and ore loss. A block grade distribution for a Zinc deposit is lognormal; the mean of raw data is 6.5 percent and the standard deviation is 7.5. The present block estimation error averages 22 percent; determine the additional metal that would be available if the average error were reduced to 10 percent, given that each block is 15,000 tonnes and annual production is 1,000 blocks.

2. It is virtually impossible to mine a tabular deposit to a predetermined simple geometric limit; there is a general tendency to "overbreak" beyond the idealized mining limit. Overbreak is commonly estimated as an average additional thickness determined by experience from mining wallrock of various breakage characteristics. A 100 m × 350 m rectangular longitudinal projection of a vertical vein deposit cuts highly altered volcanic rocks for 73 percent of its planar extent (average thickness = 1.7 m; average grade of vein = 14.3 g Au/t) and cuts competent granitic rocks for the other 27 percent of its planar extent (average thickness = 1.4 m; average grade = 18.4 g Au/t). Experience has shown that mining in volcanic rocks results in a total overbreak (both sides of the vein) of about 0.7 m, averaging 2.1 g Au/t, and that comparable dilution in the granitic rock is 0.3 m, averaging 0.0 g Au/t. Assume bulk densities of 2.7 g/cc for vein material, 3.0 g/cc for volcanic rock, and 2.9 g/cc for granitic rock. Calculate the mined tonnages and grades (i.e., including dilution) for each country rock type.

3. A near-vertical tabular deposit has a thickness distribution closely fitted by a normal distribution with mean of 2.5 m and a standard deviation of 0.9. What is the proportion of the deposit that is less than the minimum mining width of 1.3 m? What is the average thickness of that portion of the deposit below minimum mining thickness? Calculate the idealized dilution from mining that part of the deposit below minimum mining width.

17

Estimates and Reality

Time and again numbers reported as "ore reserves" have failed the practical test of full-scale mining and processing. The primary shortcoming usually is the grade of the material extracted from the ground and sent to the processing facility; the secondmost shortcoming is the inability of the mine to produce ore at the designed-for rate. . . . The cause of these problems often can be traced to improper reporting of resource/reserve numbers. In particular, insufficient attention is paid to the significant difference between a mineral resource as delineated in place and the reserves which can be extracted therefrom. (Grace, 1986b, p. 47)

Chapter 17 draws attention to comparative estimation projects in which the results of two or more estimation methods are compared, or one or more estimation methods is compared with production information. Because of the importance this text places on geostatistical methods of estimation, consideration is given to some of the more vociferous antagonists of the subject.

17.1: INTRODUCTION

The practical validity of any resource/reserve estimation technique lies in a comparison of estimates with production, a procedure commonly referred to as a *reconciliation study*. Such studies are carried out routinely throughout the industry but are rarely published; even when published, too few details are normally provided with which to undertake a true evaluation of the estimation methods. Case histories involving comparative studies of resource-estimation techniques and production are an important source of public technical information regarding the validity of particular estimation procedures. There are two types

of reconciliation studies: those based on a simulated database and those based on a real mineral deposit. Simulated data provide a means of comparing estimation methods with an idealized reality and can be conducted prior to actual production. In the end, however, estimation methods must be shown to be adequately successful in applications to real deposits. Table 17.1 is a selected list of a variety of types of geostatistical studies that incorporate elements of reconciliation between estimates and production. There are other examples scattered throughout the literature, some of which discuss the relative quality of a variety of estimation methods. In general, such comparisons show that carefully conducted geostatistical studies consistently produce estimates that are either the best or among the best relative to production.

Unfortunately, in the nontechnical literature, there are all too many examples of a basic misunderstanding of the process of estimation, which leads to misrepresentations of comparative studies. As a result, reconciliation studies involving various estimation techniques and production also can be misrepresented. It is unfortunate when ill-conceived and even misleading views of estimation results are publicized in such a fashion. Of course, all methods are subject to criticism

Table 17.1 Selected case histories involving reconciliation of geostatistical estimation techniques with production

Deposit type	Brief description	Reference
Porphyry Cu	Similkameen deposit: comparison of ordinary block kriging, "conditional probability," and polygonal estimates with production	Raymond, 1979
	Valley copper deposit: similar to Similkameen deposit	Raymond and Armstrong, 1988
Skarn Au	Comparison of ordinary block-kriging, inverse-distance, and conditional probability estimates with production	Sinclair et al., 1993
Mississippi Valley–type Pb–Zn	Comparison of conditional probability, ordinary kriging, and production	Raymond, 1984
Porphyry Mo–Cu	Comparative study of mill-head grades with kriging and inverse-distance-squared estimates	Johnstone and Blackwell, 1986; Norrish and Blackwell, 1987
Epithermal Au	Golden Sunlight: blasthole kriging versus production	Roper, 1986
	Cresson Mine: various kriging estimates versus production	Pontius and Head, 1996
Massive sulphide	Louvem deposit: compares method of sections, production, and postmortem ordinary block kriging	Vallée et al., 1982

and geostatistics is no exception (see later). Perhaps much of the concern with geostatistics arises because of the complexity of geostatistical estimation procedures relative to more traditional estimation methods. There are many methods of resource/reserve estimation, some geostatistical in nature (e.g., ordinary kriging, simple kriging, multiple indicator kriging) and some empirical (e.g., method of sections, polygonal, inverse-distance weighting). Any one method may be appropriate in a specific case. Many of these methods (and variations of them) have been described in textbooks and the technical/scientific literature and have been tested in practice over many years.

17.2: RECENT FAILURES IN THE MINING INDUSTRY

The mining literature contains numerous examples of mine failure, many of which stem from the inability of production to match reserve estimates, either in terms of grade (Fig. 17.1) or tonnage. Error in each of these quantities is inevitable, but the expectation is that appropriate estimation procedures, performed by professionals, will produce estimates that are rela-

tively close to and perhaps within 10 percent of reality. Unfortunately, most cases of serious error involve overestimation (i.e., production cannot attain the goals anticipated as quantified in earlier estimates).

Mining and exploration failures can be classed in two extreme categories: those that are outright fraud and those for which evaluation suffers from one or more flawed technical procedures. Of course, these two end member causes of failure are not mutually exclusive. Moreover, a failure might be attributable to several causes, despite our tendency to emphasize a specific cause in many cases. Clow (1991) presented a concise and particularly astute consideration of many of the Canadian gold mine failures of the 1980s. He suggests that the most important general cause of vein-type gold mine failure during that decade is the inexperience of those conducting the evaluation. Explorationists tend to be generalists without the necessary detailed background in mining geology to take into account properly all the necessary aspects of structure, continuity, wallrock conditions, number of work sites, mining methods, and production infrastructure. In addition, he voices a general comment regarding the complexity of geostatistics and the

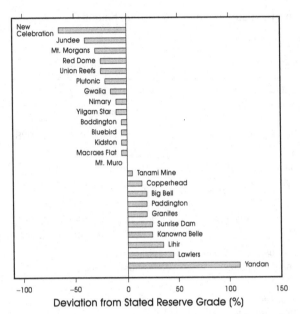

Figure 17.1: Reported deviation of production grades from estimated grades for 23 Australian mineral deposits. Note that only in a few cases are estimates within about 10 percent of production. Redrawn from Rossi (1999).

common lack of integration of geology with a geostatistical resource/reserve estimate (Clow, 1991, p. 34):

> Even the most ardent proponents of geostatistics will concede that their science (art?) encounters the most difficulty when applied to vein-type gold deposits. To compound the problem, this type of treatment [geostatistics] is very poorly understood by the average mining engineer and geologist and, as a result, the output is frequently misapplied. Despite the best efforts of the geostatistical practitioners, geology is often forgotten.

This is not so much a complaint against geostatistics per se as it is against the poor integration of geology into the estimation process. Geostatistics is not alone in having difficulty producing high-quality estimates for gold-bearing veins; all resource/reserve estimation methods have difficulty in this regard. Geostatistics has the advantage over other techniques that it is best by a quantitative criterion (i.e., the estimation variance is minimized); other methods cannot make such a claim. More generally, geostatistics takes into account the autocorrelation characteristics of a grade variable during the estimation process, a feature that other estimation methods generally cannot guarantee.

17.3: RESOURCE/RESERVE ESTIMATION PROCEDURES

Resource/reserve estimates are just that – estimates. After all, one is using sample information representing 1/100,000 to 1/1,000,000 of a deposit to determine an estimate of the metal content of the deposit. Clearly, interpretations and assumptions are necessary in arriving at tonnage and grade estimates to be used in economic evaluations and mine planning. Significant failures of estimates relative to production have been documented for many cases (e.g., King, 1982; Knoll, 1991). Unfortunately, most of the serious disparities between estimates and production are on the side of underestimation (i.e., not as much ore is found in a place as was estimated). Serious financial difficulties can ensue. One compendium of the range of errors that were encountered in a number of Australian gold deposits is indicated in Fig. 17.1.

Professional people of many backgrounds depend on resource/reserve estimates; however, not all are involved in the day-to-day procedures used to obtain these estimates. Consequently, many are unaware of the details of various methodologies or the advantages/disadvantages of any one method. The so-called traditional methods include polygonal, sectional, inverse-distance weighting, triangular, and contouring methods. A second group of geostatistical methods includes several kriging procedures such as simple kriging, ordinary kriging, and indicator kriging. Except for small deposits with small amounts of data, the application of any of these methods generally requires a significant level of computer use. Which methods are appropriate in any specific case depend on geologic features, characteristics of the variable being estimated, data density, experience of the estimator, and so on. It is important to realize that the development of the three-dimensional geometry of an orebody (geometric model) is based on geologic interpolation and extrapolation. Estimates of individual blocks within that defined geometric body are then commonly done to produce a "block model" of the deposit. The estimation method itself has no control over which estimates will be made; that decision is, in part, predetermined by the geology (i.e., the geometric model) and, in part, depends on the number and array of data available for each block that is considered for estimation. Providing the criteria regarding quantity and array of data near a block are met, a block within a zone defined geologically will be estimated by whatever method is used.

All methods of estimation produce estimates that only approach the truth. For a specific geologic character, some estimation procedures are likely to be better than others. It can be difficult to ascertain in advance of production which method of estimation is best (i.e., has an acceptably small error and can be integrated into mine planning with available personnel and facilities). Selection of the best method can be based on cross validation (e.g., Isaaks and Srivastava, 1989), bulk sampling for grade verification (e.g., John and Thalenhorst, 1991), trial mining and theoretical considerations (as discussed in various textbooks on resource/reserve estimation). Of these approaches, cross validation and theoretical considerations are cheap and easy to undertake, whereas trial mining implies a scale of operation that is very costly. Bulk sampling is an important advanced exploration procedure that can provide the necessary confidence in an estimation procedure that uses small samples at the feasibility stage. Too often, none of these procedures are applied to the question of selecting an appropriate estimation method or they are applied in poorly conceived ways. Formal commentary on any individual resource/reserve estimation procedure should be technically correct. Sadly, this is often not true, and is particularly unfortunate when incorrect or partially correct statements are made in a public venue that reaches a large number of professionals.

Simplistic views of the complex problem of resource/reserve estimation can lead to doubt and uncertainty of results. Consider the example of any "black box" approach, such as the occasional analytical scam involving a claim that a new (unexplained) analytical method finds 10 times as much gold as does the traditional fire-assay approach. Clearly, no responsible professional can accept such claims until the method is proved. Claims about technology must be backed up with appropriate verification.

It is worth recalling that high grades invariably have different continuity (geologic) characteristics than do low grades (e.g., McKinstry, 1948; Sinclair, 1995), a feature that is too often overlooked in resource/reserve estimation. How these values are considered in a mineral inventory study can be of the utmost importance because outlier samples representing less than 1 percent of the available data might account for 10 to 20 percent of the metal content of estimated resources/reserves.

17.4: GEOSTATISTICS AND ITS CRITICS

Geostatistics deals with spatial correlations of a variable; mining is but one field among its wide range of applications (e.g., forestry, fisheries, oil and gas, hydrogeology, soil science). Within the mining industry, geostatistics has found extensive applications both in resource/reserve estimation and in simulating grade distributions in mineral deposits as an aid to mine/mill planning. These applications are widely documented in established textbooks and traditional technical/scientific publications, and most authors have been reviewed by their peers. An extensive global community of consultants has been providing geostatisitical services since the mid-1970s. It is through such public examination, critical review, and practical application that the subject has evolved from concepts described initially by Matheron in the 1950s and 1960s to the much more extensive arsenal of tools now available and in use globally.

Over the years, a variety of criticisms have been leveled at geostatistics by prominent professionals with interests in the field of mineral inventory estimation. These include the following:

1. In contrast to other methods of estimation, the gain in information (i.e., estimation variance) provided by kriging is negligible. Many case histories now exist to show that this criticism is incorrect in a significant proportion of cases. Moreover, the validity of the criticism can be demonstrated only by conducting both geostatistical estimates and at least one independent estimate of mineral inventory and conducting a reconciliation with production data.

2. The method is costly. However, any method done well is generally costly. Geostatistics requires both highly trained personnel (who are still in limited supply) as well as substantial computing capability, both of which can add to costs. Any estimation done incorrectly is extremely costly. Many reserve-estimation failures in Australia (e.g., King et al., 1982) and Canada (Knoll, 1989) were related to projects that were not based on geostatistical estimation techniques, indicating a need for a theoretical as well as an empirical basis for mineral inventory estimation.

3. The subject is too complex. There is little question that geostatistical methods are complex relative to many empirical approaches to mineral inventory estimation. Understanding the principles is relatively easy, but understanding them adequately to carry out an involved geostatistical study that is well integrated with equally complex geologic features requires guidance or experience. Fortunately, the pool of capable geostatisticians is increasing as the result of a greater selection of university courses, a broad availability of short courses, a burgeoning theoretical and applied literature, and a wide range of accessible software.

4. Geostatistical procedures are too time-consuming. Geostatistics demands close attention to data quality and geologic interpretation, as does any method of estimation. The principal additional time required by geostatistics is related to semivariogram modeling, and without doubt, this can be a time-consuming undertaking. It is worth noting, however, that semivariogram modeling provides information that could improve inverse distance estimates, and in cases of significant anisotropy, invalidate any empirical method that assumes isotropy. Thus, semivariogram modeling should be a component of data evaluation even in estimation cases in which geostatistics is not to be used.

The subject has undergone scrutiny since the early 1960s by applied and theoretical scientists and a myriad of mining practitioners, quite apart from the multitude of users in fields other than mining. As a result of this scrutiny, a very small group of professionals raised concerns about certain aspects of geostatistics, concerns that have been well publicized and discussed in both peer-reviewed and non-peer-reviewed literature. These are real concerns that have been the subject of ongoing public discussion. This public forum approach has been a positive influence that has led to improvements in the understanding/application of geostatistics. Published critiques have appeared since the mid-1980s under such names as Philip and Watson (1986), Shurtz (1994), and Merks (1992, 1993), and there have been responses in the scientific/technical literature by well-known members of the geostatistical community, including Journel (1986b), Srivastava (1986), Myers (1986), Champigny (1992), Krige (1986), Deutsch (1994), and Matheron (1986). Thus far, concerns about geostatistics have been dealt with in considerable detail through a series of responses from the geostatistical community, largely in the *International Journal of Mathematical Geology*.

Philip and Watson (1986) have many problems with geostatistics: specifically, the assumptions, the semivariogram definition, the validity of linear kriging as the best interpolator, and the significance of the estimation variance. Their principal concern with assumptions relates to stationarity, which they feel is simply not possible in the context of ore deposits. However, in most geostatistical applications, only local stationarity of the grade differences is required. The use of relative semivariograms is another way in which nonstationarity is taken into account. Srivastava (1986) has pointed out that their attack on stationarity is an "attack" on all statistics. In fact, it is the model (random function) that requires stationarity; the model is simply applied to data because it proves to be an adquate representation of data.

It is a concern to Philip and Watson that the semivariogram is a global autocorrelation function, yet it is used for local estimation. This procedure is applied when there are relatively few data, and is justifiable as a best approximation. When there are abundant data, semivariograms can be determined locally and used to optimize local estimation (e.g., Johnson and Blackwell, 1986). Concerns about the treatment of outliers expressed by Philip and Watson have been taken into account in a previous discussion of this topic (Chapter 7). Philip and Watson express concern that all parameters of a semivariogram model are determined subjectively. Although this is true, it is important to remember that one is attempting to approximate a function whose form is not really known. Just as one commonly fits normal or lognormal distributions to histograms using particular assumptions, so it seems appropriate to fit models that reasonably describe the form of an experimental semivariogram. In fact, the particular model fitted is not as important as the necessity that the model describe the semivariogram adequately over the distances for which it will be used.

Philip and Watson (1986) question the advantages of kriging relative to many other interpolation procedures because (i) a global semivariogram is used to make local estimates, (ii) local trends within the data are ignored, and (iii) a kriged surface is discontinuous, requires that large matrices be inverted, and suffers from outlier problems. Some of these concerns have been dealt with previously and are of minimal importance; that local trends are ignored is simply untrue. Regarding the "discontinuous" nature of kriged results, Srivastava (1986) points out that surface discontinuities are a product of many interpolation techniques, but not of kriging theory; discontinuities arise only because of implementation procedures. Local trends are taken into account by the values of data themselves. If the local trends are on a large enough scale to be represented by the local semivariogram, they are taken into account to an even greater degree. Their concern about the weakness of kriging in the face of outlier values has led to the use/development of alternative techniques such as multiple indicator kriging and restricted kriging. Their criticism in this case has led to the implementation of different geostatistical procedures than those used previously for dealing with outliers. Note that nongeostatistical methods have equal difficulty of dealing with outlier values.

The kriging process produces the kriging variance that is recognized as an index of data configuration (Philip and Watson, 1986; Srivastava, 1986). In discussing their concern with the estimation variance, Philip and Watson (1986) emphasize that different types of continuity exist from one locality to another in a deposit. This criticism largely disappears if an orebody can be subdivided into more homogeneous domains, each characterized by its own continuity model (e.g., Krige, 1986).

The fundamental nature of the complaints made by Philip and Watson are minimized in the recognition by Journel (1986b) and Srivastava (1986), who emphasize that geostatistics is a model or methodology based on random functions; thus, it does not need a priori justification, but can be proved adequate on the basis of comparative studies with reality (cf. Krige, 1986).

Shurtz (1985, 1991, 1994, 1998) also complains vociferously about geostatistics. One of his widely publicized concerns (Shurtz, 1994, 1998) relates to the peculiar distribution of weights in the common case when strings of data (e.g., samples along a drill hole) are used to make point or block estimates by ordinary kriging. The problem that arises is that members of a string furthest from the point/block being estimated have the highest weights, certainly a feature contrary to expectation. This peculiar pattern of weights is most extreme when the continuity is greatest (e.g., zero nugget effect and long range) and becomes less extreme as the nugget effect increases. A detailed study of this topic by Deutsch (1994) indicates that the problem arises in ordinary kriging, in which the mean of the data field is unknown. The highest weights for the most distant data are explained by Deutsch as an attempt to compensate for the unknown mean value, with the most distant values being surrogates for this unknown mean. The problem can be minimized in a variety of ways, including the use of simple kriging when appropriate. When ordinary kriging is used, precautions include limiting the number of data used from any one drill hole to two, requiring data from three or more drill holes, and compositing data into longer lengths so that abundant small sample information will not be lost. In fact, this is a particularly sound reason for using composite grades for estimation purposes. It must be recognized that this question of peculiar weights is most pronounced in cases of extreme continuity that are rare features of mineral deposits (i.e., no nugget effect, very long range, and perhaps a Gaussian semivariogram model).

Shurtz (1999) also points out the mathematical instability that results from solving the kriging equations if one sample of a string (i.e., the sample nearest the point being estimated) is shifted imperceptibly off the linear trend of samples, a common situation in drill holes that have been surveyed at regular intervals (i.e., there is a slight periodic directional shift that stems from the intermittent nature of the surveying). The nearest sample can take on an unusually large weight that is either positive or negative, depending on the direction of a small departure from a linear trend of samples. This problem is avoided if the suggestions of the previous paragraph are implemented. Open discussion has led to an understanding of the phenomenon leading to the criticism, as well as suggestions to minimize or avoid the problem.

The principal complaint made by Merks (e.g., 1992, 1993) against geostatistics is that degrees of freedom are not taken into account in geostatistical theory. Of course, degrees of freedom are not essential in the development of a theory in which entire populations are considered. Hence, this complaint against geostatistics is invalid. Nevertheless, there are arguably practical situations in geostatistics in which degrees of freedom might well be considered (e.g., in the construction of an experimental semivariogram). When abundant data are available (a necessary situation in most mining geostatistics applications), it is immaterial from a practical point of view whether degrees of freedom are considered in the estimation of the semivariogram. The resulting estimates of the experimental semivariogram are affected only negligibly, because n is large. Because a kriging estimate is based on the semivariogram model and a particular data array in space, a block estimate will also be affected negligibly.

Merks (1992, p. 113) further complains, "The kriged variance violates the requirement of independence." As Chiles (1993, p. 6) observes, "the kriging error is a linear combination of point values, but since no independence is required for computing its variance, nothing is violated."

Two additional concerns relating to Merks are matters worthy of comment, partly because of how they might be interpreted by the uninitiated. A view attributed to Merks (*Northern Miner*, Sept. 28, 1998, p. 4) is that kriging "tends to inflate expectations for the continuity of mineralization between measured data points." This statement is incorrect and seems to indicate a lack of understanding of the estimation procedure. In resource/reserve estimation, the inferred continuity of mineralization is an independent geologic decision that precedes the selection/application of any estimation method. Once acceptable limits of mineralization have been defined on the basis of geology, it is possible to proceed to estimation. Clearly, it is unwise to make grade estimates where mineralization is known not to exist, regardless of the estimation method in use. A second view attributed to Merks and pertaining to kriging states, "applying the brute force of computers to fabricate more data from measured data is somewhat similar to perpetual motion" (*Northern Miner*, Sept. 28, 1998, p. 4). This loose statement gives the impression that geostatistics creates data. Of course, kriging does no such thing, nor has any responsible geostatistician ever made such a claim. Kriging provides estimates of points or blocks, just as do other estimation methods. It may be that this view pertains to simulation, but this is not evident from the context. The widely accepted process of simulation does create values with similar statistical and spatial characteristics to the grades in a deposit. A range of simulation methods exist (some of which predate the field of geostatistics), all of which are used to study a mass of data having similar statistical characteristics to the deposit. Unfortunately, most of Merks's comments against geostatistics are short on theoretical or practical support.

17.5: WHY IS METAL PRODUCTION COMMONLY LESS THAN THE ESTIMATE?

A common practical problem is that tonnage produced from a deposit is less than tonnage estimated. This problem arises because selection during mining is based on closely spaced data obtained during production, in contrast to the widely spaced exploration data used for mine planning. Ideally, if blocks originally outlined as ore were the blocks mined as ore, then production tonnage should equal the estimated tonnage, at least within reasonable error limits. However, during mining, the block on which selection is based is generally much smaller than the blocks used in preliminary evaluation. Consequently, the tonnage above cutoff changes, even when the cutoff grade is unchanged.

Estimates commonly contain an element of conditional bias (i.e., on average, high grades are overestimated and low grades are underestimated; something also known as the *regression effect*). When the entire distribution is mined, the overestimates might be compensated for by the underestimates; globally, the estimates might be unbiased. When a cutoff grade is applied to such a distribution of estimates, the low-grade part of the distribution is sent to waste. This leaves the overestimated high grades to be mined.

Consider an ideal case in which there is a small error in block estimates but no conditional bias. Even in this best-case situation, there is some misclassification of blocks relative to a cutoff grade because of the small random error in block estimates. Hence, some ore is classed as waste and some waste is classed as ore. When the cutoff grade is below the mean of the distribution for a given block estimation error, there is more ore lost to waste than waste included in ore. Thus, solely because of inherent random error in the estimation procedure, there is less metal recovered than estimated.

Outlier values can be spread far beyond their true physical extent in space by many methods of estimation that derive block estimates as weighted averages of nearby sample values. This problem has been more prevalent in the past; much more attention is now paid to defining and incorporating outliers into the estimation process. Assays that are one or two orders of magnitude larger than the great mass of data must be dealt with cautiously in order to offset the common problem of gross overestimation that might arise.

Of course, there are many other reasons why production does not attain estimates, including overly optimistic geologic definition of the extent of mineralized ground, lower than expected metal recovery in the mill, and so on. The emphasis here has been on overestimation arising from the assay data themselves and the procedure of estimation (i.e., condi-

tional bias, random error, outliers, and ore/waste classification based on a different block size for estimates and production).

17.6: PRACTICAL CONSIDERATIONS

1. Errors are inevitable in making mineral inventory estimates. The aim of professional procedures is to minimize those errors by thorough analysis of each situation, followed by the use of appropriate estimation methods for the deposit under study. Most imperative to the success of the undertaking is the thorough integration of geology into the estimation procedure.

2. Estimates in general should be done by at least two independent methods. Essential to the overall process is a detailed reconciliation of results obtained from the various estimation methods used.

3. At least one of the estimation methods used for mineral inventory estimation of a deposit should be an appropriate geostatistical approach.

4. Detailed case histories of mineral inventory estimation, including reconciliation information, are an important basis for appreciating which estimation methods are best adapted to a specific deposit type.

5. Reconciliation studies ultimately need production information with which to compare estimates. However, there commonly are significant differences between what is estimated and what is produced, so care must be taken that reconciliation conclusions are stated fairly.

17.7: SELECTED READING

Case histories such as those discussed in Table 17.1 are important background for understanding present practice.

Deutsch, C., 1994, Kriging with strings of data; Math. Geol. v. 26, no. 5, pp. 623–638.

Journel, A. G., 1986, Geostatistics: models and tools for the earth sciences; Math. Geol., v. 18, no. 1, pp. 119–140.

Myers, D. M., 1986, Matheronian geostatistics – quo vadis?: comment; Math. Geol., v. 18, no. 1, pp. 699–700.

Philip, G. M., and D. F. Watson, 1986, Matheronian geostatistics – quo vadis? Math. Geol., v. 18, no. 1, pp. 93–117.

Shurtz, R. F., 1994, Spatial least squares approximation and geologic plausibility; *in* Chung, C.-J. F. (ed.), Proc. 1994 Inter. Assoc. Math. Geol. Ann. Conf., Mont Tremblant, Quebec, Oct. 3–5, pp. 319–324.

Srivastava, R. M., 1986, Philip and Watson – quo vadunt? Math. Geol., v. 18, no. 1, pp. 141– 146.

17.8: EXERCISE

1. Case histories describing "successful" reserve estimation procedures commonly must necessarily abbreviate the abundant information in company reports, and thus generally provide incomplete comparative information for the reader's evaluation. A critical analysis of a published case history is a useful exercise for a resource/reserve student. In such an analysis, attention should be directed toward matters such as: uniformity(?) of cutoff grade through the study, variation(?) in block sizes estimated, details of each estimation method reported and compared (search radius, required data), a description of how geology was taken into account in each estimation method (e.g., adaptation of estimation method to style of mineralization, recognition of domains, estimation at domain margins, deleterious minerals), documentation of the differences in volumes estimated by various methods, and so on.

18

Resource/Reserve Classification

Every mining operation from a simple sand-and-gravel pit to the largest integrated mine-mill-concentrator-smelter-refining complex has as its foundation, a quantity of extractable natural resource – the reserves. Without reserves, there is no mine and any concentrating or extracting plant, no matter how sophisticated, is simply so much scrap metal. Without resources there are no reserves. It is remarkable, therefore, especially in this age of careful planning, high-quality engineering and state-of-the-art financial analysis, that such a fundamental and vital concept as resources/reserves is so little understood and the subject of so much confusion. (Grace, 1984, p. 1446)

Chapter 18 considers the thorny problem of classification of resources/reserves in view of the many conflicting demands on such a system. A fundamental geologic base to a classification scheme is emphasized, as is the necessity of documenting classification criteria so that classifications are easily reproducible by external auditors.

18.1: INTRODUCTION

Classification of mineral inventory estimates is necessary for several reasons, including the following:

(i) Creation of a formal inventory of the principal assets of a mining company
(ii) Documentation of assets to demonstrate potential for medium- or long-term production
(iii) Raising development funds in the speculative money market
(iv) Providing a reasonable confidence level for senior financing institutions
(v) Providing a basis for royalties, taxation, land use management, and so on.

Formal classifications were first suggested about the start of the 20th century and pertained only to ore (Table 18.1). More recently, there has been recognition that many mineral deposits or mineralized zones can be well estimated even though a sufficiently thorough study has not been done to define material as ore. This situation had led to a widely accepted trend to define both (mineral) resources and (ore) reserves as *mineral inventory*.

An abundance of recent literature on the topic of resource/reserve classification seems to depend on two principal motivations: the hope to introduce a semiquantitative or quantitative measure of uncertainty into the classification process and a hope that a set of international standards can be developed (e.g., McOuat, 1993; Riddler, 1996; Stephenson and Miskelly, 1999). For the most part, resource/reserve classification concerns relations between those with some element of day-to-day operational control of the mining property in question and other groups (share holders, senior financiers) who have supplied or will supply funds for the operation. There are different vested interests on the parts of both parties that generally are constrained or controlled to some degree by government legislation. With the increasing tendency

Table 18.1 Early definitions of mineral inventory

Source	Quality of estimate		
	Highest	Middle	Lowest
Kendall et al., 1901–2	Puritan Ore in sight, ore blocked out, ore in reserve, ore available for extraction out, probable ore in sight	Cavalier Ore reasonably expected, visible ore not blocked, ore probable but not blocked out	Others Witwatersrand drill-inferred geologic projection
Hoover, H. C., 1909	Proven ore Ore in which there is practically no risk of failure of continuity out, probable ore in sight	Probable ore Ore in which there is some risk, yet warrantable justification for assumption of continuity	Possible ore Ore that cannot be incuded in the proven–probable classes, nor definitely known or stated in any terms of tonnage
Leith, C. K., 1935 (see Taylor, 1994)	Assured ore Ore blocked out in three dimensions by mining or drilling, risk of failure remote	Prospective ore Extensions near at hand; probability high, but extent is less precise	Possible ore Presumptive evidence of ore, but of indeterminate quantity
Fennel, J. H., 1939	Ore blocked out Ore exposed on three or four sides	Probable ore Partly exposed on one or two sides	Prospective ore Not exposed
USBM/USGS, 1976	Measured ore Sampled tonnage from dimensions in workings, trenches and drill holes; tonnage or grade estimates good	Indicated ore Tonnage partly from measurements and partly projected within 20%	Inferred ore Few measurements, but some geologic knowledge presumed

Source: After Taylor (1994).

toward global operations by many mining companies, there are increasing problems with the reporting of resource/reserves because regulations from one jurisdiction to another can vary widely. This has led to a call for international standards in the classification of mineral inventory estimates (e.g., Stephenson and Miskelly, 1999), and various proposed classifications have emerged more recently that are directed toward such an end (e.g., Australian system, U.S. proposal, IMM proposal). The general framework of two classification systems are outlined in Section 1.3.4 and Fig. 1.6. An example of definitions (the Australasian system) of both reserves (proved, probable) and resources (measured, indicated, inferred) is given in Chapter 1.

Of course, classification must be viewed from two perspectives: that of the regulator requiring a standard means of reporting resources/reserves as protection for the public (including boards of directors) and that of the user (of a classification scheme), who must use specific criteria to meet the aims of two masters, viz. the regulating authority and internal company requirements. Ultimately, the classification of resources/reserves for a mineral deposit should be a reproducible undertaking. Without a clear description of the criteria and how they are applied, a classification scheme is not reproducible and is of little value. Consequently, the criteria used for classification must be spelled out clearly.

18.2: A GEOLOGIC BASIS FOR CLASSIFICATION OF MINERAL INVENTORY

There are two important stages in arriving at an acceptable level of geologic knowledge for resource/reserve classification. The first is the gathering of high-quality data, and the second is a rational interpretation of those data. A classification of resources/reserves depends first and foremost on an appreciation of mineralization continuity. Deposits are generally known at widely spaced sampling points (e.g., drill holes), surface exposures, and limited underground workings; substantial interpolation is required based on an understanding of the geologic character of a mineral deposit. The topic has been discussed at length by Sinclair and Vallée (1993), who recognize both geologic continuity and value continuity (see Section 1 and Chapter 3).

In general, geologic continuity is a present or absent condition. A mineralized zone is interpreted to exist between control points. However, even when continuity is based on thorough study and documentation, an interpretation remains subjective, perhaps dependent on the deposit model. Consequently, although arbitrary relative methods of quantifying geologic parameters exist, such as the subjective numeric scales of Blanchet and Godwin (1972) and Kilburn (1990), geologic continuity cannot be quantified in an absolute manner. Nevertheless, geologic information is the basis on which interpolations regarding geologic continuity are made; thus, geologic concepts are the underlying basis for mineral inventory estimates and their classification. Observations regarding style of mineralization and known physical extent in space of each style of mineralization (domains) are two of the important features that contribute to confidence in defining the limits to mineralized volumes within which estimates are made and classified.

No matter how quantifiable some components of the data may be, the interpretation remains subjective and might change significantly as new information is gathered or new concepts come to light. Hence, mineral inventory classification must always be considered with respect to a particular set of geologic assumptions that should be stated explicitly, with both supporting and contrary (ambiguous) evidence; justification of a model should not be left to the imagination.

The second aspect of continuity important to classification concerns grade continuity, commonly quantified by a three-dimensional autocorrelation model (Chapter 9). The distinction between geologic and grade continuity can be visualized by reference to a vein for which grades are clearly separated into high- and low-grade segments (e.g., ore shoots within a much larger vein). Such a scenario demands that the physical continuity of high grades be much less than the physical continuity of the vein. Samples represent 1/100,000 to 1/1,000,000 of the volume of a deposit. Hence, available data must be extended in some manner to nearby mineralized ground in order to make estimates of larger volumes. How far sample grades can be extended in any direction is a function of grade continuity. Grade continuity in the form of an average autocorrelation function (e.g., a semivariogram) provides quantitative insight into how grade at a data site correlates (on average) with the adjoining mineralized ground. Note that in applying the concept of grade continuity to a volume, there is a prior assumption that geologic continuity occurs (i.e., the volume is mineralized). In the past, there was often ambiguity in use of the term *continuity*; for example, grade continuity, especially high-grade continuity, has been assumed, mistakenly, to be extensive just because geologic continuity (the mineralized zone) is established with a high level of confidence. Note that the characterization of grade continuity by an autocorrelation function is independent of the grade estimation method used.

18.3: SHORTCOMINGS TO EXISTING CLASSIFICATION SYSTEMS

There are a number of reasons why existing resource/reserve classification systems have been found wanting. In particular, attempting to satisfy several masters is seen as one of the principal sources of difficulty in arriving at acceptable definitions of various categories of both resources and reserves. More specific problems include the following:

1. Classifications are not uniform from one government (or securities) jurisdiction to another. In some cases, differences result in substantial additional work to classify resources/reserves appropriately in response to jurisdictional requirements in more than one country. For example, an international mining company may be required to present resources/reserves in a particular form for taxation purposes in the country in which a mine is operating, yet may have to report them quite differently for financial reporting purposes at the head office in another country.

2. Taylor (1993, p. 146) summarizes the practical problem of resource/reserve classification that has evolved as a function of changing technology and changing emphasis of mining procedures:

   > Definitions were first devised for ore exposed in developed undergound mines, which sought recurrent or marginal profit because the interval between valuation and extraction was at most a few years and often much less. What existed was known well and what might happen to it wasn't far away. But even then, inclusion of profit in definitions was not without difficulty. Today, most primary valuation is by drilling, even when reliability may be low. Estimation of reserves is advanced too long before production, and feasibility must require prediction of a protracted and distant absolute profit. But still we use the same terms, implying falsely that earlier standards remain, though these cannot now be replicated until secondary valuation and production.

3. Although relative quality or certainty of the value of a resource/reserve is implicit in all mineral inventory classification schemes in widespread use, none contains a detailed reliability index for either global or local estimates. "[T]he category of Measured Resource does not imply 100% knowledge or 100% confidence in the estimate" (Stephenson and Stoker, 1999, p. 57).

4. Classification codes for coal inventories are traditionally distinct from those for mineral invento-

ries. There is some question as to how general a mineral inventory classification scheme should be (e.g., perhaps some industrial mineral resources need not be covered by a classification system designed for metallic mineral deposits).

5. The widespread use of geostatistics into the mineral inventory estimation scene has led to the possibility that quantitative measures might become part of a classification system (e.g., Froidevaux, 1982; Vallée, 1992; Vallée and Cote, 1992) and thus improve on the qualitative/subjective aspects of existing classification schemes.

6. Ambiguity of existing definitions of categories in a classification system leads to subjectivity and variability in practical use of terms. Definitions of various categories of resources and reserves in many jurisdictions are quite general, and their application is more dependent on traditional use and the experience of the user than on objective criteria.

7. The distinction between resources and reserves is not as rigorous in some jurisdictions as in others. In Australia, for example, "appropriate assessments, which may include feasibility studies" (JORC Code 1999, p. 8) are required in order to demonstrate which of the identified resources can be mined at a profit, and therefore classed as reserves. Other jurisdictions may be more or less demanding in this regard.

8. The limited number of resource/reserve categories (classes) means that a wide range of quality is included within each class.

18.4: FACTORS TRADITIONALLY CONSIDERED IN CLASSIFYING RESOURCES/RESERVES

Resource/reserve classification commonly involves the following considerations:

1. Geologic continuity. For each block being estimated, an evaluation must be made of the likelihood that mineralization persists through the block. This fundamental geologic decision must preempt the consideration of all other criteria

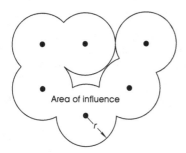

Figure 18.1: The concept of using a zone of influence around individual grades results in complex intersections and volumes that are impractical to estimate (see Readdy, 1986). The circular zones of influence shown here are cylindrical in the third dimension.

(cf. Sinclair and Vallée, 1994). Clearly, if there is no geologic continuity, there is nothing to be classified.

2. Distance from a sample site is a widely used method for resource classification (Fig. 18.1). A short radius is used to define a *cylinder* of material that is placed in a high-quality category; a larger radius adds an additional hollow cylindrical shape that is categorized in a lower class of resource. By itself, the method is inadequate. Considered with other parameters listed here, it might be useful. However, the method is widely used to produce intersecting cylindrical volumes and is awkward, if not impossible, to deal with in practice.

3. Sample density in the vicinity of each block to be estimated. The size of this *vicinity* must be defined and is very different from one deposit type to another. It is common practice to require a minimum number of samples within the search volume in order to proceed to an estimate; this minimum number should be specified. When data are abundant, it is also routine practice to specify a maximum number of data, generally those closest to the block being estimated (cf. Isaaks and Srivastava, 1989). Classic polygonal arrays (a centered sample in a polygonal volume) are an extreme and generally undesirable basis for resource/reserve classification.

4. The geometric array of data relative to each block to be estimated. There is a significant difference between *n* samples on one side of a block to

be estimated (extrapolation) and *n* samples distributed around the block to be estimated (interpolation), as shown in Fig. 18.2. Better-known categories of resources/reserves might be based only

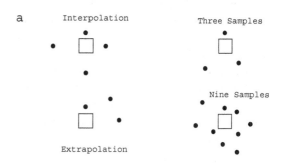

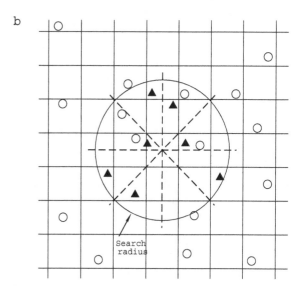

Figure 18.2: (a) Dots are sample locations relative to a block to be estimated. Interpolation intuitively provides better-quality block estimates than does extrapolation. Similarly, an increase in the number of nearby samples improves the quality of an estimate. (b) Circles and triangles are sample locations relative to a block to be estimated. Triangles are those samples within the search radius selected for estimation. Note that the nearest sample in each octant has been selected. The number of samples used in obtaining a block estimate depends on the search radius and sample density. Judicious choice of a search radius is essential. Simply enlarging the search radius to increase the number of samples does not necessarily improve the quality of an estimate adequately to change a classification. Samples can be required in three or four quadrants (or five or six octants) in order to proceed to an estimation.

on interpolation, whereas a less-precisely known category might be based on limited (but specified) extrapolation (e.g., to allow for limited extrapolation in a direction when data are limited, but a deposit is known to extend).

5. Estimated block grade must be categorized relative to a meaningful cutoff grade. In general, resource/reserve classification schemes are based on the fact that the material being considered for classification has the potential to be produced at a profit. Consequently, realistic assumptions must be made as to likely production costs so that an estimate can be made of a justifiable specified cutoff grade.

6. Various types of data might be of significantly different quality. For example, high core recovery might be necessary in areas where estimates are made with a high degree of certainty, whereas low core recovery may preclude the use of grade but may verify geologic continuity. Similarly, such features as variable sample support or reverse circulation drilling versus core drilling can affect assignment of blocks to resource categories (e.g., Stephenson and Stoker, 1999).

7. Criteria other than grade that might reject ground from a resource classification system. An extremely fine-grained character of base metal deposits could mean abnormally low metallurgic recovery such that, despite a relatively high grade, ore minerals cannot be recovered with adequate efficiency to classify the material as potentially profitable. Deleterious components (e.g., S in coal or As in gold deposits) can lead to penalties that might exceed the benefits of grade of the valuable metal (cf. Fairweather, 1997).

The forgoing considerations are easily applied to any two- or three-dimensional arrays of blocks that approximate a deposit or a mineralized domain. At the outset, it is necessary to define the block size(s) for which estimates will be made and on which classification will be based. There is no general agreement on this matter, as evidenced in the discussion by Stephenson and Stoker (1999). From a practical point of view, an appropriate block size is one that

corresponds to an SMU (the relatively small volume on which a physical separation of ore and waste is possible by the mining method to be used). More generally, classification, especially at a prefeasibility stage, might be based on a somewhat larger block size, if demanded by the available data or the estimation method.

In underground operations, individual ore blocks are commonly irregular in shape and can be as large as individual stopes. Resource/reserve classification criteria must be adapted to these less regular situations. For example, a large volume defined geologically can be approximated by a regular array of smaller blocks, as shown in Fig. 1.1, and each block can be classified using factors indicated previously. Alternatively, the array of sampling information (data density) in the block as a whole, or in large subsets of the block, can be used as the basis of classification. The classification criteria determined for such blocks should be consistent with criteria used for more regular block arrays. Figure 18.3 is an example of a vein in longitudinal section, showing blocks classified principally on the basis of relation to underground openings. A common method to define block size is to limit block extent to midway between adjoining drill sections. Such a procedure is generally inappropriate unless drill sections are very closely spaced because it results in single intersections being extended great distances, commonly with little confidence.

18.5: CONTRIBUTIONS TO CLASSIFICATION FROM GEOSTATISTICS

The increasing widespread use of geostatistics in resource/reserve estimation since the mid-1980s (Champigny and Armstrong, 1993; Kwa and Mousset-Jones, 1986) led to new parameters as an aid to resource classification (e.g., Wellmer, 1983). This arises because, in addition to providing a block estimate, geostatistical procedures provide various quantitative objective measures, including the average range of influence of a sample (range of the semivariogram) and the block-kriging variance (estimation error). Froidevaux and Roscoe (1983) suggest that the range of the semivariogram serve as the basis for

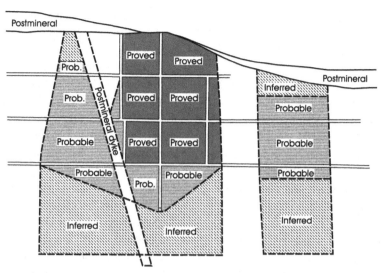

Figure 18.3: Longitudinal section of vein showing blocks of resources/reserves classified on the basis of location of workings relative to the block. Proved reserves are surrounded by workings; probable reserves are surrounded on two or three sides by workings or involve slight extrapolation from a working; unclassified resources are extrapolated from workings based on a high likelihood of geologic continuity of the vein. Modified from Rostad (1986).

resource classification in a particular case history, and define three classes as follows:

(i) Blocks that are within the sampled area and for which grade has been computed from data within the range of influence

(ii) Blocks within the sampled area, but that are either beyond the range of available data or that have only one sampled drill hole within the range

(iii) Blocks within the deposit as defined but remote from data (e.g., at the periphery of the deposit).

However, the range alone is not generally adequate as the basis of a resource classification scheme because it does not take nugget effect into account. If the structured component of the error is a small proportion of the total variability, the range is of little use as a classification criterion. Furthermore, the range of an autocorrelation function (e.g., semivariogram) commonly is not well defined with exploration data. The use of range is further complicated by the common occurrence of multiple structures and multiple models (with very different ranges depending on the anisotropy characteristics from place to place).

Royle (1977), Sabourin (1984), and Froidevaux et al. (1986) recommend that block-kriging variances

be used to classify individual blocks based on arbitrary values that separate measured, indicated, and inferred categories. Royle further standardizes this approach by basing his classification on blocks the size of the sampling grid. This latter suggestion has found little acceptance because the block estimation error is not only a function of the size of block, but also closely tied to the autocorrelation characteristics of a particular deposit (e.g., Froidevaux and Roscoe, 1983). Even with few data, it is possible to estimate the mean grade of a large volume with a fairly high degree of confidence (e.g., where the grade distribution is fairly uniform). The confidence, however, is deposit specific and should not be assumed a priori for estimation and public reporting of resources – sound geologic and sampling information is essential. Moreover, selective ore and waste mining is conducted within this large volume; hence, the average grade estimate of a large volume is of limited use for mine planning.

Blackwell (1998) demonstrates a practical use of relative-kriging variance (or relative-kriging standard deviation) as an important component of a resource classification scheme for porphyry-copper and large epithermal gold deposits. The general sequence of classification is as follows:

(i) Identify mineralized blocks (i.e., verify geologic continuity).
(ii) Identify mineralized blocks above cutoff grade.
(iii) Classify blocks above cutoff grade based on selected values of relative kriging standard deviation (RKSD). Blackwell's applications to porphyry-copper and epithermal gold deposits suggest that useful values of RKSD for classifying resources as measured, indicated, and inferred are as follows:

Measured $0.3 \leq$ Indicated $0.5 \leq$ Inferred.

Results for an epithermal gold deposit (Fig. 18.4) clearly demonstrate the advantage of RKSD over the widely used amount of data in the search volume. The simple parameter "amount of data" does not take into account the disposition of the data relative to the block being estimated. In contrast, the value of RKSD

incorporates the effects of both amount and location of data. It must be remembered that the kriging error for a block is not actually an estimate of the local variability; rather, it is a global or average error. However, the block-kriging error is a clear indicator of the number of data and the array of data relative to the block. Because it represents the interaction of these two contributors to error, the kriging error is an excellent empirical basis on which to base a classification of relative quality of block grades. Of course, the kriging error must be calibrated by selecting thresholds that provide a classification comparable to those derived from other widely used empirical criteria. Once satisfactory thresholds of the kriging error become widely accepted, they have the potential of becoming a standard against which other classification criteria can be evaluated for adequacy.

The increasing use of multiple indicator kriging (MIK) as a basis for resource/reserve estimates has led to new approaches to the problem of classification. The product of MIK is (i) an estimate of the

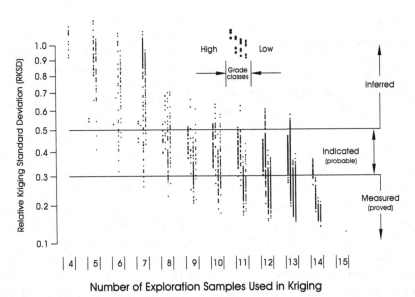

Figure 18.4: A typical example of RKSD versus number of samples used to obtain grade estimates for a large epithermal gold deposit. The closest two samples per octant within the semivariogram range were used for grade estimation. There is a broad correlation between the RKSD and the number of samples used; more samples generally improve the quality of estimates. The variation in RKSD for each number of data is a function of data geometry relative to the block being estimated. Resource categories are based on arbitrary RKSD values of less than 0.3 for measured, 0.3 to 0.5 for indicated, and more than 0.5 for inferred. These arbitrary values provide acceptable results for both epithermal gold and porphyry-copper deposits. Modified from Blackwell (1998).

probability that an SMU will be above a specified cutoff grade, and (ii) the average grade to be expected. In some cases, a large panel is estimated and the MIK output is the expected proportion of SMUs within the panel and above cutoff grade (as well as the average grade of those blocks). The problem is that the exact positions of the estimated ore blocks within the panel are not known, leading to an uncertainty that, at least superficially, this does not exist in other, more deterministic estimation procedures. Second, the MIK method does not provide error estimates for blocks, and so requires other classification criteria. The MIK method has proved useful in practice and has become popular in dealing with deposits having highly irregular grade distribution, such as large epithermal gold deposits. Consequently, some combination of number and array of data used in the estimate of a panel could be used to classify those blocks estimated to be above cutoff grade.

18.6: HISTORICAL CLASSIFICATION SYSTEMS

A concise summary of the North American history of resource/reserve classification (Taylor, 1993) is summarized in Table 18.1. A variety of classification systems throughout the English-speaking world have been reviewed recently by Vallée and McCutcheon (1997). As an example of the general nature of these classification schemes, the Australasian system (Joint Ore Reserve Committee, 1999) is reproduced here because in the absence of an international standard, it has been widely used and since the early 1990s has become a de facto international classification system. A general outline of the system is contained in Section 1.3.4 and Fig. 1.6.

A *mineral resource* is a concentration or occurrence of material of intrinsic economic interest in or on the Earth's crust in such form and quantity that there are reasonable prospects for eventual economic extraction. The location quantity, grade, geologic characteristics, and continuity of a mineral resource are known, estimated, or interpreted from specific geologic evidence and knowledge. Mineral resources are subdivided in order of increasing geologic confidence into inferred, indicated, and measured categories.

An *inferred mineral resource* is that part of a mineral resource for which tonnage, grade, and mineral content can be estimated with a low level of confidence. It is inferred from geologic evidence and assumed but not verified geologic or grade continuity. It is based on information gathered through appropriate techniques from locations such as outcrops, trenches, pits, workings, and drill holes that may be limited or of uncertain quality or reliability.

An *indicated mineral resource* is that part of a mineral resource for which tonnage, densities, shape, physical characteristics, grade, and mineral content can be estimated with a reasonable level of confidence. It is based on exploration, sampling, and testing information gathered through appropriate techniques from locations such as outcrops, trenches, pits, workings, and drill holes. The locations are too widely or inappropriately spaced to confirm geologic and/or grade continuity but are spaced close enough for continuity to be assumed.

A *measured mineral resource* is that part of a mineral resource for which tonnage, densities, shape, physical characteristics, grade, and mineral content can be estimated with a high level of confidence. It is based on detailed and reliable exploration, sampling, and testing information gathered through appropriate techniques from locations such as outcrops, trenches, pits, workings, and drill holes. The locations are spaced closely enough to confirm geologic and/or grade continuity.

The appropriate mineral resource category must be determined by a competent person or persons.

An *ore reserve* is the economically mineable part of a measured or indicated mineral resource. It includes diluting materials and allowances for losses that may occur when the material is mined. Appropriate assessments, which may include feasibility studies, have been carried out, and include consideration of and modification by realistically assumed mining, metallurgic, economic, marketing, legal, environmental, social, and governmental factors. These assessments demonstrate at the time of reporting that extraction could reasonably be justified. Ore reserves are subdivided in order of increasing confidence into probable ore reserves and proved ore reserves.

A *probable ore reserve* is the economically mineable part of an indicated and in some circumstances measured mineral resource. It includes diluting materials and allowances for losses that may occur when the material is mined. Appropriate assessments, which may include feasibility studies, have been carried out and include consideration of and modification by realistically assumed mining, metallurgic, economic, marketing, legal, environmental, social, and governmental factors. These assessments demonstrate at the time of reporting that extraction could reasonably be justified.

A *proved ore reserve* is the economically mineable part of a measured mineral resource. It includes diluting materials and allowances for losses that may occur when the material is mined. Appropriate assessments, which may include feasibility studies, have been carried out, and include consideration of and modification by realistically assumed mining, metallurgic, economic, marketing, legal, environmental, social, and governmental factors. These assessments demonstrate at the time of reporting that extraction could reasonably be justified.

The choice of an appropriate category of ore reserve is determined primarily by the classification of the corresponding mineral resource and must be made by the competent person.

Classification, in general, is accomplished by professionals with demonstrable experience in estimation projects of the type being reported. These professionals are referred to by such terms as *competent person* (Australia) and *qualified person* (Canada).

Comparable resource/reserve definitions that differ only in detail are included in a report published by the Society of Mining Engineers (Anonymous, 1999). A general comparison of various international definitions is given by Vallée and McCutcheon (1997).

18.7: THE NEED FOR RIGOR AND DOCUMENTATION

The increasing emphasis on protection of the public against outright fraud and low-quality work on mineral deposits by professional earth scientists requires that reports of mineral inventory be of a high standard. Important contributions to high-quality reports include the following:

(i) Estimates are made only when indicated by an appropriate geologic argument that is clearly documented.
(ii) Within a domain interpreted to be continuously mineralized, mineral inventory estimates can be classified using clearly documented criteria such as those in Sections 18.4 and 18.5.
(iii) The general procedure for classification should be easily reproducible by an independent auditor.

A report on mineral inventory should include the following:

(i) A brief statement regarding the basis for confidence in the geologic continuity of a mineralized zone for which resource/reserve estimates are being prepared
(ii) A concise commentary on sample density and various sample types
(iii) A justification of the block size used as a basis for classification
(iv) A concise list of criteria used to classify individual blocks, including, when possible, reference to levels of confidence for estimates
(v) Justification of the cutoff grade used to distinguish ore from waste.

With the increasing use of geostatistical estimation methods, various statistical parameters and probabilistic approaches have emerged as a basis for classification. Generally, relatively small blocks are estimated and each block is categorized individually.

Rules regarding each of these factors must be described explicitly, otherwise they are not easily subject to audit. For example, geologic continuity in many cases can be established with a high degree of certainty from a combination of outcrop patterns, sampling patterns, geophysical surveys (down-hole and surficial), and applied geochemical surveys. Within this zone of geologic continuity, individual blocks can be classified on the basis of various parameters using deterministic rules. Such a clearly defined procedure provides a sound basis for audit.

fidence in the interpretation and estimation. (Stephenson and Stoker, 1999, p. 66.)

18.8: EXAMPLES OF CLASSIFICATION PROCEDURES

1. A vertical vein (0.5 to 2 m thick) is intersected by cored drill holes. Geologic continuity was assured for a zone of dense drilling. A 10-m grid was superimposed on the plane of the vein. (a) Each 10×10 m^2 cell was estimated if it met the following criteria: a minimum number of four data occurred within a search radius of 15 m; blocks were estimated for thickness and accumulation (grade $\times$ thickness); and the average grade was determined from these estimates. Those blocks with a minimum grade of 0.1 percent $_eU_3O_8$ (i.e., 0.1 percent U_3O_8 equivalent) and a minimum accumulation of 0.01 percent were categorized as measured. (b) The criteria of (a) were relaxed by requiring only three data for block estimation and a minimum accumulation of 0.08 percent – the additional blocks estimated were classed as inferred resources.

2. Porphyry-type and large epithermal gold deposits. A resource/reserve classification scheme based on geologic continuity and RKSD for these large deposits is documented in Section 18.3.1. A more detailed description is given by Blackwell (1998).

3. A company evaluating a large open pit gold project used a combination of the number of samples and the distance to the nearest sample, generated during interpolation of block grades, to provide a block by block subdivision of the resource estimate into varying levels of confidence. This automated classification information was plotted and overlain on the interpreted geology complete with the plotted drilling data base. The project geologists and the competent person used this data as the basis for common-sense classification using all the projects geologists' and competent person's knowledge of the con-

18.9: PRACTICAL CONSIDERATIONS

1. Classification schemes are a function of individual political jurisdictions; consequently, agreement on an internationally acceptable classification is difficult and highly desirable. Classifications must adhere to the requirements of the appropriate jurisdiction.

2. Resource/reserve estimation depends first and foremost on a geologic model that provides a sound and confident expectation that a well-defined volume (deposit, domain) is mineralized throughout. Without this explicit decision regarding geologic continuity of a delimited mineralized zone, neither estimates nor classification of mineral inventory is possible.

3. Establishing a reasonable cutoff grade is required even at the resource classification stage because various jurisdictions indicate the need for classified resources to have economic potential.

4. Within a continuously mineralized zone, resources/reserves can be classified through the rigorous application of appropriate and clearly defined criteria. These criteria should be documented so they are easily reproducible by an external auditor.

5. Wherever possible, an array of uniformly sized blocks is desirable as a basis for resource/reserve classification because local criteria are then used to classify each block and because the block array approach lends itself to reproducible procedures that facilitate audit.

6. Invariably, individual cases of estimation arise that are difficult to classify within the broad framework of existing schemes. This arises because rules for classification are defined very generally and ambiguities can arise in applying them. Innovation may be necessary in such cases and should be clearly described.

7. In all cases of resource/reserve classification a clear, detailed listing of criteria used is essential.

8. Geostatistically based criteria for classification of mineral inventory are arbitrary and require standardization relative to more traditional approaches.

18.10: SUGGESTED READING

Sinclair, A. J., and G. H. Blackwell, 2000, Resource/reserve classification and the Qualified Person; Can. Inst. Min. Metall. Bull. v. 93, no. 1038, pp. 29–35.

Stephenson, P. R., and P. T. Stoker, 1999, Classification of mineral resources and ore reserves; Proceedings APCOM '99 Symp., Oct. 20–22, Colorado School of Mines, Golden Colo., pp. 55–68.

Vallée, M., 1998, Resource/reserve estimation and inventory: Can reporting definitions substitute for standards? *in* Vallée, M., and A. J. Sinclair (eds.), Quality assurance, continuous quality improvement and standards in mineral resource estimation; Expl. and Min. Geol., v. 7, nos. 1 and 2, pp. 15–24.

Vallée, M., and S. McCutcheon, 1997, Are international reporting standards feasible? Can. Inst. Min. Metall. Bull., v. 90, no. 1007, pp. 30–37.

18.11: EXERCISES

1. Comment on the usefulness of relative-kriging error as a basis for resource/reserve classification.

2. How would you incorporate the geologic character of a mineral deposit into a resource/reserve classification system? Note that *continuity* is an explicit component of many legal and quasi-legal classification schemes that have been suggested in the past. In replying to this question, it is appropriate to consider a specific deposit with which you are familiar.

3. Data density within a search radius is a criterion commonly used to classify a block as a measured, an indicated, or an inferred resource. Is this a sufficient criterion? Explain your answer in detail.

4. Describe a data selection algorithm that guarantees that all blocks to be estimated/classified will be interpolated rather than extrapolated. How can the algorithm be modified to permit limited extrapolation estimates that will be classified in a lower-quality category than the interpolated estimates?

5. It is common practice to limit the number of data to be used to estimate/classify an SMU for grade-control purposes. In some cases, the maximum number of data is 16 or 20. Is this a reasonable operating procedure?

6. Describe how a block might be classed as a measured, an indicated, or an inferred resource, based on an estimation procedure that produced a cumulative grade distribution for the block.

7. For purposes of estimation/classification, vein deposits commonly are approximated by a number of rectangular blocks on longitudinal sections. These rectangular outlines generally vary in size. Describe an approach to resource classification for such a scenario.

19

Decisions from Alternative Scenarios: Metal Accounting

Several empirical methods and attitudes [regarding gold sampling and analysis] developed in earlier generations are still alive and thriving. Many of these earlier procedures have not been reviewed in the light of modern sampling and statistical practices. Significant improvements could be made without great effort . . .(Vallée, 1992, p. 161)

Chapter 19 emphasizes the need for an open mind in the decision-making processes leading to the preparation of a mineral inventory estimation. Standard methodologies must be viewed critically, as alternatives generally exist and might be better suited to achieving estimation goals. Here, a simple mechanism of accounting is introduced, based on metal grades and an established cutoff grade as a practical means of comparing experimental results from various action scenarios. These examples are illustrative; what is important is the philosophy of critical appraisal.

19.1: INTRODUCTION

The expressions *lost grade* or *extra tonnes* are commonly used in the mineral industry to refer to metal losses or gains that result from one course of action relative to an expectation. It is not always clear that the terms are appropriate for the situations to which they are applied. Traditionally, such terms are used in reconciliations that compare or contrast estimates of contained metal with production. Clearly, it is useful to have a procedure that permits a rigorous comparison of various scenarios in terms of differences in operating profit. The simple concept of metal account-

ing is in relatively common use for such purposes, if only in an informal way.

The formalization of metal accounting is a practical and easy procedure for evaluating the results of a wide range of situations involving alternative actions, including the comparison of (i) sampling procedures and (ii) various methods of mineral inventory estimation. In general, metal accounting as used here is applied to those situations in which an operation or calculation produces a result in terms of metal grade that can be compared with a cutoff grade. Thus, the grade difference represents an operating profit (or loss) in terms of mineral or metal and, when applied to tonnage, can be transformed to a quantity of metal.

19.2: DEFINITION

A metal operating profit (loss) q_i can be defined as

$$q_i = (g_i - g_c) \cdot T \qquad (19.1)$$

where q_i is quantity of metal (profit if positive, loss if negative), g_i is the grade of an estimate or action, g_c is the cutoff grade, and T is tonnes. Units of grade determine the units of quantity of metal.

Consider a volume V of tonnage T that has been estimated by two methods to have average grades of g_1 and g_2, respectively. Substitution of these figures in Eq. 19.1 gives metal contents of q_1 and q_2, each of which can be compared with the quantity of metal eventually produced. Such a simple example of retrospective metal accounting might be part of a reconciliation study to optimize future production. However, the methodology has much wider use and is discussed here relative to several different applications, as follows:

(i) Evaluation of the impact of two blasthole sampling procedures
(ii) Evaluation of the effect of using the wrong semivariogram model for the estimation of selective mining units (SMUs)
(iii) Evaluation of the impact of block-estimation error on block classification (i.e., the combined effect of the amount of ore lost to the waste dump and the amount of dilution).

Because grades are rarely accurate to more than two or three significant figures, some caution must be exercised in how calculations are performed. Therefore, instead of calculating small differences between pairs of very large numbers (e.g., $1,000,002 - 1,000,000 = 2$), it is wiser to cast the problem in such a way that large numbers are avoided. This commonly can be achieved by dealing with differences that arise from two courses of action.

19.3: METAL ACCOUNTING: ALTERNATIVE BLASTHOLE SAMPLING METHODS

As part of a quality control program at Equity Silver Mine Ltd., 42 blasthole cuttings piles were sampled by two methods, tube sampling and channel sampling, described briefly in Fig. 5.7 and corresponding text, and in more detail elsewhere by Giroux et al. (1986). The analytical data for Ag and Cu assays are listed in Table 5.3 and in modified form in Table 19.1; only the Ag data are considered here because these were the grades used in practice to select ore and waste at the time of production. In this sampling experiment, the bulk material remaining after both types of samples had been taken was also subsampled and assayed ("Bulk" in Table 19.1) so that it was pos-

sible to estimate a weighted average grade for each cuttings pile ("Best" in Table 19.1). Thus, results of each of the two sampling methods can be compared with corresponding best estimates of cuttings piles.

For some of the comparisons reported here, each blasthole is assumed to represent a block equivalent to 360 tonnes; this is comparable to selection by polygonal estimation, as done at the mine site. So-called true values in the following discussion are the weighted mean values of tube, channel, and bulk samples of Table 19.1 (i.e., "true" = best). Equation 19.1 can be applied in a variety of ways to the information in Table 19.1 to compare real and estimated operating profits, as summarized in Table 19.2.

The information in Tables 19.1 and 19.2 warrants comment. Both tube and channel sampling methods are seen to overestimate true metal content of ore blocks. Tube sampling identifies 18 blocks as ore with an estimated mean grade of 139.2 g/t Ag, versus a true mean grade of 127.7 g/t. Channel sampling indicates a somewhat different 18 blocks as ore, estimated to average 143.3 g/t, compared with a true average of 136.1 g/t. By comparison, the "best" estimate identifies 19 blocks as ore with an average grade of 132.3 g Ag/t; this average is lower than estimates by both tube and channel sampling. Despite the relatively low grade of the ore identified by the "best" estimate, "best" returns a significantly higher "true" quantity of metal as follows:

Tube "true" $127.7 \times 18 \times 360 = 827,496$ g Ag
Channel "true" $136.1 \times 18 \times 360 = 881,928$ g Ag
Best estimate $132.3 \times 19 \times 360 = 904,932$ g Ag.

In this case, both tube and channel sampling underestimate metal that could be recovered because they fail to identify some ore blocks. However, both tube and channel samples are slightly biased on the high side and overestimate those blocks that are identified as ore. The overestimate by tube sampling is about twice that of channel sampling in terms of both grade and metal content. Hence, channel sampling is significantly better than tube sampling in two

Table 19.1 Assay data, blasthole sampling experiment, Equity Silver Mine – Cu (%), Ag (g/t)

BH no.	Tube Cu	Tube Ag	Channel Cu	Channel Ag	Bulk Cu	Bulk Ag	Best Cu	Best Ag
1	0.14	5	0.25	18	0.26	22	0.246	20.0
2	0.06	20	0.05	22	0.04	15	0.043	17.9
3	0.02	11	0.02	6	0.01	4	0.014	4.98
4	0.08	123[a]	0.07	103[a]	0.07	101	0.068	102.4[a]
5	0.03	**36**	0.06	70[a]	0.04	54	0.046	59.9[a]
6	0.69	274[a]	0.97	393[a]	0.87	389	0.879	387.5[a]
7	0.69	148[a]	0.6	163[a]	0.52	151	0.540	155.7[a]
8	0.12	10	0.1	8	0.16	20	0.132	15.0
9	0.06	76[a]	0.06	62[a]	0.05	46	0.053	53.1[a]
10	0.11	85[a]	0.15	142[a]	0.14	117	0.139	126.0[a]
11	0.03	28	0.03	32	0.04	47	0.035	40.5
12	0.02	27	0.03	27	0.04	37	0.035	32.8
13	0.28	108[a]	0.07	55[a]	0.03	50	0.051	53.5[a]
14	0.07	76[a]	0.09	98[a]	0.09	86	0.087	90.5[a]
15	0.09	126[a]	0.09	160[a]	0.12	178	0.105	169.5[a]
16	0.06	42	0.07	35	0.06	48	0.062	42.7
17	0.47	**11**	0.44	131[a]	0.38	11	0.394	58.5[a]
18	0.56	325[a]	0.43	217[a]	0.45	226	0.433	225.1[a]
19	0	0	0.02	5	0.03	7	0.025	6.0
20	0.14	**78**[a]	0.09	45	0.08	40	0.083	43.0
21	0.33	199[a]	0.24	123[a]	0.3	147	0.271	138.9[a]
22	0.07	31	0.07	28	0.05	23	0.057	25.2
23	0.04	19	0.04	18	0.03	14	0.033	15.7
24	0.04	11	0.03	10	0.03	10	0.029	10.0
25	0.04	29	0.02	18	0.03	21	0.026	20.0
26	0.42	220[a]	0.39	233[a]	0.43	268	0.403	252.8[a]
27	0.18	**52**[a]	0.11	21	0.09	16	0.097	19.0
28	0.02	301[a]	0.76	175[a]	0.79	181	0.737	181.9[a]
29	0.24	**43**	0.27	89[a]	0.21	74	0.227	79.1[a]
30	0.02	10	0.02	9	0.02	11	0.019	10.2
31	0.19	15	0.15	26	0.16	24	0.15	24.5
32	0.24	4	0.19	4	0.23	3	0.209	3.4
33	0.15	141[a]	0.25	240[a]	0.17	167	0.194	195.2[a]
34	0.06	**51**[a]	0.03	22	0.03	21	0.03	22.2
35	0.02	**16**	0.06	57[a]	0.07	66	0.063	61.1[a]
36	0.03	22	0.01	10	0.02	13	0.016	12.1
37	0.07	6	0.03	12	0.01	7	0.019	9.0
38	0.05	**28**	0.08	45	0.11	78	0.094	63.6[a]
39	0.1	**55**[a]	0.03	17	0.05	26	0.043	23.2
40	0.2	68[a]	0.19	68[a]	0.16	54	0.168	59.9[a]
41	0.04	9	0.06	18	0.04	8	0.046	12.0
42	0	2	0.01	2	0.01	2	0.009	2
MEAN[b]		139.2		143.3				132.3
TRUE MEAN[c]		127.8		136.1				

Note: **Bold type** indicates misclassified values.
[a] Grades above cutoff of 50 g/t Ag.
[b] Estimated mean value of those grades greater than cutoff of 50 g/t Ag.
[c] True mean value of those grades greater than cutoff of 50 g/t Ag (i.e., mean value based on corresponding data in "Best" column.)

Table 19.2 Metal accounting and sampling experiments, Equity Silver Mine Ltd.

Tube sampling	
Estimated operating profit 18 × (139.2 − 50) × 360	= 578,000 g Ag
True operating profit 18 × (127.7 − 50) × 360	= 503,500 g Ag
Overestimation of operating profit	= 74,500 g Ag
Lost profit (ore classed as waste) 5 × (64.4 − 50) × 360	= 25,900 g Ag
Channel sampling	
Estimated operating profit 18 × (143.3 − 50) × 360	= 604,600 g Ag
True operating profit 18 × (136.1 − 50) × 360	= 557,900 g Ag
Overestimation of operating profit	= 46,700 g Ag
Lost profit (ore classed as waste) 1 × (63.6 − 50) × 360	= 4,900 g Ag

Note: All numeric expressions are in the form of Eq. 19.1.

respects: First, overall estimates by channel sampling are much closer to true values, so productivity will be closer to expectations than would be the case with tube sampling. Second, a hidden operating loss resulting from ore erroneously identified as waste is much less with channel sampling (4,900 g Ag) than with tube sampling (25,900 g Ag). Note that of these 42 samples, 9 were misclassified by tube sampling whereas only 1 was misclassified by channel sampling. Improved metal recovery by channel sampling reflects an improvement in block classification relative to tube sampling.

If the 42 grades are assumed to be representative of material encountered during production, the metal accounting results can be transformed quickly to monetary equivalents. Assuming a production rate of 3,000 tonnes per day (tpd) mined, the 42 blocks represent 15,120 tonnes, or about 1 week of production. The difference in true annual operating profit by the two sampling methods is 52 (weeks) × 54,400 (g)/31.1 (transform to oz) = 91,000 oz Ag, equivalent to US$455,000 (assuming US$5.00 per oz Ag).

This example illustrates the need for tests of experimental design in establishing sampling methodologies for ore production. The simple transposition

of a sampling method from one deposit to another deposit is not acceptable practice without a test to demonstrate the adequacy of the sampling method in the new situation. Significant and unnecessary operating losses can result both from bias and random error, as illustrated here.

19.4: METAL ACCOUNTING: EFFECT OF INCORRECT SEMIVARIOGRAM MODEL ON BLOCK ESTIMATION

Sinclair and Giroux (1984) describe the differences in block estimates obtained in a part of the South Tail zone of Equity Silver Mine Ltd. when an incorrect semivariogram model is used to make estimates instead of the correct model (as illustrated schematically in Fig. 3.3). In their example, 25 blocks (5 × 5 × 5 m^3 – equivalent to about 360 tonnes each), as shown in Fig. 13.7 are used as a test panel. These 25 blocks are estimated by ordinary kriging using first a general semivariogram model for the entire Southern Tail zone and then a model that is specific to the area in question, in which the principal structural control is at 90 degrees to the general trend in the deposit. Block estimates are summarized in Table 19.3. With a cutoff grade of 50 g Ag/t, the operating profit under two estimation scenarios can be determined and is summarized as follows:

Operating profit using incorrect semivariogram Nine blocks were estimated to be ore, with an average grade of 65.2 g Ag/t.

Apparent operating profit
$$= 9 \times 360 (65.2 - 50) = 49,248 \, g \, Ag.$$

However, the correct semivariogram provided an average kriged block grade of 58.4 g Ag/t for the nine blocks selected by using the incorrect semivariogram model. Thus, the effective operating profit in mining those nine blocks is

$$\text{Operating profit} = 9 \times 360 (58.4 - 50)$$
$$= 27,216 \, g \, Ag.$$

That is, use of the incorrect semivariogram model provided an overestimation (about

Table 19.3 Ordinary kriging estimates for 25 blocks (each $5 \times 5 \times 5$ m^3), northern part of the South Tail zone, Equity Silver Mine

Block	Kriged block estimates (g Ag/mt)	
	Correct semivariogram	Incorrect semivariogram
1	89.8[a]	65.8[a]
2	92.0[a]	71.0[a]
3	71.0[a]	60.3[a]
4	31.7	44.0
5	22.1	26.4
6	37.6	61.0[a]
7	47.5	93.8[a]
8	53.6[a]	52.5[a]
9	55.6[a]	58.2[a]
10	50.4[a]	45.7
11	28.6	35.6
12	40.7	40.4
13	55.3[a]	50.0[a]
14	60.8[a]	62.6[a]
15	58.9[a]	58.7[a]
16	20.8	17.9
17	29.7	26.1
18	42.1	33.1
19	46.4	41.5
20	43.5	48.7
21	10.7	12.8
22	19.3	30.0
23	27.5	38.9
24	32.3	39.9
25	26.3	31.2

Note: Average estimated grade of "incorrect" ore blocks = 65.2 g/t.
Average "true" grade of "incorrect" ore blocks = 58.4 g/t.
Average "true" grade of "correct" ore blocks = 63.5 g/t.
[a] Estimate 50 g/t (cutoff grade).
Source: After Sinclair and Giroux (1984).

Table 19.4 Comparison of block estimates using incorrect and correct semivariogram models, South Tail zone, Equity Silver Mine

Incorrect semivariogram model	
Apparent operating profit	= 49,248 g Ag
$= 9 \times 360\,(65.2 - 50)$	
Effective operating profit	= 27,216 g Ag
$= 9 \times 360\,(58.4 - 50)$	
Correct semivariogram model	
"True" operating profit	= 48,600 g Ag
$= 10 \times 360\,(63.5 - 50)$	

identified by the correct semivariogram model and is

$$\text{"true" operating profit} = 10 \times 360(63.5 - 50)$$
$$= 48,600 \, \text{g Ag}.$$

Consequently, the potential operating loss by using the wrong semivariogram model for 25 blocks (about 9,000 tonnes containing 3,600 tonnes of ore) is really $(48,600 - 27,200)$ 21,400 g Ag or about 5.9 g per tonne of ore identified. Details of this example are provided in Tables 19.3 and 19.4.

Assumptions about ore continuity are too often generalized without an ongoing critical analysis as more geologic information and insight become available. Incorrect models of continuity can lead to substantial errors in reserve estimation that can be reflected in significant losses to operating profit.

19.5: METAL ACCOUNTING: EFFECT OF BLOCK ESTIMATION ERROR ON ORE/WASTE CLASSIFICATION ERRORS (AFTER SINCLAIR, 1995; POSTOLSKI AND SINCLAIR, 1998)

Block estimation error can have a dramatic impact on metal recovery. Clearly, if some blocks of ore are inadvertently classed as waste, metal is lost; similarly, if blocks of waste are included in ore, total tonnage is increased, but average grade is decreased (i.e., dilution occurs). The problem is illustrated clearly in

22,000 g Ag) for the nine blocks identified as ore, amounting to approximately 80 percent of the true operating profit for the nine blocks in question. However, use of the correct semivariogram model indicates that one of the nine blocks is actually waste and that two additional blocks identified as waste are really ore. The best estimate of metal operating profit is therefore obtained by considering the 10 blocks of ore

Table 19.5 Effect of dilution and ore loss on production grade for various errors imposed on a block-grade distribution for the Bougainville porphyry-copper mine.

	10% Error		20% Error		30% Error	
	No. of blocks	Average grade[a]	No. of blocks	Average grade[a]	No. of blocks	Average grade[a]
True ore	919	0.475	919	0.475	919	0.475
Ore rejected as waste	17.3	0.233	45.3	0.261	84.4	0.298
Ore selected	901.7	0.480	873.7	0.486	834.6	0.493
Waste selected	10.5	0.204	17.3	0.198	21.9	0.194
Production	912.2	0.477	891	0.481	856.5	0.485

[a] Average grades in wt% Cu.

Fig. 16.6. David and Toh (1989) were among the first to provide quantitative documentation of the concept of dilution due to block estimation error, using the Bougainville copper deposit as an example. A computer program has been developed (Postolski and Sinclair, 1998b) that quantifies dilution and ore losses for particular grade distributions and any average block estimation error. The methodology is described in Section 16.3.2 ; the impact of various block estimation errors on tonnage and grade are summarized for the Bougainville case in Tables 16.3 and 16.4. Information in these two tables is summarized in Table 19.5 and is presented as a metal accounting in Table 19.6.

Table 19.5 is important because it permits a comparison of the operating losses (expressed as metal) that result from various error scenarios. The differences in losses from one scenario to another provide some insight as to the potential savings if block estimation errors can be reduced. A comparable example for the Huckleberry porphyry deposit (Postolski, 1998) is summarized in Table 19.7. In this case, an improvement in block-estimation error from 20 percent to 15 percent would result in an additional profit of $205.1 - 117.7 = 87.4$ tonnes of copper from a production of 5.4 million tonnes of ore.

An important contributing factor to block-estimation errors, regardless of the method of estimation, is the combination of sampling and analytical error. This is particularly the case for gold deposits, as illustrated here for the Ortiz gold deposit, New Mexico (Springett, 1984). The GAINLOSS software by

Postolski and Sinclair (1998) was used to provide information for the Ortiz deposit comparable to Tables 16.3 and 16.4 and summarized in Table 19.8. For these estimates, the "true" Au grade distribution was taken as lognormal with a mean of 1.75 g/t (0.051 oz/st), a standard deviation of 1.23 g/t (0.036 oz/st), and a block size taken as 30 st (ca. 27 tonnes). A cutoff grade of 0.85 g/t (0.025 oz/st) was assumed for illustrative purposes. The information in Table 19.8 can be recast in metal accounting form as illustrated in Table 19.9. These results show that an improvement in block-estimation error (1 standard deviation) to 20 percent from 30 percent results in an operating profit of $(2,627 - 900) = 1,727$ additional ounces of gold (equivalent to US$492,000.00 @ $285/oz) from the mining of 100,000 blocks (3,000,000 short tons).

These examples demonstrate the impact of various levels of block-estimation error on metal losses and dilution during production. Results are deposit specific because they depend on the data distribution. However, results are fairly robust and a small change in the distribution of grades does not have a large impact on the results. The procedure is worth repeating for particular situations in order to estimate whether the effort/cost of improving low-quality estimates is worthwhile.

19.6: SUMMARY COMMENTS

Two examples illustrated here are entirely deterministic and are based on small data sets; hence, they

Table 19.6 Metal accounting summary of operating loss for block misclassification due to various levels of error, Bougainville porphyry deposit, for assumed cutoff grade = 0.215% Cu[a]

	10% Error	20% Error	30% Error
Net cost of mining waste classed as ore (tonnes of metal)	−2.24	−5.83	−9.15
Net loss of metal in ore classed as waste (tonnes of metal)	−6.23	−41.63	−140.55
Total operating loss (tonnes of metal)	**−8.47**	**−47.47**	**−149.70**

[a] For a hypothetical scenario with 1,000 blocks (2,000 tonnes each).

Table 19.7 Metal accounting summary of operating loss for block misclassification due to various levels of error, Huckleberry porphyry deposit, east zone (west domain) for assumed cutoff grade = 0.4% Cu

	10% Error	15% Error	20% Error
Net cost of mining waste classed as ore (tonnes of metal)	−23.95	−50.77	−85.16
Net loss of metal in ore classed as waste (tonnes of metal)	−29.01	−66.91	−119.91
Total operating loss (tonnes of metal)	**−52.96**	**−117.69**	**−205.07**

Table 19.8 Effect of dilution and ore loss on production grade for various errors imposed on a block grade distribution for the Ortiz gold mine, New Mexico

	10% Error		20% Error		30% Error	
	No. of blocks	Average grade[a]	No. of blocks	Average grade[a]	No. of blocks	Average grade[a]
True ore	79,569	2.05	79,569	2.05	79,569	2.05
Ore rejected as waste	1,957	0.92	4,744	1.02	8,176	1.17
Ore selected	77,612	2.08	74,825	2.12	71,393	2.15
Waste selected	1,451	0.79	2,709	0.77	3,672	0.75
Production	79,063	2.06	77,534	2.07	75,065	2.08

[a] Average grade in g Au/t.

Table 19.9 Metal accounting summary of operating loss for block misclassification due to various levels of error, Ortiz gold deposit

	10% Error	20% Error	30% Error
Net cost of mining waste classed as ore (grams of metal)	−2,152	−5,881	−10,085
(troy ounces of metal)	−69.2	−189.0	−324.2
Net loss of metal in ore classed as waste (grams of metal)	−3,920	−22,120	−71,951
(troy ounces of metal)	−126.0	−711.1	−2,313.1
Total operating loss (grams of metal)	**−6,072**	**−28,001**	**−82,035**
(troy ounces of metal)	**−195.2**	**−900.2**	**−2,627.3**

Table 19.10 Number of waste blocks mistakenly included in ore due to various levels of errors, Ortiz gold deposit

Grade interval center	Frequency in 100,000 blocks	10% Error		20% Error		30% Error	
		$P > c$	$N > c$	$P > c$	$N > c$	$P > c$	$N > c$
0.3	886.4					0.000	0.000
0.4	1996.8			0.000	0.000	0.000	0.065
0.5	3162.5	0.000	0.000	0.000	0.324	0.008	24.895
0.6	4131.3	0.000	0.016	0.016	66.228	0.080	328.617
0.7	4814.3	0.014	65.595	0.140	673.861	0.237	1139.483
0.8	5219.4	0.265	1385.163	0.377	1968.978	0.417	2178.874
Sum of misclassified waste blocks			1450.774		2709.391		3671.934
Average true grade of misclassified waste blocks			0.795		0.770		0.749

$P > c$ = proportion of waste blocks incorrectly assigned to ore. $N > c$ = number of waste blocks per 100,000 blocks, incorrectly assigned to ore.

are rough estimates and should be seen as examples that illustrate a technique while emphasizing the need for appropriate experimental design in real-life cases. The third example, Bougainville Copper Evaluation, is based on a real data distribution and the normal distribution of errors, and thus is general in nature; moreover, this example is statistical rather than deterministic and provides a general insight into the impact of errors on operating profit. Although deposit specific, the results of the Bougainville example are generally applicable to porphyry-type deposits with not-too-dissimilar grade distributions.

19.7: PRACTICAL CONSIDERATIONS

1. Metal accounting has had applications in mineral inventory studies, especially for purposes of conducting reconciliations and documenting apparent operating profits for various scenarios. The procedure has potential for extensive use in many situations related to mineral inventory in which alternative choices are possible. Any situation involves estimates by two or more methods, different levels of variability that arise due to various sampling or analytical procedures, and soon can be treated by "metal accounting," provided a reasonable estimate of cutoff grade can be made.

2. Sampling experiments are particularly amenable to evaluation by metal accounting, especially when a total mass balance can be determined, as in the Equity blasthole sampling case cited here.

3. Mineral inventory estimates by two or more methods can be compared, as illustrated for the "correct" and "incorrect" semivariogram estimates for the South Tail zone. In this example, the true value is unknown, but the magnitude of the difference between acceptable and unacceptable methods can be determined.

4. If the true grade distribution is known, the influence of error on ore selection can be determined as shown, using an extension of a published case history for the Bougainville porphyry copper deposit. The effects of both "loss of ore to waste" and "inclusion of waste in ore" are considered. In this example, the improvement in metal recovery as sampling error decreases is quantified. The cost of decreasing sampling and analytical errors can be considered in the light of the expected increase in operating profit. The general applicability of the procedure is illustrated by the Ortiz gold example.

5. Any situation in which different sampling/ analytical/estimation methods lead to different numeric results in terms of grades can generally be cast in terms of metal accounting for comparative purposes.

Table 19.11 Number of ore blocks mistakenly included in waste due to various levels of error, Ortiz gold deposit, for a cutoff grade of 0.85 g Au/t

Grade interval center	Frequency in 100,000 blocks	10% Error		20% Error		30% Error	
		$P < c$	$N < c$	$P < c$	$N < c$	$P < c$	$N < c$
0.9	5393.6	0.289	1557.788	0.391	2106.382	0.427	2300.512
1.0	5392.2	0.064	343.742	0.226	1216.937	0.308	1661.934
1.1	5265.2	0.009	49.587	0.126	661.691	0.223	1176.183
1.2	5053.5	0.001	5.624	0.069	350.555	0.164	827.727
1.3	4788.7	0.000	0.583	0.039	184.949	0.122	584.210
1.4	4494.2	0.000	0.061	0.022	98.500	0.092	415.529
1.5	4186.7	0.000	0.007	0.013	53.390	0.071	298.602
1.6	3878.1	0.000	0.001	0.008	29.589	0.056	217.056
1.7	3576.4			0.005	16.805	0.045	159.669
1.8	3286.8			0.003	9.787	0.036	118.851
1.9	3012.5			0.002	5.845	0.030	89.490
2.0	2755.4			0.001	3.576	0.025	68.126
2.1	2516.1			0.001	2.239	0.021	52.404
2.2	2294.6			0.001	1.433	0.018	40.707
2.3	2090.7			0.000	0.936	0.015	31.912
2.4	1903.6			0.000	0.623	0.013	25.233
2.5	1732.3			0.000	0.423	0.012	20.113
2.6	1575.9			0.000	0.291	0.010	16.152
2.7	1433.4			0.000	0.204	0.009	13.063
2.8	1303.6			0.000	0.145	0.008	10.633
2.9	1185.5			0.000	0.104	0.007	8.709
3.0	1078.3					0.007	7.174
3.1	980.8					0.006	5.941
3.2	892.4					0.006	4.945
3.3	812.1					0.005	4.135
3.4	739.3					0.005	3.474
3.5	673.2					0.004	2.930
3.6	613.2					0.004	2.482
3.7	558.8					0.004	2.109
3.8	509.4					0.004	1.799
3.9	464.5					0.003	1.540
4.0	423.8					0.003	1.322
4.1	386.8					0.003	1.139
Sum of misclassified ore blocks		1957.393		4744.402		8175.791	

$P < c$ = proportion of ore blocks incorrectly assigned to waste. $N < c$ = number of ore blocks per 100,000 blocks, that are incorrectly assigned to waste.

19.8: SELECTED READING

David, M., and E. Toh, 1989, Grade control problems, dilution and geostatistics: choosing the required quality and number of samples for grade control; Can. Inst. Min. Metall. Bull., v. 82, no. 931, pp. 53–60.

Postolski, T. A., and A. J. Sinclair, 1998, Quantitative estimation of dilution and ore loss resulting from block estimation errors and a specified cutoff grade; *in* Vallée, M., and A. J. Sinclair (eds.), Quality assurance, continuous quality improvement and standards in mineral resource estimation, Expl. Min. Geol. v. 7, nos. 1 and 2, pp. 91–98.

19.9: EXERCISES

1. The following information, summarized from Knudson (1992), can be presented in terms of comparative metal operating profits, estimated vs. real. Estimates for 442 blocks (420 tonnes each) above cutoff (0.5 g Au/mt) for the Cherokee gold deposit average 0.96 g Au/mt. Mill heads subsequently were found to average 0.75 g Au/mt. Calculate the overestimate as a function of apparent metal operating profit.

2. The metal accounting for silver summarized in Table 19.2 can be enhanced by adding the effects of Cu (data provided in Table 19.1). For purposes of the exercise, assume 100 percent recovery of Cu.

Appendices

Appendices

Appendix 1

British and International Measurement Systems: Conversion Factors

Distance

1 inch	= 2.54 centimeters
1 kilometer	= 0.6214 miles
1 meter	= 3.281 feet
	= 100 centimeters
1 mile	= 5,280 feet
	= 1.609 kilometers
1 millimeter	= 0.0394 inches
1 micron	= 1.0×10^{-3} millimeters

Area

1 acre	= 4,840 sq. yards
	0.404686 hectares
	4,046.86 square meters
1 hectare	= 10,000 square meters
	= 11,959.9 square yards

Circular measure

Circle	= 6.2832 radians
	= 360 degrees
1 degree	= 60 minutes
1 degree	= 17.453×10^{-3} radians
1 minute (angle)	= 0.2909×10^{-3} radians

Volume

1 cubic centimeter	= 0.061 cubic inches
1 cubic meter	= 1,000 liters
	= 35.315 cubic feet
	= 1.308 cubic yards

Mass

1 gram	= 15.432 grains
	= 5 metric carats
	= 35.274 10^{-3} ounces
	= 2.205×10^{-3} pounds
1 kilogram	= 2.2046 pounds
	= 1,000 grams

Mass

Gram/tonne	= 1 part per million
	= 0.5833 dwt (troy)/short ton
	= 0.02917 ounces (troy)/short ton
	= 0.653 dwt (troy)/long ton
Gram/cubic centimeter	= 1 tonne/cubic meter
	= 62.428 pounds/cubic foot
1 ounce (avoirdupois)	= 437.5 grains
	= 28.350 grams
1 ounce (troy)	= 20 pennyweights (dwt)
	= 480 grains
	= 31.103 grams
1 pennyweight (dwt, troy)	= 24 grains
	= 1.5552 grams
1 pound (avoirdupois)	= 7000 grains
	= 16 ounces
	= 453.59 grams
1 pound (troy)	= 12 ounces (troy)
	= 5,760 grains
	= 373.24 grams
1 stone	= 6.3503 kilograms
	= 32.6667 grams
1 long ton	= 2,240 pounds
	= 1.016 short tons
1 short ton	= 2,000 pounds
1 metric tonne	= 0.9842 long tons
	= 1.1023 short tons
	= 1,000 kilograms
	= 2,205 pounds

Appendix 2

U.S. Standard Sieves

Designation			Nominal dimensions			
National Bureau of Standards		Tyler	A.S.T.M sieve	Sieve opening		Wire diameter
Inches	Mesh number	Screen number	mm	Microns	Inches	mm
4			101.6		4.00	6.30
3.5			90.5		3.50	6.08
3			76.1		3.00	5.80
2.5			64.0		2.50	5.50
2			50.8		2.00	5.05
1 3/4			45.3		1.75	4.85
1 1/2			38.1		1.50	4.59
1 1/4			32.0		1.25	4.23
1			25.4		1.00	3.80
7/8			22.6		0.875	3.50
3/4			19.0		0.750	3.30
5/8			16.0		0.625	3.00
1/2			12.7		0.500	2.67
7/16			11.2		0.438	2.45
3/8			9.51		0.375	2.27
5/16	2 1/2	2 1/2	8.00		0.312	2.07
1/4	2 1/4	2 1/4	6.35		0.250	1.82
	3	3		5,660	0.223	1.68
	4	4		4,760	0.187	1.54
	5	5		4,000	0.157	1.37
	6	6		3,360	0.132	1.23
	7	7		2,830	0.111	1.10
	8	8		2,380	0.0937	1.000
	10	9		2,000	0.0787	0.900
	12	10		1,680	0.0661	0.810
	14	12		1,410	0.0555	0.725
	16	14		1,190	0.0469	0.650
	18	16		1,000	0.0394	0.580
	20	20		841	0.0331	0.510
	25	24		707	0.0197	0.340
	40	35		420	0.0165	0.290
	45	42		354	0.0139	0.247
	50	48		297	0.0117	0.215
	60	60		250	0.0098	0.180
	70	65		210	0.0083	0.152
	80	80		177	0.0070	0.131
	100	100		149	0.0059	0.110
	120	115		125	0.0049	0.091
	140	150		105	0.0041	0.076
	170	170		88	0.0035	0.064
	200	200		74	0.0029	0.058
	230	250		63	0.0025	0.044
	270	270		53	0.0021	0.037
	325	325		44	0.0017	0.030
	400	400		37	0.0015	0.025

Appendix 3

Drill-Hole and Core Diameters

Diamond-drill-hole diameters

	Core diameter		Hole diameter	
Size	(Decimal inches)	(mm)	(Decimal inches)	(mm)
AQ, AQ-U	1.062	26.97	1.89	48.01
BQ, BQ-U	1.432	36.37	2.360	59.94
NQ, NQ-U	1.875	47.63	2.980	75.69
HQ	2.500	63.50	3.782	96.06
PQ	3.345	84.96	4.827	122.61

Wireline drill-hole diameters

	Core diameter		Hole diameter	
Size	(Decimal inches)	(mm)	(Decimal inches)	(mm)
AQ, AQ-U	1.06	27.0	1.89	48.0
BQ, BQ-U	1.44	36.5	2.36	60.0
NQ, NQ-U	1.87	47.6	2.98	75.7
HQ	2.50	63.5	3.78	96.0

Source: Cumming (1980).

Bibliography

Adams, S. S., and B. R. Putnam III, 1992, Application of mineral deposit models in exploration: a case study of sediment-hosted gold deposits, Great Basin, western United States; *in* Annels, A. E., 2 (ed.), Case histories and methods in mineral resource evaluation; Geol. Soc., London, Spec. Pub. No. 63, pp. 1–23.

Agricola, 1556, de re metallica (Hoover translation); Mining Mag., London.

Agterberg, F. P., 1974, Geomathematics; Developments in Geomathematics 1; Elsevier Scientific Pub. Co., Amsterdam, 596 pp.

Agterberg, F. P., 1965, The technique of serial correlation applied to continuous series of element concentration values in homogeneous rocks; Jour. Geol., v. 73, pp. 142–154.

Ahrens, E. H., 1983, Practical mining geology; *in* Erickson, A. J. Jr., R. A. Metz, and D. E. Ranta (eds.), Applied mining geology: general studies, problems of sampling and grade control, ore reserve estimation, Proc. Symp. of Soc. Min. Metall. and Expl., pp. 73–82.

Allard, D., M. Armstrong, and W. J. Kleingeld, 1993, The need for a connectivity index in mining geostatistics; *in* Dimitrakopoulos, R. (ed.), Geostatistics for the next century; Kluwer Acad. Press, Proc. Int. Forum in honor of Michel David's contributions to geostatistics, Montreal, June 3–5, pp. 230–236.

Annels, A. E., 1996, Ore reserves: errors and classification, Trans. Inst. Min. Metall., v. 105, pp. A150–A156.

Annels, A. E., 1991, Mineral deposit evaluation, a practical approach; Chapman and Hall, London, 436 pp.

Annels, A. E., S. Ingram, and L. Malmstrom, 1994, Structural reconstruction and mineral resource evaluation at Zinkgruvan mine, Sweden; *in* Whateley, M. K. G., and P. K. Harvey (eds.), Mineral resource evaluation II: methods and case histories; Geol. Soc., Spec. Pub. No. 79, pp. 171–189.

Anonymous, 1999, A guide for reporting exploration information, mineral resources and mineral reserves; Soc. for Min., Metall. and Expl., January.

Anonymous, 1994, Recommendations on reserve definitions to the Canadian Institute of Mining, Metallurgy and Petroleum; Can. Inst. Min. Metall. Bull., v. 87, no. 984, pp. 20–24.

Anonymous, 1991a, Guidelines to the Australasian code for reporting of identified mineral resources and ore reserves; Prepared by the Joint Committee of the Australasian Inst. Min. Metall. and the Australian Mining Industry Council, May, 5 pp.

Anonymous, 1991b, A guide for reporting exploration information, resources and reserves (Working Party #79); Min. Eng., April, pp. 379–384.

Anonymous, 1989, Australasian code for reporting of identified mineral resources and ore reserves; report of the joint committee of the Australasian Institute of Mining and Metallurgy and Australian Mining Industry Council, February, 8 pp.

Arik, A., 1992, Outlier restricted kriging: a new kriging algorithm for handling of outlier high grade data in ore reserve estimation; *in* Kim,

Y. C. (ed.), Proc. 23rd symp. on application of computers and operations research in the minerals industry, Tucson, Ariz., pp. 181–187.

Armitage, M. G., and F. E. Potts, 1994, Some comments on the classification of resources and reserves; *in* Whateley, M. K. G., and P. K. Harvey (eds.), Mineral resource evaluation II: methods and case histories; Geol. Soc., Spec. Pub. No. 79, pp. 11–16.

Applin, K. E. S., 1972, Sampling of alluvial diamond deposits in West Africa; Trans. Inst. Min. Metall., sect. A, pp. 62–77.

Armstrong, M., 1992, Positive definiteness is not enough; Math. Geol., v. 24, no. 1, pp. 135–144.

Armstrong, M., 1984a, Improving the estimation and modeling of the variogram; *in* Verly, G., M. David, A. G. Journel, and A. Marechel (eds.), Geostatistics for natural resource characterization: Nato ASI Series C: v. 122 – Part 1, D. Reidel Pub. Co., Dordrecht, the Netherlands, pp. 1–19.

Armstrong, M., 1984b, Common problems seen in variograms; Math. Geol., v. 16, no. 3, pp. 305–313.

Armstrong, M., and A. Boufassa, 1988, Comparing the robustness of ordinary kriging and lognormal kriging: outlier resistance; Math. Geol., v. 20, no. 4, pp. 447–458.

Armstrong, M., and N. Champigny, 1989, A study on kriging small blocks; Can. Inst. Min. Metall. Bull., v. 82, pp. 128–133.

Armstrong, M., and P. Diamond, 1984, Testing variograms for positive-definiteness; Math. Geol., v. 16, no., 4, pp. 407–422.

Aspie, D., and R. J. Barnes, 1990, Infill sampling design and the cost of classification errors; Math. Geol., v. 22, no. 8, pp. 915–932.

Assibey-Bonsu, W., 1996, Summary of present knowledge on the representative sampling of ore in the mining industry; Jour. South African Inst. Min. Metall., November, pp. 289–293.

Atkinson, W. W. Jr., and A. J. Erickson, Jr., 1984, Concepts in core logging and mapping of mineral deposits: a practical example; *in* Erickson, A. J., Jr. (ed.), Applied mining geology; Soc. Min. Engs., New York, pp. 1–8.

Baafi, E. Y., Y. C. Kim, and F. Szidarovszky, 1986, On nonnegative weights of linear kriging estimation; Min. Eng., June, pp. 437–442.

Babcock, J. W., 1984, Introduction to geologic ore deposit modeling; Min. Eng., December, pp. 1631–1636.

Bacon, W. G., G. W. Hawthorn, and G. W. Poling, 1989, Gold analyses–myths, frauds and truths; Can. Inst. Min. Metall. Bull., v. 82, no. 931, pp. 29–36.

Baker, C. K., and S. M. Giacomo, 1998, Resources and reserves: their uses and abuses by the equity markets; *in* Ore reserves and finance, a joint seminar between AusIMM and ASX, Sydney, Australia, June 15.

Bancroft, B. A., and G. R. Hobbs, 1986, Distribution of kriging error and stationarity of the variogram in a coal property; Math. Geol., v. 18, no. 7, pp. 635–652.

Bardossy, A., 1988, Notes on the robustness of the kriging system; Math. Geol., v. 20, no. 3, pp. 189–204.

Bardossy, A., and G. Bardossy, 1984, Comparison of geostatistical calculations with the results of open pit mining at the Iharkut bauxite district, Hungary: a case study; Math. Geol., v. 16, no. 2, pp. 173–192.

Bardossy, A., I. Bogardi, and W. E. Kelly, 1988, Imprecise (fuzzy) information in geostatistics; Math. Geol., v. 20, no. 4, pp. 287–312.

Barnes, R. J., 1991, The variogram sill and the sample variance; Math. Geol., v. 23, no. 4, pp. 673–678.

Barnes, R. J., 1988, Bounding the required sample size for geologic site characterization; Math. Geol., v. 20, no. 5, pp. 477–490.

Barnes, R. J., and K. You, 1992, Adding bounds to kriging; Math. Geol., v. 24, no. 2, pp. 171–176.

Barnes, R. J., and T. B. Johnson, 1984, Positive kriging; *in* Verly, G., M. David, A. G. Journel, and A. Marechel (eds.), Geostatistics for natural resource characterization: NATO ASI Series C: v. 122 – Part 1, D. Reidel Pub. Co., Dordrecht, the Netherlands, pp. 231–244.

Barnes, T., 1999, Practical post-processing of indicator distributions; *in* Dagdelen, K. (ed.), Proc. 28th Symp. on Applications of computers and operations research in the minerals industry, Colorado School of Mines, Golden, Colo., pp. 227–238.

Barry, J., J. Guard, and G. Walton, 1994, Database management at the Lisheen deposit, Co. Tipperary, Ireland; *in* Whateley, M. K. G., and P. K. Harvey (eds.), Mineral resource evaluation II: methods and case histories; Geol. Soc., London, Spec. Pub. No. 79, pp. 233–239.

Becker, R. M., and S. W. Hazen, Jr., 1961, Particle statistics of infinite populations as applied to mine sampling; U.S. Bur. Mines, Rept. of Investigations 5569, 79 pp.

Bell, T. M., and M. K. G. Whateley, 1994, Evaluation of grade estimation techniques; *in* Whateley, M. K. G., and P. K. Harvey (eds.), Mineral resource evaluation II: methods and case histories; Geol. Soc., London, Spec. Pub. No. 79, pp. 67–86.

Bentzen, A., and A. J. Sinclair, 1993, P-RES – a computer program to aid in the investigation of polymetallic ore reserves; Tech. Rept. MT-9, Mineral Deposit Research Unit, Dept. of Geological Sciences, University of British Columbia, Vancouver (includes diskette), 55 pp.

Bevan, P. A., 1993, The weighting of assays and the importance of both grade and specific gravity; Can. Inst. Min. Metall. Bull., v. 86, no. 967, pp. 88–90.

Bird, H. H., 1991, Dealing with coarse gold and cutting factors or dealing with the nugget effect in practice; *in* Proc. seminar on "Sampling and Ore Reserves," Prospector and Developers Assoc. of Canada, Toronto, March 23, pp. 35–53.

Blackwell, G. H., 1998, Relative kriging errors – a basis for mineral resource classification; Expl. Min. Geol., v. 7, nos. 1 and 2, pp. 99–106.

Blackwell, G. H., 1993, Computerized mine planning for medium-size open pits; Transactions, Inst. of Min. Metall., sect. A, 102, pp. 83–88.

Blackwell, G. H., 1991, Open pit mine grade control; *in* Lawton, S. E. (ed.), Proc. Seminar on "Sampling and Ore Reserves," Prospector and Developers Assoc. of Canada, Toronto, March 23, pp. 143–149.

Blackwell, G. H., and A. J. Sinclair, 1993, QMIN, geostatistical software for mineral inventory using personal computers; *in* Elbrond, J., and X. Tang (eds.), Proc. 24th Symp. on Application of computers and operations research to the minerals industry, Montreal, v. 2, pp. 350–357.

Blackwell, G. H., and A. J. Sinclair, 1992, Geostatistical mineral inventory using personal computers; Can. Inst. Min. Metall. Bull., v. 85, no. 961, pp. 65–70.

Blackwell, G. H., M. Anderson, and K. Ronson, 1999, Simulated grades and open pit mine planning – resolving opposed positions; *in* Dagdalen, K. (ed.), Proc. 28th Symp. on Application of computers and operations research to the minerals industry, Colorado School of Mines, Golden, Colo., pp. 205–215.

Blais, R. A., and P. A. Carlier, 1968, Applications of geostatistics in ore evaluation; *in* Gill, J. E., R. A. Blais, and V. A. Haw (eds.), Ore reserve estimation and grade control, Can. Inst. Min. Metall., Spec. v. 9, pp. 41–68.

Blanchet, T., and C. I. Godwin, 1972, Geolog system for computer and manual analysis of geologic data from porphyry and other deposits; Econ. Geol., v. 76, pp. 796–813.

Bliss, J. D., G. J. Orris, and W. D. Menzie, 1987, Changes in grade, volume and contained gold during the mining life-cycle of gold placer deposits; Can. Inst. Min. Metall. Bull., v. 80, no. 903, pp. 75–80.

Brooker, P. I., 1986, A parametric study of robustness of kriging variance as a function of range and relative nugget effect for a spherical semivariogram; Math. Geol., v. 18, no. 5, pp. 477–488.

Brooker, P. I., 1985, Two-dimensional simulation by turning bands; Math. Geol., v. 17, no. 1, pp. 81–90.

Brooker, P. I., 1983, Semi-variogram estimation using a simulated deposit; Min. Eng., January, pp. 37–42.

Brooker, P. I., 1979, Kriging; Eng. and Min. Jour., September, pp. 148–153.

Brooker, P. I., 1975a, Optimal block estimation by kriging; Proc. Australasian Inst. Min. Metall., no. 253, pp. 15–19.

Brooker, P. I., 1975b, Avoiding unnecessary drilling; Proc. Australasian Inst. Min. Metall., no. 253, pp. 21–23.

Brooks, L. S., and R. C. E. Bray, 1968, A study of the tonnage and grade calculations at the Geco division of Noranda Mines; Can. Inst. Min. Metall., Spec. v. 9, pp. 177–182.

Bullis, H. R., R. A. Hureau, and P. D. Soares, 1994, Dilution measurement and control at Lupin, N.W.T.; Can. Inst. Min. Metall. Bull., v. 87, no. 982, pp. 73–79.

Burn, R. G., 1981, Data reliability in ore reserve assessments; Mining Mag., October, pp. 289–299.

Buxton, B. E., 1986, Global estimation variance; Letter to the Editor, Math. Geol., v. 18, no. 7, pp. 693–696.

Buxton, B. E., 1984, Estimation variance of global recoverable reserve estimates; in Verly, G., M. David, A. G. Journel, and A. Marechel (eds.), Geostatistics for natural resource characterization: NATO ASI Series C: v. 122 – Part 1, D. Reidel Pub. Co., Dordrecht, the Netherlands, pp. 165–183.

Caers, J., and L. Rombouts, 1995, Valuation of primary diamond deposits by extreme value statistics; Econ. Geol., v. 91, no. 5, pp. 841–854.

Calvert, A. J., and D. Livelybrooks, 1997, Borehole-radar reflection imaging at the McConnell nickel mine, Sudbury; in Gubins, A. G. (ed.), Proc. of fourth decennial international conference on mineral exploration, Prospectors and Developers Assoc., Toronto, Canada, pp. 701–704.

Caras, S., 1990, Ore reserve estimation at the feasibility study stage – the period of greatest risk; in Symposium Proceeding "Ore reserve estimates, the impact on miners and financiers," Melbourne Branch, Australasian Inst. Min. Metall., Parkville, Victoria, Australia, pp. 99–108.

Caras, S., 1987a, Gold mine evaluation – some evaluation concepts; Proc. seminar on ore reserves, Pacific Rim Congress '87, pp. 845–850.

Caras, S., 1987b, Concepts for estimating ore reserves and a comparative view of resource estimation methods; Australasian Inst. Min. Metall., Sydney Branch, Proc. Resources and Reserves Symposium, November, pp. 73–80.

Caras, S., 1984, Comparative ore reserve methodologies for gold mine evaluation; Proc. Conf. "Gold-mining, metallurgy and geology," Aus. Inst. Min. Metall., Perth and Kalgoorlie branches, October, pp. 59–70.

Chamberlain, T. C., 1965, The method of multiple working hypotheses; Science, v. 148, pp. 754–759.

Champigny, N., 1993, Use of geostatistics: current situation and future perspectives; in Bawden, W. F., and J. F. Archibald (eds.), Innovative mine design for the 21st century, Balkema, Rotterdam, pp. 193–197.

Champigny, N., 1992, Geostatistics: a tool that works; Northern Miner, May 18, p. 4.

Champigny, N., 1989, Three solitudes; Northern Miner, June, p. 55.

Champigny, N., and M. Armstrong, 1994, An overview of reserve estimation problems by an international "Groupe de reflexion"; Can. Inst. Min. Metall. Bull., v. 87, no. 977, pp. 23–25.

Champigny, N., and M. Armstrong, 1993, Geostatistics for the estimation of gold deposits. A review and survey of current practices up to 1989; Mineralium Deposita, v. 28, no. 4, pp. 279–282.

Champigny, N., and P. H. Grimley, 1990, Computer based reserve estimation and grade control: practitioners' views; Can. Inst. Min. Metall. Bull., v. 83, no. 942, pp. 75–77.

Champigny, N., and A. J. Sinclair, 1984, A geostatistical study of the Cinola gold deposit, Queen Charlotte Islands, British Columbia; Western Miner, February 1984, pp. 54–59.

Champigny, N., A. J. Sinclair, and K. G. Sanders, 1980, Specogna gold deposit of Consolidated Cinola Mines, an example of structured property exploration; Western Miner, June, pp. 35–44.

Chaouai, N.-E., and K. Fytas, 1991, A sensitivity analysis of search distance and number of samples in indicator kriging; Can. Inst. Min. Metall. Bull., v. 84, no. 948, pp. 37–43.

Chavez, Jr., W. X., 2000, Supergene oxidation of copper deposits: zoning and distribution of copper oxide minerals; Society of Economic Geologists Newsletter, no. 41, April, p. 1 and pp. 10–19.

Chou, D., and D. E. Scheck, 1984, Selecting optimum drilling locations by groups; Amer. Inst. Min. Metall. Eng., preprint 84-62, 9 pp.

Cheng, Q., 1995, The perimeter-area fractal model and its application to geology; Math. Geol., v. 27, no. 1, pp. 69–82.

Chiles, J. P., 1993, In defense of geostatistics; Eng. and Min. Jour., June, p. 6-ww.

Chung, C. F., 1984, Use of the jacknife method to estimate autocorrelation functions (or variograms); in Verly, G., M. David, A. G. Journel, and A. Marechel (eds.), Geostatistics for natural resource characterization: NATO ASI Series C: v. 122 – Part 1, D. Reidel Pub. Co., Dordrecht, the Netherlands, pp. 55–69.

Clark, I., 1999, A case study in the application of geostatistics to lognormal and quasi-lognormal problems; in Dagdelen, K. (ed.), Proc. 28th Symp. on Application of computers and operations research to the minerals industry, Colorado School of Mines, Golden, Colo., pp. 407–416.

Clark, I., 1993, Characterization of multiple mineralisation in a shear-hosted gold deposit; Elbrond, J., and X. Tang (eds.), Proc. 24th Symp. on Application of computers and operations research to the minerals industry, Montreal, v. 2, pp. 374–380.

Clark, I., 1987a, Statistical tables for mineral reserve evaluation; Geostokos Ltd., London, 32 pp.

Clark, I., 1987b, Turning the tables – an interactive approach to the traditional estimation of reserves; Jour. South African Inst. Min. Metall., v. 87, no. 10, pp. 293–306.

Clark, I., 1983, Regression revisited; Math. Geol., v. 15, no. 4, pp. 517–536.

Clark, I., 1979a, Practical geostatistics; Applied Sci. Pub. Ltd., London, 129 pp.

Clark, I., 1979b, The semivariogram – Part 1; Eng. Min. Jour., July, pp. 90–94.

Clark, I., 1979c, The semivariogram – Part 2; Eng. Min. Jour., August, pp. 92–97.

Clark, I., 1977, Roke, a computer program for non-linear least-squares decomposition of a mixture of distributions; Computers & Geosciences, v. 3, pp. 245–256.

Clark, I., 1976a, Some practical computational aspects of mine planning; in Guarascio, M., M. David, and C. Huijbregts (eds.), Advanced geostatistics in the mining industry, pp. 391–399, D. Reidel Pub. Co., Dordrecht, the Netherlands.

Clark, I., 1976b, Some auxiliary functions for the spherical model of geostatistics; Computers & Geosciences, v. 1, pp. 255–263.

Clark, I., and F. L. Clausen, 1981, Simple alternative to disjunctive kriging; Trans. Inst. Min. Metall., Sect. A, v. 90, pp. A13–A24.

Clow, G., 1991, Why gold mines fail; Northern Miner, February, pp. 31–34.

Cochrane, L. B., B. D. Thompson, and G. D. McDowell, 1998, The application of geophysical methods to improve the quality of resource and reserve estimates; in Vallée, M., and A. J. Sinclair (eds.), Quality assurance, continuous quality improvement and standards in mineral resource estimation, Expl. Min. Geol., v. 7, nos. 1 and 2, pp. 63–78.

Conolly, H. J. C., 1936, A contour method of revealing some ore structures; Econ. Geol., v. 31, pp. 259–271.

Copeland, D. J., and M. Rebagliati, 1993, El Condor's Kemess project: in transition from exploration to development; (abs) Ann. Mtg., Prospector and Developers' Assoc. Canada, Toronto.

Craig, C., 1994, Geostatistical analysis of drill data from the Main zone, Huckleberry deposit, central British Columbia; unpub. thesis, Dept. of Geological Sciences, University of British Columbia, Vancouver, 47 pp. plus appendices.

Cressie, N., 1990, The origins of kriging; Math. Geol., v. 22, no. 3, pp. 239–252.

Cressie, N., 1988, Spatial prediction and ordinary kriging; Math. Geol., v. 20, no. 4, pp. 405–422.

Cressie, N., 1985a, When are relative variograms

useful in geostatistics?; Math. Geol., v. 17, no. 7, pp. 693–702.

Cressie, N., 1985b, Fitting variogram models by weighted least squares; Math. Geol., v. 17, no. 5, pp. 563–586.

Cressie, N., 1984, Towards resistant geostatistics; *in* Verly, G., M. David, A. G. Journel, and A. Marechel (eds.), Geostatistics for natural resource characterization: NATO ASI Series C: v. 122 – Part 1, D. Reidel Pub. Co., Dordrecht, the Netherlands, pp. 21–44.

Cressie, N., and D. L. Zimmerman, 1992, On the stability of the geostatistical method; Math. Geol., v. 24, no. 1, pp. 45–60.

Crozel, D., and M. David, 1985, Global estimation variance: formulas and calculation; Math. Geol., v. 17, no. 8, pp. 785–796.

Cumming, J. D., 1980, Diamond drill handbook; J. K. Smit, Toronto, 547 pp.

Dadson, A. S., 1968, Ore estimates and specific gravity; Can. Inst. Min. Metall., Spec. v. 9, pp. 3–4.

Dagbert, M., 1987, Cut-off grades: statistical estimation and reality; Can. Inst. Min. Metall. Bull., v. 80, no. 898, pp. 73–76.

Dagbert, M., M. David, D. Crozel, and A. Desbarats, 1984, Computing variograms in folded strata-controlled deposits; *in* Verly, G., M. David, A. G. Journel, and A. Marechel (eds.), Geostatistics for natural resource characterization: NATO ASI Series C: v. 122 – Part 1, D. Reidel Pub. Co., Dordrecht, the Netherlands, pp. 1–19.

David, M., 1988, Handbook of applied advanced geostatistical ore reserve estimation; Developments in geomathematics 6, Elsevier, Amsterdam, 216 pp.

David, M., 1988, Dilution and geostatistics; Can. Inst. Min. Metall. Bull., v. 81, no. 914, pp. 29–35.

David, M., 1986, "Global estimation variance: formulas and calculation"; Letter to the Editor, Math. Geol., v. 18, no. 7, pp. 697–698.

David, M., 1977, Geostatistical ore reserve estimation; Elsevier Scientific Pub. Co., Amsterdam, 364 pp.

David, M., 1976, The practice of kriging; *in* Guarascio, M., M. David, and C. Huijbregts (eds.), Advanced geostatistics in the mining industry,

pp. 31–48, D. Reidel Pub. Co., Dordrecht, the Netherlands.

David, M., 1972, Grade–tonnage curve: use and misuse in ore-reserve estimation; Trans. Inst. Min. Metall., sect. A, pp. A129–A132.

David, M., and M. Dagbert, 1986, Computation of metal recovery in polymetallic deposits with selective mining and using equivalent metals; Can. Geol. Jour., v. 1, no. 1, pp. 34–37.

David, M., and E. Toh, 1989, Grade control problems, dilution and geostatistics: choosing the required quality and number of samples for grade control; Can. Inst. Min. Metall. Bull., v. 82, no. 931, pp. 53–60.

David, M., D. Marcotte, and M. Soulie, 1984, Conditional bias in kriging and a suggested correction; *in* Verly, G., M. David, A. G. Journel, and A. Marechel (eds.), Geostatistics for natural resource characterization: NATO ASI Series C: v. 122 – Part 1, D. Reidel Pub. Co., Dordrecht, the Netherlands, pp. 217–230.

Davis, B. M., 1987, Uses and abuses of cross-validation in geostatistics; Math. Geol., v. 19, no. 3, pp. 241–248.

Davis, B. M., 1984, Indicator kriging as applied to an alluvial gold deposit; *in* Verly, G., M. David, A. G. Journel, and A. Marechel (eds.), Geostatistics for natural resource characterization: NATO ASI Series C: v. 122 – Part 1, D. Reidel Pub. Co., Dordrecht, the Netherlands, pp. 337–348.

Davis, B. M., M. David, and Belisle, 1978, A fast method for the solution of a system of simultaneous equations – a method adapted to a particular problem; Math. Geol., v. 10, no. 4, pp. 369–374.

Davis, B. M., J. Trimble, and D. McClure, 1989, Grade control and ore selection practices at the Colosseum gold mine; Min. Eng., August, pp. 827–830.

Davis, J. C., 1986, Statistics and data analysis in geology; John Wiley and Sons, New York, 646 pp.

Deraisme, J., and C. de Fouquet, 1984, Recent and future developments of "downstream" geostatistics; *in* Verly, G., M. David, A. G. Journel, and A. Marechel (eds.), Geostatistics for natural resource characterization: NATO ASI Series C:

v. 122 – Part 2, D. Reidel Pub. Co., Dordrecht, the Netherlands, pp. 979–999.

Dent, B. M., 1937, On observations of points connected by a linear relation; Proc. Phys. Soc. London, v. 47, pt. 1, pp. 92–108.

Deutsch, C. V., 1994, Kriging with strings of data; Math. Geol., v. 26, no. 5, pp. 623–638.

Deutsch, C. V., 1993, Kriging in a finite domain; Math. Geol., v. 25, no. 1, pp. 41–52.

Deutsch, C. V., 1989, Mineral inventory estimation in vein type gold deposits: case study on the Eastmain deposit; Can. Inst. Min. Metall. Bull., v. 82, no. 930, pp. 62–67.

Deutsch, C. V., and A. G. Journel, 1998, GSLIB Geostatistical software library and user's guide; Oxford University Press, New York, 369 pp. (an earlier version is dated 1992).

Diamond, P., and M. Armstrong, 1984, Robustness of variograms and conditioning of kriging matrices; Math. Geol., v. 16, no. 8, pp. 809–822.

Diehl, P., and M. David, 1982, Classification of ore reserves/resources based on geostatistical methods; Can. Inst. Min. Metall. Bull., v. 75, no. 838, pp. 127–136.

Diering, J. A. C., 1992, ONE-D: a program for one-dimensional composite optimization; in Annels, A. E. (ed.), Case histories and methods in mineral resource evaluation; Geol. Soc., London, Spec. Pub. No. 63, pp. 185–190.

Dominy, S. C., A. E. Annels, G. S. Camm, B. W. Cuffley, and I. P. Hodkinson, 1999, Resource evaluation of narrow gold-bearing veins: problems and methods of grade estimation; Trans. Inst. Min. Metall. (Sect. A: Min. industry), v. 108, pp. 52–69.

Dowd, P. A., 1994, Optimal open pit design: sensitivity to estimated block values; in Whateley, M. K. G., and P. K. Harvey (eds.), Mineral resource evaluation II; Methods and case histories; Geol. Soc., London, Spec. Pub. No. 79, pp. 87–94.

Dowd, P. A., 1990, Structural control in stratigraphic orebodies: a constrained kriging approach; in Proc. 22nd Symp. on Applications of computers and operations research in the minerals industry, Berlin, v. II, pp. 531–540.

Dowd, P. A., 1984, The variogram and kriging: robust and resistant estimators; in Verly, G., M. David, A. G. Journel, and A. Marechel (eds.), Geostatistics for natural resource characterization: NATO ASI Series C: v. 122 – Part 1, D. Reidel Pub. Co., Dordrecht, the Netherlands, pp. 91–106.

Dowd, P. A., 1982, Lognormal kriging – the general case; Math. Geol., v. 14, no. 5, pp. 475–500.

Dowd, P. A., and M. David, 1976, Planning from estimates: sensitivity of mine production schedules to estimation methods; in Guarascio, M., M. David, and C. Huijbregts (eds.), Advanced geostatistics in the mining industry, pp. 163–183, D. Reidel Pub. Co., Dordrecht.

Dowd, P. A., and D. W. Milton, 1987, Geostatistical estimation of a section of the Perseverance nickel deposit; in Matheron, G., and M. Armstrong (eds.), Geostatistical case studies, pp. 39–67, D. Reidel Pub. Co., Dordrecht, the Netherlands.

Drummond, A. D., and C. I. Godwin, 1976, Hypogene mineralization – an empirical evaluation of alteration zoning; in Sutherland Brown, A. (ed.), Porphyry deposits of the Canadian cordillera; Can. Inst. Min. Metall., Spec. v. 15, pp. 195–205.

Dubrule, O., 1994, Estimating or choosing a geostatistical model; in Dimitrakopoulos, R. (ed.), Geostatistics for the next century, Kluwer Acad. Pub., Dordrecht, the Netherlands, pp. 3–14.

Dubrule, O., 1983, Cross validation of kriging in a unique neighborhood; Math. Geol., v. 15, no. 6, pp. 687–700.

Dunn, M. R., 1983, A simple sufficient condition for a variogram model to yield positive variances under restrictions; Math. Geol., v. 15, no. 4, pp. 553–564.

Dykes, S., H. D. Meade, A. J. Sinclair, and G. H. Giroux, 1983, Sampling and ore reserve estimation consideration in the Big Missouri precious-base metal property, northwest B. C. (abs.); Can. Inst. Min. Metall. Bull., v. 76, no. 857, pp. 47 (preprint of full paper circulated at Can. Inst. Min. Metall., Dist. Mtg., Smithers, B. C., 1983).

Elbrond, J., 1994, Economic effects of ore losses and rock dilution; Can. Inst. Min. Metall. Bull., v. 87, no. 978, pp. 131–134.

Englund, E. J., 1990, A variance of geostatisticians; Math. Geol., v. 22, no. 4, pp. 417–456.

Ewanchuck, H. W., 1968, Grade control at Bethlehem Copper; Can. Inst. Min. Metall., Spec. v. 9, pp. 302–307.

Faddies, T. B., R. E. Lippoth, and G. E. Mellor, 1984, Application of detailed geology to production, planning and ore search: Trixie mine, East Tintic district, Utah; *in* Erickson, A. J., Jr. (ed.), Applied mining geology; Amer. Inst. Min. Metall. Eng., Soc. Min. Engs., pp. 41–51.

Fairweather, M. J., 1997, Values and impurities in base metal concentrates; Can. Inst. Min. Metall. Bull., v. 90, no. 1014, pp. 91–94.

Faunes, A., J. Forkes, and J. O'Leary, 1992, Ore reserve estimation in Los Pelambres, a Chilean porphyry copper; *in* Annels, A. E. (ed.), Case histories and methods in mineral resource evaluation; Geol. Soc., London, Spec. Pub. No. 63, pp. 277–288.

Fennel, J. H., 1939, Ore reserves; Trans. Inst. Min. Metall., v. 49, p. 316.

Francois-Bongarcon, D., 1998, Extensions to the demonstration of Gy's formula; *in* Vallée, M., and A. J. Sinclair (eds.), Quality assurance, continuous quality improvement and standards in mineral resource estimation, Expl. Min. Geol., v. 7, nos. 1 and 2, pp. 149–154.

Francois-Bongarcon, D., 1998, Error variance information from paired data: applications to sampling theory; *in* Vallée, M., and A. J. Sinclair (eds.), Quality assurance, continuous quality improvement and standards in mineral resource estimation, Expl. Min. Geol., v. 7, nos. 1 and 2, pp. 161–168.

Francois-Bongarcon, D., 1993, The practice of the sampling theory of broken ore; Can. Inst. Min. Metall. Bull., v. 86, no. 970, pp. 75–81.

Francois-Bongarcon, D., 1991, Geostatistical determination of sample variances in the sampling of broken gold ores; Can. Inst. Min. Metall. Bull., v. 84, no. 950, pp. 46–57.

Froidevaux, R., 1984a, Conditional estimation variances: an empirical approach; Math. Geol., v. 16, no. 4, pp. 327–350.

Froidevaux, R., 1984b, Precision of estimation of recoverable reserves: the notion of conditional estimation variance; *in* Verly, G., M. David, A. G. Journel, and A. Marechel (eds.), Geostatistics for natural resource characterization: NATO ASI Series C: v. 122 – Part 1, D. Reidel Pub. Co., Dordrecht, the Netherlands, pp. 141–164.

Froidevaux, R., 1982, Geostatistics and ore reserve classification; Can. Inst. Min. Metall. Bull., v. 75, no. 843, pp. 77–83.

Froidevaux, R., and A. G. Journel, 1982, Prediction of local ore recoveries and ore–waste ratios at the Silver Bell uranium mine; Math. Geol., v. 14, no. 6, pp. 645–660.

Froidevaux, R., and W. E. Roscoe, 1983, The problem of objectively reporting reserves: a geostatistical re-examination; Soc. Min. Engs. of AIME, preprint no. 83–407, 6 pp.

Froidevaux, R., W. E. Roscoe, and R. I. Valiant, 1986, Estimating and classifying gold reserves at Page-Williams C zone: a case study in nonparametric geostatistics; *in* David, M., R. Froidevaux, A. J. Sinclair, and M. Vallée (eds.), Proc. symp. on "Ore reserve estimation: methods models and reality"; Can. Inst. Min. Metall., Montreal, May 10–11, pp. 280–300.

Fustos, A., 1982, Multiple cutoff evaluation of uranium resources, south limb, Elliot Lake syncline; internal report, Canmet Energy Research Prog., Ottawa, March, 28 pp. plus appendices.

Fytas, K., N.-E. Chaouai, and M. Lavigne, 1990, Gold deposit estimation using indicator kriging; Can. Inst. Min. Metall. Bull., v. 83, no. 934, pp. 77–83.

Gasparrini, C., 1983, The mineralogy of gold and its significance in metal extraction; Can. Inst. Min. Metall. Bull., v. 76, no. 851, pp. 144–153.

Gibbs, B. L., 1994, Computers: the catalyst for information accessibility; Min. Eng., June, pp. 516–517.

Giroux, G. H., and A. J. Sinclair, 1986, Geostatistics at Equity Silver Mine Ltd.: Global reserves of the South Tail zone by volume–variance relations; *in* David, M., R. Froidevaux, A. J. Sinclair, and M. Vallée (eds.), Proc. symp. on "Ore Reserve Estimation: methods, models and reality," Montreal, May 10–11, 1986, Can. Inst. Min. Metall., pp. 218–237.

Giroux, G. H., and A. J. Sinclair, 1983, Geological control of geostatistical reserve estimates, Dago zone, Big Missouri property, Westmin Resources Limited; Ann. Mtg. District 6, Can. Inst. Min. Metall., Oct. 25–29, Smithers, B. C., 18 pp.

Giroux, G. H., A. J. Sinclair, and J. H. L. Miller, 1986, Geostatistics at Equity Silver Mine Ltd.: production quality control experiments; in David, M., R. Froidevaux, A. M. Sinclair, and M. Vallée (eds.), Proc. Symp. on "Ore Reserve Estimation: methods, models and reality," Montreal, May 10–11, 1986, Can. Inst. Min. Metall., pp. 238–260.

Godwin, C. I., and A. J. Sinclair, 1979, Application of multiple regression analysis to drill-target selection, Casino porphyry copper–molybdenum deposit, Yukon Territory, Canada; Trans. Inst. Min. and Metall., v. 88B, pp. B93–B106.

Goldie, R. J., 1996, The dollar: an economic geologist's most important unit of measurement; Can. Inst. Min. Metall. Bull., v. 89, no. 997, pp. 39–41.

Goldie, R. J., and P. Tredger, 1991, Net smelter return models and their use in the exploration, evaluation and exploitation of polymetallic deposits; Geoscience Canada, v. 18, no. 4, pp. 159–171.

Goode, J. R., M. J. Davie, L. D. Smith, and C. R. Latytanzi, 1991, Back to basics: the feasibility study; Can. Inst. Min. Metall. Bull., v. 84, no. 953, pp. 53–61.

Grace, K. A., 1986a, The critical role of geology in reserve determination; in Ranta, D. E., (ed.), Applied mining geology; Amer. Inst. Min. Metall. Eng., Soc. Mining Engineers, pp. 1–7.

Grace, K. A., 1986b, Ore reserve reporting – what do the numbers mean? In Proc. Symp. on "Ore Reserve estimation: methods, models and reality"; Can. Inst. Min. Metall., Montreal, May 10–11, pp. 47–52.

Grace, K. A., 1984, Reserves, Resources and Pie-In-The-Sky; Min. Eng., October, pp. 1446–1450.

Green, W. R., and B. W. Barde, 1984, A general purpose computer program for data analysis in exploration; Can. Inst. Min. Metall. Bull., v. 77, no. 870, pp. 78–83.

Guarascio, M., 1976, Improving the uranium deposit estimation (the Novazza case); in Guarascio, M., M. David, and C. Huijbregts (eds.), Advanced geostatistics in the mining industry, pp. 351–367, D. Reidel Pub. Co., Dordrecht, the Netherlands.

Guertin, K., 1987, A correction model for conditional bias in selective mining operations; Math. Geol., v. 19, no. 5, pp. 407–424.

Guertin, K., 1984, Correcting conditional bias; in Verly, G., M. David, A. G. Journel, and A. Marechel (eds.), Geostatistics for natural resource characterization: NATO ASI Series C: v. 122 – Part 1, D. Reidel Pub. Co., Dordrecht, the Netherlands, pp. 245–260.

Guibal, D., and A. Remacre, 1984, Local estimation of the recoverable reserves: comparing various methods with the reality on a prophyry copper deposit; in Verly, G., M. David, A. G. Journel, and A. Marechel (eds.), Geostatistics for natural resource characterization: NATO ASI Series C: v. 122 – Part 1, D. Reidel Pub. Co., Dordrecht, the Netherlands, pp. 435–448.

Guney, M., M. A. Al-Marhoun, and Z. A. Nawala, 1988, Metalliferous submarine sediments of the Atlantis-II-Deep, Red Sea; Can. Inst. Min. Metall. Bull., v. 81, no. 910, pp. 33–39.

Gustafson, L. B., 1978, Some major factors of porphyry copper genesis; Econ. Geol., v. 73, pp. 600–607.

Gy, P., 1979, Sampling of particulate materials, theory and practice; Developments in geomathematics 4, Elsevier Scientific Pub. Co., Amsterdam, 431 pp.

Handley, G. A., R. W. Lewis, and G. I. Wilson, 1987, The collection and management of ore reserve estimation data; Proc. symp. on resources and reserves, Australasian Inst. Min. Metall., Sydney branch, November, pp. 27–30.

Hazen, S. W., 1968, Ore reserve calculations; Can. Inst. Min. Metall., Spec. v. 9, pp. 11–32.

Healy, C. M., 1992, Geology as a risk factor in project evaluation: its impact on reserve estimation; Expl. and Min. Geol., v. 1, no. 3, pp. 243–250.

Henley, S., 1987, Kriging – blue or pink? Math. Geol., v. 19, no. 2, pp. 155–158.

Henley, S., and J. W. Aucott, 1992, Some alternatives to geostatistics for mining and exploration; Trans. Inst. Min. Metall., v. 100, pp. A36–A40.

Hester, M. G., 1986, Geostatistical methods for the estimation of minable reserves in stratiform uranium deposits; in Ranta, D. E. (ed.), Applied mining geology: ore reserve estimation; Proc. Symp. of Amer. Inst. Min. Metall. Eng., pp. 83–90.

Hoover, H. C., and L. H. Hoover, 1950, De re metallica (translated from the first Latin edition of 1556); Dover Publ., Inc., New York, 638 pp.

Hoover, H. C., 1909, Principles of mining; McGraw-Hill Pub. Co., New York, 199 pp.

Howarth, R. J., and M. Thompson, 1976, Duplicate analysis in geochemical practice Part II. Examination of proposed method and examples of its use; Analyst, v. 101, pp. 609–709.

Houlding, S., 1991a, Computer modelling limitations and new directions – Part I; Can. Inst. Min. Metall. Bull., v. 84, no. 952, pp. 75–78.

Houlding, S., 1991b, Computer modelling limitations and new directions – Part II; Can. Inst. Min. Metall. Bull., v. 84, no. 953, pp. 46–49.

Hoy, T., G. Gibson, and N. W. Berg, 1984, Copper–zinc deposits associated with basic volcanism, Goldstream area, southwestern British Columbia; Econ. Geol., v. 79, pp. 789–814.

Huijbregts, C., 1976, Selection and grade–tonnage relationships; in Guarascio, M., M. David, and C. Huijbregts (eds.), Advanced geostatistics in the mining industry, pp. 113–135, D. Reidel Pub. Co., Dordrecht, the Netherlands.

Huijbregts, C., 1971, Reconstitution du variogramme ponctuel a partir d'un variogramme experimental regularise; Report N-244, Ecole de Mines de Paris, Centre de Geostatistique, Fontainebleau, 22 pp.

Ingamells, C. O., 1981, Evaluation of skewed exploration data – the nugget effect; Geochim. et Cosmochim. Acta, v. 45, pp. 1209–1216.

Ingamells, C. O., 1974, New approaches to geochemical analysis and sampling; Talanta, v. 21, pp. 141–155.

Ingamells, C. O., 1974, Control of geochemical error through sampling and subsampling diagrams; Geochim. et Cosmoschim. Acta, v. 38, pp. 1225–1237.

Isaaks, E. H., 1990, The application of Monte Carlo methods to the analysis of spatially correlated data; unpublished Ph.D. thesis, Stanford University, Stanford, Calif.

Isaaks, E. H., 1984, Indicator simulation: application to the simulation of a high grade uranium mineralization; in Verly, G., M. David, A. G. Journel and A. Marechel (eds.), Geostatistics for natural resource characterization: NATO ASI Series C: v. 122 – Part 2, D. Reidel Pub. Co., Dordrecht, the Netherlands, pp. 1057–1069.

Isaaks, E. H., and R. M. Srivastava, 1989, An introduction to applied geostatistics; Oxford University Press, New York, 561 pp.

Isaaks, E. H., and R. M. Srivastava, 1988, Spatial continuity measures for probabilistic and deterministic geostatistics; Math. Geol., v. 20, no. 4, pp. 313–342.

Jen, L. S., 1992, Co-products and by-products of base metal mining in Canada: facts and implications; Can. Inst. Min. Metall. Bull., v. 85, no. 965, pp. 87–92.

John, H. T., 1985, Cut-off grade calculations for an open-pit mine; Can. Inst. Min. Metall. Bull., v. 78, no. 879, pp. 73–75.

John, M., and H. Thalenhorst, 1991, Don't lose your shirt, take a bulk sample; in Proc. seminar on "Sample and Ore Reserves," Prospector and Developers Assoc. of Canada, Toronto, March 23, pp. 13–22.

Johnson, N. L., and S. Kotz, 1970, Continuous univariate distributions – 1; John Wiley and Sons, New York, 300 pp.

Johnston, T. G., and G. H. Blackwell, 1986, Short and long term open pit planning and grade control; in David, M., R. Froidevaux, A. J. Sinclair and M. Vallée (eds.), Proc. Symp. on "Ore Reserve Estimation, Methods, Models and Reality," Can. Inst. Min. Metall., Montreal, May 10–11, pp. 108–129.

Jorc Code, 1999, Australasian code for reporting mineral resources and ore reserves; Australasian Inst. Min. Metall., Sydney, Australia, 16 pp.

Journel, A. G., 1992, Comment on "Positive definiteness is not enough"; Math. Geol., v. 24, no. 1, pp. 145–147.

Journel, A. G., 1987, Geostatistics for the environmental sciences, an introduction; Project No. CR 811893, Exposure Assessment Res. Div., Environmental Monitoring Systems Laboratory, Las Vegas, Nev., 89114, 135 pp.

Journel, A. G., 1986a, Constrained interpolation and qualitative information – the soft kriging approach; Math. Geol., v. 18, no. 3, pp. 269–286.

Journel, A. G., 1986b, Geostatistics: models and tools for the earth sciences; Math. Geol., v. 18, no. 1, pp. 119–140.

Journel, A. G., 1985a, Answers to Margaret Armstrong and Robert F. Shurtz's comments on "The deterministic side of geostatistics"; Math. Geol., v. 17, no. 8, pp. 869–

Journel, A. G., 1985b, The deterministic side of geostatistics; Math. Geol., v. 17, no. 1, pp. 1–16.

Journel, A. G., 1985c, Nonparametric estimation of spatial distributions; Math. Geol., v. 15, no. 3, pp. 469–476.

Journel, A. G., 1985d, Recoverable reserves estimation – the geostatistical approach; Min. Eng., June, pp. 563–568.

Journel, A. G., 1983, Nonparametric estimation of spatial distributions; Int. Jour. Math. Geol., v. 14, no. 3, pp. 445–468.

Journel, A. G., 1979, Geostatistical simulation: methods for exploration and mine planning; Eng. Min. Jour., December, pp. 86–91.

Journel, A. G., 1976, Ore grade distributions and conditional simulations – two geostatistical approaches; in Guarascio, M., M. David, and C. Huijbregts (eds.), Advanced geostatistics in the Mining Industry, pp. 195–202, D. Reidel Pub. Co., Dordrecht, the Netherlands.

Journel, A. G., 1975, Geological reconnaissance to exploitation – a decade of applied geostatistics; Can. Inst. Min. Metall. Bull., June, pp. 1–10.

Journel, A. G., 1974, Geostatistics for conditional simulation of ore bodies; Econ. Geol., v. 69, pp. 673–687.

Journel, A. G., and F. G. Alabert, 1988, Focusing on spatial connectivity of extreme valued attributes – stochastic indicator models of reservoir heterogeneities; Soc. Petroleum Engineering, SPE 18324, pp. 621–632.

Journel, A. G., and A. Arik, 1988, Dealing with outlier high grade data in precious metals deposits; in Fytas et al. (eds.), Computer applications in the mining industry: Balkema Pub., Rotterdam, pp. 161–171.

Journel, A. G., and R. Froidevaux, 1982, Anisotropic hole-effect modeling; Math. Geol., v. 14, no. 3, pp. 217–240.

Journel, A. G., and C. Huijbregts, 1978, Mining geostatistics; Academic Press, New York, 599 pp.

Journel, A. G., and E. H. Isaaks, 1984, Conditional indicator simulation: application to a Saskatchewan uranium deposit; Math. Geol., v. 16, no. 7, pp. 685–718.

Journel, A. G., and D. Posa, 1990, Characteristic behavior and order relations for indicator variograms; Math. Geol., v. 22, no. 8, pp. 1011–1028.

Kacewicz, M., 1991, Shape prediction with a fuzzy uncertainty measure; Math. Geol., v. 23, no. 3, pp. 289–304.

Kelly, R., and K. Eldridge, 1994, ISO 9002 as the cornerstone for continuous improvement; Can. Inst. Min. Metall. Bull., v. 87, no. 982, pp. 65–68.

Kendal, J. D., 1901, Ore in sight; Trans. Inst. Min. Metall., v. 10, pp. 143–202.

Kennedy, B. A., and E. J. Wade, 1972, Feasibility studies for large open pit mines; Mining World, August, pp. 70–77.

Kermack, K. A., and J. B. S. Haldane, 1950, Organic correlation and allometry; Biometrika, v. 37, pp. 30–41.

Kilburn, L. C., 1990, Valuation of mineral properties which do not contain exploitable reserves; Can. Inst. Min. Metall. Bull., v. 83, no. 940, pp. 90–93.

Killeen, P. G., 1997, Nuclear techniques for ore grade estimation; in Gubins, A. G. (ed.), Proc. of fourth decennial international conference on mineral exploration, Prospectors and Developers Assoc., Toronto, pp. 677–684.

Killeen, P. G., and B. E. Elliott, 1997, Surveying the path of boreholes: a review of developments and methods since 1987; *in* Gubins, A. G. (ed.), Proc. of fourth decennial international conference on mineral exploration, Prospectors and Developers Assoc., Toronto, pp. 709–712.

Kim, Y. C., and E. Y. Baafi, 1984, Combining local kriging variances for short-term mine planning; *in* Verly, G., M. David, A. G. Journel, and A. Marechel (eds.), Geostatistics for natural resource characterization: NATO ASI Series C: v. 122 – Part 1, D. Reidel Pub. Co., Dordrecht, the Netherlands, pp. 185–199.

Kim, Y. C., P. K. Medhi, and I. S. Roditis, 1987, Performance evaluation of indicator kriging in a gold deposit; Min. Eng., October, pp. 947–952.

Kimura, E. T., G. D. Bysouth, and A. D. Drummond, 1976, Endako; *in* Sutherland Brown, A. (ed.), Porphyry deposits of the Canadian cordillera, Can. Inst. Min. Metall., Spec. v. 15, pp. 444–454.

King, H. F., D. W. McMahon, and G. J. Bujtor, 1982, A guide to the understanding of ore reserve estimation; Australasian Inst. Min. Metall., Supplement to Proc. No. 281, 21 pp.

King, H. F., D. W. McMahon, G. J. Bujtor, and A. K. Scott, 1985, Geology in the understanding of ore reserve estimation: an Australian viewpoint; Amer. Inst. Min. Metall. Eng., Soc. Min. Eng., preprint no. 85–355, 13 pp.

Kingston, G. A., 1992, Mineralogy in the evaluation of ore deposits; *in* Annels, A. E. (ed.), Case histories and methods in mineral resource evaluation; Geol. Soc., London, Spec. Pub. No. 63, pp. 47–59.

Knoll, K., 1989, And now the bad news; Northern Miner, v. 6, pp. 48–52.

Knudson, H. P., 1992, Blasthole samples – a source of bias? Min. Eng., March, pp. 251–253.

Koch, G. S., Jr., and R. F. Link, 1970, Statistical analysis of geological data; v. I and II, John Wiley and Sons Inc., New York.

Konya, C. J., 1996, Drilling accuracy: the key to successful blasting; Eng. Min. Jour., March, pp. 78.

Kratochvil, B., and J. K. Taylor, 1981, Sampling for chemical analysis; Analyt. Chemistry, v. 53, no. 8, pp. 925A–938A.

Krige, D. G., 1993, Variability in the gold grade and spatial structures of Witwatersrand reefs: the implications for geostatistical ore valuations; Jour. South African Inst. Min. Metall., v. 93, no. 3, pp. 79–84.

Krige, D. G., 1986, "Matheronian geostatistics – quo vadis?" by G. M. Philip and D. F. Watson; Math. Geol., v. 18, no. 5, pp. 501–502.

Krige, D. G., 1984, Geostatistics and the definition of uncertainty; Trans. Inst. Min. Metall. Sect. A, v. 93, pp. A41–A47.

Krige, D. G., 1978, Lognormal–de Wijsian geostatistics for ore evaluation; South African Inst. Min. Metall., Monograph Series, Johannesburg, South Africa, 50 pp.

Krige, D. G., 1960, On the departure of ore value distributions from the lognormal model in South African gold mines; Jour. South African Inst. Min. Metall., November, pp. 231–244.

Krige, D. G., 1951, A statistical approach to some basic mine valuation problems on the Witwatersrand; Jour. Chem., Metall. and Min. Soc. South Africa, v. 52, no. 6, pp. 119–139.

Krige, D. G., and C. E. Dohm, 1994, The role of massive grade data bases in geostatistical applications in South African gold mines; *in* Dimitrakopoulos, R. (ed.), Geostatistics for the next century, Kluwer Acad. Publ., pp. 46–54.

Krige, D. G., and P. G. Dunn, 1995, Some practical aspects of ore reserve estimation at Chuquicamata Copper Mine, Chile; Proc. 25th Symp. on Applications of computers and operations research to the minerals industry, Brisbane, Australia, July 9–14, pp. 125–133.

Krige, D. G., and E. J. Magri, 1982, Geostatistical case studies of the advantages of lognormal–de Wijsian kriging with mean for a base metal mine and a gold mine; Math. Geol., v. 14, no. 6, pp. 547–556.

Krige, D. G., and E. J. Magri, 1982, Studies of the effects of outliers and data transformation on variogram estimates for a base metal and a gold ore body; Math. Geol., v. 14, no. 6, pp. 557–564.

Krumbein, W. C., and F. Graybill, 1965, An introduction to statistical models in geology; McGraw-Hill Book Co., New York, 475 pp.

Kwa, B. L., and P. F. Mousset-Jones, 1986, Mineral reserve estimation of gold deposits – a survey of practices; *in* David, M., R. Froidevaux, A. J. Sinclair, and M. Vallée (eds.), Proc. Symp. on "Ore reserve estimation – methods, models and reality," Can. Inst. Min. Metall., Montreal, pp. 172–184.

Lahee, F. H., 1952, Field geology; McGraw-Hill Book Co., Inc., New York, 883 pp.

Lane, K. E., 1988, The economic definition of ore: cut-off grades in theory and practice; Mining Journal Books Ltd., London, 149 pp.

Lavigne, M., 1991, Geostatistical ore reserve estimation at the Lac Knife graphite deposit, Esmanville Township, Quebec; Can. Inst. Min. Metall. Bull., v. 84, no. 951, pp. 30–35.

Lawton, S. E., (ed.), 1991, Proc. Seminar on "Sampling and ore reserves"; Prosp. and Developers' Assoc. Canada, Toronto, 162 pp.

Leitch, C. H. B., X. Cheng, C. T. Hood, and A. J. Sinclair, 1991, Structural character of en echelon polymetallic veins at the Silver Queen mine, British Columbia; Can. Inst. Min. Metall. Bull., v. 85, no. 955, pp. 57–66.

Leitch, C. H. B., C. T. Hood, X. Cheng, and A. J. Sinclair, 1990, Geology of the Silver Queen mine area, Owen Lake, central British Columbia; *in* Paper 1990–1, B. C. Ministry of Energy, Mines and Petroleum Resources, Victoria, Canada, pp. 287–295.

Lemmer, I. C., 1984, Estimating local recoverable reserves via indicator kriging; *in* Verly, G., M. David, A. G. Journel, and A. Marechel (eds.), Geostatistics for natural resource characterization: NATO ASI Series C: v. 122 – Part 1, D. Reidel Pub. Co., Dordrecht, the Netherlands, pp. 349–364.

Long, S. D., 1998, Practical quality control procedures in mineral inventory estimation; *in* Vallée, M., and A. J. Sinclair (eds.), Quality assurance,

continuous quality improvement and standards in mineral resource estimation, Expl. Min. Geol., v. 7, nos. 1 and 2, pp. 117–128.

Lowell, J. D., and J. M. Guilbert, 1970, Lateral and vertical alteration-mineralization zoning in prophyry ore deposits; Econ. Geol., v. 65, pp. 373–408.

Lydon, J. W., 1984, Volcanogenic massive sulphide deposit part I: a descriptive model; Geoscience Canada v. 11, no. 4, 195–202.

Magri, E. J., and P. McKenna, 1986, A geostatistical study of diamond-saw sampling versus chip sampling; Jour. South African Inst. Min. Metall., v. 86, no. 8, pp. 335–347.

Marechal, A., 1984, Recovery estimation: a review of models and methods; *in* Verly, G., M. David, A. G. Journel, and A. Marechel (eds.), Geostatistics for natural resource characterization: NATO ASI Series C: v. 122 – Part 1, D. Reidel Pub. Co., Dordrecht, the Netherlands, pp. 385–420.

Marek, J. M., and H. E. Welhener, 1985, Cutoff grade strategy – a balancing act; Soc. Min. Eng. of Amer. Inst. Min. Metall. Eng., preprint no. 85–320, 5 pp.

Mark, D. M., and M. Church, 1974, On the misuse of regression in earth science; Math. Geol., v. 9, no. 1, pp. 63–75.

Matheron, G., 1987, A simple answer to an elementary question; Math. Geol., v. 19, no. 5, pp. 455–458.

Matheron, G., 1986, Philipian/Watsonian high (flying) philosophy; Math. Geol., v. 18, no. 5, pp. 503–504.

Matheron, G., 1976, A simple substitute for conditional expectation: the disjunctive kriging; Proc. NATO ASI, Advanced Geostatistics in the Mining Industry, D. Reidel Pub. Co., Dordrecht, the Netherlands, pp. 221–236.

Matheron, G., 1971, The theory of regionalized variables and its applications; Les cahiers du Centre de Morphologie Mathematique, Fontainebleau, No. 5, 211 pp.

Matheron, G., 1963, Principles of geostatistics; Econ. Geol., v. 58, pp. 1246–1266.

Matheron, G., 1957, Theorie lognormale de

l'echantillonage systematique des gisement; Ann. Mines, France, v. 9, pp. 566–584.

Matheron, G., and M. Armstrong (eds.), 1987, Geostatistical case studies; D. Reidel Pub. Co., Dordrecht, the Netherlands, 248 pp.

May, R. W., and W. B. Ryan, 1986, Point kriging for isopach estimation: a case study in reject assessment; *in* Ranta, D. E. (ed.), Applied mining geology: ore reserve estimation; Proc. Symp. of Amer. Inst. Min. and Metall. Engs., pp. 139–152.

McArthur, G. J., 1988, Using geology to control geostatistics in the Hellyer deposit; Math. Geol., v. 20, no. 4, pp. 343–366.

McCammon, R. B., 1968, The dendrograph: a new tool for correlation; Geol. Soc. Amer. Bull., v. 79, pp. 1663–1670.

McGaughey, W. J., and M. A. Vallée, 1997, Ore delineation in three dimensions; *in* Gubins, A. G. (ed.), Proc. of fourth decennial international conference on mineral exploration, Prospectors and Developers Assoc., Toronto, pp. 639–650.

McKelvey, V. E., 1993, Mineral resource estimates and public policy; U.S. Geological Survey Prof. Paper 820.

McKinstry, H. E., 1948, Mining geology; Prentice-Hall, Inc., New York, 680 pp.

McMahon, D., R. Morland, and R. Peck, 1990, The estimator and the user; Proc. symp on ore reserve estimates, Melbourne, Australia, March, pp. 7–9.

McMillan, W. J., T. Hoy, D. G. McIntyre, J. L. Nelson, G. T. Nixon, J. L. Hammack, A. Panteleyev, G. E. Ray, and I. C. L. Webster, 1991, Ore deposits, tectonics and metallogeny in the Canadian cordillera; British Columbia Ministry of Energy, Mines and Petroleum Resources, Paper 1991-4, 276 pp.

McOuat, J. F., 1993, Reserves – requirements for global reserve standards and practices; Eng. and Min. Jour., August, pp. ww30–33.

Merks, J. W., 1993, Abuse of statistics; Can. Inst. Min. Metall. Bull., v. 86, no. 966, pp. 40–41.

Merks, J. W., 1992, Geostatistics or voodoo science?; Northern Miner, April 20, p. A4. Also in Mining Mag., 1992, August, p. 113.

Miller, R. L., and J. S. Kahn, 1962, Statistical analysis in the geological sciences; John Wiley and Sons Inc., New York, 483 pp.

Moore, S. C., S. G. Kolb, C. McLean, and A. J. Erickson, Jr., 1984, Geologic evaluation of the Grossschloppen vein uranium deposit, West Germany; *in* Erickson, A. J., Jr., (ed.), Applied mining geology: Proc. Symp. Amer. Inst. Min. Metall. Eng., pp. 131–145.

Morris, M. D., 1991, On counting the number of data pairs for semivariogram estimation; Math. Geol., v. 23, no. 7, pp. 929–943.

Mustard, J. W., and A. C. Noble, 1986, Geological controls in geostatistical ore reserve estimation of a skarn deposit – the Mactung case study; *in* Ranta, D. E. (ed.), Applied mining geology: ore reserve estimation; Soc. Min. Engs., Littleton, Colo., pp. 71–80.

Mutton, A. J., 1997, The application of geophysics during evaluation of the Century zinc deposit; *in* Gubins, A. G. (ed.), Geophysics and geochemistry at the millenium: Proc. Exploration 97, Prospectors and Developers Assoc. of Canada, Toronto, September, pp. 599–614.

Mwenifumbo, C. J., 1997, Electrical methods for ore body delineation; *in* Gubins, A. G. (ed.), Proc. of fourth decennial international conference on mineral exploration, Prospectors and Developers Assoc., Toronto, pp. 667–676.

Myers, D. E., 1986, Matheronian geostatistics – quo vadis?: comment; Math. Geol., v. 18, no. 7, pp. 699–700.

Myers, D. E., 1983, Estimation of linear combinations and co-kriging; Math. Geol., v. 15, no. 5, pp. 633–638.

Myers, D. E., 1982, Matrix formulation of co-kriging; Math. Geol., v. 14, no. 3, pp. 249–258.

Myers, D. E., and A. Journel, 1990, Variograms with zonal anisotropies and noninvertible kriging systems; Math. Geol., v. 22, no. 7, pp. 779–786.

Noble, A. C., and D. E. Ranta, 1984, Zoned kriging – a successful union of geology and geostatistics; *in* Erickson, A. J., Jr. (ed.), Applied mining geology; Amer. Inst. Min. Metall. Eng., Soc. Min. Engs., pp. 115–128.

Norrish, N., and G. H. Blackwell, 1987, A mine operator's implementation of geostatistics; Can. Inst. Min. Metall. Bull., v. 80, no. 899, pp. 103–112.

Nowak, M. S., 1991, Ore reserve estimation, Silver Queen vein, Owen Lake, British Columbia; unpub. M.A.Sc. thesis, University of British Columbia, Vancouver, 204 pp.

Nowak, M. S., and A. J. Sinclair, 1993, Estimate of volume of felsite dykes and their position in space, Virginia deposit; internal report, Mineral Deposit Research Unit, University of British Columbia, Vancouver, report VIR92-4, 7 pp.

Nowak, M. S., A. J. Sinclair, and A. Randall, 1993, Multiple indicator kriging for block estimation at Silbak Premier mine, British Columbia; *in* Dimitrakopoulos, R. (ed.), Proc. Symp. "Geostatistics for the next century," June 3–5, Montreal, Kluwer Acad. Pub., pp. 55–63.

Nowak, M. S., R. M. Srivastava, and A. J. Sinclair, 1993, Conditional simulation, a mine planning tool for a small gold deposit; Proc. Troia Symp., Troia, Portugal, Kluwer Acad. Publ., pp. 977–987.

Nowak, M. S., A. J. Sinclair, R. M. Srivastava, and P. Holbek, 1991, Mineral inventory studies of precious metal deposits in British Columbia: Shasta gold deposit; Mineral Deposit Research Unit, Dept. of Geological Sciences, University of British Columbia, Vancouver, Tech. Rept. MT-1, 93 pp. plus appendices.

O'Donnell, N. D., and T. B. Ostrowski, 1986, Ore reserve estimation at Syncrude Canada Ltd.; *in* Ranta, D. E. (ed.), Applied mining geology: ore reserve estimation, Soc. Min. Engs., Littleton, Colo., pp. 121–135.

O'Dowd, R. J., 1991, Conditioning of coefficient matrices of ordinary kriging; Math. Geol., v. 23, no. 5, pp. 721–758.

Olea, R. A., 1984, Sampling design optimization for spatial functions; Math. Geol., v. 16, no. 4, pp. 369–392.

O'Leary, J., 1994a, Mining project finance and the assessment of ore reserves; *in* Whateley, M. K. G., and P. K. Harvey (eds.), Mineral resource evaluation II: methods and case histories; Geol. Soc., London, Spec. Pub. No. 79, pp. 129–139.

O'Leary, J., 1994b, Cia Minera Los Pelambres: a project history; *in* Whateley, M. K. G., and P. K. Harvey (eds.), Mineral resource evaluation II:

methods and case histories; Geol. Soc., London, Spec. Pub. No. 79, pp. 249–263.

Omre, H., 1984, The variogram and its estimation; *in* Verly, G., M. David, A. G. Journel, and A. Marechel (eds.), Geostatistics for natural resource characterization: NATO ASI Series C: v. 122 – Part 1, D. Reidel Pub. Co., Dordrecht, the Netherlands, pp. 107–125.

Owens, O., and W. P. Armstrong, 1994, Ore reserves – the 4 C's; Can. Inst. Min. Metall. Bull., v. 87, no. 979, pp. 52–54.

Pakalnis, R., and S. Vongpaisal, 1993, Mine design, an empirical approach; *in* Bawden, W. F., and J. F. Archibald (eds.), Innovative mine design for the 21st century, Balkema Publ., Rotterdam, pp. 455–467.

Pakalnis, R. C., R. Poulin, and J. Hadjigeoriou, 1995, Quantifying the cost of dilution in underground mines; Min. Eng., December, pp. 1136–1141.

Pakalnis, R., R. Poulin, and S. Vongpaisal, 1995, Quantifying dilution for underground mine operations; preprint, Can. Inst. Min. Metall., Ann. Gen. Mtg., Halifax, N. S., May 14–18, 8 pp. plus diagrams.

Pan, G., 1998, Smoothing effect, conditional bias and recoverable reserves; Can. Inst. Min. Metall. Bull. v. 91, no. 1019, pp. 81–86.

Pan, G., 1995, Practical issues of geostatistical reserve estimation in the mining industry; Can. Inst. Min. Metall. Bull., v. 88, no. 993, pp. 31–37.

Pan, G., 1994, Restricted kriging: a link between sample value and sample configuration; Math. Geol., v. 26, no. 1, pp. 135–155.

Pan, G., and A. Arik, 1993, Restricted kriging for mixture of grade models; Math. Geol., v. 25, no. 6, pp. 713–736.

Pan, G., D. Gaard, K. Moss, and T. Heiner, 1993, A comparison between cokriging and ordinary kriging: case study with a polymetallic deposit; Math. Geol., v. 25, no. 3, pp. 377–398.

Parker, H. M., 1991, Statistical treatment of outlier data in epithermal gold deposit estimation; Math. Geol., v. 23, no. 2, pp. 175–200.

Parker, H. M., 1984, Trends in geostatistics in the mining industry; *in* Verly, G., M. David, A. G. Journel, and A. Marechel (eds.), Geostatistics for natural resource characterization: Nato ASI Series C: v. 122 – Part 2, D. Reidel Pub. Co., Dordrecht, the Netherlands, pp. 915–934.

Parker, H. M., 1979, The volume–variance relationship: a useful tool for mine planning; Eng. and Min. Jour., October, pp. 106–123.

Parker, H. M., A. G. Journel, and W. C. Dixon, 1979, The use of conditional lognormal probability distribution for the estimation of open-pit ore reserves in stratabound uranium deposits – a case study; in Proc. 16th Symp. on Applications of computers and operations research in the minerals industry, Tucson, Ariz., October, 28, Soc. Min. Engs., New York, pp. 133–148.

Parrish, I. S., 1997, Geologist's Gordian knot: to cut or not to cut; Min. Eng., April, pp. 45–49.

Parrish, I. S., 1995, Incremental ore and reserves – a paradox; Min. Eng., November, 986 pp.

Parrish, I. S., 1993, Tonnage factor – a matter of some gravity; Min. Eng., March, pp. 1268–1271.

Parrish, I. S., 1990, Reserve audits: what they are, what they should cover, how they are used; Min. Eng., November, pp. 1247–1248.

Pasieka, A. R., and Sotirow, G. V., 1985, Planning and operational cutoff grades based on computerized net present value and net cash flow; Can. Inst. Min. Metall. Bull., v. 78, no. 878, pp. 47–54.

Patterson, J. A., 1959, Estimating ore reserves follows logical steps; Eng. Min. Jour., September, pp. 111–115.

Peterson, U., and D. McMillan, 1992, Electron microprobe mineral analysis: applications in exploration and mine development; Min. Eng., February, pp. 139–143.

Petruk, W., 1987, Applied mineralogy in ore dressing; in Yarar, B., and Z. M. Dogan (eds.), Mineral processing design, NATO ASI Series E, No. 122, pp. 2–36.

Philip, G. M., and D. F. Watson, 1987, An open letter to G. Matheron; Math. Geol., v. 19, no. 5, pp. 451–453.

Philip, G. M., and D. F. Watson, 1986, Geostatistics and spatial data; Math. Geol., v. 18, no. 5, pp. 505–508.

Philip, G. M., and D. F. Watson, 1986, Matheronian geostatistics – quo vadis?; Math. Geol., v. 18, no. 1, pp. 93–118.

Phillips, R. J., 1968, Trend surface analysis in mining; in Proc. Symp. on Decision-making in mineral exploration; Vancouver, pp. 28–33.

Pitard, F. F., 1994, Exploration of the nugget effect; in Dimitrakopoulos, R. (ed.), Proc. Symp. on "Geostatistics for the next century," Montreal, Kluwer Acad. Publ., pp. 124–136.

Pitard, F. F., 1989, Pierre Gy's sampling theory and sampling practice, v. 1: heterogeneity and sampling, 214 pp., v. II: Sampling correctness and sampling practice; CRC Press, Boca Raton, Fla., 247 pp.

Poliquin, M. J., 1994, A geostatistical procedure for mineral inventory at the Aurora gold deposit, Mineral County, Nevada; unpubl. thesis, Dept. of Geological Sciences, University of British Columbia, 64 pp. plus appendices.

Pontius, J. A., and J. A. Head, 1996, Cresson mine: case history of a rapidly evolving mining project; Min. Eng., January, pp. 26–30.

Postolski, T. A., and A. J. Sinclair, 1998a, Geology as a basis for refining semi-variogram models for porphyry-type deposits; in Vallée, M., and A. J. Sinclair (eds.), Quality assurance, continuous quality improvement and standards in mineral resource estimation; Expl. Min. Geol., v. 7, nos. 1 and 2, pp. 45–50.

Postolski, T. A., and A. J. Sinclair, 1998b, Quantitative estimation of dilution and ore loss resulting from block estimation errors and a specified cut-off grade; in Vallée, M., and A. J. Sinclair (eds.), Quality assurance, continuous quality improvement and standards in mineral resource estimation; Expl. Min. Geol., v. 7, nos. 1 and 2, pp. 91–98.

Puente, C. E., and R. L. Bras, 1986, Disjunctive kriging, universal kriging, or no kriging: small sample results with simulated fields; Math. Geol., v. 18, no. 3, pp. 287–306.

Purdie, J. J., 1968, Mineral reserve calculations at Lake Dufault; Can. Inst. Min. Metall., Spec. v. 9, pp. 183–192.

Quinlan, T., and T. A. Leahey, 1979, Geostatistical ore reserve estimation for the Warrego mine, Northern Territory; Australasian Inst. Min. Metall.,

Sydney branch, Proc. Symp. on "Estimation and statement of mineral resources," October, 15 pp.

Radlowski, Z. A., and A. J. Sinclair, 1993, GYSAMPLE, a computer program for the evaluation of sample reduction schemes; Min. Dep. Res. Unit, Dept. Geological Sciences, University of British Columbia, Vancouver, Tech. Rept. MT-6, 7 pp. plus appendices and diskette.

Ranta, D. E., A. D. Ward, and M. W. Ganster, 1984, Ore zoning applied to geologic reserve estimation of molybdenum deposits; in Erickson, A. J., Jr., R. A. Metz, and D. E. Ranta (eds.), Applied mining geology; Soc. Min. Engs., New York, pp. 83–114.

Ravenscroft, P. J., 1992, Risk analysis for mine scheduling by conditional simulation; Trans. Inst. Min. Metall., v. 101, pp. A104–A108.

Ravenscroft, P. J., 1992, Recoverable reserve estimation by conditional simulation; in Annels, A. E. (ed.), Case histories and methods in mineral resource evaluation; Geol. Soc., London, Spec. Pub. No. 63, pp. 289–298.

Ravenscroft, P. J., and M. Armstrong, 1990, Kriging of block models – the dangers re-emphasized; Proc. 22nd Symp. on Applications of computers and operations research in the minerals industry, Berlin, v. II, pp. 577–587.

Raymond, G. F., 1984, Geostatistical applications in tabular style lead-zinc ore at Pine Point, Canada; in Verly, G., M. David, A. G. Journel, and A. Marechel (eds.), Geostatistics for natural resource characterization: NATO ASI Series C: v. 122 – Part 1, D. Reidel Pub. Co., Dordrecht, the Netherlands, pp. 469–483.

Raymond, G. F., 1982, Geostatistical grade estimation of Mount Isa's copper orebodies; Proc. Aust. Inst. Min. Metall., No. 284, pp. 17–39.

Raymond, G. F., 1979, Ore estimation in an erratically mineralized orebody; Can. Inst. Min. Metall. Bull., v. 72, no. 806, pp. 90–98.

Raymond, G. F., and W. P. Armstrong, 1988, Sample bias and conditional probability ore reserve estimation at Valley; Can. Inst. Min. Metall. Bull., v. 81, no. 911, pp. 128–136.

Readdy, L. A., 1986, An example of classification of reserve types by historical input to geostatistics and other considerations in defining ore; in Ranta, D. A. (ed.), Applied mining geology: ore reserve estimation; Soc. Min. Engs., Littleton, Colo., pp. 45–51.

Rech, W. D., E. B. Jensen, G. Hauk, and D. R. Stewart, 1993, Using geostatistical software to manage panel caving operations; Min. Eng., March, pp. 239–243.

Reedman, J. H., 1979, Techniques in mineral exploration; Applied Science Publ. Ltd., London, 533 pp.

Rendu, J.-M., 1997, Practical geostatistics at Newmont Gold Company, a story of adaptation; SME Ann. Mtg., Denver, Colo., Feb. 24–27, Soc. for Min., Metall. and Expl. Inc., preprint 97–152, 6 pp.

Rendu, J.-M., 1986, How the geologist can prevent a geostatistical study from running out of control some suggestions; in Ranta, D. E. (ed.), Applied mining geology: ore reserve estimation; AIME, Soc. Min. Engs., pp. 11–17.

Rendu, J.-M., 1984, Geostatistical modelling and geological controls; Trans. Inst. Min. Metall. sect. B, pp. B166–B172.

Rendu, J.-M., 1980, Kriging for ore valuation and mine planning; Eng. Min. Jour., January, pp. 114–120.

Rendu, J.-M., 1979a, Kriging, logarithmic kriging and conditional expectation: comparison of theory with actual results; in Proc. 16th Symp. on Applications of computers and operations research in the minerals industry, Tucson, Ariz., October 28, Soc. Min. Engs., New York, pp. 199–212.

Rendu, J.-M., 1979b, Normal and lognormal estimation; Math. Geol., v. 11, no. 4, pp. 407–422.

Rendu, J.-M., 1978, An introduction to geostatistical methods of mineral evaluation; South African Inst. Min. Metall. Monograph Series: Geostatistics 2, 84 pp.

Rhys, D. A., and C. I. Godwin, 1992, Preliminary structural interpretation of the Snip mine; in Geological fieldwork 1991, B. C. Ministry of Energy, Mines and Petroleum Resources, Rept. 1992-1, pp. 549–554.

Richards, D. G., and E. Sides, 1991, Evolution of ore-reserve estimation strategy and methodology at Neves–Corvo copper–tin mine, Portugal; Trans. Inst. Min. Metall. (Sect. B: Appl. earth sci.), v. 100, pp. B192–B208.

Riddler, G. P., 1996, Towards an international classification of reserves and resources; Australasian Inst. Min. Metall. Bull., no. 1, February, pp. 31–39.

Riddler, G. P., 1994, What is a mineral resource? in Whateley, M. K. G., and P. K. Harvey (eds.), Mineral resource evaluation II: methods and case histories; Geol. Soc., London, Spec. Pub. No. 79, pp. 1–10.

Ripley, B. D., and M. Thompson, 1987, Regression techniques for the detection of analytical bias; Analyst, v. 112, pp. 377–383.

Rivoirard, J., 1987, Two key parameters when choosing the kriging neighbourhood; Math. Geol., v. 19, no. 8, pp. 851–856.

Rivoirard, J., 1987, Computing variograms on uranium data; in Matheron, G., and M. Armstrong (eds.), Geostatistical case studies, D. Reidel Pub. Co., Dordrecht, the Netherlands, pp. 1–22.

Rivoirard, J., 1984, Looking for a kriging plan in a stockwork deposit; in Verly, G., M. David, A. G. Journel, and A. Marechel (eds.), Geostatistics for natural resource characterization: NATO ASI Series C: v. 122 – Part 2, D. Reidel Pub. Co., Dordrecht, the Netherlands, pp. 951–963.

Roberts, R. G., and P. A. Sheahan, 1988, Ore deposit models; Geol. Assoc. Canada, Reprint Series 3, 194 pp.

Rogers, R. S., 1998, Forensic geology and mineral exploration projects; Expl. Min. Geol., v. 7, nos. 1 and 2, pp. 25–28.

Rombouts, L., 1995, Sampling and statistical evaluation of diamond deposits; Jour. Explor. Geochem., v. 53, pp. 351–367.

Rombouts, L., 1987, Evaluation of low grade/ high value diamond deposits; Mining Mag., September, pp. 217–220.

Rossi, M. E., 1999, Uncertainty and risk models for decision-making processes; in Dagdalen, K. (ed.), Proc. 28th Symp. on Applications of computers and operations research in the minerals industry,

Oct. 20–22, Colorado School of Mines, Golden, Colo., pp. 185–195.

Rossi, M. E., and H. M. Parker, 1994, Estimating recoverable reserves: is it hopeless?; Proc. Symp. "Geostatistics for the next century," Montreal, June 3–5, Kluwer Acad. Publ., pp. 259–276.

Rossi, M. E., and J. C. Vidakovich, 1999, Using meaningful reconciliation information to evaluate predictive models; Soc. Min. Engs., preprint 99–20, 8 pp.

Rostad, O. H., 1986, Geologic and allied considerations in ore reserve estimates for vein-type deposits; in Ranta, D. E. (ed.), Applied mining geology: ore reserve estimation; AIME, Soc. Min. Engs., pp. 61–71.

Roper, M. W., 1986, Geostatstics in planning and production, Golden Sunlight mine, Whitehall, Montana; in David, M., R. Froidevaux, A. J. Sinclair, and M. Vallée (eds.), Proc. Symp. "Ore reserve estimation: methods, models and reality," Can. Inst. Min. Metall., Montreal, May 10–11, pp. 163–171.

Rowe, R. G., and R. G. Hite, 1984, Applied geology: the foundation for mine design at Exxon Minerals Company's Crandon deposit; in Erickson, Jr., A. J. (ed.), Applied mining geology, Soc. Min. Engs., AIME, pp. 9–27.

Royle, A. G., 1981, Optimization of assay cutoff orebodies; Trans. Inst. Min. Metall., Sect. A., pp. A55–A60.

Royle, A. G., 1979, Why geostatistics?; Eng. Min. Jour., May, pp. 92–101.

Royle, A. G., 1977, How to use geostatistics for ore reserve classification; Eng. Min. Jour., February, pp. 52–55.

Royle, A. G., and M. J. Newton, 1972, Mathematical models, sample sets and ore reserve estimation; Trans. Inst. Min. Metall., sect. A, pp. 78–85.

Royle, A. G., M. J. Newton, and V. K. Sarin, 1972, Geostatistical factors in design of mine sampling programmes; Trans. Inst. Min. Metall., sect. A, pp. A82–A88.

Sabourin, R., 1984, Application of a geostatistical method to quantitatively define various categories of resources; in Verly, G., M. David, A. G.

Journel, and A. Marechel (eds.), Geostatistics for natural resource characterization: NATO ASI Series C: v. 122 – Part 1, D. Reidel Pub. Co., Dordrecht, the Netherlands, pp. 201–215.

Sabourin, R., 1983, Geostatistics as a tool to define various categories of resources; Math. Geol., v. 15, no. 1, pp. 131–144.

Sahin, A., and A. A. Abdul-Latif, 1990, Geostatistical evaluation of the Abu Tartur phosphate deposit, Western Desert, Egypt; Can. Inst. Min. Metall. Bull., v. 83, no. 934, pp. 56–61.

Sandefur, R. L., and D. C. Grant, 1980, Applying geostatistics to roll front uranium in Wyoming; Eng. Min. Jour., February, pp. 90–96.

Sangster, D. F., 1995, Presidential perspective; SEG newsletter, no. 20, January, p. 4.

Schaap, W., 1981, Effects of mineral grain size and ore hardness on mill-dump cutoff grades; Trans. Inst. Min. Metall. sect. A, v. 90, pp. A27–A33.

Schaap, W., and J. D. St. George, 1981, On weights of linear kriging estimator; Trans. Inst. Min. Metall. sect. A, v. 90, pp. A25–A227.

Schofield, N., 1988, Ore reserve estimation at the Enterprise mine, Pine Creek, Northern Territory, Australia. Part I: structural and variogram analysis; Part II: The multigaussian kriging model; Can. Inst. Min. Metall. Bull., v. 81, no. 909, pp. 56–61 and 62–66.

Schwarz, F. P., Jr., S. M. Weber, and A. J. Erickson, Jr., 1984, Quality control of sample preparation at the Mount Hope molybdenum prospect, Eureka County, Nevada; in Erickson, A. J., Jr. (ed.), Applied mining geology; AIME, Soc. Min. Engs., pp. 175–187.

Scoble, M. J., and A. Moss, 1994, Dilution in underground bulk mining: implications for production management; in Whateley, M. K. G., and P. K. Harvey (eds.), Mineral resource evaluation II: methods and case histories; Geol. Soc., London, Spec. Pub. No. 79, pp. 95–108.

Shahkar, A., 1995, A geostatistical study of the C 7-04 stope, Giant mine, Royal Oak Mines, Inc., Yellowknife, N.W.T; unpub. thesis, Dept. of Geological Sciences, University of British Columbia, 49 pp. plus appendices.

Shaw, D. M., 1964, Interpretation geochimique des elements en trace dans les roches cristalline; Masson et Cie., Paris.

Sheahan, P. A., and M. E. Cherry, 1993, Ore deposit models volume II; Geol. Assoc. Canada Reprint Series 6, 154 pp.

Shurtz, R. F., 1998, Kriging merely geostatistical gibberish; Northern Miner, Dec. 26, pp. 5.

Shurtz, R. F., 1994, Spatial-least squares approximation and geological plausibility; in Chung, C.-J. F. (ed.), Proc. 1994 International Association for Mathematical Geology Annual Conference, Mont Tremblant, Quebec, Oct. 3–5, pp. 319–324.

Shurtz, R. F., 1991, Comment: a study of "probabilistic" and "deterministic" geostatistics; Math. Geol., v. 23, no. 3, pp. 443–479.

Shurtz, R. F., 1985, A critique of A. Journel's "The deterministic side of geostatistics"; Math. Geol., v. 17, no. 8, pp. 861–868.

Sichel, H. S., 1966, The estimation of means and associated confidence limits for small samples from lognormal populations; Jour. South African Inst. Min. Metall., v. 66, pp. 106–123.

Sichel, H. S., 1952, New methods in the statistical evaluation of mine sampling data; Trans. Inst. Min. Metall., pp. 261–288.

Sichel, H. S., W. J. Kleingeld, and W. Assibey-Bonsu, 1992, A comparative study of three frequency-distribution models for use in ore evaluation; Jour. South African Inst. Min. Metall., v. 92, no. 4, pp. 91–99.

Sides, E. J., 1994, Quantifying differences between computer models of orebody shapes; in Whateley, M. K. G., and P. K. Harvey (eds.), Mineral resource evaluation II: methods and case histories; Geol. Soc., London, Spec. Pub. No. 79, pp. 109–121.

Sides, E. J., 1992a, Reconciliation studies and reserve estimation; in Annels, A. E. (ed.), Case histories and methods in mineral resource evaluation, Geol. Soc., London, Spec. Pub. No. 63, pp. 197–218.

Sides, E. J., 1992b, Modeling of geological discontinuities for reserve estimation purposes at Neves-Corvo, Portugal; in Pflug, R., and J. W. Harbaugh (eds.), Computer graphics in geology:

three dimensional modeling of geological structures and simulating geologic processes, Lecture notes in earth sciences, v. 41, pp. 213–228, Springer-Verlag, Berlin.

Sillitoe, R. H., 1993, Gold-rich porphyry copper deposits – geological model and exploration implications; in Kirkham, R. V., W. D. Sinclair, R. I. Thorpe, and J. M. Duke (eds.), Mineral deposit modeling: Geol. Assoc. Canada, Spec. Paper 40, pp. 465–478.

Sinclair, A. J., 2001, High-quality geology, axiomatic to high quality resource/reserve estimates; Can. Inst. Min. Metall. Bull., v. 94, no. 1049, pp. 37–41.

Sinclair, A. J., 1998a, Geological controls in resource/reserve estimation; in Vallée, M., and A. J. Sinclair (eds.), 1998, Quality assurance, continuous quality improvement and standards in mineral resource estimation; Expl. Min. Geol., v. 7, nos. 1 and 2, pp. 29–44.

Sinclair, A. J., 1998b, Exploratory data analysis: a precursor to resource/reserve estimation; in Vallée, M., and A. J. Sinclair (eds.), Quality assurance, continuous quality improvement and standards in mineral resource estimation; Expl. Min. Geol., v. 7, nos. 1 and 2, pp. 77–90.

Sinclair, A. J., 1995, Selected topics in mineral inventory estimation; notes for short course given in conjunction with the British Columbia and Yukon Chamber of Mines, Vancouver, 220 pp.

Sinclair, A. J., 1991, A fundamental approach to threshold estimation in exploration geochemistry: probability plots revisited; Jour. Expl. Geochem. v. 41, pp. 1–22.

Sinclair, A. J., 1986, Statistical interpretation of soil geochemical data; in Fletcher, W. K., S. J. Hoffman, M. B. Mehrtens, A. J. Sinclair, and I. Thomson, Exploration Geochemistry: design and interpretation of soil surveys; Soc. Econ. Geol., Reviews in Econ. Geol., v. 3, pp. 97–115.

Sinclair, A. J., 1978, Sampling a mineral deposit for feasibility studies and metallurgical testing; in Mular, A. L., and R. B. Bhappu (eds.), Proc. Symp. on mineral processing plant design, American Inst. Min. Metall., pp. 115–134.

Sinclair, A. J., 1976, Applications of probability graphs in mineral exploration; Spec. v. 4, Assoc. Expl. Geochem., 95 pp.

Sinclair, A. J., 1974, Selection of thresholds in geochemical data using probability graphs; Jour. Geochem. Expl., v. 3, pp. 129–149.

Sinclair, A. J., and G. J. Blackwell, 2000, Resource/reserve classification and the qualified person; Can. Inst. Min. Metall. Bull., v. 93, no. 1038, pp. 29–35.

Sinclair, A. J., and A. Bentzen, 1998, Evaluation of errors in paired analytical data by a linear model; in Vallée, M., and A. J. Sinclair (eds.), Quality assurance, continuous quality improvement and standards in mineral resource estimation; Expl. Min. Geol., v. 7, nos. 1 and 2, pp. 167–174.

Sinclair, A. J., and Deraisme, 1976, A 2-dimensional geostatistical study of a skarn deposit, Yukon Territory, Canada; in Guarascio, M., M. David, and C. Huijbregts (eds.), Advanced geostatistics in the mining industry, pp. 369–379, D. Reidel Pub. Co., Dordrecht, the Netherlands.

Sinclair, A. J., and J. Deraisme, 1974, A geostatistical study of the Eagle copper vein, northern British Columbia; Can. Inst. Min. Metall. Bull., v. 67, pp. 131–142.

Sinclair, A. J., and G. H. Giroux, 1984, Geological controls of semi-variograms in precious metal deposits; in Verly, G., M. David, A. G. Journel, and A. Marechel (eds.), Geostatistics for natural resource characterization: NATO ASI Series C: v. 122 – Part 2, D. Reidel Pub. Co., Dordrecht, the Netherlands, pp. 965–978.

Sinclair, A. J., and Postolski, T. A., 1999, Geology – a basis for quality control in resource/reserve estimation of porphyry-type deposits; Can. Inst. Min. Metall. Bull., v. 92, no. 1027, pp. 37–44.

Sinclair, A. J., and O. J. Tessari, 1981, Vein geochemistry, an exploration tool in Keno Hill camp, Yukon Territory, Canada; Jour. Geochem. Explor., v. 14, pp. 1–14.

Sinclair, A. J., and M. Vallée, 1994, Reviewing continuity: an essential element of quality control for deposit and reserve estimation; Expl. Min. Geol., v. 3, no. 2, pp. 95–108.

Sinclair, A. J., and M. Vallée, 1993, Improved sampling control and data gathering for improved mineral inventories and production control; *in* Dimitrakopoulos, R. (ed.), Proc. Symp. on "Geostatistics for the next century," June 3–5, Montreal, Kluwer Acad. Pub., The Netherlands, pp. 323–329.

Sinclair, A. J., and L. J. Werner, 1978, Geostatistical investigation of the Kutcho Creek chrysotile deposit, northern British Columbia; Math. Geol., v. 10, no. 3, pp. 273–288.

Sinclair, A. J., E. T. Lonergan, and W. E. McConechy, 1983, Geostatistics at the feasibility stage, Golden Sunlight deposit, Montana; Nev. Bur. Mines and Geol., Rept. 36, pp. 165–172.

Sinclair, A. J., M. S. Nowak, and Z. A. Radlowski, 1993, Geostatistical estimation of dilution by barren dykes at Snip gold mine and Virginia porphyry Cu–Au deposit; *in* Elbrond, J., and X. Tang (eds.), Proc. 24th Symp. on Application of computers and operations research to the minerals industry, Oct. 31–Nov. 3, Montreal, v. 2, pp. 438–444.

Sinclair, A. J., Z. A. Radlowski, and G. Raymond, 1994, Mineral inventory of a gold-bearing skarn, the Nickel Plate mine, Hedley, British Columbia; *in* Dimitakopoulos, R. (ed.), Proc. Symp. on "Geostatistics for the next century," June 3–5, 1993, Montreal, Kluwer Acad. Press, The Netherlands, pp. 64–72.

Sinclair, A. J., Z. A. Radlowski, R. M. Srivastava, and A. Samis, 1993, Geostatistical estimation of grade and dilution, Snip gold mine, northern British Columbia; *in* Bawden, W. F., and J. F. Archibald (eds.), "Mining into the 21st century," Kingston, Ont., August 23–26, 1993, pp. 217–224.

Sketchley, D. A., 1998, Gold deposits: establishing sampling protocols and monitoring quality control; *in* Vallée, M., and A. J. Sinclair (eds.), Quality assurance, continuous quality improvement and standards in mineral resource estimation; Expl. Min. Geol., v. 7, nos. 1 and 2, pp. 129–138.

Smee, B., 1997, Quality control procedures in mineral exploration; Northern Miner, June 23, p. 4.

Smith, C. G., 1991, Theoretical estimation of in situ bulk density of coal; Can. Inst. Min. Metall. Bull., v. 84, no. 949, pp. 49–52.

Smith, J., 1991, How companies value properties; Can. Inst. Min. Metall. Bull., v. 84, no. 953, pp. 50–52.

Smith, L. D., 1994, Checklist for economic evaluations of mineral deposits; Can. Inst. Min. Metall. Bull., v. 87, no. 983, pp. 32–37.

Smith, M., 1999, Gold ore reference materials: a quality assurance tool for geologists; Assoc. Explor. Geochemists, Explore, no. 102, pp. 8–10.

Soares, A., 1992, Geostatistical estimation of multiphase structures; Math. Geol., v. 24, no. 2, pp. 149–160.

Soares, A., 1990, Geostatistical estimation of orebody geometry: morphological kriging; Math. Geol., v. 22, no. 7, pp. 787–802.

Soares, A., 1984, Local estimation of the block recovery functions; *in* Verly, G., M. David, A. G. Journel, and A. Marechel (eds.), Geostatistics for natural resource characterization: NATO ASI Series C: v. 122 – Part 1, D. Reidel Pub. Co., Dordrecht, the Netherlands, pp. 485–493.

Solow, A. R., 1993, On the efficiency of the indicator approach in geostatistics; Math. Geol., v. 25, no. 1, pp. 53–58.

Solow, A. R., 1990, Geostatistical cross-validation: a cautionary note; Math. Geol., v. 22, no. 6, pp. 637–639.

Solow, A. R., 1986, Mapping by simple indicator kriging; Math. Geol., v. 18, no. 3, pp. 335–352.

Soregaroli, A. E., and W. I. Nelson, 1976, Boss Mountain; *in* Sutherland Brown, A. (ed.), Porphyry deposits of the Canadian cordillera; Can. Inst. Min. Metall., Spec. v. 15, pp. 432–443.

Springett, M., 1993, Solving open pit grade control problems; *in* Bawden, W. F., and J. F. Archibald (eds.), Innovative mine design for the next century; Balkema Pub., Rotterdam, pp. 225–228.

Springett, M., 1989, Biasing unbiased data: mine versus mill; *in* Weiss, A., (ed.), Proc. 21st Symp. on Application of computers and operations research to the minerals industry, pp. 286–296.

Springett, M., 1984, Sampling practices and problems; in Erickson, A. J., Jr. (ed.) Applied mining geology; Amer. Inst. Min. Metall. Eng., Soc. Min. Eng., pp. 189–195.

Springett, M., 1983, Sampling and ore reserve estimation for the Ortiz gold deposit, New Mexico; Nev. Bur. Mines and Geol., Rept. 36, pp. 152–164.

Srivastava, R. M., 1987, Minimum variance or maximum profitability; Can. Inst. Min. Metall. Bull., v. 80, no. 901, pp. 63–68.

Srivastava, R. M., 1986, Philip and Watson–quo vadunt?; Math. Geol., v. 18, no. 1, pp. 141–146.

Srivastava, R. M., and E. H. Isaaks, 1991, Comment: A study of "probabilistic" and "deterministic" geostatistics by R. F. Shurtz; Math. Geol., v. 23, no. 3, pp. 481–496.

Srivastava, R. M., and H. M. Parker, 1988, Robust measures of spatial continuity; in Armstrong, M. (ed.), Proc. third international geostatistics congress; D. Reidel Pub. Co., Dordecht, the Netherlands, pp. 295–308.

Stanley, C. R., 1998, NUGGET: PC-software to calculate parameters for samples and elements affected by the nugget effect; in Vallée, M., and A. J. Sinclair (eds.), Quality assurance, continuous quality improvement and standards in mineral resource estimation, Expl. Min. Geol., v. 7, nos. 1 and 2, pp. 139–148.

Stanley, C. R., 1987, PROBPLOT: an interactive computer program to fit mixtures of normal (or lognormal) distributions with maximum likelihood procedures; Assoc. Expl. Geochem., Spec. v. 14, 40 pp. (includes diskette).

Stephenson, P. R., and N. Miskelly, 1999, Reporting standards and the JORC code; Australasian Inst. Min. Metall., Monograph 23.

Stephenson, P. R., and P. T. Stoker, 1999, Classification of mineral resources and ore reserves; in Dagdelen, K. (ed.), Proc. 28th Symp. on Applications of computers and operations research in the minerals industry, Oct. 20–22, Colorado School of Mines, Golden, Colo., pp. 55–68.

Stoiser, L. R., 1986, Exploration and geologic evaluation of the Los Bronces copper–molybdenum deposit, Chile; Min. Eng., v. 38, no. 3, pp. 187–194.

Stone, J. G., 1986, Contact dilution in ore reserve estimation; in Ranta, D. E. (ed.), Proc. Symp. on Applied mining geology: ore reserve estimation; Amer. Inst. Min. Metall. Eng., Soc. Min. Eng., Alburquerque, New Mexico, pp. 153–166.

Stone, J. G., and P. G. Dunn, 1994, Ore reserve estimates in the real world; Soc. Econ. Geol., Spec. Pub. No. 3, 150 pp.

Sutphin, D., and J. D. Bliss, 1990, Disseminated flake graphite and amorphous graphite deposit models: an analysis using grade and tonnage models; Can. Inst. Min. Metall. Bull., v. 83, no. 940, pp. 85–89.

Swan, A. R. H., and M. Sandilands, 1995, Introduction to geological data analysis; Blackwell Science, London, 446 pp.

Swanson, C. O., 1945, Probabilities in estimating the grade of gold deposits; Trans. Can. Inst. Min. Metall., v. XLVIII, pp. 323–350.

Switzer, P., and H. M. Parker, 1975, The statistics of selective mining; Tech. Rept. No. 75, Dept. of Statistics, Stanford University, Stanford, Calif., 21 pp.

Szidarovszky, E., Y. Baafi, and Y. C. Kim, 1987, Kriging without negative weights; Math. Geol., v. 19, no. 6, pp. 549–559.

Taylor, H. K., 1994, Ore reserves, mining and profit; Can. Inst. Min. Metall. Bull., v. 87, no. 983, pp. 38–46.

Taylor, H. K., 1993, Reserve inventory practice-reply; Can. Inst. Min. Metall. Bull., v. 86, no. 968, pp. 146–147.

Taylor, H. K., 1992, The guide to the evaluation of gold deposits; integrating deposit evaluation and reserve inventory practices – discussion; Can Inst. Min. Metall. Bull., v. 85, no. 964, pp. 76–77.

Taylor, H. K., 1991, Ore reserves – the mining aspects; Trans. Inst. Min. Metall., v. 100, pp. 146–158.

Taylor, H. K., 1986, The war against "ore", lets end it; in David, M., et al. (eds.), Proc. Symp. on "Ore Reserve Estimation, Methods, Models and Reality," Can. Inst. Min. Metall., Montreal, May 10–11, pp. 32–46.

Taylor, H. K., 1985, Cutoff grades – some further reflections; Trans. Inst. Min. Metall., v. 94, sect. A, pp. A204–A216.

Taylor, H. K., 1972, General background theory of cutoff grades; Trans. Inst. Min. Metall., v. 81, sect. A, pp. A160–A179.

Tessari, O. J., and A. J. Sinclair, 1980, Metal and mineral zoning models and their practical importance; Keno Hill–Galena Hill camp, Yukon Territory; Western Miner, October, pp. 52–66.

Thompson, M., and R. J. Howarth, 1978, A new approach to the estimation of analytical precision; Jour. Geochem. Expl., v. 9, pp. 23–30.

Thompson, M., and R. J. Howarth, 1976, Duplicate analysis in geochemical practice Part I. Theoretical approach and estimation of analytical reproducibility; Analyst, v. 101, pp. 690–698.

Thompson, M., and R. J. Howarth, 1973, The rapid estimation and control of precision by duplicate determinations; Analyst, v. 98, no. 1164, pp. 153–160.

Till, R., 1974, Statistical methods for the earth scientist–an introduction; McMillan Press Ltd., London, 154 pp.

Till, R., 1971, Are there geochemical criteria for differentiating reef and non-reef carbonates? Bull. Amer. Assoc. Petroleum Geol., v. 55, pp. 523–528.

USBM/USGS, 1976, Principles of the mineral resource classification system of the U.S. Bureau of Mines and the U.S. Geological Survey; Geological Survey Bull., 1450-A, 5 pp.

Valente, J., L. C. Cardoso, M. Cardoso, and E. Silva, 1997, Ore reserve estimation of alluvial diamond deposits; in Baafi, E. Y., and N. A. Schofield (eds.), Geostatistics Wollongong '96, v. 2, pp. 842–850.

Vallée, M., 1999, Quality assurance: a tool to maximize returns on exploration, deposit appraisal and mining investments; Can. Inst. Min. Metall. Bull., v. 92, no. 1027, pp. 45–54.

Vallée, M., 1998a, Sampling quality control; in Vallée, M., and A. J. Sinclair (eds.), Quality assurance, continuous quality improvement and standards in mineral resource estimation; Expl. Min. Geol., v. 7, nos. 1 and 2, pp. 107–116.

Vallée, M., 1998b, Quality assurance, continuous quality improvement and standards; in Vallée, M., and A. J. Sinclair (eds.), Quality assurance, continuous quality improvement and standards in mineral resource estimation; Expl. Min. Geol., v. 7, nos. 1 and 2, pp. 1–14.

Vallée, M., 1998c, Resource/reserve estimation and inventory: Can reporting definitions substitute for standards? in Vallée, M., and A. J. Sinclair (eds.), Quality assurance, continuous quality improvement and standards in mineral resource estimation; Expl. Min. Geol., v. 7, nos. 1 and 2, pp. 15–24.

Vallée, M., 1993, Mineral deposit evaluation and reserve inventory practice–reply; Can. Inst. Min. Metall. Bull., v. 86, no. 968, pp. 147–148.

Vallée, M., 1992, Guide to the evaluation of gold deposits; Can. Inst. Min. Metall., Spec. v. 45, 299 pp.

Vallée, M., and D. Cote, 1992, The guide to the evaluation of gold deposits; integrating deposit evaluation and reserve inventory practices; Can. Inst. Min. Metall. Bull., v. 85, no. 947, pp. 50–61.

Vallée, M., and D. Cote, 1992, The guide to the evaluation of gold deposits; integrating deposit evaluation and reserve inventory practices–reply; Can. Inst. Min. Metall. Bull., v. 85, no. 964, pp. 77–78.

Vallée, M., and S. McCutcheon, 1997, Are international reporting standards feasible? Can. Inst. Min. Metall. Bull., v. 90, no. 1007, pp. 30–37.

Vallée, M., and A. J. Sinclair (eds.), 1998, Quality assurance, continuous quality improvement and standards in mineral resource estimation; Expl. Min. Geol., v. 7, nos. 1 and 2, 180 pp.

Vallée, M., and A. J. Sinclair, 1998, Quality control of resource/reserve estimation – where do we go from here? Can. Inst. Min. Metall. Bull., v. 91, no. 1022, pp. 55–57.

Vallée, M., and A. J. Sinclair, 1997, Efficient resource and reserve estimation depends on high quality geology and evaluation procedures; Can. Inst. Min. Metall. Bull., v. 90, no. 1012, pp. 76–79.

Vallée, M., and A. J. Sinclair, 1993, Quality management methods for more reliable estimations of deposits and reserves; in Elbrond, J., and X. Tang (eds.), Proc. 24th Symp. on Application of computers and operations research in the minerals industry, Oct. 31–Nov. 3, Montreal, v. 2, pp. 461–468.

Vallée, M., J. M. Belisle, and M. David, 1982, Kriging as a tool to avoid overestimation of grade in sulphide orebodies; Tran. Inst. Min. Metall., v. 91, Sect. A, pp. 1–3.

Vallée, M., M. Dagbert, and D. Cote, 1993, Quality control requirements for more reliable mineral deposit and reserve estimates; Can. Inst. Min. Metall. Bull., v. 86, no. 969, pp. 65–75.

Vallée, M., J. Gilbert, and M. Dagbert, 1986, A case study of gold sampling: developing the SOQUEM-New Pascalis gold deposit; in David, M., R. Froidevaux, A. J. Sinclair, and M. Vallée (eds.), Proc. Symp. on "Ore reserve estimation: methods, models and reality," Can. Inst. Min. Metall., Montreal, pp. 195–217.

Verly, G., 1984, The block distribution given a point multivariate normal distribution; in Verly, G., M. David, A. G. Journel, and A. Marechel (eds.), Geostatistics for natural resource characterization: NATO ASI Series C: v. 122 – Part 1, D. Reidel Pub. Co., Dordrecht, the Netherlands, pp. 495–515.

Verly, G., and J. Sullivan, 1985, Multigaussian and probability krigings – application to the Jerritt Canyon deposit; Min. Eng., June, pp. 568–574.

Walpole, R. E., and R. H. Myers, 1978, Probability and statistics for engineers and scientists; McMillan Pub. Co., New York, 580 pp.

Watson, D. F., and G. M. Philip, 1986, A derivation of the Isted formula for average mineral grade of a triangular prism; Math. Geol., v. 18, no. 3, pp. 329–334.

Wellman, F. W., 1993, Classification of ore reserves based on geostatistical and economical parameters – discussion; Can. Inst. Min. Metall. Bull., v. 86, no. 972. pp. 78–79.

Wellmer, F. W., 1983, Classification of ore reserves by geostatistical methods; Erzmetall, v. 6, no. 7/8, pp. 315–321.

Whiting, B. H., and A. J. Sinclair, 1990, Variations of gold content with depth, San Antonio gold mine; Can. Inst. Min. Metall. Bull., v. 83, no. 941, pp. 43–46.

White, W. H., A. A. Brookstrom, R. J. Kamilli, M. W. Ganster, R. P. Smith, D. E. Ranta, and R. C. Steininger, 1981, Character and origin of Climax-type molybdenum deposits; Econ. Geol., 75th Anniv., v. pp. 270–316.

Wignall, T. K., and J. de Geoffroy, 1987, Statistical models for optimizing mineral exploration; Plenum Press, New York, 432 pp.

Wober, H. H., and P. J. Morgan, 1993, Classification of ore reserves based on geostatistical and economical parameters; Can. Inst. Min. Metall. Bull., v. 86, no. 966, pp. 73–76.

Wober, H. H., and P. J. Morgan, 1993, Classification of ore reserves based on geostatistical and economical parameters – reply; Can. Inst. Min. Metall. Bull., v. 86, no. 972, pp. 79.

Wolfe, J. A., and J. A. Niederjohn, 1962, Statistical control of an exploration program; Min. Eng., November, pp. 54–58.

Wong, J., 1997, Borehole-radar reflection imaging at the McConnell nickel deposit, Sudbury; in Gubins, A. G. (ed.), Proc. of fourth decennial international conference on mineral exploration, Prospectors and Developers Assoc., Toronto, pp. 697–700.

Wortman, D., 1993, Mineral deposit evaluation and reserve inventory practice-discussion; Can. Inst. Min. Metall. Bull., v. 86, no. 968, pp. 144–146.

Wright, J. H., 1984, Method for calculating tonnage and grade in a longhole sublevel stope; Min. Eng., January, pp. 70–71.

Wunderlich, A. D., 1990, Beware the gold scams; Min. Eng., August, pp. 994–995.

Yalcin, E., and A. Unal, 1992, The effect of the variogram estimation on pit-limit design; Can. Inst. Min. Metall. Bull., v. 85, no. 961, pp. 47–50.

Yfantis, E. A., G. T. Flatman, and J. V. Behar, 1987, Efficiency of kriging estimations for square, triangular and hexagonal grids; Math. Geol., v. 19, no. 3, pp. 183–206.

Zhu, H., and A. G. Journel, 1991, Mixture of populations; Math. Geol., v. 23, no. 4, pp. 647–671.

Zimmerman, D. L., 1993, Another look at anisotropy in geostatistics; Math. Geol., v. 25, no. 4, pp. 453–470.

Index

Accounting – *see* Metal Accounting
Accumulation, 222–223, 278
Accuracy, 14–15, 39, 108
Affine Correction, 232, 260
Alteration, 33, 47, 49, 53, 60, 165
Analyses – *see* Assay
Analytical Error, 107, 198, 342
Analytical Methods, 107, 125–127
Anisotropy, 61, 64, 96, 161, 183, 192, 193, 199–204, 208, 278, 313
Area of Influence – *see* Search radius
Arithmetic Mean, 77, 228
Assay, 88, 107, 124–129, 148, 272
Assay Mapping, 57
Atomic Weight, 298
Audit, 4, 334
Aurora Gold Vein, 155, 156, 159
Autocorrelation, 11, 71, 72, 73, 96, 107, 139, 160, 167, 182–183, 185–186, 189, 192, 196, 212, 215–216, 218, 268, 318, 327, 331
Auxiliary Function, 181, 188–190

Back Transform, 176
Bedding, 35, 47, 148, 149
Bench, 148–150, 280
Besshi Type Deposit, 47–48
Bias – *see* Error, systematic
Binomial Distribution, 88
Bivariate Methods, 147, 155–160
Bivariate Normal Distribution, 287, 288
Blasthole Cuttings, 117–120, 141, 338
Blasthole Drilling, 141, 269
Blasthole Sampling, 99–100, 280, 307, 338–340
Blending, 285
Block, 11–14, 268, 330
Block Array/Model, 2, 45–46
Block Size, 219, 222, 247, 269–271, 274, 306, 330
BLUE, Best Linear Unbiased Estimator, 215
Boss Mountain Deposit, 37, 50

Bougainville (Porphyry Copper Deposit), 86, 258–261, 307–311, 342–344
Boundary
    General, 41, 50, 56, 268
    Gradational, 41–45, 55, 57, 274
    Hard – *see* Boundary, Sharp
    Interpretation, 38–45
    Ore/Waste, 5–7, 38–46, 56, 57, 272–274
    Sharp, 41, 55, 57, 273
    Soft – *see* Boundary, Gradational
Box Plots, 152, 210
Breccia/Brecciated, 37, 66
Brenda Porphyry Deposit, 141, 317
Broken Hill Deposit, 54, 55
Bulk Density – *see* Density, Bulk
Bulk Sample – *see* Sampling, Bulk
By-product, 159, 251–253

Calibration, 108
Capping – *see* Cutting
Cash Flow, 6–8
Casino Porphyry Deposit, 164, 165
Cell Size, 82, 83, 152
Central Limit Theorem, 134
Central Tendency, 77
Change of Support, 189, 223, 232, 275, 285
Channel Sample, 104, 105, 338
Chain of Custody, 128
Chi Square Method, 151
Chip Sample, 104, 105
Chuquicamata Porphyry Deposit, 50
Cinola Gold Deposit, 173, 174, 245, 278, 279
Class Interval, 80, 81
Classification, Ore/Waste, 5–7, 287, 307–311, 323, 334, 341–344
Classification System, 8, 10, 59, 325–336
Climax Molybdenum Mine, 22, 173
Clustering, 152
Co-product, 159, 251–253
Coefficient of Correlation – *see* Correlation, Coefficient
Coefficient of Variation, 80, 87, 89, 176, 235

Colosseum Gold Mine, 118
Competent Person, 9, 334–335
Composite, 62, 149–151, 165, 280, 321
Composition of Terms, 244–245
Concentrate, 23, 24
Conditional Bias, 97–98, 116, 183, 218, 219, 224, 233–236, 240, 255, 265, 269, 274, 323
Conditional Probability, 276, 277
Conditional Simulation, 3, 4, 69, 72–73, 191, 284–293
Confidence Limits, 97
Connectivity, 6
Conolly Diagram, 62, 64
Contact, 35, 274, 305
Contact Dilution – *see* Dilution, External
Contamination, 106, 126
Continuity
    General, 7–9, 15, 31, 59–75, 115, 154, 160–162, 177, 255, 306–307, 322, 341
    Geologic, 8, 59–63, 67, 70, 328
    Grade – *see* Continuity Value
    Domains, 65–66, 72, 235
    Value, 59, 63–65, 68–71, 176, 192, 276, 327, 341
Continuous Distribution, 83–90, 152, 153, 255, 256, 266
Contour(ing), 12, 13, 20–22, 27, 51, 53, 161–162, 191, 239, 276, 312–313
Contoured Plots, 22, 62, 161, 162, 313
Copper Queen Deposit, 34
Core/Cuttings Sample, 106
Core Recovery/Loss, 35, 106
Correlation – *see also* Autocorrelation
    Coefficient, 43, 94, 95, 99, 155–159, 233, 269–271
    Graphic Display, 158–159
    Q-Mode, 94
    Simple Linear, 43, 94–96, 99, 137, 155–159, 233, 269–271
    Spatial, 192–214, 289, 319
Correlogram, 94, 96, 186, 192, 212
Covariance, 80, 94, 96, 185–186, 212

Covariogram, 96, 186, 192, 212
Crandon Massive Sulphide Deposit, 33
Cresson Mine, 317
Cross-Validation, 218, 221–224, 228, 319
Cumulative Distribution, 90–94, 231, 244
Curved Space, 210–211
Cutoff Grade, 5–7, 112–113, 232, 239, 245, 52–253, 255–266, 277, 308, 323, 337
Cutting, 71–72, 167–170, 178, 281

Daisy Creek Cu Prospect, 154
Dago Sulphide Deposit, 263–266
Data
    Amount – see Sample Size
    Arrays – see Sampling Array
    Bivariate, 147, 155–159
    Corroboration, 128
    Distribution – see individual distribution types
    Duplicate, 107–108, 111–112, 116–118, 120
    Editing, 147, 149, 170–172
    Evaluation, 83–100, 125–127, 129–139, 146–166, 171–177
    Input, 148, 149
    Location, 34, 108, 149, 237–239, 269, 282
    Multivariate, 147, 160–165, 298
    Quality, 79, 98–100, 107–108, 125–127, 129–139, 147
    Strings, 236–237, 321
    Univariate, 147, 151–155
Data Organization, 33–36, 51, 52, 147, 148
Declustering, 152, 263
Degrees of Freedom, 79, 322
Delaunay Triangulation, 19
Deleterious (substance, mineral), 53, 330
Density
    Bulk, 39, 77, 248–249, 251–252, 281, 294–300, 304, 306
    Determination, 295
    Errors, 77, 295–297
    Mineralogy Impact, 295–296
    Porosity Impact, 296
Dendrograph, 158, 159
Deposit Types, 31, 37, 45, 56, 60–62, 317
Detection Limit, 110
Dilution
    Barren Dike, 311–314
    External, 282, 301–306
    General, 9–10, 12, 273, 301–314, 342
    Internal, 306–311
Dispersion, 78–80, 81, 100, 149, 150, 181, 189, 232, 275, 306
Disseminations, 172
Distal VMS Deposit, 47
Dizon Porphyry Deposit, 54, 55
Domains, 37, 44, 50, 53, 55–56, 65–66, 140, 156, 172, 176, 244, 268, 321

Drill(ing)
    Core, 33–34, 105–108, 151, 314
    General, 33–35, 37, 105–108, 141–142, 181, 224, 269, 272, 279
    Twinned Holes, 129
Duplicate Analyses/Data, 39, 88, 99, 110–112, 117, 119, 120, 132–139, 147, 171

Eagle Copper Vein, 173, 177, 244–245
Eastmain Gold Deposit, 247
Endako Molybdenum Mine, 33, 36, 73
Equity Silver Mine, 61, 117–120, 125, 126, 255–256, 276, 338, 339
Equivalent Grade, 253, 264
Error
    Accumulation, 245
    Analytical, 39, 79, 110, 122–124, 198, 265, 342
    Block Estimation, 308–311, 332
    Computer-Derived, 39
    Classification, 108–110
    Data Base, 107
    Deliberate, 104, 127–129, 319, 334
    Density, 248–249, 251, 295–297
    Estimation, 181, 182, 190, 286, 314
    Extension, 181, 182
    General, 14–15, 113, 146, 286, 290, 291
    Geologic, 34–36, 38–40, 198
    Geometric, 39–45, 274
    Global, 39, 238, 247, 269
    Kriging, 217, 220, 247, 286
    Local, 39, 114, 268
    Ore Margins – see Error, Geometric
    Random, 10, 14, 108
    Sample Location – see also Data Location, 114
    Sampling, 39, 79, 99, 107, 122–124, 198, 265, 342
    Standard, 79
    Subsampling, 79, 121–124
    Surface Measure, 249–251
    Systematic, 14, 88, 108–110, 112–114, 128–129, 130–131, 133, 134, 140–141
    Thickness, 245, 247, 251
    Thompson-Howarth, 111–112
    Tonnage, 113, 248, 296–297
    Total, 79
    Variance, 79, 181, 182, 186, 187, 216, 238
    vs Concentration, 110–112, 134
Estimation Method
    Contouring, 16, 20–21
    Inverse Distance, 16, 19–20, 21, 182, 222, 290, 291
    General, 16–22, 136–139, 182, 313, 318–319, 339
    Geostatistical – see Kriging
    Global, 3, 152, 227, 242–254

Local, 3, 182–183, 255, 268–282, 311
    Method of Sections, 16, 17
    Nearest Neighbor, 18, 19, 222
    Ore Zoning, 22
    Polygonal, 16, 17–19, 182, 221, 262
    Triangular, 16, 19
Evaluation, 73, 146–166, 255
Expected Value, 185
Exploration – see Mineral Exploration
Exploration Information, 3, 4, 10, 31
Exploration Target, 3, 258
Exponential Model, 197
Extrapolation, 57, 183, 184, 219, 329

F-Function, 187, 213, 228, 243, 259, 275, 306
F-test, 152
Falconbridge Nickel Mines, 97, 98
Faulting, 32
Feasibility Study, 2, 9, 115
Feedback, 39
File Design, 147–149
Finance/Financial Analysis, 286, 325
Fire Assay, 124
Fixed Bias, 133, 134, 140
Folding, 32
Fracture Density, 32, 35
Fraud – see Salting, 334
Fundamental Error, 122

GAINLOSS Software, 307, 309, 335
Gangue Minerals, 51, 53
Gaussian Distribution – see Normal Distribution, 83
Gaussian Semivariogram, 197
Geco Mine, 295
Genetic Model – see Ore Deposit Model
Geochemical Contrast, 43–45
Geologic
    Boundary/Margin, 150
    Coding, 33, 149
    Domain, 37, 47, 50, 55–56, 114, 150
    Estimate – see In Situ Estimate
    Framework, 36, 37
    Continuity, 8, 59–63, 67, 70, 328
    History, 31, 36, 107
    Information, 32, 33, 35, 38, 40, 148
    Interpretation, 8, 31, 32, 38, 40, 61
    Mapping, 31–37
    Model(ing) – see Geometric Model
    Resource, 3
    Structure, 33, 35
    Variables, 32, 35, 165
Geology
    Deposit, 33, 34, 37–39, 40, 48–49, 65, 67, 70
    Detailed, 33–35, 37–39
    General, 31, 36–37

Errors, 39–45
    Mine Planning, 35
    Scale, 32
Geometric Error, 39–45, 274
Geometric Mean, 87, 228
Geometric Model, 15, 39–45, 114
Geophysics, 106, 165, 295, 299
Geostatistics
    Benefits, 183, 190–191
    Classification of Resources/Reserves,
        330–333
    Critics, 319–322
    Introduction, 22–23, 181–191
    General, 23, 181–191, 268
    Terminology, 185
Giant Gold Mine, 238, 246, 247
Global Estimation – *see* Estimation,
    Global
Golden Sunlight Mine, 65, 150, 210, 247,
    248, 317
Goldstream Deposit, 47–49
Grab Sample, 104
Graca Orebody, 40
Grade, 140–143, 235
Grade Control, 1, 2, 140–142, 268, 286
Grade–Tonnage Curve, 18, 152, 255–266,
    275
Grade Overestimation, 167
Grade Profile, 63, 71, 140, 160, 273, 274,
    284
Grain Count, 88–90, 296, 300
Grain Size, 88, 89
Graphic Display Software, 38
Greedy Algorithm, 238, 239
Grid, 249–250, 272
GSLIB Software, 69, 287, 292
Gy's Sampling Constant, 55, 122–124
Gy's Sampling Equation, 104, 107,
    122–125

h-Scattergrams, 171, 208, 209
Heterogeneity, 104, 123
Histogram
    Cumulative, 82, 263
    Errors, 131, 132, 221
    Examples, 78, 81, 113, 131, 132, 153,
        222, 223, 267, 303, 306
    General, 78, 80–83, 129–132, 152, 303
    Grade–tonnage, 257–258
    Quality Control, 129–132
    Raw (Naïve), 81, 152
    Unbiased, 152, 153, 255, 257–259, 266,
        275, 303
Hole Effect, 195, 211–212
Homogeneity, 62, 121, 185
Homoscedasticity, 164
Horse, 312
Horsetail Vein, 65

Huckleberry (Porphyry Deposit), 64–65,
    114, 171, 278, 342–343
Hypogene Mineralization, 50

Illegal Activity – *see* Salting, 104
In Situ Estimate, 3
Incremental Ore, 7
Indicator Kriging – *see* Kriging, Indicator
Inflection Point, 154, 173
Information Effect, 306
Interpolation, 32, 38, 57, 183, 184, 219, 327,
    329
Interpretation, 31, 327
Inverse Distance Weighting, 19–21, 161,
    221, 222, 280
Isopachs, 48, 49, 62
Isotropy, 61, 64, 161, 200, 201, 278

J&L Massive Sulphide Deposit, 111
Jerritt Canyon, 173, 177

Kemess Porphyry Deposit, 65
Keno #18 Vein, 156–159
Krige's Relation, 97, 98, 189, 245
Kriging
    Data Strings, 236–237, 321
    Error, 216, 220, 331–332
    General, 142, 183, 190, 215–216,
        218–219, 271, 280, 314, 322
    Indicator, 178, 229–233, 311–313
    Lognormal, 228–229
    Multiple Indicator, 178, 230–233, 262,
        263, 281, 285, 292, 322, 323
    Negative Weights, 169, 219, 224–226,
        240
    Ordinary, 216–217, 221–222, 235, 239,
        246, 290, 291
    Punctual, 215
    Restricted, 178, 227–228, 240, 281
    Simple, 215, 217–218, 235, 239
    Small Blocks, 269–271
    Strings of Samples, 236, 237, 240
    Universal, 215
    Variance, 331–332
    Weights, 150
Kuroko Type Deposit, 46–47
Kurtosis, 80

Lac Knife Graphite Deposit, 151
Lag, 186, 193–195, 196, 198
Lagrange Multiplier, 216, 228, 235
Lake Dufault Mine, 296, 297
Leaching, 50
Least Squares, 136, 182, 190
Liberation, 54
Liberation Factor, 123
Linear Regression Model – *see* Regression
Linear Sample, 105–106

Linear Semivariogram Model, 252–253
Local Estimation, 268–282
Location Error (of data), 108
Location of Estimates, 278–279
Log Transform, 176
Lognormal Distribution, 85, 86–88, 92–93,
    152, 252, 253, 258, 303, 307
Louvem Massive Sulfide Deposit, 317
Lowell-Guilbert Model, 49–50
Lupin Gold Mine, 304

Mactung Tungsten Deposit, 295–296
Margin – *see* Boundary
Matrix, Illconditioned, 221
Mean, Arithmetic, 77
Mean, Geometric, 228
Median, 78
Metal Accounting, 337–346
Metamorphism, 37, 51
Method of Sections, 17
Method of Triangles, 19
Mill Efficiency, 285
Mine Failure, 14, 15, 317–318
Mine Planning/Design, 2, 36, 256, 279, 286,
    313, 318
Mine Profitability, 1
Mine Revenue, 23–26
Mineral Exploration, 3, 256, 276
Mineral Inventory, 2–16, 46, 146–149, 177,
    181–183, 242, 320, 325–326
Mineral Inventory Concepts, 4–15
Mineral Zoning, 33, 46–50, 53–54, 295
Mineralization, 32, 50, 66, 280
Mineralogic Factor, 123
Mineralography, 52
Mineralogy, 51–55, 298, 299
Minimum Mining Width, 302–304
Mining, Underground, 35
Mining Software, 26–27
Misclassification, 5, 307, 308, 323
Mississippi Valley-Type Deposits, 317
Mode, 78
Molar Weight, 296
Mount Hope Molybdenum Deposit, 118
Mount Isa Copper Mine, 150, 205, 206
Moving Average, 160
Multiple Populations, 91–93, 154–155, 170,
    172–178, 198, 209, 232, 286
Multiple Regression – *see* Regression
Multiple Domains, 163, 247–248
Multiple Working Hypotheses, 31, 46
Multivariate Data Analysis, 162–165

Nearest Neighbor Estimation, 18, 19, 222
Negative Weights, 167, 169–170, 197, 219
Nested Structures, 192, 205–207
Net Smelter Return, 24–26
Neves Corvo Copper Tin Mine, 38, 39, 40

Nickel Plate Mine (inc. Mascot, Bulldog
    Pit), 69–72, 122, 131–132, 154, 155,
    172, 314
Nonbias Conditions, 77
Normal Distribution, 81, 83–86, 88, 91, 258,
    303, 307–308
Nugget Effect, 107, 183, 184, 186, 194,
    197–199, 272, 306, 321

Octant Search, 219, 226, 237, 329
Omnidirectional Semivariogram, 201
Operating Cost, 6, 278
Operating Profit, 6, 337, 340
Optimize Data Locations, 237–239
Order Relations, 232, 233
Ordinary Kriging (OK) – see Kriging,
    Ordinary
Ore
    Free-Milling, 54
    General, 4–5,
    Incremental, 7
    Loss, 273
    Mineral, 51
    Refractory, 54
Ore Deposit Model, 45–51
Ore Zoning Method, 22
Operating Profit, 337, 340
Ore/Waste Margin, 5–7, 38–44, 56, 272–274
Ortiz Gold Deposit, 118, 121, 139,
    342–345
Overbreak – see Dilution, External
Outlier, 95, 136, 146, 147, 149, 156, 159,
    167–180, 192, 198, 208–209, 224, 225,
    226, 227–228, 232, 235, 281–282, 319,
    320, 323
Overestimation – see Conditional Bias, 5,
    18, 118, 317, 318

P-RES Software, 159, 167, 175
Page-Williams Mine, 232
Paired Data, 99–100, 129–139, 194, 195
Panel Sample, 105, 106
Parameters, 76, 83, 92, 93, 94
Particle Count, 51, 52
Particle Shape Factor, 123
Partitioning, 77, 78, 80, 91–94, 173–176
Percentile, 174, 176
Pit Optimization, 275
Point Count, 296
Point Sample/Estimate, 11–13, 104, 105,
    212, 268
Poisson Distribution, 88–90, 116
Polygonal Estimation, 17–19, 221, 222
Population, 76
Porosity, 294, 296, 299
Porphyry-Type Deposit, 33, 49–50, 65,
    140–143, 260
Positive Definite Function, 196, 218

Punctual Estimate – see Point
    Sample/Estimate
Precision, 14–15, 108–109, 129–139
PROBPLOT Software, 175
Probability Density Function – see
    individual distribution types
Probability Graph, 70, 90–94, 153–155,
    172–177, 179
Proportional Effect, 65, 152, 160, 192, 199,
    204, 205
Proportional Bias, 132, 133, 134, 136
Pulp, 121

Quadrant Search, 219, 237, 329
Qualified Person, 9, 334
Quality Control
    Assays, 124–127
    Duplicate Data, 88, 99, 110–112,
        132–139
    Global Bias, 130–132
    Histogram, 131, 132
    Linear Model, 133–139
    Scatter Plot, 133
    Time Plots, 126, 127
    Value Plots, 126, 128

Radius of Influence, 183–184, 186, 194,
    197–199, 313, 329
Rand Type Gold Deposit, 87, 94, 218
Random Sampling – see also Sampling, 113,
    114, 141
Random Stratified Sampling – see also
    Sampling, 113
Random Function, 183–185, 193, 286
Random Variable – see Variable, Random
Range (of influence) – see Radius of
    Influence
Rare Grain Effect, 88
Reconciliation, 108, 129, 142–143, 220, 299,
    304, 316, 323
Recoverable Reserves, 256, 257, 271,
    274–276
Recoverable Resources, 282
Recovery, 35, 51, 253, 273, 323
Regionalization, 11
Regionalized Variable, 10–12, 90, 192, 215
Regression
    Duplicate Data, 132–139
    Model, 94, 135, 136, 234
    Multiple, 162, 164–165, 249, 251–252,
        298
    Reduced Major Axis (RMA), 98–100,
        137
    Simple Linear, 97–98, 111–112, 159, 160
Regression Effect, 323
Regular Sampling – see also Sampling, 113,
    249, 250
Regularization, 11, 199, 212–213

Reject, 121
Relative Error, 108, 163
Replicate Analyses – see Data, Duplicate
Representativity, 63, 115–116
Reproducibility – see Precision
Resource – see also Classification
    General, 274, 333
    Global, 242–254
    Grade-Tonnage, 255
    Inferred, 242
Resource/Reserve, 2, 8–10, 59, 242, 268,
    318, 325
Reserve – see also Classification
    General, 8–10, 333
    Grade-Tonnage, 256
Revenue – see Mine Revenue
Riffle, 121, 141
RMA – see Regression, RMA

Safety Line, 122–123, 124
Salting, 127–129
Sample
    Contamination, 106, 126
    Density, 115, 116, 329
    General, 76–77, 104–108
    Location, 108, 167, 269
    Reduction, 120–124
    Representativity, 115
    Size, 115, 116, 150, 306, 322
    Statistical, 76
    Support, 11, 148–149, 165, 183, 212, 259,
        306
    Type, 104–106, 330
Sample Reduction Diagram, 121, 122
Sampling
    Array, 113–116, 183, 199, 207, 243–244,
        246–247, 286, 329
    Bulk, 319
    Chain of Custody, 128
    Experiments, 116–120
    Gy's Constant, 122–124
    Large Lots, 115, 118, 120
    Methods, 104–107, 338–340, 344
    Personnel, 141–142
    Protocol, 121–129, 253
    Safety Line, 122, 123, 124
    Security, 127
    Scanners, 141
    Standards, 126, 128
    Theory, 79, 104, 120–124
San Antonio Mine, 54
Scam, 319
Scatter Diagram, 5, 97, 99, 120, 126, 128,
    136, 159–160, 171, 205, 222–223, 251,
    277, 291
Screen Effect, 225–226
Search Radius (Ellipse, Ellipsoid, Volume),
    219, 329

Selective Mining Unit, 13–14, 55, 62, 150, 151, 187, 232, 245, 256, 259–262, 264, 268, 275, 285, 306, 330, 333

Semivariogram
  Anisotropy, 201–204, 212, 276
  Cloud, 171
  Experimental, 192, 193–204
  General, 107, 150, 186, 192–213, 232, 243, 245, 259, 287, 289, 290
  Models, 184, 187, 189, 195, 196–207, 211, 229, 276–278, 282, 287, 288, 292, 306, 320, 322
  Incorrect Model, 279, 340–341
  Nested Structures, 192, 205–207
  Relative, 176, 192, 199, 204, 205, 208, 221, 320
  Robustness, 209, 210
  In Curved Space, 192, 210, 211
  Regularization, 139
  Proportional Effect, 204–205

Shasta JM Zone, 68, 69, 71, 131, 132, 171, 251–252, 287

Sichel's *t*-Estimator, 86–88

Significant Figures, 14

Silbak Premier Gold Mine, 99, 131, 132, 136, 138

Silver Queen Deposit, 66–68, 71, 221–223, 303–304

Sill, 107, 198

Silver Stack Deposit, 131, 132

Similkameen Porphyry Deposit, 205, 206, 253, 280, 288–290, 317

Simulation
  Applications, 289–291
  Conditional, 3, 142, 236, 284–293, 285
  Sequential Indicator, 292
  Sequential Gaussian, 286–287

Size Range Factor, 123–124

Skarn Deposit – *see also* Nickel Plate Mine, 279–280, 295

Skewed Distributions, 198

Skewness, 80

Sleeper Gold Mine, 173, 177

Smelter Contract, 14, 23–26

Smoothing, 150, 261, 269, 274, 280

SMU – *see* Selective Mining Unit

Snip Gold Mine, 163–164, 311–313

Spatial Dependence – *see* Autocorrelation

Specific Gravity, 248, 249, 294, 296, 299

Specimen, 104

Standards, Analytical, 108, 127, 128

Standard Deviate, 84

Standard Deviation, 79, 94, 99, 199, 303

Standard Error of the Mean, 79

Standard Normal Distribution, 84, 303

Standard Variables, 98

Stationarity, 185, 193, 218, 219, 235, 320

Stockpiling, 285

Stockwork, 35

Stoichiometry, 295

Stope, 247, 279, 280

Stratified Sampling – *see* Sampling, Array

Strip Ratio, 6, 7

Structure, 35–36, 193

Subpopulations, 71–73, 152, 174, 175, 177

Sudbury Nickel Deposits, 97, 98, 159, 160, 249, 298

Support – *see* Sample, Support

Sunnyside Gold Mine, 173, 174, 176, 177

Survey, Geochemical/Geophysical, 32, 51

Systematic Error, 14

t-Test, 152

Target Hardening, 128

Thompson-Howarth Method, 110–112

Threshold, 92, 93, 167, 173–176, 229, 247, 281, 287, 292

Transform, 98, 164, 204, 228, 229, 230, 287, 288, 292

Trend, 146, 156, 159, 161, 321

Triangular Diagram, 163–164

Triangular Estimation, 16, 19

Truncated Distribution, 112

Turning Bands, 286

Umpire Lab, 129

Uniform Grid, 82

Univariate Procedures, 151–155

Universe, 76

Uranium, 74, 164, 175

Valley Copper Mine, 140, 141, 205, 206, 277, 317

Value Continuity, 59, 63–65, 68–71, 139, 140, 172, 341

Variable
  Dependent, 97, 99
  Independent, 97, 99
  Random, 185
  Regionalized, 10–12, 190, 215

Variance
  Analytical, 79
  Dispersion, 181, 189, 245, 259, 262, 269–271, 284, 306
  Estimation, 216, 238
  Extension, 186–188, 243
  General, 79, 185, 198, 199, 235, 243
  Sampling, 79
  Kriging, 307, 330
  Subsampling, 79
  Two–Population, 80

Variogram – *see* Semivariogram, 96, 186, 192

Variogram Cloud, 208, 209

Variography, 189, 192, 279

Vein, 32, 33, 34, 37, 50, 66–69, 71, 331

Vein Density, 32

Virginia Porphyry Deposit, 152, 161, 162, 311, 313

VMS/Volcanic Massive Sulphide Deposit, 46, 47, 51

Volume-Variance Relation, 152, 189, 232, 245–246, 258–264, 266, 275–276, 306

Voronoi Tessellation, 18

Warrego Gold Deposit, 202

Waste, 314

Weathering, 50

Weight(ing), 77, 78, 216, 218, 257, 321

Weight, Negative, 169–170, 224–226

Weighted Average, 149, 231, 257, 281, 303

Whitehorse Copper Mine, 279–280

Woodlawn Deposit, 40, 51, 53

ZERROR Software, 112

Zonation – *see* Zoning Patterns

Zone, 37–39, 152, 274, 299, 327

Zone (Area, Range, Volume) of Influence – *see also* Radius of Influence, 183–184, 329

Zoning Patterns – *see also* Mineral Zoning, 48–49, 53, 54, 248, 295, 299